Lecture Notes in Artificial Intel

Subseries of Lecture Notes in Computer Science

LNAI Series Editors

Randy Goebel
  University of Alberta, Edmonton, Canada
Yuzuru Tanaka
  Hokkaido University, Sapporo, Japan
Wolfgang Wahlster
  DFKI and Saarland University, Saarbrücken, Germany

LNAI Founding Series Editor

Joerg Siekmann
  DFKI and Saarland University, Saarbrücken, Germany

Luís Correia   Luís Paulo Reis
José Cascalho (Eds.)

Progress in Artificial Intelligence

16th Portuguese Conference
on Artificial Intelligence, EPIA 2013
Angra do Heroísmo, Azores, Portugal, September 9-12, 2013
Proceedings

 Springer

Volume Editors

Luís Correia
University of Lisbon
Laboratory of Agent Modelling (LabMAg)
Campo Grande, 174-016 Lisbon, Portugal
E-mail: luis.correia@di.fc.ul.pt

Luís Paulo Reis
University of Minho
Artificial Intelligence and Computer Science Lab. (LIACC)
Campus de Azurém, 4800-058 Guimarães, Portugal
E-mail: lpreis@dsi.uminho.pt

José Cascalho
University of the Azores
Information Technology and
Applied Mathematics Center (CMATI) and LabMAg
Campus de Angra do Heroísmo
9700-042 Angra do Heroísmo, Azores, Portugal
E-mail: jmc@uac.pt

ISSN 0302-9743 e-ISSN 1611-3349
ISBN 978-3-642-40668-3 e-ISBN 978-3-642-40669-0
DOI 10.1007/978-3-642-40669-0
Springer Heidelberg New York Dordrecht London

Library of Congress Control Number: 2013946556

CR Subject Classification (1998): I.2, J.4, H.3, H.5.2, I.5, I.4, F.4.1, K.4

LNCS Sublibrary: SL 7 – Artificial Intelligence

Typesetting: Camera-ready by author, data conversion by Scientific Publishing Services, Chennai, India

Printed on acid-free paper

Springer is part of Springer Science+Business Media (www.springer.com)

Preface

The Portuguese Conference on Artificial Intelligence is a long standing international forum organized to present and discuss research in Artificial Intelligence (AI) in a broad scope. It covers themes strongly rooted in AI while simultaneously keeping an eye on newly-developed fringes, with the program organized as a set of thematic tracks. By maintaining a clear international character it is a well-known and attractive event, for the most recent research in the area, with over one hundred participants. The 2013 edition was no exception.

The conference was organized under the auspices of the Portuguese Association for Artificial Intelligence (APPIA – http://www.appia.pt) and, as in previous editions, the program was organized as a set of thematic tracks dedicated to specific themes of AI. A total of 12 main tracks were initially selected following the guidelines of the Organizing and Advisory Committees and 10 tracks were selected to take place at EPIA 2013, as the result of the submission and review processes. Each track was coordinated by an Organizing Committee composed of, at least, two researchers in the specific field coming from distinct institutions. An international Program Committee (PC) was created for each of the tracks, composed by experts in the corresponding research areas.

This volume gathers a selection of the best papers submitted to the 16th Portuguese Conference on Artificial Intelligence, EPIA 2013, held in Angra do Heroísmo, Azores, Portugal. In this edition, total of 157 contributions were received coming from 31 Countries (Argentina, Belgium, Brazil, Canada, China, Colombia, Czech Republic, Finland, France, Germany, Hungary, Iran, Ireland, Italy, Japan, Mexico, Nepal, Netherlands, New Zealand, Norway, Pakistan, Poland, Portugal, Romania, Singapore, South Africa, Spain, Sweden, Tunisia, United Arab Emirates, United Kingdom, and United States). Each submission was reviewed by at least 3, and on average 4, international (PC) members. For these proceedings 45 papers were accepted resulting in an acceptance rate under 30%.

Forming the volume we adopted an organization in sections corresponding to the thematic tracks of the conference. The General AI (GAI) section groups papers stemming from symbolic oriented AI areas, where a theoretical stance is prevalent. Artificial Intelligence in Transportation Systems (AITS) is a section dedicated to the application of AI to transportation and mobility systems. Its contents are oriented towards a rather practical perspective. Artificial Life and Evolutionary Algorithms (ALEA) section includes papers in the wide area of Artificial Life (ALife) while particularly emphasizing Evolutionary Algorithms (EA). Therefore, optimization, evolutionary models, and artificial ecologies are the core of this section. Ambient Intelligence and Affective Environments (AmIA) presents research work in intelligent proactive tools to assist people with their day-to-day activities, especially indoors. Affect and social behavior are common research subjects in this domain. The track of Computational Methods in Bioinformatics and Systems Biology (CMBSB) is also an application oriented area, where AI techniques are used in Systems Biology problems. Intelligent Robotics (IROBOT) is one of the most long standing areas of application of AI. It has also had an important return to the field on account of being a privileged interface of intelligent systems with the physical world. Knowledge Discovery and Business Intelligence (KDBI) has become a significant area of AI, dedicated to the extraction of knowledge from large amounts of data envisaging decision support for management in industry and services. Multi-Agent Systems: Theory and Applications (MASTA) covers one of the most representative AI themes, Multi-Agent Systems (MAS), with all the involved aspects such as theories, architectures, coordination, negotiation, organization, and applications. Social Simulation and Modelling (SSM) may be considered as a spin-off of MAS with a special focus on simulation and synthesis of social behaviors in order to understand real social systems. Text Mining and Applications (TEMA) comprehends another key area of AI consisting on approaches to deal with human language gathering both symbolic and statistic approaches.

The conference program included 4 invited talks. As a discipline AI has always been supported by new findings both in the core areas as well as extending, theory and application, to new areas, deepening its developments into scenarios of increasing complexity. The four invited speakers, Maria Fasli, Carla Gomes, Ana Paiva, and Kenneth Stanley, provided the participants an updated image in a variety of current research areas of AI.

Two special tracks completed EPIA 2013 program, the well established Doctoral Symposium on AI and the Challenges initiative, whose contributions integrate another book. The Challenges special track had the main objective of empowering possible research opportunities and new research projects in which problems related to islands and archipelagos were the target of potential AI applications. This special track has been dubbed "Azores Scientific Workspace Meets Artificial Intelligence" and had the contribution of António Menezes, João Luis Gaspar, Gui M. Menezes and Paulo Borges as Challenges invited speakers.

The idea of organizing this conference came initially from a challenge made by Helder Coelho pointing out that Azores was one of the regions in Portugal that had never welcomed EPIA. We accepted the challenge and went on with the proposal of organizing the 16th EPIA in Azores as conference and program chairs. This endeavor would not have happened without the contribution of the local organizers Luís Gomes and Hélia Guerra who accepted the idea with the condition of having the venue in Angra do Heroísmo. The local organizing team was completed with Pedro Cardoso.

We must express our appreciation for the work of the track organizing chairs that sustained an important part of the scientific arrangements, and for the mission of the advisory committee, program committees and all the reviewers, without whose work the conference would not have been possible. An acknowledgment is due to Governo Regional dos Açores (Azores Local Government) and Fundação Luso Americana (FLAD) for their financial support. We thank Orlando Guerreiro, Filomena Ferreira and Rui Carvalho for their organization support in situ, and the team of Centro para a Ciência de Angra do Heroísmo (CCAh).

This conference was only possible with the joint effort of people from different institutions. The main contribution came from University of Azores, CMATI and Biodiversity Group/CITA-A. Other partners were University of Lisbon, University of Minho, LabMAg and LIACC. We acknowledge the contribution of Centro para a Ciência de Angra do Heroísmo, Câmara Municipal de Angra do Heroísmo, Museu de Angra do Heroísmo and Banco Espírito Santo dos Açores. A final word goes to Springer for their assistance in producing this volume.

July 2013   Luís Correia
                                                             Luís Paulo Reis
                                                             José Cascalho

Organization

16th Portuguese Conference on Artificial Intelligence, EPIA 2013,
Angra do Heroísmo, Azores, Portugal, September 2013

Conference and Program Co-chairs

Luís Correia   University of Lisbon, Portugal
Luís Paulo Reis   University of Minho, Portugal
José Cascalho   University of the Azores, Portugal

Advisory Committee

Alípio Jorge   University of Porto, Portugal
Amílcar Cardoso   University of Coimbra, Portugal
Arlindo Oliveira   Technical University of Lisbon, Portugal
Carlos Ramos   Polytechnic Institute of Porto, Portugal
Ernesto Costa   University of Coimbra, Portugal
Eugénio Oliveira   University of Porto, Portugal
Gaël Dias   University of Caen Basse-Normandie, France
Helder Coelho   University of Lisbon, Portugal
José Carlos Maia Neves   University of Minho, Portugal
José Gabriel Pereira Lopes   New University of Lisbon, Portugal
José Júlio Alferes   New University of Lisbon, Portugal
Luis M. Rocha   University of Indiana, USA
Luis Seabra Lopes   University of Aveiro, Portugal
Manuela Veloso   Carnegie Mellon University, USA
Miguel Calejo   APPIA, Portugal
Pavel Brazdil   University of Porto, Portugal
Pedro Barahona   New University of Lisbon, Portugal
Pedro Rangel Henriques   University of Minho, Portugal
Salvador Abreu   University of Évora, Portugal

AmIA Track Chairs

Paulo Novais   University of Minho, Portugal
Ana Almeida   ISEP - Polytechnic Institute of Porto, Portugal
Sara Rodríguez González   University of Salamanca, Spain
Goreti Marreiros   ISEP - Polytechnic Institute of Porto, Portugal

AmIA Program Committee

Pedro Alves Nogueira	LIACC - University of Porto, Portugal
César Analide	University of Minho, Portugal
Javier Bajo	Pontifical University of Salamanca, Spain
Carlos Bento	University of Coimbra, Portugal
Lourdes Borrajo	University of Vigo, Spain
Juan Botia	University of Murcia, Spain
Antonio Fernández Caballero	University of Castilla-La Mancha, Spain
Antonio Camurri	University of Genova, Italy
Amílcar Cardoso	University of Coimbra, Portugal
Davide Carneiro	University of Minho, Portugal
Juan Corchado	University of Salamanca, Spain
Ângelo Costa	University of Minho, Portugal
Ricardo Costa	Polytechnic of Porto, Portugal
Laurence Devillers	LIMS-CNRS, France
Eduardo Dias	New University of Lisbon - UNL, Portugal
João Dias	INESC-ID Lisbon, Portugal
Anna Esposito	Second University of Naples, Italy
Florentino Fdez-Riverola	University of Vigo, Spain
Lino Figueiredo	Polytechnic University of Porto, Portugal
Diego Gachet	European University of Madrid, Spain
Hatice Gunes	Imperial College London, UK
Eva Hudlicka	Psychometrix Associates Blacksburg, USA
Javier Jaen	Polytechnic University of Valencia, Spain
Rui José	University of Minho, Portugal
Vicente Julián	Polytechnic University of Valencia, Spain
Kostas Karpouzis	ICCS, Greece
Boon Kiat Quek	National University of Singapore
José M. Molina	University Carlos III of Madrid, Spain
Luis Macedo	University of Coimbra, Portugal
José Machado	University of Minho, Portugal
Stacy Marsella	University of Southern California, USA
José Neves	University of Minho, Portugal
Andrew Ortony	NorthWestern University, USA
Ana Paiva	INESC-ID, Portugal
Frank Pollick	University of Glasgow, UK
Carlos Ramos	Polytechnic of Porto, Portugal
Emilio S. Corchado	University of Salamanca, Spain
Ricardo Santos	ISEP - Polytechnic Institute of Porto, Portugal
Ichiro Satoh	National Institute of Informatics, Japan
Dante Tapia	University of Salamanca, Spain
Arlette van Wissen	VU University Amsterdam, The Netherlands

AITS Track Chairs

Rosaldo Rossetti · University of Porto, Portugal
Matteo Vasirani · École Polytechnique Fédérale de Lausanne,
Switzerland
Cristina Olaverri · Technische Universität München - TUM,
Germany

AITS Program Committee

Ana Almeida · ISEP - Polytechnic Institute of Porto, Portugal
Constantinos Antoniou · National Technical University of Athens,
Greece
Elisabete Arsénio · LNEC, Portugal
Federico Barber · Universitat Politècnica de Valencia (UPV),
Spain
Ana Bazzan · Federal University of Rio Grande do Sul -
UFRGS, Brazil
Carlos Bento · University of Coimbra, Portugal
Eduardo Camponogara · Federal University of Santa Catarina, Brazil
António Castro · TAP Air Portugal and LIACC, Portugal
Hilmi Berk Celikoglu · Istanbul Technical University, Turkey
Hussein Dia · AECOM, Australia
Alberto Fernández · University Rey Juan Carlos, Spain
Adriana Giret · Universitat Politècnica de Valencia, Spain
Maite López Sánchez · University of Barcelona, Spain
Paulo Leitão · Polytechnic Institute of Bragança, Portugal
Ronghui Liu · University of Leeds, UK
Jorge Lopes · Brisa, Portugal
José Manuel Menendez · Universidad Politécnica de Madrid, Spain
Jeffrey Miller · University of Alaska Anchorage, USA
Luís Nunes · ISCTE - University Institute of Lisbon,
Portugal
Sascha Ossowski · Universidad Rey Juan Carlos, Spain
Luís Paulo Reis · University of Minho, Portugal
Francisco Pereira · University of Coimbra, Portugal
Michael Rovatsos · University of Edinburgh, UK
Miguel A. Salido · Universitat Politècnica de Valencia, Spain
Javier Sanchez Medina · University of Las Palmas de Gran Canaria,
Spain
J'urgen Sauer · University of Oldenburg, Germany
Thomas Strang · German Aerospace Center, Germany
Agachai Sumalee · Hong Kong Polytechnic University, China
Shuming Tang · Inst. of Autom., Shandong Acad. of Sci., China
José Telhada · University of Minho, Portugal

Harry Timmermans	Eindhoven University of Technology, The Netherlands
Fausto Vieira	IT/FCUP, Portugal
Giuseppe Vizzari	University of Milano-Bicocca, Italy
Fei-Yue Wang	Chinese Academy of Sciences, China
Geert Wets	Hasselt University, Belgium
Danny Weyns	Linnaeus University, Sweden

ALEA Track Chairs

Leonardo Vanneschi	ISEGI - New University of Lisbon, Portugal
Sara Silva	INESC-ID Lisbon, Portugal
Francisco Baptista Pereira	Polytechnic Institute of Coimbra, Portugal

ALEA Program Committee

Stefania Bandini	University of Milano-Bicocca, Italy
Helio J.C. Barbosa	Lab. Nacional de Computação Científica, Brazil
Daniela Besozzi	University of Milan, Italy
Christian Blum	University of the Basque Country, Spain
Stefano Cagnoni	University of Parma, Italy
Philippe Caillou	INRIA, France
Mauro Castelli	ISEGI - New University of Lisbon - UNL, Portugal
Luís Correia	University of Lisbon, Portugal
Ernesto Costa	CISUC - University of Coimbra, Portugal
Carlos Cotta	University of Málaga, Spain
Ivanoe De Falco	ICAR - CNR, France
Antonio Della Cioppa	University of Salerno, Italy
Anikó Ekárt	Aston University, UK
Carlos M. Fernandes	University of Granada, Spain
James A. Foster	University of Idaho, USA
António Gaspar-Cunha	University of Minho, Portugal
Carlos Gershenson	Universidad Nacional Autónoma de Mexico, Mexico
Mario Giacobini	University of Torino, Italy
Jin-Kao Hao	University of Angers, France
Arnaud Liefooghe	Université Lille 1, France
Fernando Lobo	University of Algarve, Portugal
Penousal Machado	CISUC - University of Coimbra, Portugal
Ana Madureira	ISEP - Polytechnic Institute of Porto, Portugal
Luca Manzoni	University of Milano-Bicocca, Italy

Pedro Mariano   University of Lisbon, Portugal
Rui Mendes   University of Minho, Portugal
Telmo Menezes   CAMS - CNRS, France
Alberto Moraglio   University of Birmingham, UK
Shin Morishita   Yokohama National University, Japan
Luís Paquete   CISUC - University of Coimbra, Portugal
Marc Schoenauer   INRIA, France
Roberto Serra   University of Modena and Reggio Emilia, Italy
Anabela Simões   Polytechnic Institute of Coimbra, Portugal
Ricardo H.C. Takahashi   Federal University of Minas Gerais, Brazil
Leonardo Trujillo Reyes Instituto Tecnológico de Tijuana, Mexico

CMBS Track Chairs

José Luis Oliveira   University of Aveiro, Portugal
Miguel Rocha   University of Minho, Portugal
Rui Camacho   FEUP - University of Porto, Portugal
Sara Madeira   IST - Technical University of Lisbon, Portugal

CMBS Program Committee

Alexessander Alves   Imperial College, UK
Paulo Azevedo   University of Minho, Portugal
Chris Bystroff   Rensselaer Polytechnic Institute, USA
João Carriço   IMM/FCUL - University of Lisbon, Portugal
Alexandra Carvalho   IT/IST - Technical University of Lisbon,
                           Portugal
André Carvalho   University of São Paulo, Brazil
Vítor Costa   University of Porto, Portugal
Francisco Couto   University of Lisbon, Portugal
Fernando Diaz   University of Valladolid, Spain
Inês Dutra   Federal University of Rio de Janeiro - UFRJ,
                           Brazil
Florentino Fdez-Riverola University of Vigo, Spain
Nuno Fonseca   EMBL-EBI, UK
Alexandre Francisco   IST - Technical University of Lisbon, Portugal
Ana Teresa Freitas   IST - Technical University of Lisbon, Portugal
Daniel Glez-Pena   University of Vigo, Spain
Ross King   University of Wales, Aberystwyth, UK
Stefan Kramer   Technische Universitat Munchen (TUM),
                           Germany
Marcelo Maraschin   Federal University of Santa Catarina, Brazil
Sérgio Matos   University of Aveiro, Portugal

Rui Mendes	University of Minho, Portugal
Luis M. Rocha	University of Indiana, USA
Marie-France Sagot	Inria Rhône-Alpes, Université Claude Bernard, Lyon I, France
Mário Silva	IST - Technical University of Lisbon, Portugal
Jorge Vieira	IBMC - University of Porto, Portugal
Susana Vinga	INESC-ID, Portugal

GAI Track Chairs

Luís Correia	University of Lisbon, Portugal
Luís Paulo Reis	University of Minho, Portugal
José Cascalho	University of the Azores, Portugal

GAI Program Committee

Salvador Abreu	University of Évora, Portugal
José Júlio Alferes	New University of Lisbon - UNL, Portugal
César Analide	University of Minho, Portugal
Luis Antunes	University of Lisbon, Portugal
Reyhan Aydogan	Bogazici University, Turkey
João Balsa	University of Lisbon, Portugal
Pedro Barahona	New University of Lisbon - UNL, Portugal
Fernando Batista	INESC, Portugal
Carlos Bento	University of Coimbra, Portugal
Lourdes Borrajo	University of Vigo, Spain
Pavel Brazdil	University of Porto, Portugal
Stefano Cagnoni	University of Parma, Italy
Philippe Caillou	INRIA, France
Miguel Calejo	APPIA, Portugal
Rui Camacho	University of Porto, Portugal
Amílcar Cardoso	University of Coimbra, Portugal
Margarida Cardoso	ISCTE - University Institute of Lisbon, Portugal
João Carreia	University of Coimbra, Portugal
Carlos Carreto	Polytechnic Institute of Guarda, Portugal
Alexandra Carvalho	IT/IST - Technical University of Lisbon, Portugal
André Carvalho	University of São Paulo, Brazil
Cristiano Castelfranchi	ISTC-CNR, Italy
Gladys Castillo	University of Aveiro, Portugal
António Castro	TAP Air Portugal and LIACC, Portugal
Daniel Castro Silva	University of Coimbra, Portugal

Marc Cavazza	University of Teesside, UK
Luis Cavique	University Aberta, Portugal
Hilmi Berk Celikoglu	Istanbul Technical University, Turkey
Hélder Coelho	University of Lisbon, Portugal
Juan Corchado	University of Salamanca, Spain
João Cordeiro	University of Beira Interior, Portugal
Paulo Cortez	University of Minho, Portugal
Ernesto Costa	CISUC - University of Coimbra, Portugal
José Costa	Federal University of Rio Grande do Norte - UFRN, Brazil
Vítor Costa	University of Porto, Portugal
Walter Daelemans	University of Antwerp, Belgium
Eric de La Clergerie	INRIA, France
Yves Demazeau	Laboratoire d'Informatique de Grenoble, France
Gael Dias	University of Caen, France
Frank Dignum	Utrecht University, The Netherlands
Antoine Doucet	University of Caen, France
Amal El Fallah	Pierre-and-Marie-Curie University - UPMC, France
Mark Embrechts	Rensselaer Polytechnic Institute, USA
Anna Esposito	Second University of Naples, Italy
Carlos Ferreira	ISEP - Polytechnic University of Porto, Portugal
João Gama	University of Porto, Portugal
Adriana Giret	Universitat Politècnica de Valencia, Spain
Paulo Gomes	University of Coimbra, Portugal
Paulo Gonçalves	Polytechnic Institute of Castelo Branco, Portugal
Gregory Grefenstette	CEA, France
Alejandro Guerra-Hernández	University Veracruzana, Mexico
Jin-Kao Hao	University of Angers, France
Samer Hassan	Complutense University of Madrid, Spain
Pedro Henriques	University of Minho, Portugal
Eva Hudlicka	Psychometrix Associates Blacksburg, USA
Javier Jaen	Polytechnic University of Valencia, Spain
Alípio Jorge	University of Porto, Portugal
Vicente Julián	Polytechnic University of Valencia, Spain
Kostas Karpouzis	ICCS, Greece
João Leite	New University of Lisbon - UNL, Portugal
Augusto Loureiro Da Costa	Federal University of Bahia, Brazil
José M. Molina	University Carlos III of Madrid, Spain
Gonzalo Méndez	Complutense University of Madrid, Spain

Paulo Trigo   ISEL - Lisbon Polytechnic Institute, Portugal
Leonardo Trujillo Reyes   Instituto Tecnológico de Tijuana, Mexico
Paulo Urbano   University of Lisbon, Portugal
Marco Vala   IST and INESC-ID, Portugal
Wamberto Vasconcelos   University of Aberdeen, UK
Matteo Vasirani   EPFL, France
Manuela Veloso   Carnegie Mellon University, USA
Rosa Vicari   Federal University of Rio Grande do Sul -
                                   UFRGS, Brazil
Aline Villavicencio   Federal University of Rio Grande do Sul -
                                   UFRGS, Brazil
Neil Yorke-Smith   American University of Beirut, Lebanon

IROBOT Track Chairs

Nuno Lau   University of Aveiro, Portugal
António Paulo Moreira   University of Porto, Portugal
Carlos Cardeira   IST - Technical University of Lisbon, Portugal

IROBOT Program Committee

César Analide   University of Minho, Portugal
Kai Arras   University of Freiburg, Germany
Stephen Balakirsky   Nat. Inst. Stand. and Techn., USA
Reinaldo Bianchi   University Center of FEI, Brazil
Carlos Carreto   Polytechnic Institute of Guarda, Portugal
Xiaoping Chen   University of Science and Technology, China
Luís Correia   University of Lisbon, Portugal
Alexandre da Silva Simões   UNESP, Brazil
Jorge Dias   University of Coimbra, Portugal
Marco Dorigo   Université Libre de Bruxelles, Belgium
Paulo Gonçalves   Polytechnic Institute of Castelo Branco,
                                   Portugal
José Luis Gordillo   Tecnológico de Monterrey, Mexico
Anna Helena Costa   EPUSP, Brazil
Axel Hessler   TU Berlin, Germany
Fumiya Iida   ETHZ, Switzerland
Nicolas Jouandeau   Université Paris 8, France
Augusto Loureiro Da Costa   Federal University of Bahia, Brazil
André Marcato   Federal University of Juíz de Fora, Brazil
Luis Moreno   University Carlos III of Madrid, Spain
Luis Mota   ISCTE - University Institute of Lisbon,
                                   Portugal
Angélica Muñoz   INAOE, Mexico

António José Neves   University of Aveiro, Portugal
Urbano Nunes   University of Coimbra, Portugal
Paulo Oliveira   Instituto Superior Técnico - University of
                               Lisbon, Portugal
Anibal Ollero   University of Seville, Spain
Fernando Osório   University of São Paulo, Brazil
Armando J. Pinho   University of Aveiro, Portugal
Daniel Polani   University of Hertfordshire, UK
Mikhail Prokopenko   CSIRO ICT Centre, Australia
Luís Paulo Reis   University of Minho, Portugal
Josemar Rodrigues de Souza   University of Estado da Bahia, Brazil
Sanem Sariel Talay   Istanbul Technical University, Turkey
André Scolari   Federal University of Bahia, Brazil
Luis Seabra Lopes   University of Aveiro, Portugal
Saeed Shiry Ghidary   Amirkabir University, Iran
Armando Sousa   University of Porto, Portugal
Guy Theraulaz   CRCA, France
Flavio Tonidandel   University Center of FEI, Brazil
Paulo Urbano   University of Lisbon, Portugal
Manuela Veloso   Carnegie Mellon University, USA

KDBI Track Chairs

Paulo Cortez   University of Minho, Portugal
Luís Cavique   University Aberta, Portugal
João Gama   University of Porto, Portugal
Nuno Marques   New University of Lisbon, Portugal
Manuel Filipe Santos   University of Minho, Portugal

KDBI Program Committee

António Abelha   University of Minho, Portugal
Carlos Alzate   KU Leuven, Belgium
Fernando Bacao   New University of Lisbon - UNL, Portugal
Karin Becker   Federal University of Rio Grande do Sul -
                               UFRGS, Brazil
Orlando Belo   University of Minho, Portugal
Albert Bifet   University of Waikato, New Zealand
Agnes Braud   University Robert Schuman, France
Rui Camacho   University of Porto, Portugal
Margarida Cardoso   ISCTE - University Institute of Lisbon,
                               Portugal
André Carvalho   University of São Paulo, Brazil
Gladys Castillo   University of Aveiro, Portugal

Pedro Castillo-Ugr   University of Granada, Spain
Ning Chen   Institute of Engineering of Porto, Portugal
José Costa   Federal University of Rio Grande do Norte -
   UFRN, Brazil

Marcos Domingues   University of São Paulo, Brazil
Mark Embrechts   Rensselaer Polytechnic Institute, USA
Carlos Ferreira   ISEP - Polytechnic University of Porto,
   Portugal

Mohamed Gaber   University of Portsmouth, UK
Paulo Gomes   University of Coimbra, Portugal
Alípio Jorge   University of Porto, Portugal
Stéphane Lallich   University Lyon 2, France
Luis Lamb   Federal University of Rio Grande do Sul -
   UFRGS, Brazil

Phillipe Lenca   Telecom Bretagne, France
Stefan Lessmann   University of Hamburg, Germany
José Machado   University of Minho, Portugal
Armando Mendes   University of the Azores, Portugal
Filipe Pinto   Polytechnic Institute of Leiria, Portugal
Bernardete Ribeiro   University of Coimbra, Portugal
Fátima Rodrigues   Institute of Engineering of Porto, Portugal
Joaquim Silva   New University of Lisbon - UNL, Portugal
Murate Testik   Hacettepe University, Turkey
Aline Villavicencio   Federal University of Rio Grande do Sul -
   UFRGS, Brazil

Leandro Krug Wives   Federal University of Rio Grande do Sul -
   UFRGS, Brazil

Yanchang Zhao   Australia Government, Australia

MASTA Track Chairs

Paulo Trigo   ISEL - Lisbon Polytechnic Institute, Portugal
António Pereira   University of Porto, Portugal
Daniel Castro Silva   University of Coimbra, Portugal
Manuel Filipe Santos   University of Minho, Portugal

MASTA Program Committee

Pedro Abreu   University of Coimbra, Portugal
Diana Adamatti   Federal University of Rio Grande - FURG,
   Brazil
César Analide   University of Minho, Portugal
Luís Antunes   University of Lisbon, Portugal
Reyhan Aydogan   Bogazici University, Turkey

Andrea Omicini	University of Bologna, Italy
Santiago Ontañón Villar	IIIA-CSIC, Spain
Luís Paulo Reis	University of Minho, Portugal
Ana Paula Rocha	University of Porto, Portugal
António Rocha Costa	Federal University of Rio Grande do Sul - UFRGS, Brazil
Juan Antonio Rodriguez	IIIA-CSIC, Spain
Rosaldo Rossetti	University of Porto, Portugal
Jordi Sabater-Mir	IIIA-CSIC, Spain
Marija Slavkovik	University of Liverpool, UK
Jordan Srour	The American University of Beirut
Paolo Torroni	University of Bologna, Italy
Joana Urbano	Instituto Superior Miguel Torga, Portugal
Paulo Urbano	University of Lisbon, Portugal
Wamberto Vasconcelos	University of Aberdeen, UK
Matteo Vasirani	EPFL, France
Laurent Vercouter	École Nationale Supérieure des Mines de Saint-étienne, France
Matt Webster	University of Liverpool, UK
Neil Yorke-Smith	American University of Beirut, Lebanon

SSM Track Chairs

Luis Antunes	University of Lisbon, Portugal
Jaime Sichman	University of São Paulo, Brazil
Jorge Louçã	ISCTE - University Institute of Lisbon, Portugal
Maciej Latek	George Mason University, USA

SSM Program Committee

Frederic Amblard	University Toulouse 1, France
Pedro Andrade	INPE, Brazil
João Balsa	University of Lisbon, Portugal
Pedro Campos	FEUP - University of Porto, Portugal
Amílcar Cardoso	University of Coimbra, Portugal
Cristiano Castelfranchi	ISTC-CNR, Italy
Shu-Heng Chen	National Chengchi University, China
Hélder Coelho	University of Lisbon, Portugal
Nuno David	ISCTE - University Institute of Lisbon, Portugal
Paul Davidsson	Blekinge Institute of Technology, Sweden
Julie Dugdale	Grenoble Informatics Laboratory (LIG), France

Nigel Gilbert	University of Surrey, UK
Nick Gotts	Macaulay Institute, UK
Deffuant Guillaume	Cemagref, France
Samer Hassan	Complutense University of Madrid, Spain
Rainer Hegselmann	University of Bayreuth, Germany
Cesareo Hernandez	University of Valladolid, Spain
Wander Jager	University of Groningen, The Netherlands
Pedro Magalhães	ICS - University of Lisbon, Portugal
Jean-Pierre Muller	CIRAD, France
Akira Namatame	National Defense Academy, Japan
Juan Pavón	Complutense University of Madrid, Spain
Juliette Rouchier	Greqam/CNRS, France
Jordi Sabater-Mir	IIIA-CSIC, Spain
David Sallach	Argonne National Lab., University of Chicago, USA
Keith Sawyer	Washington University St. Louis, USA
Oswaldo Teran	University of Los Andes, Colombia
Takao Terano	University of Tsukuba, Japan
José Tribolet	IST - Technical University of Lisbon, Portugal
Klaus Troitzsch	University of Koblenz, Germany
Harko Verhagen	Stockholm University, Sweden
Nanda Wijermans	Stockholm University, Sweden

TEMA Track Chairs

Joaquim Silva	New University of Lisbon, Portugal
Vitor Rocio	University Aberta, Portugal
Gaël Dias	University of Caen Basse-Normandie, France
José Gabriel Pereira Lopes	New University of Lisbon, Portugal

TEMA Program Committee

Helena Ahonen-Myka	University of Helsinki, Finland
João Balsa	University of Lisbon, Portugal
Fernando Batista	INESC, Portugal
Mohand Boughanem	University of Toulouse III, France
António Branco	University of Lisbon, Portugal
Pavel Brazdil	University of Porto, Portugal
Ricardo Campos	Polytechnic Institute of Tomar, Portugal
Guillaume Cleuziou	University of Orléans, France
João Cordeiro	University of Beira Interior, Portugal

Bruno Cremilleux	University of Caen, France
Francisco Da Camara Pereira	University of Coimbra, Portugal
Walter Daelemans	University of Antwerp, Belgium
Eric de La Clergerie	INRIA, France
Antoine Doucet	University of Caen, France
Elena Elloret	University of Alicante, Spain
Marcelo Finger	University of São Paulo, Brazil
Pablo Gamallo	University of Santiago de Compostela, Spain
Brigitte Grau	CNRS-LIMSI, France
Gregory Grefenstette	CEA, France
Diana Inkpen	University of Helsinki, Finland
Adam Jatowt	University of Kyoto, Japan
Nattiya Kanhabua	University of Hannover, Germany
Zornitsa Kozareva	Information Sciences Institute, Univ. Southern California, USA
Mark Lee	University of Birmingham, UK
João Magalhães	New University of Lisbon - UNL, Portugal
Belinda Maia	University of Porto, Portugal
Nuno Marques	New University of Lisbon - UNL, Portugal
André Martins	IST - Technical University of Lisbon, Portugal
Paulo Quaresma	University of Évora, Portugal
Irene Rodrigues	University of Évora, Portugal
Sriparna Saha	Indian Institute of Technology Patna, India
Antonio Sanfilippo	Pacific Northwest National Laboratory, USA
Isabelle Tellier	Université Paris 3 - Sorbonne Nouvelle, France
Manuel Vilares Ferro	University of Vigo, Spain
Aline Villavicencio	Federal University of Rio Grande do Sul - UFRGS, Brazil
Katerzyna Wegrzyn-Wolska	ESIGETEL, France
Pierre Zweigenbaum	CNRS-LIMSI, France

Additional Reviewers

Alexandre Miguel Pinto	Jie Jiang
Amir H. Razavi	João Cunha
Ana Lopes	João Silva
Antonio Goncalves	Jonathan de Andrade Silva
Baris Gokce	Jose A. Mocholi
Brígida Mónica Faria	Lea Canales
Daniel Cabrera-Paniagua	Luca Manzoni
Eliseo Ferrante	Luís Nunes
Guibing Guo	Luís Santos
Gustavo Pessin	Maksims Fiosins
Isabel Moreno	Marco Nobile

Matthias Knorr
Milagros Fernández Gavilanes
Nuno Gonçalves
Nuno Laranjeiro
Paolo Cazzaniga
Pedro Torres
Rafael Ferrari

Rui Prada
Rui Gomes
Sara Silveira
Tiago Trindade Ribeiro
Victor Darriba
Zhe Zhao

Table of Contents

Computational Methods in Bioinformatics and Systems Biology

General Artificial Intelligence

Intelligent Robotics

Knowledge Discovery and Business Intelligence

Multi-Agent Systems: Theory and Applications

Social Simulation and Modelling

Text Mining and Applications

Studying Stress on e-Learning Users

Davide Carneiro, Sérgio Gonçalves, Paulo Novais, and José Neves

CCTC/Department of Informatics, University of Minho
Braga, Portugal
{dcarneiro,pjon,jneves}@di.uminho.pt, sergiomcgoncalves@gmail.com

Abstract. E-Learning, much like any other communication processes, has been significantly shaped by technological evolution. In its original form, e-Learning aimed to bring the education closer to people, making it more modular and personalized. However, in reality, we observe that it represents a separation between student and teacher, simplifying this relationship to the exchange of "text-based messages", leaving aside all the important contextual richness of the classroom. We are addressing this issue by devising a contextual layer for e-Learning platforms. Particularly, in this paper we describe a solution to convey information about the level of stress of the students so that the teacher can take better and more informed decisions concerning the management of the learning process.

Keywords: Stress, Behavioural Analysis, Moodle.

1 Introduction

In traditional teaching, the relationship between teachers and students is a close one. The almost daily access to the teacher results in a sense of proximity, of accompaniment: they feel that the teacher cares. This is essential for the student's motivation. Teachers, on the other hand, benefit from this proximity by having a constant update on the state of the students, on their worries, on their feedback concerning each topic. All this contextual information, much of it analysed in an unconscious way, allows the teacher to intuitively assess his students and steer his methodologies and strategies in order to optimize success.

However, when a student attends an electronic course or makes use of an e-Learning platform, the interaction between student and teacher, without all these non-verbal interactions, is poorer. Thus, the assessment of the feelings, the state and the attitudes of the student by the teacher becomes more difficult. In that sense, the use of technological tools for teaching, with the consequent teacher-student and student-student separation, may represent a risk as a significant amount of context information is lost. Since students' effectiveness and success in learning is highly related and can be partially explained by their mood while doing it, such issues should be taken into account when in an e-Learning environment. In a traditional classroom, the teacher can detect and even foresee that some negative situation is about to occur and take measures accordingly to mitigate it. When in a virtual environment, similar actions are impossible.

L. Correia, L.P. Reis, and J. Cascalho (Eds.): EPIA 2013, LNAI 8154, pp. 1–12, 2013.

The lack of certain aspects of communication in Virtual Environments (VEs) that include our body language, tone of voice or gestures has been studied by research (see for example [1–4]). The general lesson to learn from here is that human communication is a rich multi-modal process that cannot, by any means, be reduced to simple words without a significant loss in performance [5].

Hover, stress and emotions, in particular, can play an important (usually negative) role in education [7, 9]. In that sense, its analysis in an e-Learning environment assumes greater importance. Generally, stress assessment is done either through questionnaires (an easily misleading approach and certainly not a dynamic one) or through physiological sensors (very accurate but invasive to the point of making them impractical in e-Learning).

In [10], the authors studied emotion in users of e-Learning platforms. They do it using four physiological sensors: hearth rate, skin conductance, blood volume pressure and EEG brainwaves. Despite its accuracy, this approach will never result in a real-life application due to its invasiveness. Other less invasive approaches also exist. [6] contains an overview of different applications of the so-called Affective Computing [8] field in e-Learning. They point out the use of facial expression and speech recognition as potential ways to detect emotional states. However, facial recognition requires a dedicated camera of some quality to be placed in front of the user otherwise it will be inaccurate. It is thus also invasive. Speech recognition, on the other hand, is less invasive but is also much more prone to error, being difficult to develop an accurate speech model given that each individual has his own speech rhythm, tone, pitch or intonation, aspects that are much cultural-dependent.

With these issues in mind, in this paper we present a novel approach to assess the level of stress of students on an e-Learning platform. It is characterized by being non-invasive, personal and transparent. Our objective is indeed to assess the level of stress of students by analysing their behaviour when using the e-Learning platform, i.e., their interaction patterns while using the mouse and the keyboard. A total of 12 features are extracted and analysed that fully describe the way students use these peripherals when under the effect of stress.

2   Experimental Study

In order to analyse the effect of stress on the use of such platforms, we studied specifically the behaviour of students while performing evaluation activities. In this section it is described the experimental study that was conducted with this aim, involving 74 students.

During the process of data collection, two scenarios were set up. In Scenario A, an activity was performed whose main objective was simply to assess the student's knowledge on the selected topics, with no effect on their marks. The activity was performed without any constraints; the students were simply requested to answer the questions that were provided on Moodle.

In a posterior phase, the students were requested to participate on Scenario B. Under this scenario they had to perform another activity, now with some constraints. The students were told by the teacher that this activity would be effectively used for their evaluation, with an impact on the final score. They were also given access passwords and the notion that they would be responsible for managing their activity to maximize their score given the time available. While the students were performing the activity, the teacher insisted regularly on the importance of their performance on their final score and on the decreasing amount of time available.

When analysing the behaviours of the students during the execution of the activities under the two scenarios, the teacher noted that the students were visibly more stressed on the second scenario. After brief talk with some of them, their main feedback concerned their disappointment for not completing the activity or for not answering each question with the needed attention. The most frequent factors for the feeling of unnacomplishment included: (1) the time limit; (2) the noise on the room; (3) the constant interventions of the teacher; (4) the existence of a password for initiating the activity; and (5) the weight of their performance on the final score.

Despite the two activities being very similar in terms of contents and difficulty, it was observed by the teacher that, for similar questions, students would often answer incorrectly or in an incomplete way.

2.1   Methodology and Data Collection

Six different classes participated in the study, which resulted in 12 different moments of data collection: one for each class and for each scenario. The study involved a total of 74 students, 49 boys and 25 girls, aged between 13 and 17.

The data gathered was analysed in order to determine statistically significant differences between scenarios A and B. This analysis was performed with the data of each student individually and with all the data collected. Measures of central tendency and variability were calculated for all variables of interest. Provided that most of the distributions are not normal, the Mann-Whitney-Wilcoxon Statistical test was used to test whether there are actual differences between the distributions of the data for the two scenarios. The data analysis was performed using Wolfram Mathematica, Version 8.0.

In order to collect the data used for analysing the behaviour of the students a very simple logger was developed that was kept running throughout the duration of the activities. This logger was not visible for the student and had no effect on the usability of the Moodle platform. The main objective of the logger was to collect data on how, in each of the activities, the student used the mouse and the keyboard, for later analysis.

The logger listens to system events concerning the mouse and the keyboard. It generates a list of the following events:

– MOV, timestamp, posX, posY - an event describing the movement of the mouse, in a given time, to coordinates (posX, posY);

- MOUSE_DOWN, timestamp, [Left|Right], posX, posY - this event describes the first half of a click (when the mouse button is pressed down), in a given time. It also describes which of the buttons was pressed (left or right) and the position of the mouse in that instant;
- MOUSE_UP, timestamp, [Left|Right], posX, posY - an event similar to the previous one but describing the second part of the click, when the mouse button is released;
- MOUSE_WHEEL, timestamp, dif - this event describes a mouse wheel scroll of amount *dif*, in a given time;
- KEY_DOWN, timestamp, key - describes a given *key* from the keyboard being pressed down, in a given time;
- KEY_UP, timestamp, key - describes the release of a given *key* from the keyboard, in a given time;

A different log is built for each student under each scenario. The data collected allows to build information about the following features:

- Key Down Time - the timespan between two consecutive KEY_DOWN and KEY_UP events, i.e., for how long was a given key pressed (in milliseconds).
- Time Between Keys - the timespan between two consecutive KEY_UP and KEY_DOWN events, i.e., how long did the individual took to press another key (in milliseconds).
- Velocity - The distance travelled by the mouse (in pixels) over the time (in milliseconds). The velocity is computed for each interval defined by two consecutive MOUSE_UP and MOUSE_DOWN events.
- Acceleration - The velocity of the mouse (in pixels/milliseconds) over the time (in milliseconds). A value of acceleration is computed for each interval defined by two consecutive MOUSE_UP and MOUSE_DOWN events.
- Time Between Clicks - the timespan between two consecutive MOUSE_UP and MOUSE_DOWN events, i.e., how long did it took the individual to perform another click (in milliseconds).
- Double Click Duration - the timespan between two consecutive MOUSE_UP events, whenever this timespan is inferior to 200 milliseconds.
- Average Excess of Distance - this feature measures the average excess of distance that the mouse travelled between each two consecutive MOUSE_UP and MOUSE_DOWN events (in pixels).
- Average Distance of the Mouse to the Straight Line - in a few words, this feature measures the average distance of the mouse to the straight line defined between two consecutive clicks (in pixels).
- Distance of the Mouse to the Straight Line - this feature is similar to the previous one in the sense that it will compute the distance to the straight line between two consecutive MOUSE_UP and MOUSE_DOWN events. However, it returns the sum rather than its average value during the path (in pixels).

- Signed Sum of Angles - with this feature the aim is to determine if the movement of the mouse tends to "turn" more to the right or to the left (in degrees).
- Absolute Sum of Angles - this feature is very similar to the previous one. However, it seeks to find only how much the mouse "turned", independently of the direction to which it turned.
- Distance between clicks - represents the total distance travelled by the mouse between two consecutive clicks, i.e., between each two consecutive MOUSE_UP and MOUSE_DOWN events.

3 Results

Here, a detailed feature-by-feature analysis is performed considering the data collected from all the participants. This aims to identify behaviours that are, with a significant measure of support, common to all the participants. Given the length of the features, only the most interesting ones were selected.

In order to perform this analysis we made a prior individual analysis in which we computed the mean and median value of each feature, for each individual. These values were then combined into a single dataset and analysed together.

Key Down Time. When analysing the average time that a key is pressed down while typing, the main conclusion is that a stressed student tends to press the keys during a smaller amount of time. While on the baseline data the mean duration of this feature is of 102.85 ms, under stress the mean value is of 97.8 ms. This same trend was observed in 70.5% of the students. Concerning the median, it decreases in average from 98.5 ms to 96.2 ms, showing a decreasing tendency in 68.9% of the cases analysed. However, this does not necessarily indicates that the student writes faster when under stress, only that he spends less time pressing the keys.

When observing the significance of the differences between the baseline and stressed distributions, for each student, only in 31% of the cases are the differences statistically significant. However, the trend of decreasing time while the key is down does exist.

Time between Keys. In this feature the time spent between pressing two consecutive keys is analysed, which defines the typing velocity. While without stressors, a student spends in average nearly 3 seconds between each two consecutive keys pressed (2904.86 ms). While under stress, this time increases to 5202.19 ms. Moreover, 60% of the students evidence this increase in the average and 83.6% evidence an increase in the median, from 449.156 ms in average to 1979.51 ms.

This denotes that the student writes at a slower pace, when under stress. Statistically significant differences between the baseline and the stressed data for each student were observed in 54% of the cases.

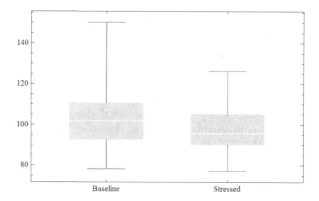

Fig. 1. The time (in milliseconds) during which a key remains down while typing tends to decrease when students are under stress

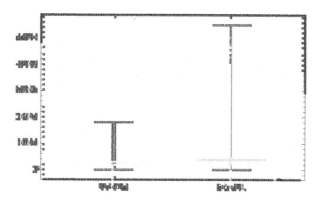

Fig. 2. The time spent between each two consecutive keys being pressed tends to increase when under stress

Acceleration. The initial expectation was that when under stress, students would have faster and more sudden movements, sometimes even involuntary or arbitrary. However, the results point the other way around: the acceleration measured on the mouse is smaller when the students are under stress.

The mean value of the acceleration between each two consecutive clicks in the baseline data is $0.532 \ px/ms^2$, decreasing to $0.449 \ px/ms^2$, which represents a difference of $-0.083 \ px/ms^2$. This decreasing tendency in the mean value of the acceleration was observed in 77.42% of the students. If we consider the value of the median of the acceleration, it is of $0.2 \ px/ms^2$ in the baseline data and of $0.169 \ px/ms^2$ in the stressed data. 87.1% of the students evidence a decrease in the median of the acceleration. This points out a remarkable tendency that can be generalized to a large number of students.

Concerning the statistical significance of this data, significant differences between the baseline and the stressed data have been observed in 77% of the

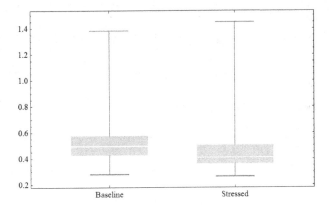

Fig. 3. Two box plots detailing the differences on the acceleration between the baseline and stressed data. In general, stressed students have smaller values of acceleration.

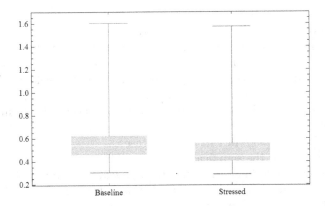

Fig. 4. Similarly to the acceleration, the value of the velocity of the mouse decreases with stress

students, i.e., not only were they affected but the differences due to the effects were statistically significant.

Velocity. Similarly to acceleration, an increase in the velocity was expected due to stress. However, the opposite tendency was observed: stressed students move their mouse slower. A decrease in the mean value of the velocity between each two clicks was observed in 77.4% of the students, from 0.58% px/ms to 0.49 px/ms. The difference in the median was even more striking, decreasing in 90.3% of the students, from 0.22 px/ms to 0.189 px/ms. Similarly to acceleration, a large number of students showed this same tendency. Moreover, significant statistical differences between the calm and the stressed data have been observed in 81% of the students.

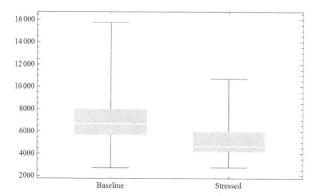

Fig. 5. Two box plots detailing the differences on the time spent between clicks, for the baseline and stressed data. Generally, stressed students spend less time between clicks.

Time between Clicks. The rate at which the students use the mouse to click is affected by stressors and a marked tendency can also be identified: when under stress, students spend less time between each consecutive click. While without stressors each student spends roughly 7 seconds without clicking (7033 ms), under stress this time decreases nearly 2 seconds to 5104 ms. This tendency of decreasing was observed in 80.6% of the students. Concerning the median, its value is of 3149.18 ms for the baseline data, decreasing to 2349.61 on the stressed data. The median decreases in 74.2% of the cases.

However, concerning the significance of the differences for each student, only 32% of the students evidence statistically significant differences between the baseline and stressed data. This points out that, although the tendency does exist, it might not be a such marked one.

Average Distance of the Mouse to the Straight Line. The average distance of the pointer to the straight line defined by two consecutive clicks also tends to decrease with stress, meaning that stressed students become more efficient in the way they move the mouse, moving in more straight lines between their objectives. The mean value of this feature for all students while without stressors was of 59.85 pixels, decreasing to 44.51 pixels when under stress, a decrease of 25.63% in the average distance. 85.48% of the students evidence this same behaviour. Similarly, the median decreases for 82.26% of the students, from and average of 30.14 to 16.63 pixels.

Distance of the Mouse to the Straight Line. This feature is related to the previous one, expect that it measures the total excess of the distance to the straight line between each two clicks, rather than its average value. Thus being, the results observed are in line with the previous ones. The sum of the distances of the mouse to the closest point in the straight line between the two clicks is in

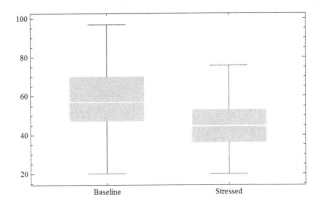

Fig. 6. These two box plots show that when a student is stressed, he moves the mouse with more precision as he minimizes the distance to the straight line that is defined by each two consecutive clicks

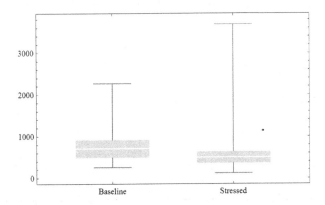

Fig. 7. These two box plots allow to reach a conclusion in line with the previous feature

average 782.03 pixels without stressors, decreasing to 549.752 pixels when under stressors. 87.1% of the students behave like this while under stress. The value of the median also decreases in average from 241.1 pixels to 104.07 pixels, with 80.65% of the students showing a decrease in its value.

Absolute Sum of Angles. Here we analysed the absolute sum of angles, i.e., "how much" the mouse turned rather than "to which side" the mouse turned more. Without stress, between each two clicks the mouse turned in average 8554.4°, while under stress this value decreased to 5119-75°, which represents a decrease of 764.64° between each two clicks, in average. Moreover, 69.35% of the students decrease the amount of how much they turn their mouses. The value of the median also decreases from 6598.54° to 3134.04°.

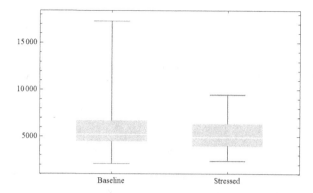

Fig. 8. These box plots show that, in line with previous conclusions, stressed students move in a more straight line, curving less

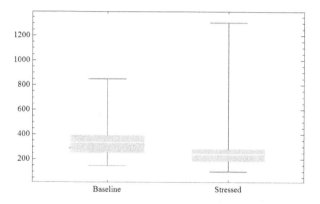

Fig. 9. These plots show that, despite some outlier, students tend to move their mouse less when they are under stress: they become more precise

Distance between Clicks. The total distance travelled by the mouse between each two consecutive clicks also shows a consistent decrease due to stress. In average, the mouse of a student that was not stressed travelled 342.61 pixels between each to consecutive clicks. This value decreased by 92 pixels to 250.64 pixels while under stress, a decrease of 27%, in average. 85.5% of the students evidence this behaviour. The median value also decreases, for 87.1% of the students, from 186.5 to 103.5 pixels.

4   Conclusion and Future Work

This paper started by analysing current drawbacks of online communication tools, specifically in the context of e-Learning. Interestingly enough, e-Learning tools, that started with the objective of bringing teacher and students closer together, actually end up having the opposite effect. The teacher may even be more accessible, easily reachable on the other end of the screen. However, there is also a significant sense of distance that is nowadays hard to overcome.

Besides other issues identified such as lack of motivation by the part of the students, what we are really concerned is for the lack of important contextual information. Generally, when a teacher analysis the state of his students, he takes into consideration more information than just the evaluation results. He builds, partially in an intuitive way, a representation of the student that includes the subjects that he is more comfortable at, a measure of this degree of comfort, among other issues. And he does so by having a daily access to pieces of information such as the attitudes, the actions, the reactions, the behaviours inside and outside the classroom or the occasional small-talk. All this important information is absent from an e-Learning platform. The teacher must thus assess the state of his students merely by the results of the activities performed, which is a far more poor approach.

In this paper we detailed an approach that aims to bridge this gap by providing the teacher with real-time access to context information about his students. Particularly, we have focused on the study of stress as this is one of the key factors that influence the performance of the student.

The main conclusion achieved is that students, when under stress, tend to behave in a more efficient way, decreasing the number of unnecessary actions: they are more focused on their task. However, some students were also found that behave the other way around: they become less efficient and more sloppy when under stress. This points out that: 1) generic models can be developed that can, to a large extent, shape the response of students to stress; and 2) personalized models should not be disregarded as stress is a very individual phenomenon, with each student having his own particular response.

The results achieved show not only that stress does have an effect on the interaction patterns and also that this influence can be quantified and measured. Nevertheless, the features related with the mouse show much more satisfying results than the ones related with the keyboard. We assume that this can be related with the nature of the experiment: students were not required to write much, they mostly had to click on the correct answers. The data collected from the mouse is thus much more than the data collected from the keyboard. In future work we will study the use of the keyboard in more detail in order to build a more detailed stress model.

We are also now starting a similar study to measure the effects of fatigue on the use of the same peripherals. The main goal of this study is, similarly, to determine if the effects of fatigue can be measured and quantified. Our ultimate goal is to develop multi-modal non-invasive classifiers that can act in real time to provide the best and most accurate description of the state of the student to the teacher. This will bring the students closer to the teacher, allowing him to better understand their difficulties and the topics in which they are more at ease. Ultimately, with access to this information, the teacher will be able to adjust his teaching strategies and methodologies to shape such changes, making the learning process a more efficient one.

Acknowledgements. This work is part-funded by ERDF - European Regional Development Funds through the COMPETE Programme (operational programme for competitiveness) and by National Funds through the FCT - Fundação para a Ciência e a Tecnologia (Portuguese Foundation for Science and Technology) within project FCOMP-01-0124-FEDER-028980 (PTDC/EEI-SII/1386/2012). The work of Davide Carneiro is also supported by a doctoral grant by FCT (SFRH/BD/64890/2009).

References

1. Alsina-Jurnet, I., Gutiérrez-Maldonado, J.: Influence of personality and individual abilities on the sense of presence experienced in anxiety triggering virtual environments. International Journal of Human-Computer Studies 68(10), 788–801 (2010)
2. Beale, R., Creed, C.: Affective interaction: How emotional agents affect users. International Journal of Human-Computer Studies 67(9), 755–776 (2009)
3. Dodds, T.J., Mohler, B.J., Bülthoff, H.H.: Talk to the Virtual Hands: Self-Animated Avatars Improve Communication in Head-Mounted Display Virtual Environments. PLoS ONE 6(10), e25759 (2011)
4. Hudlicka, E.: To feel or not to feel: The role of affect in human-computer interaction. International Journal of Human-Computer Studies 59(1-2), 1–32 (2003)
5. Jaimes, A., Sebe, N.: Multimodal human-computer interaction: A survey. Computer Vision and Image Understanding 108(1-2), 116–134 (2007)
6. Lin, H., Pan, F., Wang, Y.: Affective Computing in E-learning, E-learning, pp. 118–128. InTech (2010) ISBN 978-953-7619-95-4
7. Palmer, S., Cooper, C., Thomas, K.: Creating a Balance: Managing Stress. British Library, London (2003)
8. Picard, R.W.: Affective Computing. The MIT Press (2000)
9. Rodrigues, M., Fdez-Riverola, F., Novais, P.: Moodle and Affective Computing - Knowing Who's on the Other Side. In: ECEL 2011 - 10th European Conference on E-learning, Brighton, UK, pp. 678–685 (2011); ISBN: 978-1-908272-22-5
10. Shen, L., Wang, M., Shen, R.: Affective e-Learning: Using "Emotional" Data to Improve Learning in Pervasive Learning Environment. Educational Technology & Society 12(2), 176–189 (2009)

Multi-agent System for Controlling a Cloud Computing Environment

Fernando de la Prieta[1], María Navarro[1], Jose A. García[1],
Roberto González[2], and Sara Rodríguez[1]

[1] University of Salamanca, Computer Science and Automatic Control Department
Plaza de la Merced s/n, 37007, Salamanca, Spain
{fer,mar90ali94,rgonzalezramos,srg,jalberto}@usal.es
[2] Centro de Innovación Tecnológica CENIT, Salamanca, Spain
jgarccor@insags.com

Abstract. Nowadays, a number of computing paradigms have been proposed, of which the latest one is known as Cloud computing. Cloud computing is revolutionizing the services provided through the Internet, and is continually adapting itself in order to maintain the quality of its services. In this paper is proposes a cloud platform for storing information and files by following the cloud paradigm. Moreover, a cloud-based application has been developed to validate the services provided by the platform.

Keywords: Cloud Computing, muti-agent system, agent-based cloud computing, cloud storage, utility computing.

1 Introduction

The term "Cloud Computing" defined the infrastructure as a "Cloud" from which businesses and users are able to access applications from anywhere in the world on demand. Thus, the computing world is rapidly transforming towards developing software for millions to consume as a service, rather than to run on their individual computers. As a result, the number of both closed and open source platforms has been rapidly increasing [2]. Although at first glance this may appear to be simply a technological paradigm, reality shows that the rapid progression of Cloud Computing is primarily motivated by economic interests that surround its purely computational or technological characteristics [1].Since user requirements for cloud services are varied, service providers have to ensure that they can be flexible in their service delivery while keeping the users isolated from the underlying infrastructure.

Nowadays, the latest paradigm to emerge is that of Cloud computing which promises reliable services delivered through next-generation data centers that are built on virtualized compute and storage technologies. Cloud computing platforms has properties of clusters or grids environments, with its own special attributes and capabilities such strong support for virtualization, dynamically composable services with Web Service interfaces, value added services by building on Cloud compute, application

L. Correia, L.P. Reis, and J. Cascalho (Eds.): EPIA 2013, LNAI 8154, pp. 13–20, 2013.
© Springer-Verlag Berlin Heidelberg 2013

services and storage. On this last point, it is important to note that information storage is not performed in the same way today as it was in the past. During the incipient stages of computer sciences, information was stored and accessed locally in computers. The storage process was performed in different ways: in data files, or through the use of database management systems that simplified the storage, retrieval and organization of information, and were able to create a relationship among the data. Subsequently, data began to be stored remotely, requiring applications to access the data in order to distribute system functions; database system managers facilitated this task since they could access data remotely through a computer network. Nevertheless, this method had some drawbacks, notably that the users had to be aware of where the data were stored, and how they were organized. Consequently, there arose a need to create systems to facilitate information access and management without knowing the place or manner in which the information was stored, in order to best integrate information provided by different systems.

This paper presents a Cloud architecture developed in the +Cloud system [5] to manage information. +Cloud is a Cloud platform that makes it possible to easily develop applications in a cloud. Information access is achieved through the use of REST services, which is completely transparent for the installed infrastructure applications that support the data storage. In order to describe the stored information and facilitate searches, APIs are used to describe information, making it possible to search and interact with different sources of information very simply without knowing the relational database structure and without losing the functionality that they provide. Finally, in order to validate the functionality of the services proposed by the +Cloud platform, the paper presents Warehouse 3.0, a cloud storage application.

This paper is structured as follows: the next section provides an overview of the +Cloud platform; Section 3 presents the cloud-based application developed to validate the services provided by the platform: Warehouse 3.0; and finally some conclusions are shown.

2   +Cloud Platform

The platform is composed by a layered structure that coincides with the widely accepted layered view of cloud-computing [3]. This platform allows services to be offered at the PaaS (Platform as a Service) and SaaS (Software as a Service) levels.

The SaaS (Software as a Service) layer is composed of the management applications for the environment (control of users, installed applications, etc.), and other more general third party applications that use the services from the PaaS (Platform as a Service) layer. At this level, each user has a personalized virtual desktop from which they have access to their applications in the Cloud environment and to a personally configured area as well. The next section presents the characteristics and modules of PaaS Layer in +Cloud and +Cloud in greater detail. Both the PaaS and SaaS layers are deployed using the internal layer of the platform, which provides a virtual hosting service with automatic scaling and functions for balancing workload. Therefore, this platform does not offer an IaaS (Infrastructure as a Service) layer.

The virtual and physical resources are managed dynamically. To this end, a virtual organisation of intelligent agents, that is off the topic covered in this paper, and that monitor and manage the platform resources is used [4][5].

The *PaaS Layer* provides its services as APIs, offered in the form of REST web services. The most notable services among the APIs are the identification of users and applications, a simple non-relational database, and a file storage service that provides version control capabilities and emulates a folder-based structure.

The services of the Platform layer are presented in the form of stateless web services. The data format used for communication is JSON, which is more easily readable than XML and includes enough expression capability for the present case. JSON is a widely accepted format, and a number of parsing libraries are available for different programming languages. These libraries make it possible to serialize and de-serialize objects to and from JSON, thus facilitating/simplifying the usage of the JSON-based APIs.

The *FSS (File Storage Service)* provides an interface to a file container by emulating a directory-based structure in which the files are stored with a set of metadata, thus facilitating retrieval, indexing, searching, etc. The simulation of a directory structure allows application developers to interact with the service as they would with a physical file system. A simple mechanism for file versioning is provided. If version control is enabled and an existing file path is overwritten with another file, the first file is not erased but a new version is generated. Similarly, an erased file can be retrieved using the "restore" function of the API. In addition to being organized hierarchically, files can be organized with taxonomies using text tags, which facilitates the semantic search for information and makes the service more efficient. The following information is stored for each file present in the system: (i) Its virtual path as a complete name and a reference to the parent directory. (ii) Its length or size in bytes. (iii) An array of tags to organize the information semantically. (iv) A set of metadata. (v) Its md5 sum to confirm correct transfers and detect equality between versions. (vi) Its previous versions.

Web services are implemented using the web application framework Tornado[1] for Python. While Python provides excellent maintenance and fast-development capabilities, it falls short for intensive I/O operations. In order to keep file uploads and downloads optimized, the APIs rely on the usage of the Nginx[2] reverse proxy for the actual reads and writes to disk. The actual file content is saved in a distributed file system so that the service can scaled, and the workload is distributed among the frontend servers by a load balancer. The structure of the service allows migrating from one distributed file system to another without affecting the client applications.

File metadata and folder structure are both stored in a MongoDB[3] database cluster, which provides adequate scalability and speed capabilities for this application. Web service nodes deploy Tornado and Nginx as well as the distributed file system clients

---

[1] http://www.tornadoweb.org/

[2] http://nginx.org/

[3] http://www.mongodb.org/

Table 1. Restfull web services exposed by FSS

REST Web Call	Description
PutFile	creates a new file (or a new version of an existing file) in response to a request containing the file and basic metadata (path, name and tags) in JSON, structured in a standard multipart request.
Move	changes the path of a file or a folder
Delete	deletes a file. Can include an option to avoid the future recovery of the file, erasing it permanently
GetFolderContents	returns a JSON array with a list of the immediate children nodes of a specific directory.
GetMetadata	returns the metadata set of a file or directory providing its identifier or full path.
GetVersions	returns the list of all the recoverable versions of a file.
DownloadFile	returns the content of a file (a specific older version can be specified).
Copy	creates a copy of a file or a recursive copy of a folder.
CreateFolder	creates a new folder given its path.
DeleteVersion	permanently deletes a specific version of a file.
Find	returns a list of the children nodes of a folder (recursively).
GetConfiguration	retrieves the value of a configuration parameter for the application.
SetConfiguration	sets the value of a configuration parameter (e.g. enabling or disabling version control)
GetSize	retrieves the size of a file. If a folder path is passed, then the total size of the folder is returned.
RestoreVersion	sets an older version of a file as the newest.
Undelete	restores a file.

(GlusterFS[4]/NFS), and the access to the MongoDB cluster that can be located either within or exterior to the nodes.

The *OSS (Object Storage Service)* is a document-oriented and schemaless database service, which provides both ease of use and flexibility. In this context, a document is a set of keyword-value pairs where the values can also be documents (this is a nested model), or references to other documents (with very weak integrity enforcement). These documents are grouped by collections, in a manner similar to how tuples are grouped by tables in a relational database. Nevertheless, documents are not forced to share the same structure. A common usage pattern is to share a subset of attributes among the collection, as they represent entities of an application model. By not needing to define the set of attributes for the object in each collection, the migration between different versions of the same application and the definition of the relationships among the data become much easier. Adding an extra field to a collection is as easy as sending a document with an extra key. A search on that key would only retrieve objects that contain it. The allowed types of data are limited to the basic types present

---

[4] http://www.gluster.org/

in JSON documents: strings, numbers, other documents and arrays of any of the previous types.

As with the FSS, the web service is implemented using Python and the Tornado framework. By not managing file downloads or uploads, there is no need to use the reverse proxy that manages them in every node; therefore Nginx is used only to balance the workload at the entry point for the service.

Table 2. Restfull web services exposed by OSS

REST Web Call	Description
Create	creates a new object inside a collection according to the data provided. It returns the created object, adding the newly generated identifier. If the collection does not exist, it is created instantly.
Retrieve	retrieves all objects that match the given query.
Update	updates an object according to the data provided (the alphanumeric identifier of the object must be provided).
Delete	deletes all objects that match the given query.

The *Identity Manager* is in charge of offering authentication services to both customers and applications. Among the functionalities that it includes are access control to the data stored in the Cloud through user and application authentication and validation. Its main features are:

- Single sign-on web authentication mechanism for users. This service allows the applications to check the identity of the users without implementing the authentication themselves.
- REST calls to authenticate application/users and assign/obtain their roles in the applications within the Cloud.

3   Warehouse

Every user has a root folder that contains all their data. The information stored by a user can be shared with other users through invitations for specific folders or using the user's "Public" folder. It is also possible to create groups of users, which work in a similar way to e-mail groups, in order to allow massive invitations. The user interface is updated asynchronously by using WebSockets. The changes made by a user over a shared resource are automatically displayed in the browsers of the other users. The application has syntactic and semantic search capabilities that are applied to different types of files (text, images or multimedia) due to the extraction and indexing of both the textual content and the metadata present in those files. Furthermore, the search results are presented next to a tag cloud that can be used to refine the searches even more. Finally, the application allows users to retrieve and manipulate different versions of their files. This function is powered by the mechanisms present in the underlying file storage API that has been previously described.

Fig. 1. Snapshot of the user interface

The contents of the files and the file system structure are stored in the FSS. Additional information is necessary to establish relationships between the data and to maintain the folder-sharing logic. This extra information is stored in the OSS. Due to the scalability and high-performance of the APIs, the application can execute tasks that will mainten the referential integrity of its model and the high number of recursive operations that are necessary to move and copy folders.When a user creates a folder, three properties are assigned to it automatically: (i) Host user: keeps permissions over the folder. A number of operations are reserved for this user: move, delete, rename and cancel sharing; (ii) A list of invited users, initially empty; and finally, (iii) A list of users with access privileges to the file, initially containing only the host user.

This tool makes intensive usage of both the file and object storage services and it serves the purpose of being the first real-application test for the developed APIs. Warehouse is the first non-native application that has been developed for the +Cloud platform. Using last-generation standards such as HTML5 and WebSockets[5], the tool allows storing and sharing information using the cloud environment. The user interface is shown at Fig. 1. The available mechanisms for file uploading include HTML5's drag&drop technique. The sharing algorithms are capable of multi-level folder sharing: a children folder of one shared folder can be shared with another list of users. This second group of users will only be allowed to navigate the most-nested folder. The actions related to folder sharing include:

- Invite: adds a user to the list of invited users.
- Accept or decline invitation: If the invitation is accepted, the user is added to the list of access-allowed users. Otherwise, the target user is removed from the list of invited users.
- Leave folder: the user that leaves the folder is removed from the list of access-allowed users. If the host user leaves the folder, the folder will be moved to

---

[5] http://www.websocket.org/

another user's space and that user will be the new host. If there is more than one user remaining, the current host must choose which user will be the new host.

- Turn private: this operation can only be executed by the host user, and deleting all invitations and resetting the access list.
- Move: if the host moves the file, the other users will see a change in the reference to the shared folder. If the operation is done by another user, then only the reference of that user is modified (no move operation is performed).
- Delete: only the host can execute this operation. The shared folder can be moved to the space of another user, or be completely removed.

The next figure depicts the layered architecture of the application Warehouse 3.0. There are three layers, interface, control and model: (i) the interface layer is developed using HTML5 and jQuery, (ii) the control layer is developed using WebSocket for automatically updating runtime information among all users Warehouse using the system, and finally, (iii) the persistence layer that implemented a DAO pattern specially developed to manage the persistence of information OSS (Object Storage System) and files FSS (File Storage System) in the Cloud Computing environment.

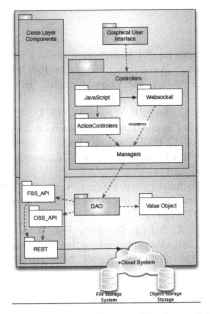

Fig. 2. Layered Architecture Warehouse 3.0

4   Conclusions

Cloud computing is a new and promising paradigm delivering computing services. As Clouds are designed to provide services to external users, providers need to be compensated for sharing their resources and capabilities. In this paper, we have proposed architecture, +Cloud, that has made it possible to store information in applications without having previously established a data model. The storage and retrieval of information is done transparently for the applications, and the location of the data and the storage methods are completely transparent to the user. JSON can define information that is stored in the architecture, making it possible to perform queries that are more complete than those allowed by other cloud systems. This characteristic makes it possible to change the infrastructure layer of the cloud system, facilitating the scalability and inclusion of new storage systems without affecting the applications. In conclusion, we have presented various Cloud efforts in practice from the service-oriented perspective to reveal its emerging potential for the creation of third-party services to enable the successful adoption of Cloud computing, such as object and file storage infrastructure.

Acknowledgements. This research has been supported by the project *SOCIEDADES HUMANO-AGENTE: INMERSION, ADAPTACION Y SIMULACION.* TIN2012-36586-C03-03.(Ministerio de Ciencia e Innovación. Proyectos de Investigación Fundamental No Orientada). Spain.

References

1. Buyya, R., Yeo, C.S., Venugopal, S.: Market-oriented cloud computing: Vision, hype, and reality for delivering it services as computing utilities. In: 10th IEEE International Conference on High Performance Computing and Communications, HPCC 2008, pp. 5–13. IEEE (September 2008)
2. Peng, J., Zhang, X., Lei, Z., Zhang, B., Zhang, W., Li, Q.: Comparison of several cloud computing platforms. In: 2nd International Symposium on Information Science and Engineering, ISISE 2009, pp. 23–27. IEEE Computer Society (2009)
3. Mell, P., Grance, T.: The Nist Definition of Cloud Computing. In: NIST Special Publication 00-145, pp. 1–3. NIST (2011)
4. Heras, S., De la Prieta, F., Julian, V., Rodríguez, S., Botti, V., Bajo, J., Corchado, J.M.: Agreement technologies and their use in cloud computing environments. Progress in Artificial Intelligence 1(4), 277–290 (2012)
5. De la Prieta, F., Rodríguez, S., Bajo, J., Corchado, J.M.: A multiagent system for resource distribution into a Cloud Computing environment. In: Demazeau, Y., Ishida, T., Corchado, J.M., Bajo, J. (eds.) PAAMS 2013. LNCS, vol. 7879, pp. 37–48. Springer, Heidelberg (2013)

Application of Artificial Neural Networks to Predict the Impact of Traffic Emissions on Human Health

Tânia Fontes[1], Luís M. Silva[2,3], Sérgio R. Pereira[1], and Margarida C. Coelho[1]

[1] Centre for Mechanical Technology and Automation / Department of Mechanical Engineering, University of Aveiro, Aveiro, Portugal
[2] Department of Mathematics, University of Aveiro, Aveiro, Portugal
[3] INEB-Instituto de Engenharia Biomédica, Campus FEUP (Faculdade de Engenharia da Universidade do Porto), Rua Dr. Roberto Frias, s/n, 4200-465 Porto, Portugal
{trfontes,lmas,sergiofpereira,margarida.coelho}@ua.pt

Abstract. Artificial Neural Networks (ANN) have been essentially used as regression models to predict the concentration of one or more pollutants usually requiring information collected from air quality stations. In this work we consider a Multilayer Perceptron (MLP) with one hidden layer as a classifier of the impact of air quality on human health, using only traffic and meteorological data as inputs. Our data was obtained from a specific urban area and constitutes a 2-class problem: above or below the legal limits of specific pollutant concentrations. The results show that an MLP with 40 to 50 hidden neurons and trained with the cross-entropy cost function, is able to achieve a mean error around 11%, meaning that air quality impacts can be predicted with good accuracy using only traffic and meteorological data. The use of an ANN without air quality inputs constitutes a significant achievement because governments may therefore minimize the use of such expensive stations.

Keywords: neural networks, air quality level, traffic volumes, meteorology, human health protection.

1 Introduction

Artificial Neural Networks (ANN) are powerful tools inspired in biological neural networks and with application in several areas of knowledge. They have been commonly used to estimate and/or forecast air pollution levels using pollutant concentrations, meteorological and traffic data as inputs. Viotti et al. (2002) proposed an ANN approach to estimate the air pollution levels in 24-48 hours for sulphur dioxide (SO_2), nitrogen oxides (NO, NO_2, NO_X), total suspended particulate (PM_{10}), benzene (C_6H_6), carbon monoxide (CO) and ozone (O_3). Since then, a lot of studies have been made based on the prediction of one or more pollutants. Nagendra and Khare (2005) demonstrate that ANN can explain with accuracy the effects of traffic on the CO dispersion. Zolghadri and Cazaurand (2006) predict the average daily concentrations

L. Correia, L.P. Reis, and J. Cascalho (Eds.): EPIA 2013, LNAI 8154, pp. 21–29, 2013.
© Springer-Verlag Berlin Heidelberg 2013

of PM_{10} but to improve the results they suggest the use of traffic emissions data. On the other hand, Chan and Jian (2013) demonstrate the viability of ANN to estimate $PM_{2.5}$ and PM_{10} concentrations. They verified that the proposed model can accurately estimate not only the air pollution levels but also to identify factors that have impact in those air pollution levels. Voukantsis et al. (2011) also analyzed the $PM_{2.5}$ and PM_{10} concentrations. However, they propose a combination between linear regression and ANN models for estimating those concentrations. An improved performance in forecasting the air quality parameters was achieved by these authors when compared with previous studies. The same conclusions were obtained in a study conducted by Slini et al. (2006) to forecast the PM_{10} concentrations. Nonetheless, Slini et al. (2006) stress the need to improve the model by including more parameters like wind profile, opacity and traffic conditions. Cai et al. (2009) uses an ANN to predict the concentrations of CO, NO_x, PM and O_3. The ANN predicts with accuracy the hourly air pollution concentrations with more than 10 hours in advance. Ibarra-Berastegi et al. (2008) make a more embracing analysis and proposes a model to predict five pollutants (SO_2, CO, NO, NO_2, and O_3) with up to 8 hours ahead.

Literature review shows that ANN have been essentially used as regression models using pollutant concentrations in two ways: (i) first, as input variables to the model; (ii) and second, as variable(s) to be predicted (usually with up to h hours ahead). This means that information of such concentrations has to be collected, limiting the applicability of such models only to locals where air quality stations exist. In this work we propose two modifications. First, we rely just on meteorological and traffic data as input variables to the ANN model, eliminating the use of pollutant concentrations and consequently, the need for air quality stations. Second, we use an ANN as a classifier (and not as a regression model) of the air quality level (below or above to the human health protection limits, as explained in the following section). Such a tool will provide the ability to predict the air quality level in any city regardless of the availability of air quality measurements.

2   Material and Methods

2.1   The Data

Hourly data from 7 traffic stations, a meteorological station and an air quality station, located in a congested urban area of Oporto city (Portugal), were collected for the year 2004 (Figure 1). Traffic is monitored with sensors located under the streets and the meteorological station (at 10'23.9''N and 8° 37''21.6''W) was installed according to the criteria of the World Meteorological Organization (WMO, 1996). Table 1 presents the traffic and meteorological variables used as inputs in the ANN model while Table 2 shows the air quality pollutants measured as well as the respective limits of human health protection according to the Directive 2008/50/EC. These pollutant

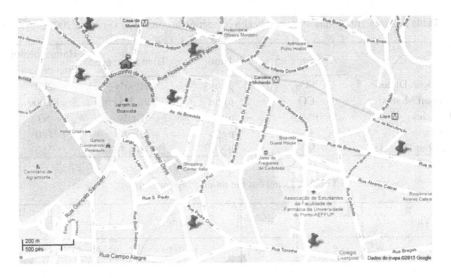

Fig. 1. Study domain (Oporto - Boavista zone): traffic station; air quality station. The meteorological station is located 2 km away from this area.

concentrations are just used to build our classification problem as follows: an instance belongs to class 1 if all the pollutant concentrations are below those limits and to class 2 otherwise.

For the year 2004, is expected that 8,784 hourly instances with data from all of the stations were available. However, due to the gaps and missing data existing in the historical records and also those produced after data pre-processing, the total number of available instances for this year is 3,469. Thus, we have a two-class problem organized in a data matrix with 3,469 lines corresponding to the number of instances and 8 columns corresponding to the 7 input variables and 1 target variable, coded 0 (class 1) and 1 (class 2). Table 3 shows the sampling details of each meteorological and air quality variables.

Table 1. Input variables for the ANN

Variables		Abbreviation	Units
Hour		H	-
Month		M	-
Traffic volumes		V	Vehicles
	Wind speed	WS	m/s
Meteorological	Wind direction	WD	°
variables	Temperature	T	°C
	Solar radiation	SR	W/m^2

Table 2. Air quality limits of human health protection according to Directive 2008/50/EC

	Abbreviation	Units	Time reference	Human health limit protection
Nitrogen Dioxide	NO_2	$\mu g/m^3$	Hourly	200
Carbon Monoxide	CO	$\mu g/m^3$	Octo-hourly	10,000
Particles	PM_{10}	$\mu g/m^3$	Daily	50
Ozone	O_3	$\mu g/m^3$	Hourly	180
Sulphur Dioxide	SO_2	$\mu g/m^3$	Hourly	350

Table 3. Details of the monitoring instruments

Variables		Reaction time	Accuracy	Range	Technique
Meteo-rology	WS	n.a.	±0.1 m/s	0-20 m/s	Anemometer
	WD	n.a.	±4°	0-360°	Anemometer
	T	30-60 s	±0.2 °C	−50 to 70 °C	Temperature sensor
	SR	10 µs	±1%	0-3,000 W.m^{-2}	Pyranometer
Air quality	NO_2	< 5s	n.a.	0-500 µg.m^{-3}	Chemiluminescence analyzer
	CO	30 s	1%	0 ~ 0.05-200 ppm	Infrared photometry
	PM_{10}	10-30 s	n.a.	0 ~ 0.05-10,000 µg/m^3	Beta radiation
	O_3	30 s	1.0 ppb	0-0.1~10 ppm	UV photometric
	SO_2	10 s	0.4 ppb	0-0.1	UV fluorescence

## 2.2	The Model

We considered the most common architecture of an ANN, the Multilayer Perceptron (MLP). In general, an MLP is a nonlinear model that can be represented as a stacked arrangement of layers, each of which is composed of processing units, also known as neurons (except for the input layer, which has no processing units). Each neuron is connected to all the neurons of the following layer by means of parameters (weights) and computes a nonlinear signal of a linear combination of its inputs. Each layer serves as input to the following layer in a forward basis. The top layer is known as output layer (the response of the MLP) and any layer between the input and output layer is called hidden layer (its units are designated hidden neurons). In this work we restricted to the case of a single hidden layer. In fact, as Cybenko (1989) shows, one hidden layer is enough to approximate any function provided the number of hidden neurons is sufficient. The MLP architecture is 7: n_{hid}:1, that is, it has 7 inputs, corresponding to the 7 variables described in Table 1 (in fact, normalized versions of those variables, see Section 2.3), n_{hid} hidden neurons and one output neuron. Figure 2 depicts the architecture used.

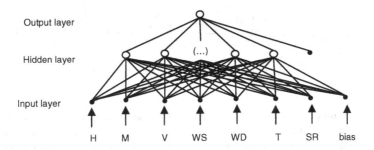

Fig. 2. MLP architecture used: 7-n_{hid}-1 with n_{hid} varying from 5 to 50 in steps of 5

In a formal way, the model can be expressed as:

$$
\begin{aligned}
y &= \varphi_2\left(\sum_{j=1}^{n_{hid}} w_j^{(2)} h_j + b^{(2)}\right) \\
&= \varphi_2\left(\sum_{j=1}^{n_{hid}} w_j^{(2)} \varphi_1\left(\sum_{k=1}^{7} w_{kj}^{(1)} x_k + b_j^{(1)}\right) + b^{(2)}\right),
\end{aligned}
\tag{1}
$$

where:

$w_j^{(2)}$	–	Weight connecting hidden neuron j to the output neuron;
h_j	–	Output of hidden neuron j;
$b^{(2)}$	–	Bias term connected to the output neuron;
$w_{kj}^{(1)}$	–	Weight connecting input k to hidden neuron j;
x_k	–	k-th input variable;
$b_j^{(1)}$	–	Bias term connected to hidden neuron j;
n_{hid}	–	Number of hidden neurons

and φ_1 and φ_2 are the hyperbolic tangent and sigmoid activation functions, respectively, responsible for the non-linearity of the model:

$$
\varphi_1(a) = \frac{e^{2a} - 1}{e^{2a} + 1}
\tag{2}
$$

$$
\varphi_2(a) = \frac{1}{1 + e^{-a}}
\tag{3}
$$

Parameter (weights) optimization (also known as learning or training) is performed by the batch backpropagation algorithm (applying the gradient descent optimization), through the minimization of two different cost functions: the commonly used Mean Square Error (MSE) and the Cross-Entropy (CE) (Bishop, 1995), expressed as

$$\text{MSE} = \frac{1}{n} \sum_{i=1}^{n} (t_i - y_i)^2, \qquad (4)$$

$$\text{CE} = - \sum_{i=1}^{n} y_i \log(y_i) + (1 - y_i) \log(1 - y_i). \qquad (5)$$

Here, n is the number of instances, $t_i \in \{0,1\}$ is the target value (class code) for instance i and $y_i \in [0,1]$ is the network output for instance i. The MLP predicts class 1 (code 0) whenever $y_i \leq 0.5$ and class 2 (code 1) otherwise. We have also used an adaptive learning rate with an initial value of 0.001 (Marques de Sá et al., 2012).

For a comprehensive approach on ANN please refer to Bishop (1995) or Haykin (2009).

2.3   Experimental Procedure

The experimental procedure was as follows. For each number n_{hid} of hidden neurons tested, we performed 30 repetitions of:

1. Randomization of the whole dataset;
2. Split in training, validation and test sets (50%, 25% and 25% respectively of the whole dataset) maintaining class proportions;
3. Normalization of these sets, such as to have inputs with zero mean and unit standard deviation (validation and test sets are normalized using the parameters of the training set);
4. Training the MLP (initialized with small random weights) during 50,000 epochs.

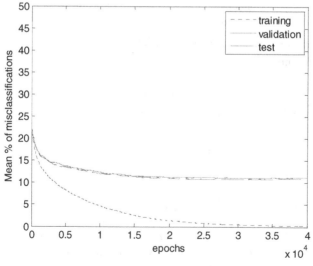

Fig. 3. Training, validation and test set mean misclassifications during the learning process of an MLP with 45 hidden neurons and using the cross-entropy cost

To assess the accuracy of the trained models, we keep track of the training, validation and test set misclassifications during the learning process. Graphs such as the one shown in Figure 3 where produced and used as follows: the validation error is used to perform "early stopping" by choosing the number of epochs m where its mean error is minimum. The mean misclassification error and standard deviation over the 30 repetitions at epoch m are then recorded for the training, validation and test sets (see Section 3). Computations were performed using MATLAB (MathWorks, 2012).

3   Results and Discussion

Table 4 presents the estimates of the mean (over 30 repetitions) misclassification errors and standard deviations both for the MSE and CE cost functions. As explained in the previous section, these records correspond to the epoch number (also given in Table 4) where the mean validation set error was smaller.

Table 4. Mean misclassification errors (in %) for different number of hidden neurons and different cost functions

n_{hid}	Epochs (m)	MSE Mean error (standard deviation)			Epochs	CE Mean error (standard deviation)		
		Train	Validation	Test		Train	Validation	Test
5	49,900	33,00(4.32)	33.97(3.82)	33.63(4.29)	31,800	18.39(0.81)	19.30(1.53)	19.46(1.47)
10	50,000	22.49(2.15)	25.29(2.01)	25.74(2.55)	25,700	15.05(0.87)	17.77(1.58)	16.96(1.23)
15	50,000	14.59(2.28)	19.46(1.53)	19.60(2.90)	10,300	12.70(0.72)	16.47(1.53)	16.07(1.14)
20	50,000	10.18(1.74)	17.10(1.55)	17.06(1.38)	35,600	9.85(0.89)	15.34(1.35)	15.40(1.29)
25	50,000	7.06(1.34)	15.54(1.15)	14.94(1.46)	19,000	7.93(0.77)	13.99(1.31)	14.40(1.33)
30	49,900	4.56(0.72)	13.78(1.21)	13.47(0.95)	21,400	5.28(0.80)	13.07(1.44)	13.03(1.03)
35	49,700	4.25(1.13)	13.70(1.57)	13.84(1.08)	40,000	1.83(0.68)	11.43(1.13)	11.88(0.08)
40	49,800	9.76(1.17)	13,67(1.63)	12.86(1.12)	23,700	1.64(1.06)	11.33(0.97)	11.28(0.08)
45	47,800	9.79(0.63)	13.27(1.30)	13.46(1.32)	30,000	4.60(0.38)	11.07(1.13)	10.77(1.05)
50	49,700	9.47(0.57)	13.08(1.07)	13.14(1.28)	30,500	3.70(0.36)	11.02(1.22)	10.68(1.08)

At a first glance we may observe that the use of different cost functions has important impacts in the results. In fact, CE revealed to be a more efficient cost function in the sense that for the same architectures, a better generalization error is achieved with the use of fewer training epochs when compared to MSE. This is in line to what is known about CE for which several authors reported marked reductions on convergence rates and density of local minima (Matsuoka and Yi, 1991; Solla et al., 1988).

Experimental results show that an MLP with 40 to 50 hidden neurons and trained with CE is able to achieve a mean error around 11%. The standard deviations are also small showing a stable behavior along the 30 repetitions. This suggests that the air quality level can be predicted with good accuracy using only traffic and meteorological data. This is significant as governments may therefore minimize the use of expensive air quality stations.

If the algorithm is applied in a densely urban network of traffic counters the air quality levels could be quickly obtained with a high spatial detail. This represents an important achievement because the implementation of this tool can contribute to assess the potential benefits of the introduction of an actuation plan to minimize traffic emissions in a real-time basis.

4   Conclusions

In this paper, a multilayer perceptron with one hidden layer was applied to automate the classification of the impact of traffic emissions on air quality considering the human health effects. We found that with a model with 40 to 50 hidden neurons and trained with the cross-entropy cost function, we may achieve a mean error around 11% (with a small standard deviation) which we can consider as a good generalization. This demonstrates that such a tool can be built and used to inform citizens in a real-time basis. Moreover, governments can better assess the potential benefits of the introduction of an actuation plan to minimize traffic emissions as well as reducing costs by minimizing the use of air quality stations. Future work will be focused on the use of more data from urban areas, as well as data from other environmental types like suburban and rural areas. We will also seek for accuracy improvements by applying other learning algorithms to this data, such as support vector machines and deep neural networks.

Acknowledgements. This work was partially funded by FEDER Funds through the Operational Program "Factores de Competitividade – COMPETE" and by National Funds through FCT–Portuguese Science and Technology Foundation within the projects PTDC/SEN-TRA/115117/2009 and PTDC/EIA-EIA/119004/2010. The authors from TEMA also acknowledge the Strategic Project PEst-C/EME/UI0481/2011.

References

1. Bishop, C.M.: Neural Networks for Pattern Recognition. Oxford University Press (1995)
2. Cai, M., Yin, Y., Xie, M.: Prediction of hourly air pollutant concentrations near urban arterials using artificial neural network approach. Transportation Research Part D: Transport and Environment 14, 32–41 (2009)
3. Chan, K.Y., Jian, L.: Identification of significant factors for air pollution levels using a neural network based knowledge discovery system. Neurocomputing 99, 564–569 (2013)
4. Cybenko, G.: Aproximation by superpositios of a sigmoidal function. Math. Control Signals System 2, 303–314 (1989)
5. Directive 2008/50/EC: European Parliament and of the Council of 21 May 2008 on ambient air quality and cleaner air for Europe entered into force on 11 June 2008
6. Haykin, S.: Neural Networks and Learning Machines, 3rd edn. Prentice Hall (2009)

7. Ibarra-Berastegi, G., Elias, A., Barona, A., Saenz, J., Ezcurra, A., Argandona, J.D.: From diagnosis to prognosis for forecasting air pollution using neural networks: Air pollution monitoring in Bilbao. Environmental Modelling & Software 23, 622–637 (2008)
8. Marques de Sá, J., Silva, L.M., Santos, J.M., Alexandre, L.A.: Minimum Error Entropy Classification. SCI, vol. 420. Springer (2012)
9. MathWorks, MATLAB and Statistics Toolbox Release 2012, The MathWorks, Inc., Natick, Massachusetts, United States (2012)
10. Matsuoka, K., Yi, J.: Backpropagation based on the logarithmic error function and elimination of local minima. In: Proceedings of the 1990 IEEE International Joint Conference on Neural Networks (1991)
11. Nagendra, S.S.M., Khare, M.: Modelling urban air quality using artificial neural network. Clean Techn. Environ. Policy 7, 116–126 (2005)
12. Slini, T., Kaprara, A., Karatzas, K., Moussiopoulos, N.: PM10 forecasting for Thessaloniki, Greece. Environmental Modelling & Software 21, 559–565 (2006)
13. Solla, S., Levin, E., Fleisher, M.: Accelerated learning in layered neural networks. Complex Systems 2(6), 625–639 (1988)
14. Viotti, P., Liuti, G., Di, P.: Atmospheric urban pollution: applications of an artificial neural network (ANN) to the city of Perugia. Ecological Modelling 148, 27–46 (2002)
15. Voukantsis, D., Karatzas, K., Kukkonen, J., Rasanen, T., Karppinen, A., Kolehmainen, M.: Intercomparison of air quality data using principal component analysis, and forecasting of PM10 and PM2.5 concentrations using artificial neural networks. Thessaloniki and Helsinki, Science of The Total Environment 409, 1266–1276 (2011)
16. Zolghadri, A., Cazaurang, F.: Adaptive nonlinear state-space modeling for the prediction of daily mean PM10 concentrations. Environmental Modelling & Software 21, 885–894 (2006)
17. WMO, Guide to meteorological instruments and methods of observation, 6th edn., World Meteorological Organization, No. 8 (1996)

Outdoor Vacant Parking Space Detector for Improving Mobility in Smart Cities

Carmen Bravo, Nuria Sánchez*, Noa García, and José Manuel Menéndez

Grupo de Aplicación de Telecomunicaciones Visuales, Universidad Politécnica de Madrid
Av. Complutense, 30, Madrid - 28040, Spain
{cbh,nsa,ngd,jmm}@gatv.ssr.upm.es

Abstract. Difficulty faced by drivers in finding a parking space in either car parks or in the street is one of the common problems shared by all the big cities, most of the times leading to traffic congestion and driver frustration. Exploiting the capabilities that Computer Vision offers, an alternative to those ITS commercial solutions for parking space detection that rely on other sensors different from cameras is presented. The system is able to detect vacant spaces and classify them by the type of vehicle that could park in that area. First of all, an approximate inverse perspective transformation is applied for 2D to 3D reconstruction of parking area. In addition, feature analysis based on Pyramid Histogram of Oriented Gradients (PHOG) is carried out on every parking zone within the parking area. Experiments on real scenarios show the excellent capabilities of the proposed system with independence of camera orientation in the context.

Keywords: vehicle detection, video surveillance, outdoor parking.

1   Introduction

Large scale car parks with hundred/thousands of spaces are more and more common in cities nowadays. However, still very often, especially during busy hours, motorists have trouble parking cars driving around with no guarantee to park.

Increasingly gaining popularity, the vacant parking space detection has been implemented by using various technologies: RFID sensors [1], ultrasonic sensors [2][3], laser scanners [4][5], short-range radar networks [6], and those relying on computer vision [7-14]. With independence of the technology being used, parking guidance systems share the common objective of ensuring easy and time effective operation of parking usage.

Most of the existing solutions to detect parking availability in real environments, rely either on a network of active wireless sensors [15] or ultrasonic ones like the commercial products [16][17]. While they are able to detect in real-time where and when cars are parked, these solutions require a dedicated sensor (installed under the pavement surface or attached to a ceiling on its top) per each individual parking space to be monitored.

---

* Corresponding author.

L. Correia, L.P. Reis, and J. Cascalho (Eds.): EPIA 2013, LNAI 8154, pp. 30–41, 2013.
© Springer-Verlag Berlin Heidelberg 2013

Computer vision based approaches have received particular interest for their application outdoor because of presence of surveillance cameras already installed in most of the parking lots. In this case, images from a single camera can be processed to analyze a wider area of the parking lot. This allows controlling several parking spaces using just one sensor, at the same time installation and maintenance costs associated to other types of sensors are reduced.

Although more sensitive to variations of lighting conditions, weather, shadow and occlusion effects, as well as to camera resolution and perspective distortion, detection methods based on cameras and video processing offer a lot of possibilities for creating simple, easy-to-use, and cost-effective outdoor parking spot detection solutions [13]. Based on this technique, many parking space detection algorithms have already been proposed.

Some of them [7][12], proposed the use of visual surveillance to detect and track vehicle movements in an area. Relying on a well-known background subtraction approach, system in [7] is able to recognize simple events in which vehicles can be involved at a parking lot ('stopped vehicle'), at the same time the position of the vehicle is identified in the image. This approach requires the real-time interpretation of image sequences. Other authors [9][10] apply also frame differencing followed by segmentation to locate vacant parking spaces. A different approach to differentiate whether a parking space is occupied or not relies on the classification of the state of a particular region of interest (ROI), i.e. determining whether the ROI is a part of a vehicle or ground surface respectively. In this regard, Dan [11] firstly characterizes each ROI by a feature vector whose components define the textures observed in that region of the image, to train later a general SVM classifier to detect the status of each single parking space. This approach, although more effective, still requires a big amount of training. Based also on ROI characterization, True [14] proposes the combination of color histogram and interest points in vehicles based on Harris corner detection, while Chen et al. [13] select edge density features for classification. Authors in [24] introduced a new descriptor, PHOG, which is based on the analysis of the histograms of oriented gradients in an image but weighting each edge orientation according to its magnitude, so that more edge information is taken into account. This allows a more flexible representation of the spatial layout of local image shape. Results show increased performance compared to other classifiers previously proposed.

A potential downside of these methods is the fact that they analyze the status of each parking space individually, requiring the segmentation in the image of every parking space manually, so that the assistance of a human operator is needed in any case. This is a drawback for the application of this kind of systems to monitor other types of parking areas, different from wide open parking lots, where parking spaces are usually not delimited by lines (e.g. side parking in streets). Additionally, most of the methods do not use to provide the mechanism for their adaptation to a broad set of scenarios each characterized by a different orientation of vehicles and zones of interest in the image due to variations of camera configuration during the installation process. Thus, to the author's knowledge, there is no camera-based solution robust and flexible enough for its consideration as part of a real deployment using surveillance cameras already installed at parking lots and streets.

With this motivation, a novel procedure of vacant parking space detection using a monocular camera is presented in this paper. It relies on the correlation of PHOG features [24] extracted both online and during the initial training phase, followed by the processing of a Dissimilarity-Graph which results as a dedicated geometrical car model slides across the zone. Finally, the distance between adjacent detected cars is used to classify a region as vacant parking spot.

Based on the appearances of the parking area and vehicles in each of the zones in the scene, a simple yet effective method to implement approximate inverse perspective transformation is proposed to reconstruct their 3D geometry beforehand. This approach is particularly used to fit in the image such a geometrical car model over the zone where vehicles are parked in order to find correspondences.

The rest of the paper is organized as follows: Theoretical basis of the system are discussed in Section 2. In Section 3, experimental setup is presented and results on different parking configurations are compared. Section 4 concludes with the summary and future work.

2   System Overview

Following the approach proposed by Pecharromán et al. in [12], our system is divided into independent modules as shown in Figure 1 which separate the low-level of processing, where context is modeled as well as target objects detected, from the high-level, where inference about activity in the parking lot is made.

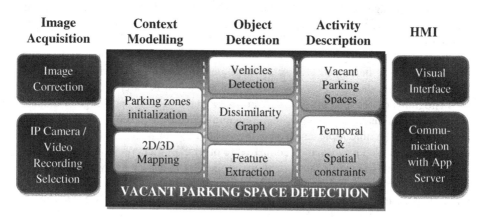

Fig. 1. System overview

First of all, the assistance of a human operator is required for the initialization of parking zones and the calibration of the contextual objects (both area and vehicles). Images from an IP Camera are processed following a bottom-up approach and using individual modules which provides the system with enough flexibility for future improvement. After image correction is applied, the vacant parking space detection starts. Once detection is completed, context-based spatio-temporal analysis is applied.

Adding intelligence in the last step aids the system to filter possible errors in the previous stage. Details of the methodology can be found in following Section. Finally, the system sends relevant information about vacant and occupied spaces in each zone to an App server in charge of providing a service for parking finding to drivers.

2.1   Unconstrained Characterization of Parking Areas

A stream of images is firstly provided to the system by an IP Camera located in the proper position covering its field of view the parking lot. Only in the case that the optic of the camera introduces aberrations (e.g. straight lines in the image become apparently curved), distortion must be corrected before the image can be processed [19].

We are particularly interested in radial distortion, specifically that known as barrel distortion, where image magnification decreases with the distance from the optical axis. Based on Brown's model [20] in the XY system coordinate, if (x,y) is the distorted point, taking into account only the first order term of radial distortion, the corrected points (x',y') will be calculated as:

$$x' = x + k_1 \cdot x \cdot (x^2 + y^2), \qquad y' = y + k_1 \cdot y \cdot (x^2 + y^2), \tag{1}$$

being k_1 positive since the camera used presents barrel distortion.

Once the image has been corrected, the scene needs to be calibrated in order to reconstruct the 3D geometry of the scene, which provides orthogonal information besides the appearance of the object models of interest (in this case, parking area and car models within the zone the area is divided). Most of approaches for 3D reconstruction algorithms count on associating corresponding points in different views to retrieve depth, thus assuming Lambertian surface [18]. However, this is not true for a general application where usually just a sequence of single monocular images from a parking lot is available for its processing with no information about camera calibration parameters.

In this paper, a projective mapping from 3D-World coordinates to pixel coordinates on image plane assuming a Pinhole Camera Model [23] is obtained. The classic technique mostly used requires the use of a dedicated structured object [21]. Auto-Camera Calibration [22] is also possible, although it requires multiple unstructured images quite difficult to obtain when using static surveillance cameras.

The pinhole camera is the simplest, and the ideal, model of camera function. It is usually modeled by placing the image plane between the focal point of the camera and the object. This mapping of three dimensions onto two dimensions is called a perspective projection, being the relationship between World coordinates (X, Y, Z) and pixel coordinates (x, y) on image plane:

$$x = X * \frac{f}{Z}, \qquad y = Y * \frac{f}{Z} \tag{2}$$

where f is the focal length (see Figure 2).

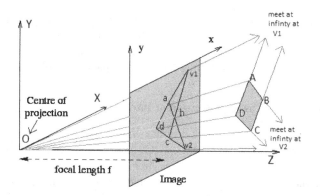

Fig. 2. 2D/3D Mapping using Pin Hole Camera Model

As shown in the Figure 2, suppose *ABCD* represents a rectangular ground plane, which projects polygon *abcd* on image. Although **AD** and **BC** are parallel, they intersect in infinity at a vanishing point which is projected on the image plane and can be found at the intersection of **ad** and **bc** marked as *v1* in Figure 2. Similarly, a second vanishing point *v2* can be found. If the distance between *a* and *A* is very small compared to distances *v1-V1* and *v2-V2*, the 3D point corresponding to *v1*, *V1*, can be approximated applying Equation (2). Similarly, *V2* can be obtained.

Vector algebra can be applied to move any point in 3D space towards the vanishing points at the same time the approximate inverse perspective projection of such point over 2D space can be obtained. Points *ABCD* can be moved proportionately toward or away from center of projection *O*, while their projection will still fall on *abcd* respectively. Direction perpendicular to the ground plane *ABCD* can be found easily by the cross product of the direction vectors along vanishing points *V1* and *V2*. This transformation has shown to be a good approximation for reconstructing object models in the scene, i.e. parking area model that covers the entire scene.

The last step in the initialization process is the establishment of each zone within the entire parking scene. Three points in the image must be selected manually clockwise. The first two points define the orientation of the car within the area, while the second and third point describes the direction of the global zone being extracted.

2.2   From Feature Extraction to the Detection of Vehicles in the Image

Once the system is initialized, a second module makes use of knowledge from context modeled in the first module to analyze features extracted from the zone of interest and thus detect vehicles in the image.

First of all, a moving cuboidal car appearance model is placed over the parking zone in 3D coordinate system to create the database of training samples of occupied spaces. Two kinds of features were considered for feature analysis: Edge Density [13] and Pyramidal Histogram of Orientation Gradients (PHOG) [24].

Once the system has been trained, the cuboidal car model is moved again on the parking zone and new features are again extracted in order to compare them with training samples. This analysis results in a Dissimilarity Graph which is subsequently processed in order to discriminate vacant spots from the occupied ones in the parking lot being monitored.

Edge Density Features

To extract these features, each side of the car model needs firstly to be divided into blocks as shown in Figure 3. The edge density of a block region is given by:

$$\text{Edge density} = \frac{1}{\text{Block Area}} \sqrt{\sum_{block} \left(g_v^2 + g_h^2 \right)} \qquad (3)$$

where g_v and g_h are the vertical and horizontal responses to the Sobel operator at each pixel within the block.

Original test scenario   Car model on test
                                 vehicle sample

Side 1

Edges map

Fig. 3. Edge density feature extraction

Only the most visible sides of the car model, divided into smaller blocks, are used for calculating the corresponding edge density feature vector.

In order to find correspondences between test and training samples, a Covariance distance metric is used as proposed in [25]. This dissimilarity measurement is based on the correlation among the edge density corresponding to different blocks from the car model:

$$d(C_{test}, C_{train}) = \sqrt{\sum_{i=1}^{n} \ln^2 \lambda_i (C_{test}, C_{train})} \qquad (4)$$

where C_{train} is the covariance matrix of the feature vectors in the training set and C_{test} is an approximated covariance matrix calculated from a single test sample as indicated in equation (4). Finally, λ_i represents their i^{th} Eigen-value.

$$C_{test} = (x - m_i) \cdot (x - m_i)^t \qquad (5)$$

m_i is the mean of the feature vectors in the training set, and x is the test feature vector where the distance obtained is added to similarity vector. The similarity vector is processed as described in the next section.

PHOG Features

PHOG features have been described in [24] by Bosch et al. These features represent an image by its local shape and the spatial layout of the shape. The descriptor consists of a histogram of orientation gradients over each image block, at each resolution level. The PHOG method uses L levels, which determine the number of blocks taken into account for analysis. In each level, the image is divided into 2^L blocks for each dimension. The local shape is captured by the distribution over edge orientations within a region *(L=0)*, while the spatial layout by tiling the image into regions at multiple resolutions *(L>0)*.

Two different PHOG approaches are tested. In the first one, the descriptor is computed by using edge orientation $\Theta(x,y)$ over each image block, while in the second approach, the contribution of each edge orientation is weighted according to its magnitude $m(x,y)$. Magnitude and orientation of the gradient on a pixel (x,y) are calculated as:

$$m(x, y) = \sqrt{g_x(x,y)^2 + g_y(x,y)^2} \qquad (6)$$

$$\theta(x, y) = \arctan\left(\frac{g_x(x,y)}{g_y(x,y)}\right) \qquad (7)$$

where $g_x(x,y)$ and $g_y(x,y)$ are image gradient along x and y directions, respectively.

The resulting PHOG vector for each level is concatenated in order to obtain a final feature vector. A $L=3$ has been used in our experiments.

In this case, a Chi-square distance χ^2 between a single test sample vector, X_{test}, and a single PHOG feature vector of the training set, X_{train}, is used for analyzing similarity:

$$\chi^2 = \sum_{i=1}^{n} \frac{(X_{test,i} - X_{train,i})^2}{(X_{test,i} + X_{train,i})}, \qquad (8)$$

denoting as i the ith element of the feature vector of length n.

At every model location, the concatenated PHOG feature is compared with the training samples according to equation (8) and the minimum distance is added to the similarity vector. This similarity vector is processed as described in the following section in order to locate vacant parking spaces in the scene.

2.3   Vacant Space Detection

Occupied spaces by vehicles are used as training samples, and a similarity vector D_{car} represents similarity with the car training samples at each location. When represented as a graph, the x-axis represents every location at which the car model is placed and the y-axis contains the similarity measured (Figure 4). D_{car} has lower values in regions where cars are located.

Figure 4 shows what is known as Dissimilarity Graph, i.e. graphical representations of Chi-Square Distance between test and training samples when car model is displaced from beginning to end of a parking zone. The use of gradients orientations in the zone requires that differences between consecutive maximums and minimum must be greater than certain threshold to be detected as vehicle in the image. It can be shown that the incorporation of magnitude information produces a graph in which cars can be classified with an automatic absolute threshold.

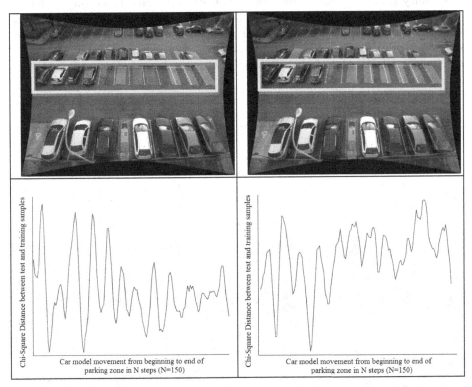

Fig. 4. Comparison between Dissimilarity-Graph using PHOG features, using edge orientations only (left) and edge orientations weighted by magnitude (right) for the second zone in the image. Marked in red those locations detected as cars in the image.

3   Experiments

In order to analyze the performance of our system for both kinds of features we have tested the proposed methods using several video sequences. The sequences were captured using two IP cameras: (1) an AXIS-211 CCD camera with 640 × 480 resolution and frame rate up to 30 fps; (2) AXIS P3346/-VE with 2048 × 1536 resolution and frame rate up to 30 fps. Once the radial distortion is corrected, the image is resized to 640 × 480. Two sceneries were tested. Camera (1) monitors Parking C while camera (2) monitors Parking B. The training dataset was constructed from a group of 15 static images. Each zone within the parking is trained individually. For the car dataset, around 30 samples of every side of the single-car model were extracted. Parking C was tested with frames covering around 6 hours of one day. In contrast, Parking B was tested also with two sequences with frames that cover up to 6 hours. The sequences were recorded at 10 meters of height from a building in front of the parking scene on days with different lighting and weather conditions. The experiments were carried out on a PC with a 3.4 GHz i7 processor and 8GB RAM.

Accuracy (ratio of samples that the system has been able to classify correctly among all samples) is the main indicator used to evaluate performance. Results shown in Table 1 compare results of the proposed approach with the edge density one, where only edge density features were used; the proposed approach providing clearly better performance. Visual results from both parking areas are shown in Figures 5 and 6.

Table 1. Rates using our method based on PHOG features and those relying on edge density features

	Accuracy		
	Parking B Cloudy day	Parking B Sunny day	Parking C
PHOG-based	0. 937	0.821	0.878
Edge density	0.658	0.523	0.853

Fig. 5. Vacant space detection visual results at 'Parking C' context at UPM's premises

Fig. 6. Vacant space detection visual results at 'Parking B' context at UPM's premises

The computational cost for the proposed method depends directly on the number of sides of the car model to be studied. Under these conditions, and using a system like the specified above, one parking space for which only one side is extracted takes up to 2 seconds to be fully classified.

4   Conclusions and Future Work

In this paper, a system for scene reconstruction as well as vacant space detection in a parking lot is proposed. Experiments performed in real parking scenarios show good detection results achieved with the proposed method.

However, further work is needed to make system, on one hand, more robust to possible errors due to manual configuration during initialization process. Tests carried out on two different parking scenes demonstrate that the higher the number of points taken in each direction the better. The selection of points delimiting the whole parking area contributes to the reduction of the extrapolation error.

Besides, if the number of vehicle training samples is reduced too much, the dissimilarity graph is deteriorated and the detection is harder to achieve. It seems to be necessary the introduction of a process able to update progressively the vehicle database that will be used during the classification process always that a new possible candidate is available for testing. In any case, more tests on other parking scenes seem to be needed before conclusions on the suitable minimum number of initial training samples that need to be extracted.

Now, a first stable version has been achieved. Current work at different lighting conditions shows that the proposed method can be used with good accuracy under these circumstances although further testing and adjustment is needed in this sense. Improvements should be made in order to achieve higher rates at several extreme weather conditions, in presence of shadows projected either by cars in the vicinity of a parking space or by buildings closed to the parking area.

On one hand, the incorporation of an adaptive process for establishing the threshold used during classification of vehicles that lead to a more automatic system is also foreseen. On the other hand, temporal analysis and the incorporation of more information about context around occupied spaces should filter, at a higher-level in the processing chain, some of the errors obtained due to misclassification. For instance, in some parking lots the white painted layout present on the parking floor could be used for a precise analysis of each parking space locally. It would probably help calibrating the camera. Finally, the use of a multi-camera system would enhance detection resulting after applying fusion at a higher level of abstraction.

The final objective is to increase the robustness of the proposed system in a broad set of outdoor parking spaces, not only in parking lots but also in the streets.

Acknowledgements. This work has been performed in the framework of the WePark and FOTsis projects. WePark has received research funding from the Spanish Industry, Tourism and Commerce Council and ERDF funds (European Regional Development Fund) under the Project reference TSI-020100-2011-57. The FOTsis project is partly financed by the European Commision in the context of the 7th Framework Programme (FP7) for Research and Technological Development (Grant Agreement no. 270447).

References

1. Swedberg, C.: SF Uses Wireless Sensors to Help Manage Parking. RFID Journal (2007)
2. Degerman, P., Pohl, J., Sethson, M.: Hough transform for parking space estimation using long range ultrasonic sensors. SAE Paper. Document Number: 2006-01-0810 (2006)
3. Satonaka, H., Okuda, M., Hayasaka, S., Endo, T., Tanaka, Y., Yoshida, T.: Development of parking space detection using an ultrasonic sensor. In: 13th World Congress on Intelligent Transportation Systems and Services (2006)
4. Jung, H.G., Cho, Y.H., Yoon, P.J., Kim, J.: Integrated side/rear safety system. In: 11th European Automotive Congress (2007)
5. Schanz, A., Spieker, A., Kuhnert, D.: Autonomous parking in subterranean garages: a look at the position estimation. In: IEEE Intelligent Vehicle Symposium, pp. 253–258 (2003)
6. Gorner, S., Rohling, H.: Parking lot detection with 24 GHz radar sensor. In: 3rd International Workshop on Intelligent Transportation (2006)
7. Foresti, G.L., Micheloni, C., Snidaro, L.: Event classification for automatic visual-based surveillance of parking lots. In: 17th International Conference on Pattern Recognition, vol. 3, pp. 314–317 (2004)
8. Wang, X.G., Hanson, A.R.: Parking lot analysis and visualization from aerial images. In: 4th IEEE Workshop Applications of Computer Vision, pp. 36–41 (1998)

9. Lee, C.H., Wen, M.G., Han, C.C., Kou, D.C.: An automatic monitoring approach for unsupervised parking lots in outdoors. In: 39th Annual International Carnahan Conference, pp. 271–274 (2005)
10. Masaki, I.: Machine-vision systems for intelligent transportation systems. In: IEEE Conference on Intelligent Transportation System, vol. 13(6), pp. 24–31 (1998)
11. Dan, N.: Parking management system and method. US Patent, Pub. No.: 20030144890A1 (2003)
12. Pecharromán, A., Sánchez, N., Torres, J., Menéndez, J.M.: Real-Time Incidents Detection in the Highways of the Future. In: 15th Portuguese Conference on Artificial Intelligence, EPIA 2011, Lisbon, pp. 108–121 (2011)
13. Chen, L., Hsieh, J., Lai, W., Wu, C., Chen, S.: Vision-Based Vehicle Surveillance and Parking Lot Management Using Multiple Cameras. In: 6th International Conference on Intelligent Information Hiding and Multimedia Signal Processing, Washington, DC, pp. 631–634 (2010)
14. True, N.: Vacant Parking Space Detection in Static Images, Projects in Vision & Learning, University of California (2007)
15. SFPark project, http://sfpark.org/ (accessed May 2013)
16. SiPark SSD car park guidance system, Siemens AG (2011)
17. IdentiPark, Nortech Internacional (2013)
18. Kang, S.B., Weiss, R.: Can We Calibrate a Camera Using an Image of a Flat, Textureless Lambertian Surface? In: Vernon, D. (ed.) ECCV 2000, Part II. LNCS, vol. 1843, pp. 640–653. Springer, Heidelberg (2000)
19. Torres, J., Menendez, J.M.: A practical algorithm to correct geometrical distortion of image acquisition cameras. In: IEEE International Conference on Image Processing, vol. 4, pp. 2451–2454 (2004)
20. Brown, D.C.: Decentering distortion of lenses. In: Photogrommetric Eng. Remore Sensing, pp. 444–462 (1966)
21. Zhang, Z.: A flexible new technique for camera calibration. IEEE Transactions on Pattern Analysis and Machine Intelligence 22(11), 1330–1334 (2000)
22. Faugeras, O.D., Luong, Q.-T., Maybank, S.J.: Camera Self-Calibration: Theory and Experiments. In: 2nd European Conference on Computer Vision, pp. 321–334. Springer, London (1992)
23. Hartley, R., Zisserman, A.: Multiple View Geometry in computer vision. Cambridge University Press, Cambridge (2003)
24. Bosch, A., Zisserman, A., Munoz, X.: Representing shape with a spatial pyramid kernel. In: 6th ACM International Conference on Image and Video Retrieval, pp. 401–408. ACM, New York (2007)
25. Förstner, W., Moonen, B.: A metric for covariance matrices. Technical Report, Department of Geodesy and Geoinformatics, Stuttgart University (1999)

Paying the Price of Learning Independently in Route Choice

Ana L.C. Bazzan

PPGC / Instituto de Informática – UFRGS
Caixa Postal 15.064, 91.501-970, Porto Alegre, RS, Brazil
`bazzan@inf.ufrgs.br`

Abstract. In evolutionary game theory, one is normally interested in the investigation about how the distribution of strategies changes along time. Equilibrium-based methods are not appropriate for open, dynamic systems, as for instance those in which individual drivers learn to select routes. In this paper we model route choice in which many agents adapt simultaneously. We investigate the dynamics with a continuous method (replicator dynamics), and with learning methods (social and individual). We show how the convergence to one of the Nash equilibria depends on the underlying learning dynamics selected, as well as on the pace of adjustments by the driver agents.

1 Introduction and Related Work

The number of large metropolitan areas with more than ten million inhabitants is increasing rapidly, with the number of these so-called mega-cities now at more than 20. This increase has strong consequences to traffic and transportation. According to the keynote speaker of the IEEE 2011 forum on integrated sustainable transportation systems, Martin Wachs, *mobility is perhaps the single greatest global force in the quest for equality of opportunity* because it plays a role in offering improved access to other services. Congestion is mentioned as one of the major problems in various parts of the world, leading to a significant decrease in the quality of life, especially in mega-cities of countries experiencing booming economies. Mobility is severely impacted with 4.8 billion hours of travel delay that put the cost of urban congestion in USA alone at 114 billion dollars (`www.its.dot.org`). Moreover environmental costs must be considered.

One important part in the whole effort around intelligent transportation systems (ITS), in which artificial intelligence (AI) plays a role, relates to efficient management of the traffic. This in turn can be achieved, among others, by an efficient assignment of routes, especially if one thinks that in urban scenarios a considerable important part of the network remains sub-utilized whereas jams occur in a small portion of the network.

The conventional process of modeling assignment is macroscopic, based on equilibrium. In the traffic engineering literature, the Nash equilibrium is referred as Wardrop equilibrium [1]. This is convenient as it has sound mathematical grounds. Static traffic assignment models have mathematical properties such as

L. Correia, L.P. Reis, and J. Cascalho (Eds.): EPIA 2013, LNAI 8154, pp. 42–53, 2013.
© Springer-Verlag Berlin Heidelberg 2013

existence and uniqueness of equilibrium. Each link in the transportation network can be described by the so-called volume-delay functions expressing the average or steady-state travel time on a link. However, equilibrium in traffic networks are often inefficient at collective level. Important results regarding this relate to the price of anarchy [2–4] and Braess paradox [5]. Given these results, one may try to improve the load balancing by routing drivers. However, whereas it is possible to do this in particle systems (e.g., data packages in communication networks), this is not the case in social systems.

Other drawbacks of equilibrium-based methods are that they assume stationarity, and rationality plus common knowledge. Regarding the former, it is not clear what equilibria can say about changes in the environment. Regarding rationality and common knowledge, it is well-known that both conceptually and empirically this argument has many problems. Just to mention one of them, in games with more than one equilibria, even if one assumes that players are able to coordinate their expectations using some selection procedure, it is not clear how such a procedure comes to be common knowledge.

Therefore, evolutionary game theory (EGT) offers alternative explanations to rationality and stationarity by focusing on equilibrium arising as a long-run outcome of a process in which populations of bounded rational agents interact over time. See for instance [6] for an extensive discussion about the various approaches based on learning in games. However, we note that one of the EGT approaches, the replicator dynamics (RD), presents some problems regarding its justification in decentralized or multiagent encounters. Thus an explanation for the replicator could be that there is an underlying model of learning (by the agents) that gives rise to the dynamics.

Some alternatives have been proposed in this line. Fudenberg and Levine [6] refer to two interpretations that relate to learning. The first is social learning, a kind of "asking around" model in which players can learn from others in the population. The second is a kind of satisfying-rather-than-optimizing learning process in which the probability of a determined strategy is proportional to the payoff difference with the mean expected payoff. This second variant has been explored, among others, by [7, 8]. In particular, in the stimulus-response based approach of Börgers and Sarin, the reinforcement is proportional to the realized payoff. In the limit, it is shown that the trajectories of the stochastic process converge to the continuous RD. This is valid in a stationary environment. However, this does not imply that the RD and the stochastic process have the same asymptotic behavior when the play of *both* players follow a stimulus-response learning approach.

We remark that [6] specifically mention a two agent or two population game, but the same is true (even more seriously), when it comes to more. The reasons for this difference are manifold. First, Börgers and Sarin's main assumption is "...an appropriately constructed continuous time limit", i.e., a gradual (very small) adjustment is made by players between two iterations of the game. More specifically, the RD treats the player as a population (of strategies). By the construct of the continuous time of [7], in each iteration, a random sample of the population is taken to play the game. Due to the law of large numbers, this sample represents the whole population. However, in the discrete learning process, at each time, only one strategy is played by each agent. Moreover, the outcome of each of these interactions affects the probabilities with which the strategies are used in the next time step. This implies that the discrete learning model evolves stochastically, whereas the equations of the RD are deterministic. Also, there is the fact that players may be stuck in suboptimal strategies because they are all using a learning mechanism, thus turning the problem non-stationary.

These facts have as consequences that other popular dynamics in game-theory as, e.g., best response dynamics, which involves instantaneous adjustments to best replies, depict difference in the asymptotic behavior. Theoretical results regarding an extension by [8] were verified with experiments in 3 classes of 2×2 games. The differences in behavior of the continuous/discrete models were verified, with matching pennies converging to a pure strategy (thus, not a Nash equilibrium), while the RD cycles. This suggests that other dynamics, e.g., based on less gradual adjustments may lead to different results in other games as well.

In summary, there are advantages and disadvantages in using the discussed approaches and interpretations, i.e., the continuous, analytical variant of RD, and learning approaches such as best response, social learning, and stimulus-response based models.

In this paper we aim at applying different approaches and compare their performance. Specifically, we use three of them: the analytical equations of the RD, a kind of social learning where dissatisfied agents ask their peers, and individual Q-learning. We remark that other multiagent reinforcement learning approaches are not appropriate for this problem as they either consider perfect monitoring (observation of other agents' actions, as in [9]), or modeling of the opponent (as in fictitious play), or both. [1] In our case this is not possible given the high number of agents involved and the unlikelihood of encounters happening frequently among the same agents.

Here, the scenario is an asymmetric population game that models traffic assignmen. Populations with different sets of actions interact and the payoff matrix of the stage game is also asymmetric. We note that, most of the literature refers to homogeneous population, two actions, rendering the game symmetric. We are interested in the trajectory of a population of decision-makers with very little knowledge of the game. Indeed, they are only aware of the payoff or reward received, thus departing from the assumption of knowledge of payoff matrix and rationality being common knowledge, frequently made in GT.

In the next section we introduce our method and the formalization of route choice as population games. Experiments and their analysis follow in Section 3. The last section concludes the paper and discusses some future directions.

---

[1] Since a comprehensive review on learning in repeated games is not possible here, we refer the reader to [10, 11] and references therein for a discussion on the assumptions made in the proposed methods.

2   Methods

2.1   Formalization of Population Games

Population games are quite different from the games studied by the classical GT because population-wide interaction generally implies that the payoff to a given member of the population is not necessarily linear in the probabilities with which pure strategies are played. A population game can be defined as follows.

- (**populations**) $\mathcal{P} = \{1, ..., p\}$: society of $p \geq 1$ populations of agents, where $|p|$ is the number of populations
- (**strategies**) $\mathcal{S}^p = \{s_1^p, ..., s_m^p\}$: set of strategies available to agents in population p
- (**payoff function**) $\pi(s_i^p, \mathbf{q}^{-\mathbf{P}})$

Agents in each p have m^p possible strategies. Let n_i^p be the number of individuals using strategy s_i^p. Then, the fraction of agents using s_i^p is $x_i^p = \frac{n_i^p}{N^p}$, where N^p is the size of p. $\mathbf{q}^{\mathbf{P}}$ is the m^p-dimensional vector of the x_i^p, for $i = 1, 2, ..., m^p$. As usual, $\mathbf{q}^{-\mathbf{P}}$ represents the set of $\mathbf{q}^{\mathbf{P}}$'s when excluding the population p. The set of all $\mathbf{q}^{\mathbf{P}}$'s is $\mathbf{q}$. Hence, the payoff of an agent of population p using strategy s_i^p while the rest of the populations play the profile $\mathbf{q}^{-\mathbf{P}}$ is $\pi(s_i^p, \mathbf{q}^{-\mathbf{P}})$.

Consider a (large) population of agents that can use a set of pure strategies $\mathcal{S}^p$. A population profile is a vector σ that gives the probability $\sigma(s_i^p)$ with which strategy $s_i^p \in \mathcal{S}^p$ is played in p.

One important class within population games is that of symmetric games, in which two random members of a *single* population meet and play the stage game, whose payoff matrix is symmetric. The reasoning behind these games is that members of a population cannot be distinguished, i.e., two meet randomly and each plays one role but these need not to be the same in each contest. Thus the symmetry. However, there is no reason to use a symmetric modeling in other scenarios beyond population biology. For instance, in economics, a market can be composed of buyers and sellers and these may have asymmetric payoff functions and/or may have sets of actions whose cardinality is not the same. In the route choice game discussed here, asymmetric games correspond to multi-commodity flow (more than one origin-destination pair).

Before we present the particular modeling of asymmetric population game, we introduce the concept of RD. The previously mentioned idea that the composition of the population of agents (and hence of strategies) in the next generations changes with time suggests that we can see these agents as replicators. Moreover, an evolutionary stable strategy (ESS) may not even exist, given that the set of ESSs is a possibly empty subset of the set of Nash equilibria computed for the normal form game (NFG). In the RD, the rate of use of a determined strategy is proportional to the payoff difference with the mean expected payoff, as in Eq. 1.

$$\dot{x}_i^p = (\pi(s_i^p, \mathbf{x}^{\mathbf{P}}) - \bar{\pi}(\mathbf{x}^{\mathbf{P}})) \times x_i^p \qquad (1)$$

The state of population p can be described as a vector $\mathbf{x}^{\mathbf{P}} = (x_i^p, ..., x_m^p)$. We are interested in how the fraction of agents using each strategy changes with time,

Table 1. Available routes in the three populations

route	description	length
G0	$B1 \to C1 \to C3 \to E3 \to E5$	7
S0	$B1 \to F1 \to F2 \to E2 \to E5$	9
B0	$B1 \to F1 \to F4 \to E4 \to E5$	9
G1	$A2 \to A3 \to E3 \to E5$	7
S1	$A2 \to A5 \to E5$	10
B1	$A2 \to A6 \to F6 \to F4 \to E4 \to E5$	13
G2	$D5 \to D3 \to E3 \to E5$	5
S2	$D5 \to D4 \to C4 \to C5 \to E5$	5

Table 2. Payoff matrices for the three-player traffic game; payoffs are for player 0 / player 1 / player 2 (the three Nash equilibria in pure strategies are indicated in boldface)

G2

	G1	S1	B1
G0	1/1/4	**5/6/7**	5/1/7
S0	3/4/6	4/6/8	4/1/8
B0	5/5/7	**5/6/8**	4/0/9

S2

	G1	S1	B1
G0	4/4/8	7/4/6	7/1/8
S0	4/6/8	5/4/6	5/1/8
B0	**5/7/8**	5/4/6	4/0/8

i.e., the derivative $\frac{dx_i^p}{dt}$ (henceforth denoted $\dot{x}_i^p$). In Eq. 1, $\bar{\pi}(\mathbf{x^P})$ is the average payoff obtained by p:

$$\bar{\pi}(\mathbf{x^P}) = \sum_{i=1}^{m} x_i^p \pi(s_i^p, \mathbf{x^P})$$

Obviously, to analytically compute this average payoff, each agent would have to know all the payoffs, which is quite unrealistic in real scenarios.

Henceforth, in order to illustrate the approach and introduce the evaluation scenario, we refer to a specific instance. However this has some important properties: non-symmetry and presence of several equilibria.

In the three-population game considered here, to avoid confusion we use the term "player" with its classical interpretation, i.e., the decision-makers of the normal form game (NFG). Because this game is played by randomly matched individuals, one from each population, we call these individuals "agents". Thus player refers to a population of agents.

The way the three populations interact determines their reward functions. For the sake of illustration, it is assumed that the three populations of agents use a road network Γ to go from their respective origins, and that there is a single destination (typical morning peak).

Each agent in each p has some alternative routes or strategies $\mathcal{S}^p$. These are named after the following reasoning: G means greedy selection (G is the most preferred because this route yields the highest utility *if not shared with other populations*); S means second preferred alternative; and B means border route

Table 3. Five Nash equilibria for the three-player traffic game

profile	G0 x_0^0	S0 x_1^0	B0 $(1 - x_0^0 - x_1^0)$	G1 x_0^1	S1 x_1^1	B1 $(1 - x_0^1 - x_1^1)$	G2 x_0^2	S2 $(1 - x_0^2)$	payoff
σ_a	1	0	0	0	1	0	1	0	5/6/7
σ_b	0	0	1	0	0	1	1	0	5/6/8
σ_c	0	0	1	1	0	0	0	1	5/7/8
σ_d	0	0	1	$\frac{2}{3}$	$\frac{1}{3}$	0	$\frac{3}{4}$	$\frac{1}{4}$	5/5.5/ ≈ 7.3
σ_e	≈ 0.474	0	≈ 0.526	≈ 0.386	≈ 0.614	0	≈ 0.352	≈ 0.648	5/$\approx$ 4.7/$\approx$ 6.8

(a route that uses the periphery of Γ). Populations $p = 0$ and $p = 1$ have strategies $\mathcal{S}^0 = \{G0, S0, B0\}$ and $\mathcal{S}^1 = \{G1, S1, B1\}$. $p = 2$ has only two strategies: $\mathcal{S}^2 = \{G2, S2\}$. Combining all these sets, there are 18 possible assignments of routes.

Each agent selects a route and the payoff obtained is a function of the delay on the route taken. The delay in each route is the sum of delays on each link in the route. These are given by a volume-delay function (VDF). The VDF used in the present paper considers the number of agents using each link. Specifically, it adds 1 unit each time an agent uses a given link. This way, a link has cost 0 if no agent uses it; cost 1 if one agent uses it; and so forth. The only exception is a bottleneck link b that belongs to the greedy routes G0, G1, and G2. Link b does not accommodate all agents. Thus if too many agents use it at the same time, each receives a penalty of 1 unit. Hence, considering the 18 combinations of routes that appear in Table 1, costs depend on length of routes and how routes share the Γ. The maximum cost is incurred by agents in $p = 1$ when the following combination of route choices is made: B0 / B1 / G2. This cost is 13. In order to deal with maximization (of payoff) rather than cost minimization, costs are transformed in payoffs. The highest cost of 13 is transformed in reward zero and so on. Payoffs computed this way are given in Table 2. Note that strategy B1 is dominated or $p = 1$, thus the learning models tested here must be able to recognize this.

Values in Table 2 represent an arbitrary assignment of utility of the three players involved, based on the topology of Γ as explained. The utility function $u(.)$ that underlies Table 2 is however equivalent to any other $\hat{u}(.)$ if $\hat{u}(.)$ represents identical preferences of the players, and $u(.)$ and $\hat{u}(.)$ differ by a linear transformation of the form $\hat{u}(.) = A \times u(.) + B$, $A > 0$. Of course equivalence here refers to the solution concept, i.e., a qualitative, not quantitative concept. Equivalent game models will make the same prediction or prescription.

For the three-population game whose payoffs are given in Table 2, there are five Nash equilibria. All appear in Table 3. In this table, columns 2–4 specify $\mathbf{x^0}$ (fraction of agents selecting each strategy s_i^0 in population $p = 0$), columns 5–7 specify $\mathbf{x^1}$ and the last two columns specify $\mathbf{x^2}$. For example, for the first equilibrium (profile σ_a), because $x_0^0 = 1$, $x_1^1 = 1$, and $x_0^2 = 1$, all agents in $p = 0$ select action G0, all agents in $p = 1$ select S1 and all agents in $p = 2$ select G2. Regarding the mixed strategy profile σ_d, all agents in $p = 0$ select action B0 (because $x_0^0 = x_1^0 = 0$), whereas in $p = 1$, $\frac{2}{3}$ of agents select G1 and $\frac{1}{3}$ select S1. In $p = 2$, $\frac{3}{4}$ of agents select G2 and $\frac{1}{4}$ select S2. Profiles σ_b, σ_c, and σ_e can be similarly interpreted.

It must also be noticed that in asymmetric games, all ESS are pure strategies (for a proof see, e.g., [12]). Thus only σ_a, σ_b, and σ_c are candidates for ESS. Besides, clearly, among σ_a, σ_b, and σ_c, the first two are (weakly) Pareto inefficient because σ_c is an outcome that make all agents better off.

2.2   Replicator Dynamics, Social and Individual Learning

As mentioned, the continuous RD model, which is hard to justify, can be reproduced with some forms of learning. To compare the performance of these learning models, we first formulate the continuous RD for our specific three-population game. The equation for $\dot{x}_0^0$ (others are similar), derived from Eq. 1, is: $\dot{x}_0^0 = x_0^0(-x_0^2 x_0^1 - 2x_0^2 - 4x_0^1 + 3 + x_0^0 x_0^2 x_0^1 + 2x_0^2 x_0^0 + 4x_0^0 x_0^1 - 3x_0^0 + x_0^2 x_1^1 + 2x_1^0 x_0^1 - x_1^0 - x_1^1 + x_0^0 x_1^1 + x_1^0 x_1^1)$

The three Nash equilibria that need to be investigated are those in pure strategies (σ_a, σ_b, and σ_c). We have analytically checked that only σ_c is an ESS (by means of the divergence operator, to find out where all derivatives are negative). This way, it was verified that the only Nash equilibria where all derivatives are negative is σ_c.

Now we turn to the learning models. In both models reported below, in each time step, agents from each population p play g games in which payoffs are as in Table 2.

For the social learning, we use one of the possibilities mentioned in [6], which is based on an ask-around strategy. This of course involves at least some sampling of other agents' rewards. However, it does not involve sophisticated modeling as perfect monitoring. It works as follows: when dissatisfied with their own rewards, some agents ask around in their social network and eventually change strategy. To replicate this behavior, we use an ask-around rate p_a: at each time step, with probability p_a each agent in the population p copies the strategy of a better performing acquaintance. The higher p_a, the more "anxious" is the agent (i.e., the faster it gets dissatisfied). We recall that, according to [7], it is expected that if the adjustment is not gradual, there may be no convergence to the behavior of the continuous RD.

For the individual learning, no ask-around strategy is used. Rather, agents learn using individual Q-learning (Eq. 2), thus assessing the value of each strategy by means of Q values.

$$Q(s,a) \leftarrow Q(s,a) + \alpha\,(r + \gamma\,max_{a'}\,Q(s',a') - Q(s,a)) \qquad (2)$$

For action selection, ε-greedy was used. In line with the just mentioned issue of gradual adjustments, and from [8], we know that the value of ε is key to reproduce the behavior of the continuous RD.

3   Experiments and Results

In this section we discuss the numerical simulations of the learning based approaches and compare them with the continuous RD, from which we know the

Nash equilibria, and that the only ESS is profile σ_c. With the learning models, we are interested in investigating issues such as what happens if each population p starts using a given profile σ^p in games that have more than one equilibrium. To which extent the rate p_a shifts this pattern?

The main parameters of the model, as well as the values that were used in the simulations are: $\mathcal{P} = \{0, 1, 2\}$, $N^0 = N^1 = N^2 = 300$, $g = 10,000$, $\alpha = 0.5$; ε, Δ (number of time steps) and p_a were varied. In all cases, at step 0, agents select strategies from a uniform distribution of probability.

We first discuss the social learning. Because five variables (strategies) are involved, it is not possible to show typical (2d) RD-like plots that depict the trajectory of these variables. Therefore, as an alternative to show the dynamics, we use heatmaps. In the plots that appear in Figures 1 to 3, heatmaps are used to convey the idea of the intensity of the selection of each of the 18 joint actions (represented in the vertical axis) along time (horizontal axis), with $\Delta = 1000$. Due to an internal coding used, the 18 joint actions are labeled such that 0, 1 and 2 mean the selection of G, S, or B respectively. The order of the triplet is such that the first digit indicates the action of $p = 2$, the second digit is for the action of $p = 1$, and the third digit is for $p = 0$. In particular, the three Nash equilibria (σ_a, σ_b, and σ_c) are represented as 010 (G2-S1-G0), 012 (G2-S1-B0), and 102 (S2-G1-B0).

In the heatmaps, to render them cleaner we just use shades of gray color (instead of hot colors as usual). In any case, the darker the shade, the more intense one joint action is selected. Thus we should expect that the three Nash equilibria correspond to the darker strips.

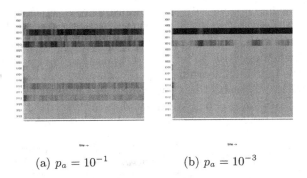

(a) $p_a = 10^{-1}$ (b) $p_a = 10^{-3}$

Fig. 1. Evolution of dynamics for different p_a

In Figure 1 we show how the selection evolves along time, for relatively high p_a. Figure 1(a) is for $p_a = 10^{-1}$, i.e., each agent asks around with this rate. It is possible to see that although σ_a (010) is clearly selected more frequently, other joint actions also appear often, as, e.g. 012. Interestingly, their counterparts 110 and 112, which differ from 010 and 012 by $p = 2$ selecting S2 instead of G2, also appear relatively often. This indicates that agents in $p = 2$ try to adapt

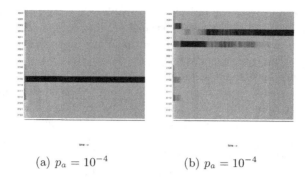

(a) $p_a = 10^{-4}$ (b) $p_a = 10^{-4}$

Fig. 2. Evolution of dynamics for $p_a = 10^{-4}$

to the other two populations. In the end the performance is poor because this co-adaptation is disturbed by the high rate of changes by the social agents.

This overall picture improves a little bit with the reduction in p_a. When $p_a = 10^{-2}$ (not shown) and $p_a = 10^{-3}$, (Figure 1(b)) the convergence pattern is clearer but still it is not possible to affirm that one profile has established.

When we decrease the rate to $p_a = 10^{-4}$ (Figure 2), it is possible to observe that one of the two cases occur: either profile 102 (σ_c) establishes right in the beginning (Figure 2(a)), or there is a competition between 010 and 012, with one or the other ending up establishing. For $p_a = 10^{-5}$ the pattern is pretty much the same as for $p_a = 10^{-4}$.

With further decrease in p_a, the time needed to either 010 or 012 establish decreases, if 102 has not already set. For instance, comparing Figure 3(b) to Figure 2(b), one sees that profile σ_a (010) established before in the former case.

A remark is that the dominated strategy B1, in $p = 1$, is quickly discarded in all cases, even when $p_a = 10^{-1}$.

Regarding the individual learning, experiments were run with different values of α and change in ε. It seems that α has much less influence in the result than ε. Thus we concentrate on $\alpha = 0.5$. We show plots for ε starting at 1.0 with decay

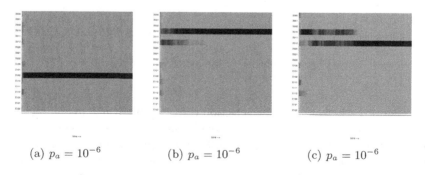

(a) $p_a = 10^{-6}$ (b) $p_a = 10^{-6}$ (c) $p_a = 10^{-6}$

Fig. 3. Evolution of dynamics for $p_a = 10^{-6}$

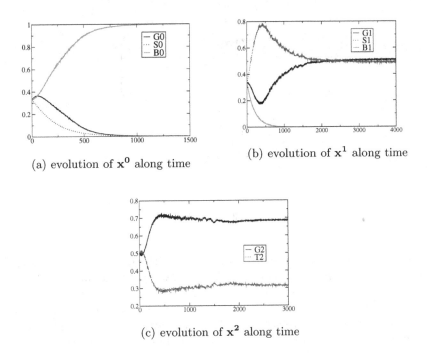

(a) evolution of $\mathbf{x^0}$ along time

(b) evolution of $\mathbf{x^1}$ along time

(c) evolution of $\mathbf{x^2}$ along time

Fig. 4. Evolution of each $\mathbf{x^p}$ (y axis) along time (x axis) using QL

of 0.995 each time step, which, at the end of $\Delta = 5000$ turns ε of the order of 10^{-11}. Figure 4 depicts the evolution of each component of each vector $\mathbf{x^p}$ (i.e., state of each population in terms of strategies selected) along time. As seen in the three plots, B0 establishes for population $p = 0$, G1 and S1 are selected with near the same probabilities in $p = 1$, and G2 converges to nearly 0.7. This pattern is not a Nash equilibrium. However, the payoffs for the three population of agents are: ≈ 5 ($p = 0$), ≈ 5.4 ($p = 1$), and ≈ 7.3 ($p = 2$), which are very close to σ_d (see Table 3). Note that strategy B1, dominated, is quickly discarded (Figure 4(b)).

How agents have converged to this behavior is better explained by the trajectories of the probabilities to play each strategy, for each population. Due to the number of variables, it is not possible to plot them all together. Even a 3d plot, where one could see at least one variable per population, was rendered not informative. Thus we opted to show the behavior of selected variables in a pairwise fashion, namely B0 x G1, B0 x G2, G1 x G2 (Figure 5). It is possible to see that the fraction of agents using G2 (x_0^2), the black line, vertical component, has reached $\approx \frac{3}{4}$ (as in profile σ_d), but this was not stable and there was a shift to lower values, which has influenced also the fraction of agents using G1 (blue circles, vertical), where we also see a shift.

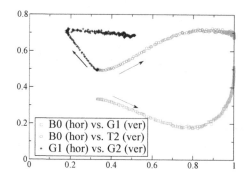

Fig. 5. Trajectories: B0 x G1, B0 x S2, G1 x G2

In short, some conclusions can be drawn from these simulations. First, simultaneous learning by the agents does not always lead to a Nash equilibrium when there are more than one of them, much less to the ESS computed for the corresponding RD of the static NFG. Whereas there is only one ESS among the three Nash equilibria in pure strategies (σ_c), depending on the p_a rate, and on the exploration rate ε, any of the three Nash equilibria (in pure strategies) may establish, or agents may be stuck at a sub-optimal state, even if, as in the case shown here, this state is very close to one of the Nash equilibria. This is in line with the result in [7], which prescribes gradual adjustments. The profile σ_c does establish fast (and does not shift) if it sets at all. When this is not the case, there is a competition between the other two. This competition is determined by agents in $p = 2$: from the payoff matrix (Table 2), one can see that only agents in $p = 2$ have different payoffs in profiles σ_a and σ_b. This leads to an important remark: agents in $p = 2$ have an asymmetric role in this game (not only due to the fact that they have less options), given that the establishment of profile σ_c is also related to a choice by agents in $p = 2$. Thus, in real-world traffic networks, such a study may prove key to determine which fraction of drivers to target when controlling traffic by means of route guidance.

4   Conclusion and Future Work

This paper contributes to the effort of analyzing route choices among population of agents. It shows that the use of models that assume stationarity may fail. An alternative approach is provided that helps the analysis of the dynamics of the RD of the static game, as well as the dynamics provided by two learning methods. In the case of the social learning, convergence is achieved depending on the rate of experimentation in the populations. Thus, anxious drivers may be stuck at sub-optimal equilibrium. For the individual Q-learning, results can be extrapolated to cases in which drivers tend to make too much experimentation (e.g., in response to broadcast of route information). In the case illustrated here, agents converge to a solution close to a Nash equilibrium, which is not an ESS. We remark that the use of a standard macroscopic modeling method (common practice in traffic engineering) would not have provided such insights.

Although the scenario used as illustration considers three populations only, each having a few actions, we claim that this is not an unrealistic simplification. In fact, in the majority of the situations a traffic engineer has to deal with, there is a small number of commodities (origin-destination pairs) that really matter, i.e., end up collecting (thus representing) several sources and sinks. Regarding the number of actions, it is equally the case that in the majority of the real-world cases drivers do not have more than a handful of options to go from A to B.

A future direction is to explicitly consider information broadcast to agents, in order to have a further way to model action selection.

Acknowledgement. The author is partially supported by CNPq. This project was funded by FAPERGS.

References

1. Wardrop, J.G.: Some theoretical aspects of road traffic research. Proceedings of the Institute of Civil Engineers 1, 325–362 (1952)
2. Koutsoupias, E., Papadimitriou, C.: Worst-case equilibria. In: Meinel, C., Tison, S. (eds.) STACS 1999. LNCS, vol. 1563, pp. 404–413. Springer, Heidelberg (1999)
3. Papadimitriou, C.H.: Algorithms, games, and the internet. In: Proc. of the 33rd ACM Symp. on Theory of Computing (STOC), pp. 749–753. ACM Press (2001)
4. Roughgarden, T., Tardos, É.: How bad is selfish routing? J. ACM 49, 236–259 (2002)
5. Braess, D.: Über ein Paradoxon aus der Verkehrsplanung. Unternehmensforschung 12, 258 (1968)
6. Fudenberg, D., Levine, D.K.: The Theory of Learning in Games. The MIT Press (1998)
7. Börgers, T., Sarin, R.: Learning through reinforcement and replicator dynamics. Journal of Economic Theory 77, 1–14 (1997)
8. Tuyls, K., Hoen, P.J., Vanschoenwinkel, B.: An evolutionary dynamical analysis of multi-agent learning in iterated games. Autonomous Agents and Multiagent Systems 12, 115–153 (2006)
9. Claus, C., Boutilier, C.: The dynamics of reinforcement learning in cooperative multiagent systems. In: Proceedings of the Fifteenth National Conference on Artificial Intelligence, pp. 746–752 (1998)
10. Panait, L., Luke, S.: Cooperative multi-agent learning: The state of the art. Autonomous Agents and Multi-Agent Systems 11, 387–434 (2005)
11. Shoham, Y., Powers, R., Grenager, T.: If multi-agent learning is the answer, what is the question? Artificial Intelligence 171, 365–377 (2007)
12. Webb, J.N.: Game Theory – Decisions, Interaction and Evolution. Springer, London (2007)

On Predicting the Taxi-Passenger Demand: A Real-Time Approach

Luis Moreira-Matias[1,2,3], João Gama[2,4], Michel Ferreira[1,5], João Mendes-Moreira[2,3], and Luis Damas[6]

[1] Instituto de Telecomunicações, 4200-465 Porto, Portugal
[2] LIAAD-INESC TEC, 4200-465 Porto, Portugal
[3] Dep. de Engenharia Informática, Faculdade de Engenharia, U. Porto, 4200-465 Porto, Portugal
[4] Faculdade de Economia, U. Porto 4200-465 Porto, Portugal
[5] Dep. de Ciência de Computadores, Faculdade de Ciências, U. Porto, 4169-007 Porto, Portugal
[6] Geolink, Lda., Avenida de França, 20, Sala 605, 4050-275 Porto, Portugal
{luis.m.matias,jgama,joao.mendes.moreira}@inescporto.pt,
michel@dcc.fc.up.pt, luis@geolink.pt

Abstract. *Informed driving* is becoming a key feature to increase the sustainability of taxi companies. Some recent works are exploring the data broadcasted by each vehicle to provide live information for decision making. In this paper, we propose a method to employ a learning model based on historical GPS data in a real-time environment. Our goal is to predict the spatiotemporal distribution of the Taxi-Passenger demand in a short time horizon. We did so by using learning concepts originally proposed to a well-known online algorithm: the *perceptron* [1]. The results were promising: we accomplished a satisfactory performance to output the next prediction using a short amount of resources.

Keywords: taxi-passenger demand, online learning, data streams, GPS data, auto-regressive integrated moving average (ARIMA), perceptron.

1 Introduction

The rising cost of fuel has been decreasing the profit of both taxi companies and drivers. It causes an unbalanced relationship between passenger demand and the number of running taxis, thus decreasing the profits made by companies and also the passenger satisfaction levels [2]. S. Wong presented a relevant mathematical model to express this need for equilibrium in distinct contexts [3]. An equilibrium fault may lead to one of two scenarios: (Scenario 1) excess of vacant vehicles and excessive competition; (Scenario 2) larger waiting times for passengers and lower taxi reliability [3,4]. However, a question remains open: Can we guarantee that the taxi spatial distribution over time will always meet the demand? Even when the number of running taxis already does?

The taxi driver mobility intelligence is an important factor to maximize both profit and reliability within every possible scenario. Knowledge about where the

L. Correia, L.P. Reis, and J. Cascalho (Eds.): EPIA 2013, LNAI 8154, pp. 54–65, 2013.

services (i.e. the transport of a passenger from a pick-up to a drop-off location) will actually emerge can be an advantage for the driver - especially when there is no economic viability of adopting random cruising strategies to find their next passenger. The GPS historical data is one of the main variables of this topic because it can reveal underlying running mobility patterns. Today, the majority of the public transportation networks have vehicles equipped with this technology. This kind of data represents a new opportunity to learn/predict relevant patterns while the network is operating (i.e. in real-time).

Multiple works in the literature have already explored this kind of data successfully with distinct applications like smart driving [5], modeling the spatiotemporal structure of taxi services [6–8] or building intelligent passenger-finding strategies [9, 10]. Despite their useful insights, the majority of techniques reported are tested using offline test-beds, discarding some of the main advantages of this kind of signal. In other words, they do not provide any live information about the passenger location or the best route to pick-up one in this specific date/time while the GPS data is mainly a data stream. In this work, we are focused into predicting the short-term spatiotemporal distribution of the taxi-passenger demand by using machine learning algorithms capable of learning and predicting in a real-time environment.

One of the most recent advances on this topic was presented by Moreira-Matias *et. al* in [4, 11]. They propose a discrete time series framework to predict the event count (i.e. number of services) for the next P-minutes with a periodicity τ of 30 minutes. This framework handles three distinct types of memory: 1) short term (ARIMA - AutoRegressive Integrated Moving Average [12]), 2) mid term and 3) long term one (both based in time-varying poisson models [13]). This model presented three main contributions facing the existing literature [4]:

1. It builds accurate predictions on a stream environment(i.e. using a real-time test bed);
2. Part of the model is able to *forget* some past data by summarizing it into *sufficient* statistics;
3. It is able to update itself on a short amount of time[1] reusing the last real event count to learn about the novelty thereby introduced;

However, such approach presents two relevant limitations: 1) it just produces predictions each 30 minutes while the decision process is made in real-time (i.e. can we guarantee that a prediction made at 8:00am is still *informative* at 8:20am?); 2) the ARIMA weights are fitted (i.e. re-calculated using an offline learning process) and not updated before each prediction by reusing the entire time series of recent event counts plus the most recent one. A research question arises from this analysis: **Can this model handle with a real-time granularity** (i.e. to build predictions per each period of 5, 2, 1 minute or even *on demand*)?

---

[1] An averaged processing time of 99.77 seconds is reported in [4] to update the predictive model and to build the next prediction.

In fact, this framework is divided into four components that work according to distinct computational models: 1,2) while the two time varying Poisson models are incremental learning methods, 3) the ARIMA model employed is an offline one. Finally, the 4) ensemble method used is an online learning. To ease the interpretation of the system characteristics, some definitions about computational learning methods are presented below.

- **Offline Learning:** a method able to learn a predictive model from a *finite* set of instances where the post-training queries do not improve its previous training [14];

- **Incremental Learning:** a method able to learn and update its predictive model *as long as* the true labels of the input samples are known (i.e. a stepwise method where each step uses one or more samples) [15];

- **Online Learning:** an *incremental* learning method which is able to update the model *every time* a true label of a newly arrived sample is known (i.e. it learns from one instance at time) [14];

- **Real-Time Learning:** an online process able to operate in *real-time* (i.e. to use the last sample *true label* to update the predictive model *before* the next sample arrives) [16];

In this paper, we propose a way to minimize the limitations previously described as much as possible by 1) constantly updating the historical time series aggregation of events and by 2) proposing an *incremental* framework to update the ARIMA weights using just the most recent real event count. We tested our improved model by using two distinct case studies: A) a large-sized fleet of 441 vehicles running in the city of Porto, Portugal and B) a small-sized fleet of 89 vehicles running in the cities of Odivelas and Loures, Portugal. While case study A corresponds to a Scenario 1 city - where each vehicle waits on average 44 minutes to pick-up a passenger - case study B is a Scenario 2 one, where the expected waiting time to pick-up a passenger is just 21 minutes.

As input, we considered the data about the services dispatched in each stand over the time. As output, we produced predictions about the demand to arise in the next 30 minutes ($P = 30$) with a periodicity $\tau = 5$ minutes. The test-bed ran continuously for 9 and 6 months for case studies A and B, respectively. However, we just produced outputs for the last four and two months of data, respectively. The results demonstrated the usefulness of our contribution by reducing the typical processing time to each in more than 40% (37.92 seconds) maintaining a satisfying performance - a maximum aggregated error of 24% was accomplished on both case studies.

The remainder of the paper is structured as follows. Section 2 revises the predictive model, describing the extension hereby proposed. The third section describes how we acquired and preprocessed the dataset used as well as some statistics about it. The fourth section describes how we tested the methodology in two concrete scenarios: firstly, we introduce the experimental setup and metrics used to evaluate our model; then, the obtained results are detailed. Finally, conclusions are drawn as well as our future work.

2 Methodology

The model previously proposed in [4, 11] is mainly an incremental one - it just keeps sufficient statistics about the input data constantly updated using the majority of the recently arrived samples. Since the present time series of historical data is a discrete one, it is not easy to propose a true real-time predictive model. However, we can accomplish a good approximation by reducing the prediction periodicity τ from 30 to 5 minutes. The main challenge relies into doing this without increasing the needs on computational resources. In this section, we firstly revisit the model definition. Secondly, we detail how we can maintain an aggregation without recalculating the entire time series in each period. Finally, we propose an *incremental* ARIMA model.

2.1 A Review on the Predictive Model

The following model is deeply described in section II in [4]. Let $S = \{s_1, s_2, ..., s_N\}$ be the set of N taxi stands of interest and $D = \{d_1, d_2, ..., d_j\}$ be a set of j possible passenger destinations. Consider $X_k = \{X_{k,0}, X_{k,1}, ..., X_{k,t}\}$ to be a discrete time series (aggregation period of P-minutes) for the number of demanded services at a taxi stand k. Our goal is to build a model which determines the set of service counts $X_{k,t+1}$ for the instant $t + 1$ per each taxi stand $k \in \{1, ..., N\}$. To do so, we propose three distinct short-term prediction models and a well-known data stream ensemble framework to use them all. We formally describe those models along this section.

Time-Varying Poisson Model. Consider the probability to emerge n taxi assignments in a determined time period - $P(n)$ - following a **Poisson Distribution**. We can define it using the following equation

$$P(n; \lambda) = \frac{e^{-\lambda}\lambda^n}{n!} \tag{1}$$

where λ represents the rate (averaged number of the demand on taxi services) in a fixed time interval. However, in this specific problem, the rate λ is not constant but time-variant. So, we adapt it as a function of time, i.e. $\lambda(t)$, transforming the Poisson distribution into a non homogeneous one. Let λ_0 be the average (i.e. expected) rate of the Poisson process over a full week. Consider $\lambda(t)$ to be defined as follows

$$\lambda(t) = \lambda_0 \delta_{d(t)} \eta_{d(t),h(t)} \tag{2}$$

where $\delta_{d(t)}$ is the relative change for the weekday $d(t)$ (e.g.: Saturdays have lower day rates than Tuesdays); $\eta_{d(t),h(t)}$ is the relative change for the period

$h(t)$ on the day $d(t)$ (e.g. the peak hours); $d(t)$ represents the weekday 1=Sunday, 2=Monday, ...; and $h(t)$ the period in which time t falls (e.g. the time 00:31 is contained in the period 2 if we consider 30-minutes periods).

Weighted Time Varying Poisson Model. The model previously presented can be faced as a time-dependent average which produces predictions based on the long-term historical data. However, it is not guaranteed that every taxi stand will have a highly regular passenger demand: actually, the demand at many stands can often be **seasonal**. To face this specific issue, we propose a weighted average model based on the one presented before: our goal is to increase the relevance of the demand pattern observed in the previous week by comparing it with the patterns observed several weeks ago (e.g. what happened on the previous Tuesday is more relevant than what happened two or three Tuesdays ago). The weight set ω is calculated using a well-known time series approach to these kind of problems: the Exponential Smoothing [17].

AutoRegressive Integrated Moving Average Model. The AutoRegressive Integrated Moving Average Model (ARIMA) is a well-known methodology to both model and forecast univariate time series data. A brief presentation of one of the simplest ARIMA models (for non-seasonal stationary time series) is enunciated below. For a more detailed discussion, the reader should consult a comprehensive time series forecasting text such as Chapters 4 and 5 in [12].

In an autoregressive integrated moving average model, the future value of a variable is assumed to be a linear function of several past observations and random errors. We can formulate the underlying process that generates the time series (taxi service over time for a given stand k) as

$$\begin{aligned} R_{k,t} = \kappa_0 + \phi_1 X_{k,t-1} + \phi_2 X_{k,t-2} + ... + \phi_p X_{k,t-p} \\ + \varepsilon_{k,t} - \kappa_1 \varepsilon_{k,t-1} - \kappa_2 \varepsilon_{k,t-2} - ... - \kappa_q \varepsilon_{k,t-q} \end{aligned} \tag{3}$$

where $R_{k,t}$ and $\{\varepsilon_{k,t}, \varepsilon_{k,t-1}, \varepsilon_{k,t-2}, ...\}$ are the actual value at time period t and the *Gaussian white noise'* error terms observed in the past signal, respectively; $\phi_l (l = 1, 2, ..., p)$ and $\kappa_m (m = 0, 1, 2, ..., q)$ are the model parameters/weights while p and q are positive integers often referred to as the order of the model.

Sliding Window Ensemble Framework. How can we combine them all to improve our prediction? In the last decade, regression and classification tasks on streams attracted the community attention due to its drifting characteristics. The ensembles of such models were specifically addressed due to the challenge related with. One of the most popular models is the weighted ensemble [18]. The model we propose below is based on this one.

Consider $M = \{M_1, M_2, ..., M_z\}$ to be a set of z models of interest to model a given time series and $F = \{F_{1t}, F_{2t}, ..., F_{zt}\}$ to be the set of forecasted values to the next period on the interval t by those models. The ensemble forecast E_t is obtained as

$$E_t = \sum_{i=1}^{z} \frac{F_{it} * (1 - \rho_{iH})}{\Upsilon}, \Upsilon = \sum_{i=1}^{z} (1 - \rho_{iH}) \tag{4}$$

where ρ_{iH} is the error of the model M_i in the periods contained in the time window $[t - H, t]$ (H is a user-defined parameter to define the window size) while compared with the real service count time series. As the information is arriving in a continuous manner for the next periods $t, t + 1, t + 2, ...$ the window will also **slide** to determine how the models are performing in the **last H periods**.

2.2   How Can We Update the Time Series Aggregation without a High Computational Effort?

In this paper, we propose to use the model previously described to build predictions with a periodicity τ for the next period of P-minutes (where $P >> \tau \wedge P$ mod $\tau = 0$). However, it requires a *newly* calculated discrete time series each τ-minutes. How can we do this calculation without a high computational effort? One of the main ways to do so is to perform an **incremental discretization** [19]. An event count X_t in an interval $[t, t + P]$ will be very *similar* to the count X_{t+1} in the interval $[t + \tau, t + P + \tau]$ (as much as $\tau \sim 0$). We can formulate it as

$$X_{t+1} = X_t + X'_{[t+P, t+P+\tau]} - X'_{[t, t+\tau]} \tag{5}$$

where X' represents both the continuous event count on the first τ-minutes of the interval $[t, t + P]$ and on the τ-minutes immediately after the same period. By generalization, we can define two discrete time series of services demand on a taxi stand k as $X_k = \{X_{k,0}, X_{k,1}, ..., X_{k,t}\}$ and $Y_k = \{Y_{k,0}, Y_{k,1}, ..., Y_{k,t'}\}$ (where $t' \geq t$) using granularities of P and τ minutes, respectively. Let X'_k be the discrete time series needed to predict the event count on the interval $[t', t' + \tau]$. We can define the event count at the time period $[t', t' + P]$ as following

$$X'_{k,t'} = \begin{cases} X'_{k,t'-1} + Y_{k,t'+\theta-1} - Y_{k,t'-1}, \theta = P/\tau \text{ if } t' > t \\ X_{k,t} \text{ if } t' = t \end{cases} \tag{6}$$

We take advantage of the *additive* characteristics of both time series to rapidly calculate a new series of interest maintaining two aggregation levels/layers: P and τ. An illustrative example about how this series can be calculated is presented in Fig. 1.

2.3   An *Incremental* ARIMA Model

The ARIMA model relies on calculating the present event count using a linear combination of previous samples. In eq. 3 the $\phi_l(l = 1, 2, ..., p)$ and $\kappa_m(m = 0, 1, 2, ..., q)$ are the model weights. Such weights usually need to be fitted using the entire historical time series every time we build a new prediction (i.e. an offline learning process). This operation can represent a high computational cost if we employ it at such a large scale as we do here.

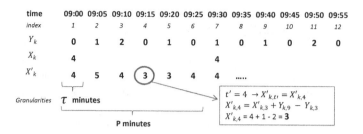

Fig. 1. An example about how can we *additively* calculate one term of the series $X'_{k,t}$

To overcome this issue, we propose to use the *delta rule* to update these weights recursively instead of re-calculating them iteratively as we did so far. The *delta rule* is a gradient descent learning rule commonly used for updating weights of online learning algorithms. It was firstly introduced in [1] to update the weights of the artificial neurons in a single-layer *perceptron*. This rule consists of updating the weights by increasing/decreasing them using a *direct proportion* of the difference between the predicted and the real output. Consider $R = \{R_{k,1}, R_{k,2}, ..., R_{k,t}\}$ to be a time series with the number of services predicted for a taxi stand of interest k in the period $[1, t]$ and $X = \{X_{k,1}, X_{k,2}, ..., X_{k,t}\}$ the number of services actually emerged in the same conditions. Let $w_{k,t} = \{w_{k,t,1}, w_{k,t,2}, ..., w_{k,t,z}\}$ be a set of z weights of a predictive model of interest (like ϕ and κ in the ARIMA one) used to calculate $R_{k,t}$. The update set $\Delta w_{k,t} = \{\Delta w_{k,t,1}, ..., \Delta w_{k,t,j}\}$ can be calculated as follows

$$\Delta w_{k,t,j} = \beta(R_{k,t} - X_{k,t})w_{k,t,j} \tag{7}$$

where β is an user-defined proportionally constant which sets how reactive the model should be and $j \in \{1, ..., z\}$. This way, the ARIMA weights can be *incrementally* updated.

3   Data Acquisition and Preprocessing

In case study A, we focused on the event data stream of a taxi company (which contains 441 running vehicles) operating in Porto, Portugal between Aug. 2011 and April 2012. This Scenario 1 city is the center of a medium size area (consisting of 1.3 million habitants) which contains 63 taxi stands. In case study B, we considered the data broadcasted by the 89 vehicles of a company running in the cities of Odivelas and Loures, Portugal from June to Dec. of 2012. These two cities are part of outer Lisbon - the urban area surrounding the capital of Portugal which has more than 3 million inhabitants. Two maps containing the spatial distribution of the taxi stands of each case study are presented in Fig. 2.

In this section, we describe the data acquisition process and the preprocessing applied to it.

Fig. 2. Taxi Stand spatial distribution over the case studies A and B, respectively

3.1 Data Acquisition

The data was continuously acquired using the telematics installed in the vehicles. These two taxi centrals usually run in one out of three 8h shifts: midnight to 8am, 8am-4pm and 4pm to midnight. Each data chunk arrives with the following six attributes: (1) TYPE – relative to the type of event reported and has four possible values: *busy* - the driver picked-up a passenger; *assign* – the dispatch central assigned a service previously demanded; *free* – the driver dropped-off a passenger and *park* - the driver parked at a taxi stand. The (2) STOP attribute is an integer with the ID of the related taxi stand. The (3) TIMESTAMP attribute is the date/time in seconds of the event and the (4) TAXI attribute is the driver code; attributes (5) and (6) refer to the LATITUDE and LONGITUDE corresponding to the acquired GPS position.

Our study only uses as input/output the services obtained directly at the stands or those automatically dispatched to the parked vehicles (more details in the section below). We did so because the passenger demand at each taxi stand is the main feature to aid the taxi drivers' decision.

3.2 Preprocessing and Data Analysis

As preprocessing, two time series of taxi demand services aggregated were built: one with a periodicity of P-minutes and other with τ minutes. There are three types of accounted events: (1) *busy* set directly at a taxi stand; (2) *assign* set directly to a taxi parked at a taxi stand and (3) *busy* set while a vacant taxi is cruising. We consider both a type 1 and type 2 event as service demanded. However, for each type 2 event, the system receives a *busy* event a few minutes later – as soon as the driver effectively picked-up the passenger – this is ignored by our system.

Table 1 details the number of taxi services demanded per daily shift and day type in the two case studies. Additionally, we could state that, in both cases, the central service assignment is 24% of the total service (*versus* the 76% of

Table 1. Taxi Services Volume (Per Daytype/Shift)

Case Study	Daytype Group	Total Services Emerged	Averaged Service Demand per Shift		
			0am to 8am	8am to 4pm	4pm to 0am
A	Workdays	957265	935	2055	1422
	Weekends	226504	947	2411	1909
	All Daytypes	1380153	1029	2023	1503
B	Workdays	247694	267	831	531
	Weekends	57958	275	920	674
	All Daytypes	354304	270	826	559

the one demanded directly in the street) while 77% of the service demanded directly in the street is demanded in the stand (and 23% is assigned while they are cruising). In case study A, the average waiting time (to pick-up passengers) of a taxi parked at a taxi stand is 42 minutes while the average time for a service is only 11 minutes and 12 seconds. Such low ratio of busy/vacant time reflects the current economic crisis in Portugal and the inability of the regulators to reduce the number of taxis in Porto. It also highlights the importance of our recommendation system, where the shortness of services could be mitigated by getting services from the competitors. Conversely, the average waiting time in case study B is just 21 minutes.

4   Results

4.1   Experimental Setup

We used a H-sized sliding window to measure the error of our model before each new prediction about the service count of the next period (the metrics used to do so are defined in section 4.2). Each data chunk was transmitted and received through a socket. The model was programmed using the R language [20].

The aggregation period of 30-minutes was maintained ($P = 30$) and a new time series with an aggregation period of 5-minutes ($\tau = 5$) was created according the definition presented in 2.2. Both the ARIMA model (p, d, q values and seasonality) and the weight set ϕ and κ were firstly set (and updated each 24h) by learning/detecting the underlying model (i.e. autocorrelation and partial autocorrelation analysis) running on the historical time series curve of each stand during the last two weeks. To do so, we used an automatic time series function in the [forecast] R package - *auto-arima* – and the *arima* function from the built-in R package [stats]. The weight set is then *incrementally* updated for each 24h period according with the eq. 7.

The time-varying Poisson averaged models (both weighted and non-weighted) were also updated every 24 hours. A sensibility analysis carried out with data previous to the one used on these experiments determined the optimal values for the parameters α, β and H as 0.4, 0.01 and 4 (i.e. it represents a sliding window of 20 minutes), respectively. The hardware configuration is equivalent to the one used in [4, 11].

4.2   Evaluation Metrics

We used the data obtained from the last four and two months to evaluate our framework of case studies A and B, respectively. A well-known error measurement was employed to evaluate our output: the Symmetric Mean Percentage Error (*sMAPE*) [21]. However, this metric can be too intolerant with small magnitude errors (e.g. if two services are predicted at a given period for a taxi stand of interest but no one actually emerges, the error measured during that period would be 1). To produce more accurate statistics about series containing very small numbers, we can add a Laplace estimator [22] to the previous definition of (*sMAPE*) in [21]. In this case, we will do it by adding a constant c to the denominator (i.e.: originally, it was added to the numerator to estimate a success rate [22]). Therefore, we can re-define $sMAPE_k$ as follows

$$sMAPE_k = \frac{1}{t} \sum_{i=1}^{t} \frac{|R_{k,i} - X_{k,i}|}{R_{k,i} + X_{k,i} + c} \tag{8}$$

where c is a user-defined constant. To simplify the estimator application, we will consider its most used value: $c = 1$ [22].

This metric is focused just on one time series for a given taxi stand k. However, the results presented below use an weighted average of the error as evaluation measure. The weight of each stand error is the number of services emerged in the stand during the test period.

4.3   Results

The error measured for each model in the two case studies considered is presented in Table 2. The results are firstly presented per shift and then globally. The overall performance is good: the maximum value of the error using the ensemble was **25.90%** during the evening shift. The sliding window ensemble is always the best model in every shift and case study considered. The models just present

Table 2. Error Measured on the Models using *sMAPE*

Case Study	Model	Periods			
		00h−08h	08h−16h	16h−00h	24h
A	Poisson Mean	27.67%	24.29%	25.27%	25.32%
	W. Poisson Mean	27.27%	24.62%	25.66%	25.28%
	ARIMA	28.47%	24.80%	25.60%	26.21%
	Ensemble	**24.86%**	**23.14%**	**24.07%**	**23.77%**
B	Poisson Mean	28.66%	21.10%	23.34%	23.99%
	W. Poisson Mean	26.27%	22.01%	24.32%	23.83%
	ARIMA	31.88%	21.10%	23.63%	25.53%
	Ensemble	**25.90%**	**19.97%**	**22.09%**	**21.80%**

slight discrepancies within the defined shifts. Our model took - in average - 37.92 seconds to build the next prediction about the spatiotemporal distribution of the demand by all stands.

5   Final Remarks

In this paper, we proposed a method to apply a complex learning model [4,11] to build predictions about the taxi passenger demand in a real-time environment. We did so by extending the typical definition of an ARIMA model [12] to an *incremental* one using **the delta rule** - a rule firstly introduced in the perceptron algorithm [1] which is able to update its weights step by step.

We tested this approach using two case studies with distinct scenarios in Portugal: A) in the city of Porto, where the number of vehicles is larger than the demand and B) in Odivelas and Loures, which have the opposite situation. Our model was able to produce predictions about the spatiotemporal distribution of the demand during the next 30 minutes $P = 30$ with a periodicity τ of 5 minutes. The results demonstrated the relevance of our contribution: we maintained the aggregated error ratio lower than 24% and 22% in the case studies A and B, respectively. On the other hand, we were able **to reduce the typical computational time** used to build each prediction **by 40%** (from the 99.77 seconds firstly proposed in [4] to 37.92 seconds accomplished by the present framework).

Acknowledgements. The authors would like to thank Geolink, Lda. and its team for the data supplied. This work was supported by the VTL: "Virtual Traffic Lights" and KDUS: "Knowledge Discovery from Ubiquitous Data Streams" projects under the Grants PTDC/EIA-CCO/118114/2010 and PTDC/EIA-EIA/098355/2008 grants.

References

1. Rosenblatt, F.: The perceptron: a probabilistic model for information storage and organization in the brain. Psychological Review 65(6), 386 (1958)
2. Schaller, B.: Entry controls in taxi regulation: Implications of us and canadian experience for taxi regulation and deregulation. Transport Policy 14(6), 490–506 (2007)
3. Wong, K., Wong, S., Bell, M., Yang, H.: Modeling the bilateral micro-searching behavior for urban taxi services using the absorbing markov chain approach. Journal of Advanced Transportation 39(1), 81–104 (2005)
4. Moreira-Matias, L., Gama, J., Ferreira, M., Mendes-Moreira, J., Damas, L.: Online predictive model for taxi services. In: Hollmén, J., Klawonn, F., Tucker, A. (eds.) IDA 2012. LNCS, vol. 7619, pp. 230–240. Springer, Heidelberg (2012)
5. Yuan, J., Zheng, Y., Zhang, C., Xie, W., Xie, X., Sun, G., Huang, Y.: T-drive: driving directions based on taxi trajectories. In: Proceedings of the 18th SIGSPATIAL International Conference on Advances in Geographic Information Systems, pp. 99–108. ACM (2010)

6. Deng, Z., Ji, M.: Spatiotemporal structure of taxi services in Shanghai: Using exploratory spatial data analysis. In: 2011 19th International Conference on Geoinformatics, pp. 1–5. IEEE (2011)
7. Liu, L., Andris, C., Biderman, A., Ratti, C.: Uncovering taxi drivers mobility intelligence through his trace. IEEE Pervasive Computing 160, 1–17 (2009)
8. Yue, Y., Zhuang, Y., Li, Q., Mao, Q.: Mining time-dependent attractive areas and movement patterns from taxi trajectory data. In: 2009 17th International Conference on Geoinformatics, pp. 1–6. IEEE (2009)
9. Li, B., Zhang, D., Sun, L., Chen, C., Li, S., Qi, G., Yang, Q.: Hunting or waiting? Discovering passenger-finding strategies from a large-scale real-world taxi dataset. In: 2011 IEEE International Conference on Pervasive Computing and Communications Workshops (PERCOM Workshops), pp. 63–68 (March 2011)
10. Lee, J., Shin, I., Park, G.: Analysis of the passenger pick-up pattern for taxi location recommendation. In: Fourth International Conference on Networked Computing and Advanced Information Management (NCM 2008), pp. 199–204. IEEE (2008)
11. Moreira-Matias, L., Gama, J., Ferreira, M., Damas, L.: A predictive model for the passenger demand on a taxi network. In: 15th International IEEE Conference on Intelligent Transportation Systems (ITSC), pp. 1014–1019 (2012)
12. Cryer, J., Chan, K.: Time Series Analysis with Applications in R. Springer, USA (2008)
13. Ihler, A., Hutchins, J., Smyth, P.: Adaptive event detection with time-varying poisson processes. In: Proceedings of the 12th ACM SIGKDD International Conference on Knowledge Discovery and Data Mining, pp. 207–216. ACM (2006)
14. Burke, E., Hyde, M., Kendall, G., Ochoa, G., Ozcan, E., Woodward, J.: A classification of hyper-heuristic approaches. In: Handbook of Metaheuristics, vol. 146, pp. 449–468. Springer, US (2010)
15. Chalup, S.: Incremental learning in biological and machine learning systems. International Journal of Neural Systems 12(06), 447–465 (2002)
16. Huang, G.B., Zhu, Q.Y., Siew, C.K.: Real-time learning capability of neural networks. IEEE Transactions on Neural Networks 17(4), 863–878 (2006)
17. Holt, C.: Forecasting seasonals and trends by exponentially weighted moving averages. International Journal of Forecasting 20(1), 5–10 (2004)
18. Wang, H., Fan, W., Yu, P., Han, J.: Mining concept-drifting data streams using ensemble classifiers. In: Proceedings of the Ninth ACM SIGKDD International Conference on Knowledge Discovery and Data Mining, pp. 226–235. ACM (2003)
19. Gama, J., Pinto, C.: Discretization from data streams: applications to histograms and data mining. In: Proceedings of the 2006 ACM Symposium on Applied Computing, pp. 662–667. ACM (2006)
20. R Core Team: R: A Language and Environment for Statistical Computing. R Foundation for Statistical Computing, Vienna, Austria (2012)
21. Makridakis, S., Hibon, M.: The m3-competition: results, conclusions and implications. International Journal of Forecasting 16(4), 451–476 (2000)
22. Jaynes, E.: Probability theory: The logic of science. Cambridge University Press (2003)

An Ecosystem Based Model for Real-Time Generative Animation of Humanoid Non-Player Characters

Rui Filipe Antunes and Frederic Fol Leymarie

Goldsmiths, University of London, United Kingdom
rui.antunes@gold.ac.uk, ffl@gold.ac.uk

Abstract. In this paper a novel approach to a decentralized autonomous model of agency for general purpose Non-Player Characters (NPCs) is presented: the AI model of Computational Ecosystems. We describe the technology used to animate a population of gregarious humanoid avatars in a virtual world. This artistic work is an ethnographic project where a population of NPCs inhabit the virtual world and interact autonomously among themselves as well as with an audience of outsiders (human observers). First, we present the background, motivation and summary for the project. Then, we describe the algorithm that was developed to generate the movements and behaviors of the population of NPC "story-tellers". Finally, we discuss some of the critical aspects of this implementation and contextualize the work with regards to a wider usage in computer games and virtual worlds.

1 Introduction

Animation of crowds on the historical site of Petra, Jordan, the Pennsylvania train station of New York city [1] or of visitors in theme parks [2] are good examples of a developing area of research that looks at modeling virtual spaces inhabited by communities of humanoid avatars that self-organize and interact autonomously. Commercial videogames such as The Sims [3] or Grand Theft Auto [4] do also share similar goals with great success in terms of mainstream appeal. Traditionally the algorithms that model crowds follow one of three main techniques when individuals: (i) are represented as particles subject to physical forces [5]; (ii) are represented as states of cells in cellular automata [6]; (iii) follow a rule–based scheme [7]. These techniques have the merit of modeling in a realistic way the macro–level behavior of the crowd, including the features of its spatial flow. Recently, more attention has been put at the micro–level, which is centered on an individual's behavior within a multitude. Some works attempt to recreate the spontaneity of behaviors visible in small groups of a few dozens of individuals. Some works also attempt to implement human moods [8] or add traits of personality to the NPCs [9]. Pelechano *et al.* for instance model a group of humanoids interacting socially at a cocktail party [10]. Good surveys of the field of crowd animation up to 2008 can be found in the books of Pelechano *et al.* [11] and Thalmann *et al* [12].

We aim at producing the visual effects of a lively community made of small interacting groups of gregarious individuals: *i.e.* groups of 2 to 10 individuals, Fig. 1. Motivated

L. Correia, L.P. Reis, and J. Cascalho (Eds.): EPIA 2013, LNAI 8154, pp. 66–77, 2013.
© Springer-Verlag Berlin Heidelberg 2013

by an artistic work that draws on European colonialism which requires animating a humanoid population driven by *predatorial* behaviors, we are looking, in particular, at systems where agents are organized in the hierarchical structure of a food-chain while trading token units of energy and biomass as a way of promoting community dynamics. A *Computational Ecosystem* (CE) is a generative model that provides for complex environment simulations rich in the heterogeneity and spontaneity we are targeting.

This paper is organized in three main sections. The first of these describes the motivations for the artwork *Where is Lourenço Marques?*, the catalyst for our present investigation (a more detailed description of the motivations is available in a recent manuscript of R. Antunes [13]). Section two presents technical details about the CE, describing the AI model that animates each of the individuals in the community. Finally, in section three, we present a discussion on the CE as a generative technology: First this discussion looks at EvoArt and how this approach stems from EvoArt traditions; and then we progress towards a critical analysis of our specific implementation.

(a)   *(b)*

Fig. 1. *(a)* Interacting story-tellers. *(b)* Story-teller interacting with the audience.

1.1 Where Is Lourenço Marques?

In the landscape of colonial Portugal, Lourenço Marques was the proud capital of the province of Mozambique. This work is dedicated to those who lived and inhabited that time and place, in a city which since its independence in 1975, became known as Maputo. Many of the inhabitants of Lourenço Marques left Mozambique for socio-economical and political reasons during the process of decolonization. The work we describe here is a representation of that city that builds up on the subjective mediation of the memories of those who experienced it in the last period of its colonial times. More than the representation of the architectural space, the work is described as an ethnographic exercise meant to address the experience of living in the city back then. This memorial takes expression in a 3D virtual world and is available at http://www.lourencomarques.net. A short video is provided at http://www.youtube.com/watch?v=yQJ5VydiEqE.

From Interviews to Avatar 'Story-Tellers'. The elaboration of the project started from a process of interviews with a community who left the city in those dramatic days, and now lives in Portugal. Their accounts and shared material form a memorabilia, from which an artistic and subjective reconstruction of the city was built in a 3D virtual environment (using Unity3D as a supporting platform). Our focus in this communication is on a community of humanoid avatars which roam autonomously the city and interact with each others as well as with the human audience. These avatars are the agent "story-tellers" (Fig. 1). When the user selects (points or clicks on) any one of these avatars, it interrupts its current activity and approaches the camera. Then, animated and gesticulating as if they were speaking, such an agent streams an audio testimony of one of the oral accounts recorded during the interview process. Thus, this population of avatars assists in the task of bringing the experiences of the respondents. Each of the individuals in this population is the bearer of an excerpt from an interview, functioning as the carrier and mediator of real-life human stories. The audience is thus implicitly invited to seek-out these story-tellers, 'chasing' virtual characters through the city, in order to listen to their stories.

The Allegory of the Computational Ecosystem. The model of artificial intelligence (AI) used to animate this population of characters is a computational ecosystem (CE). This model of AI is based on a community of autonomous agents trading units of energy and organized as a simulation of a food–chain. Each of the agents emulates a simplified form of the life cycle of generic carbon–based life forms. The community evolves through processes inspired by Mendelian genetics. Characteristics such as the size or speed of an agent pass from parents to children when couples reproduce in a process that evokes sexual reproduction. Energy is required for the activities of these individuals: including moving, running, or simply breathing. The population competes for energy and space, and the dynamics of energy transfer occur in *predatory* acts. In particular, when the energy level of an individual is too low (*i.e.* below a pre-specified threshold), it is considered dead and removed from the community as a result.

The motivation for using a CE to animate such a population came about to illustrate by means of an hyperbole the social situation then lived in the colonial city. In the tradition of literary allegory, the behavior of the population carries a second level of narrative. Social groups become alike *species* in the virtual city/ecosystem. Social interactions of these story-tellers in the virtual world are dictated by their performance in the habitat. However, instead of the usual primal actions of animals that attack and eat each other, what is offered to the audience are animations of the interactions of humanoids gesturing in apparent conversation. Each action in the ecosystem is translated into an animation of a corresponding gesture. For example, when two individuals interact, if one is to "attack" the other, the movement of the arms of the winning individual will be more expressive than the movements of its victim.

The Computational Ecosystem as an AI Engine. CEs have been used as generative engines in diverse contexts such as audio-visual installations [14], music [15] or choreography of avatars in virtual worlds [16]. We innovate by the use of the complexity of a CE as an AI to coordinate the social movements of humanoid NPCs.

By design, the observable behaviors of the population corresponds to a visual representation of the ecosystem inner-workings. The spatial layout of the virtual city provides a support for the visualization of the underlying multi-agent state space, with a direct connection between the current state of the ecosystem and the configuration of the avatars on the layout. Each NPC is an individual belonging to the population and we refer to each one as an 'avatar'. The birth of an avatar is represented by a new NPC appearing next to its parents, while its death is represented by its "dematerialization" in thin air. Each interaction during the lifetime of an avatar is translated into a series of movements or gesticulations while being constrained to the surface of the world. The CE establishes correspondences between states and movements and actions performed by each avatar. For example, the action of an avatar feeding in the virtual ecosystem might correspond to a certain gesture in the animation, while its escape after a fight with another creature will correspond to a different form of gesticulation. In this work we have defined a set of *nine base animations*. This small set is deemed sufficient for a prototype to allow to explore and demonstrate the potential of CEs in animating virtual populations of humanoid characters. The set includes eight different arm movements (or gesticulations) used during conversations and one basic animation for walking. In the elaboration of the behaviours of the avatars we implemented a system which is described next.

2   Technical Description

To animate our CE, we have combined a number of techniques with a proven record of successful results in animating populations of multiple individuals. These are: (1) a *hormonal system* as suggested by Daniel Jones in his framework for swarms [17], (2) a *metabolic system* specifying the diet of our avatars based on the model of Saruwatari *et al.* [18], and (3) a *classifier system* following on John Holland's guidelines [19], which drives the behaviors of our avatars. We now describe our implementation of each of these three main techniques.

2.1   The Hormonal System

Individual behaviors are influenced by the hormone-like system from Jone's work on swarms [17]. This is composed of five variables: (a) testosterone — increases with age and crowdness, it decreases upon giving birth, and causes an increase in the likelihood of reproduction; (b) adrenaline — increases with overcrowding, decreases as a result of internal regulation overtime and causes a greater rate and variance of movement; (c) serotonin — increase with 'day' cycles, decreases during 'night' cycles, decreases as result of hunger; and causes a greater social attraction towards other agents; (d) melatonin — increases during 'night' cycles, and decreases during 'day' cycles and also decreases rate of movement; (e) leptin — increases upon eating, decreases steadily at all other times, causes downward regulation of serotonin when depleted, and causes greater attraction to food.

The Blueprint Descriptor. These hormones are initially configured by the genetic descriptor containing a blueprint for the physical attributes of the agent. This is a string with 15 binary digits, where different sections of the string code for a set of six specific features: (a) age (**gAge**) — rate at which the agent ages; (b) introspection (**gIntr**)

— the level of gregariousness; (c) hormone cycles (**gCycl**) — the strength or speed of the hormone cycle (*e.g.* how much will it increase per day); (d) hormone uptakes (**gUptk**) — the intake during an hormone cycle (*e.g.* how much will the hormone increase when an avatar meets another one); (e) body chemistry (**gChem**) — chemical components present in the body; (f) Metabolism (**gMetab**) — what chemicals can be 'digested'. In avatars from the *initial* population, these features are determined randomly. In a Mendelian-like process of diploid reproduction, each subsequent generation inherits such information from their parents, using crossover operators on Gtypes. In the process noise is added to the information to mimic mutations (with a small probability $Pm = 0.05$ of effective mutation on each bit).

2.2   The Metabolic System

To determine the dietary specifications we use the framework of Saruwatari *et al.* [18] based on two strings, the first of which describes the body's constitution, while the second describes the avatar's metabolism. Potential preys are those whose constitution-string matches the predator's metabolic-string. Saruwatari *et al.* have shown this simple mechanism potentially leads to the emergence of complex multi-trophic food-chains with variable depths.

The first part of the dietary specification is given by the string of 3 binary digits present in the Gtype section **gChem**. This represent the avatar's body 'chemical' composition. The first digit codes for the presence of the chemical A, the second for chemical B and the third for chemical C. Each avatar is equipped with chemical repositories that are direct translations from the composition-Gtype **gChem**. Take for instance an avatar with **gChem** of "101". The only chemicals present in the repository will be A and C. Another avatar with Gtype "010" will only have repository B active. When an hypothetical avatar (X) preys another avatar (Y), X will extract from the chemical repositories of Y the existing content filling in its own repositories. Illustrating carbon-based food-web entropy, in the process 90% of the value is wasted. Each repository only carries a maximum capacity which is directly related with the avatar's body size. The body size is given by a direct translation of the binary value of **gChem**. These chemical attributes play an essential role in the ecosystem, since these determine part of the dietary specification in the community and thus the interactions of the individuals.

The second part of the dietary specification is provided by the component that defines the avatar's metabolism: **gMetab**, *i.e.* what chemicals can the individual 'digest'. For instance an avatar with **gMetab** 100, will be able to prey individuals whose **gChem** codes the A component: 100, 110, 111, 101. Consequently the combination **gChem - gMetab** establish the fundamental interactions in the predator-prey relationships. This predator-prey mechanism of matching the metabolic-string with the composition-string provides an interaction space of size 8 x 8, which was wide enough for the current work.

The Metabolic Rules.   The metabolic system emulates a simplified food-energy conversion. Besides the chemical-repositories, and the hormonal system, avatars have one main structuring variable contributing to their behavior: energy. This is generated from the chemical repository of the avatar. To simulate a conversion from mass to energy, we have defined an arbitrary chemical reaction which requires three chemical units to produce one unit of energy (*e.g.* $2A + 1B \implies 1\,energy\,unit$). Energy can be spent when breathing or performing actions in the world such as moving, attacking, fleeing, preying, eating or mating. Below a certain energetic level an avatar needs to generate more from its chemical-repositories. Below a certain threshold the avatar starts to feel hungry, activating the internal sensor. When the energy level reaches the value 0 the avatar dies and is removed.

2.3   Behavior of Avatars via a Classifier

Avatar's behavior is characterized in three stages: perception, decision, and action. Each avatar is equipped with: a) sensors for external contact which are triggered by the proximity with other avatars, and b) internal sensors described by energy and the system of hormones. Depending on these inputs, the avatar will decide what to do based on the system described below and, in accordance, it will perform some kind of action. This mechanism is implemented with a classifier system inspired by the description provided by John Holland [19], which was used in the well-known Echo system [20] and also inspired the famous artwork Eden [21].

The Action-Selection Mechanism.   During the process of perception the system inspects the state of the hormonal variables and the level of energy, as well as if the body of this avatar is entering in contact with any other avatars. When any of these variables is active, such as when (i) the energy or leptin levels are below pre-selected thresholds, (ii) the testosterone, adrenalin or serotonin levels are above some pre-fixed thresholds, (iii) the avatar is touching another body, then an action-message is generated. This message takes the form of a string of length 6, and is composed from the grammar set $\{0, 1, \#\}$, identifying which sensor is active — binary values indicating the active and inactive states and # functioning as a wildcard. This length provides a space of 64 mapping possibilities, and consequently grants a high degree of freedom to model a significant number of behaviors.

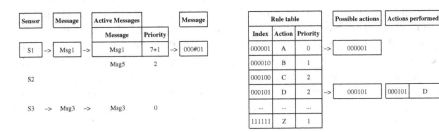

Fig. 2. *Left:* Illustration of a situation where the active messages Msg1 and Msg3 are generated due to the simultaneous activation of sensors S1 and S3. The list of current active-messages is updated with the two new messages. Msg1 increases the priority of the existing entry for Msg1; On its hand Msg3 updates the list with a new entry with priority 0. *Right:* An illustration of the process of transformation from message to action. The message 000#01 finds two compatible indices matching in the table of rules (000001 and 000101). From these, the action of the one with highest priority is performed.

The Active Messages List. This is a list with messages waiting to be processed. If a message is new on the list, it will be inserted with an assigned priority of 0. If, on the contrary, the message already exists, this means that the same sensor has already triggered one or more messages but these have yet to be processed. If that is the situation then the priority of the existing message will be increased by 1 (Fig 2 Left). During the decision stage, the message with highest priority from this list will be removed and processed against the table of rules to generate actions. When two or more messages have the same priority, the oldest in the list is selected.

The Table of Rules. This table describes a set of actions and their indices. The rules are initially similar for all avatars. Each rule contains three parameters: index, action and priority. The index is again composed from the grammar $\{0, 1, \#\}$ and is used to match the rules to the corresponding active message being processed.

Multiple rules can match one single message. Since the character # functions as a wildcard this means that any value can be accepted in the particular character of the index where the wildcard is located. However, as each of the rules has assigned a specific priority (initialized with a random value), from all the candidate rules (those which indices matching a message) only the one with highest priority will be selected. In case of ties, the oldest candidate takes precedence.

The action to perform is coded on the second section of the rule. This is an alpha-numeric code which is translated into a procedural action such as an instruction to prey on any avatar that is within a certain distance, or an instruction to move towards the closest avatar, *etc.* (Fig 2 Right).

The Reward System. The priority of the rules is updated according to the consequences of the actions performed in the world. This translates into the avatar recognizing which actions are advantageous or not. An *ad hoc* reward is attributed to some of the possible actions such as eating when hungry, or victory or defeat in battle. If for instance the selected action is to feed, this implies a positive reward. On the contrary, being hit implies a negative value. The reward can be associated not only with the current rule which has triggered an event, but also preceding rules leading to the current action. Each avatar has a FIFO (First In First Out) memory stack which stores the last five rules performed. This block of memory is also rewarded accordingly, with the rules being credited in a decremental way corresponding to the time they have been present on memory. For instance, when an avatar manages to prey, the rule for the action which has triggered the preying event is rewarded with 5. The immediate rule-action prior to that one is rewarded with 4; the anterior with 3, and so on. When a new event occurs and a new action is performed, the oldest rule-action is removed from the stack.

Generation of New Rules. As an outcome of reproduction, a newborn inherits from the current rule–table of the parents. To constitute the new rule–table, the rules with top priority from both progenitors are inherited in a 50/50 proportion. Each of the indices of the new set of rules then is perturbed by a process of (possible) mutations. These indices may suffer a transformation resulting from four possible attempts for digit mutation, each with a success probability of 50%. We set the mutation rate at such a high probability to ensure rapid variability in behaviors.

The Mapping of Behaviors. As mentioned earlier, actions are encoded in the second part of each rule. Each of these rules might trigger new messages or generate some physical action for the agent such as move or talk. To render visible each physical action on the virtual world one associated animation needs to be played. The relationship between the animations and the actions is rigidly defined *a priori*. For instance, for the rule triggering the action 'eat', the animation associated with 'eating' is performed (Figure 3). These animations were selected as interesting and playful in illustrating conversations via gesticulating in well defined separate sets of body movements. This type of contextual adaptation is similar to work by Petty *et al.*: they use a lookup operation on predefined tables to assign the actions to be performed in relation to the current 'moods' of the agents [22]. In similitude, we use actions dictated by the ecosystem states which are assigned to behaviors of performing agents, where each such behavior is transposed in a visualization via our pre-defined tables.

The continuous dynamics of the virtual world is generated by the on-going displacements of the avatars. These movements can be of three types: (i) in the direction of an arbitrary 'preferencial location', *i.e.* a randomly selected layout point set of coordinates which, once reached, is re-initialised with a new target position; (ii) in the direction of other characters as a consequence of the internal needs as determined by the classifier system under the influence of the hormonal and metabolic sub–systems; and (iii) moving towards a member of the public in response to being selected for interaction (*i.e.* telling a story).

In the next section, we contextualize our work in relation to Evolutionary Art. Then we provide a critical analysis of the current implementation.

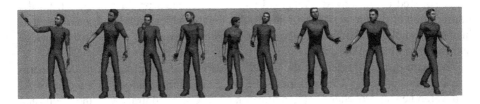

Fig. 3. Still images of the animations corresponding to each of the actions performed by the agent. From left to right: 1- Eat a prey; 2- attempt to mate with a partner that is not ready; 3- reproduction or play alone; 4- successful mating; 5- attempt to mate but no compatible individuals exist on the vicinity; 6- loosing an attack; 7- play with happy mate; 8- victorious attack; 9- walking (move to mate, wander, move to prey).

3   Discussion

The *visualization of the internal dynamics* of a functioning ecosystem is proposed as a way to structure and coordinate the animation of a population of humanoid avatars. This is exemplified with a population of NPCs simulating conversational behavior in a virtual world. The result is a populated landscape where individuals roam through the city, ignoring some of the members of the crowd while joining with others in small groups and performing visibly expressive dialogues.

The generative process that we use here finds direct affiliations with the processes and methods from Evolutionary Art (EvoArt) [23]. The foundations of EvoArt, establish its roots upon a methodological approach borrowed from Evolutionary Computing. In the classic procedure of Genetic algorithms, a syntactic element is transformed into its semantic representation. With traditional EvoArt, similarly, an encoded blueprint (the genotype - Gtype) is converted to its imagetic or audible representation (the phenotype - Ptype). The complexity of this process of conversion from Gtype to Ptype is open to the artistic creativity, and the linearity and the distance involved on this process of transformation differs widely among the artists. The diversity of the outcomes this simple methodology entails is illustrated for example by computational evolutionary art pioneers Latham and Rooke [24]: William Latham produces 3D morphologies based on a process of shape deformation, while Steven Rooke generates abstract imagery based on a language of mathematical and visual operators.

One of the established ways the Gtype-Ptype paradigm has been explored is by means of applying this idea to whole populations of interacting autonomous agents in CEs. As mentioned earlier, CE are communities formed by multiple autonomous individuals which are organized in hierarchical food-chains and trade amongst them units of energy. With these systems, individuals are as well represented and structured by the information initially written in their Gtypes, and later this information is also transformed into some form of phenotypic representation. However, in addition to this process, the autonomy of the individuals, which is so characteristic from CEs, generates an interesting dynamics of self-organization and emergence with cyclical changes of density. CEs used in EvoArt explore this process of self-organization and emergence, as main mechanisms to generate heterogeneity and novelty. In some cases the Gtypes are directly sonified or visualized. Wakefield and Ji for instance produce sounds from the transcription of the Gtypes [25]. In other cases the dynamics itself has been deemed as interesting enough to be visualized or sonified. The sonifications produced by Eldridge and Dorin result from the actions and status of the agents in the virtual space [15].

The main principle orienting the present research is the exploration of the generative features implicit in CEs and the use of the ecosystem paradigm as a model of AI to develop new approaches for the animation of NPCs. In contrast with the traditional approach visualizing the information of the Gtypes, in this work the information carried on the Gtype of the individuals describes their features such as body-composition and dietary constraints. We then use this information to determine the location of this particular individual in the dynamic food-chain. What is visualized instead are the ephemeral states of the individuals in the ecosystem: the actions such as the exchange of energy or their multiplication, for instance. As it happens with the Gtype-Ptype paradigm, the linearity and the distance between the syntactic and the semantic levels in the process of conversion can vary. For each action there is a corresponding symbol in the internal system and there is a posterior interpretation and visualization of that particular symbol. During this process of interpretation, converting symbols into visualizations, there is scope for artistic creativity. For instance, in considering the action 'attack', rather than using the animation of an animal fighting, we might instead use a specific short piece being played by an orchestra or use a specific choreographed sequence to be performed by a dancer.

Our work explores these ideas in the animation of a population of gregarious humanoids simulating conversational behavior in a virtual world. This investigation has been driven by the design specifications for the artwork. These required the existence of an autonomous population of humanoid avatars to play the role of story–tellers in the virtual world. However, the same specifications did also require the behavior of this community of humanoids to be the result of the dynamics of an ecosystem where individuals would seek out each other with predatory intentions. A CE, a system of agents organized in a hierarchical structure (of a food-chain) and trading token units (of energy and biomass) generates this dynamics. The potential complexity of crowd interaction is revealed by the spontaneity of group formation of avatars engaging in heterogeneous gesticulations during their conversations.

This work has shown some interesting results based on a limited set of predefined movements. The system was tested in a public display at the closing ceremony of the Watermans Festival of Digital Art 2012, andat the Tin Shed Gallery, both in London, with positive feedback from the audience. We obtained twenty four responses from anonymous users. These responded to a questionnaire where 80% said the group formation and the talking movements in the simulation appeared to obey to an internal and coherent logic, whereas 65% said the behaviors to be believable in a cartoonish way.

To build our model of AI we have put together a set of established and efficient techniques for the agency of populations — Holland's classifier system to generate a hierarchy of behavioral preferences, Jones' hormone-framework to provide biological "moods", and Saruwatu's model of dietary specifications to establish interactions in the food-chain (§2). The complexity of the system could however be reduced, *e.g.* by removing the hormone based component. Events could also be regulated by the energy level only. Our current design of a CE generates more noise in the system and consequent variability, than such simpler implementations. Also, it provides a CE platform which is closer to a biological modelling, which we propose is of greater interest for an artwork which has its sources in human socio-history (of recent European colonialism). Nevertheless, our approach to a CE implementation is currently not computationally tractable for large crowds, when relying on the limited computer power of a single typical PC (our case and a typical situation when performing exhibits at small venues). Note that this type of constraint is bound to evolve with the greater use and access of cloud computing capacities, with much higher performance delivered at low cost.

Another aspect to notice, is that there is an implied semantic resulting from interactions of physical gestures which in our case was ignored. The resulting conversation is not immediately intelligible from the point of view of the actions in the ecosystem. Also, the small-number of pre-generated movements, as well as the lack of sophisticated blending is limiting and hides the richness that could be effectively reached. An increased set of animations would make the system more flexible. To explore this potential further, it would be interesting to consider gestures and poses which more subtly reflect the internal states of the avatars. This could be enriched with a wider set of animations, which would reflect such nuances. Moreover, in contrast with our current deterministic approach of defining a limited set of animations, it would be interesting to break the animations into smaller bits. The use of the CE was mainly for exploring its feature as a generator of heterogeneous behaviors. In terms of emergence and novelty this work would become richer with the incorporation of elements of a language of movements on which the CE could act. This would create the conditions for procedural movements with potential emergence of unexpected movements, perhaps rich in subtleties more in accord with a virtual ecosystem and evolutionary paradigm. Other interesting possibilities would be of exploring the sonification of these internal states of the agents. Another criticism we can address on the present outcome is that the avatars appear and disappear as if by magic, in thin air. One easy way to circumvent this issue is to create an upper level of operation where the character, after being born, instead of appearing next to its parents emerges opening the door of the closest building, and in a similar process moves to an arbitrary building in order to die out of sight.

Combining a crowd approach (based on avatars) and a genuine ecosystem approach, such as the one discussed here, might offer new domains of exploratory possibilities. Further exploration on hybrids where the ecosystem paradigm is combined with a human crowd behaviour approach could have relevance to NPCs in computer games, particularly in Massively Multiplayer Online (MMO) games where maintaining scripted behavior for hundreds of NPCs is a big overhead. The advantage of the ecosystem approach to the animation of NPCs, over other established methods of population simulation such as those based on particles or cellular automata, is the inherent nature of the ecosystem as a generative system. The motivations for the behaviors of the individuals in the space are provided by the framework. The behavior system allows flexible representation of complex scenarios, and is relatively easily adaptable to different buildings or different numbers of individuals. Additionaly the ecosystem paradigm provides natural flutuations in the population density which might be interesting from the point of view of the realism of the simulation.

Our aim with this development is to look at the ecosystem paradigm as a generative methodology in a wider context than in the traditional form of explicit and direct visualization of ecologies where 'carnivores eats herbivores'. The great advantage of this framework is its high level of openness. This was something previously suggested by those exploring this technology in sonic and visual artistic developments. We have modeled a population of gregarious NPCs showing the potential for spontaneous and conversational behaviors. Our approach took advantage of one of the fundamental properties of CEs: by relying on the variety and spontaneity of the elementary behaviors, the autonomy and self-organization of the agents generates ever-changing heterogeneous patterns at the global scale of the community.

Acknowledgements. This research is supported in part in the form of a PhD studentship to Mr. Antunes by Fundação para a Ciência e Tecnologia from Portugal, contract reference SFRH / BD / 61293 / 2009. We also thank the anonymous reviewers for their helpful comments and constructive criticism.

References

1. Shao, W., Terzopoulos, D.: Populating reconstructed archaeological sites with autonomous virtual humans. In: Gratch, J., Young, M., Aylett, R. (eds.) IVA 2006. LNCS (LNAI), vol. 4133, pp. 420–433. Springer, Heidelberg (2006)
2. Huerre, S.: Agent-based crowd simulation tool for theme park environments. In: 23rd Inter. Conf. on Comp. Anim. and Social Agents (CASA), St-Malo, France (2010)

3. Electronic Arts: The Sims, http://gb.thesims3.com/
4. Rockstar Games: Grand Theft Auto 5., http://www.rockstargames.com/grandtheftauto/
5. Helbing, D.: A Fluid Dynamic Model for the Movement of Pedestrians. Complex Systems 6, 391–415 (1992)
6. Banerjee, B., Abukmail, A., Kraemer, L.: Advancing the layered approach to agent-based crowd simulation. In: Wksh on Princip. Adv. and Distrib. Simul. (PADS), pp. 185–192 (2008)
7. Reynolds, C.W.: Flocks, Herds and Schools: A Distributed Behavioral Model. ACM SIGGRAPH Comp. Graphics 21(4), 25–34 (1987)
8. Ahn, J., Wohn, K., Oh, S.: Optimized Solution Simplification for Crowd Anim. Comp. Anim. and Virtual Worlds 17, 155–165 (2006)
9. Durupinar, F., Allbeck, J., Pelechano, N., Badler, N.: Creating Crowd Variation with the OCEAN Personality Model. In: Proc. of the 7th Inter. Joint Conf. on Autonomous Agents and Multiagent Syst., pp. 1217–1220 (2008)
10. Pelechano, N., Stocker, C., Allbeck, J., Badler, N.: Being a Part of the Crowd: Towards Validating VR Crowds Using Presence. In: Proc. Auton. Agents and Multiagent Systems (AAMAS), pp. 136–142. ACM Press (2008)
11. Pelechano, N.: Virtual Crowds: Methods, Simulation, and Control. Synth. Lect. on Comp. Graphics and Anim. Morgan and Claypool Publish. (2008)
12. Thalmann, D., Musse, S.: Crowd Simulation. Springer (2007)
13. Antunes, R.F.: Where is Lourenço Marques?: A Mosaic of Voices in a 3D Virtual World. Leonardo Electronic Almanac 18 (Touch and Go) (3), 114–121 (2012)
14. Dorin, A.: Pandemic - Generat. Software Installation, Exhib.: Bouillants 4, Vern-sur-Seiche, Brittany, France, G., Allin, L. Dupuis (artistic dir.) (April 22-May 20 2012)
15. Eldridge, A., Dorin, A.: Filterscape: Energy recycling in a creative ecosystem. In: Giacobini, M., Brabazon, A., Cagnoni, S., Di Caro, G.A., Ekárt, A., Esparcia-Alcázar, A.I., Farooq, M., Fink, A., Machado, P. (eds.) EvoWorkshops 2009. LNCS, vol. 5484, pp. 508–517. Springer, Heidelberg (2009)
16. Antunes, R.F., Leymarie, F.F.: Generative choreography: Animating in real-time dancing avatars. In: Machado, P., Romero, J., Carballal, A. (eds.) EvoMUSART 2012. LNCS, vol. 7247, pp. 1–10. Springer, Heidelberg (2012)
17. Jones, D.: AtomSwarm: A framework for swarm improvisation. In: Giacobini, M., Brabazon, A., Cagnoni, S., Di Caro, G.A., Drechsler, R., Ekárt, A., Esparcia-Alcázar, A.I., Farooq, M., Fink, A., McCormack, J., O'Neill, M., Romero, J., Rothlauf, F., Squillero, G., Uyar, A.Ş., Yang, S. (eds.) EvoWorkshops 2008. LNCS, vol. 4974, pp. 423–432. Springer, Heidelberg (2008)
18. Saruwatari, T., Toqunaga, Y., Hoshino, T.: ADIVERSITY: Stepping Up Trophic Levels. In: Brooks, R.A., Maes, P. (eds.) Proc. of the 4th Inter. Workshop on the Synthesis and Simul. of Living Systems, pp. 424–429 (1994)
19. Holland, J.: Hidden Order: How Adaptation Builds Complexity. Helix Books (1996)
20. Forrest, S., Jones, T.: Modeling Complex Adaptive Syst. with Echo. In: Stonier, R., Yu, X. (eds.) Complex Systems: Mechanisms of Adapt., pp. 3–21. IOS Press (1994)
21. McCormack, J.: Eden: An evolutionary sonic ecosystem. In: Kelemen, J., Sosík, P. (eds.) ECAL 2001. LNCS (LNAI), vol. 2159, pp. 133–142. Springer, Heidelberg (2001)
22. Petty, M., McKenzie, F., Gaskins, R.: Requirements, Psychological Models and Design Issues in Crowd Modelling for Military Simul. In: Proc. of the Huntsville Simul. Conf. (2003)
23. Bentley, P., Corne, D.: Creative Evol. Systems. Academic Press (2002)
24. Whitelaw, M.: Metacreation: Art and Artificial Life. MIT Press (2004)
25. Wakefield, G., Ji, H(H.): Artificial Nature: Immersive World Making. In: Giacobini, M., et al. (eds.) EvoWorkshops 2009. LNCS, vol. 5484, pp. 597–602. Springer, Heidelberg (2009)

An Efficient Implementation of Geometric Semantic Genetic Programming for Anticoagulation Level Prediction in Pharmacogenetics

Mauro Castelli[1,2], Davide Castaldi[3], Ilaria Giordani[5], Sara Silva[2,4],
Leonardo Vanneschi[1,2,3], Francesco Archetti[3,5], Daniele Maccagnola[3]

[1] ISEGI, Universidade Nova de Lisboa, 1070-312 Lisboa, Portugal
[2] INESC-ID, IST / Universidade Técnica de Lisboa, 1000-029 Lisboa, Portugal
[3] D.I.S.Co., Università degli Studi di Milano-Bicocca, 20126 Milano, Italy
[4] CISUC, Universidade de Coimbra, 3030-290 Coimbra, Portugal
[5] Consorzio Milano Ricerche, 20126 Milano, Italy

Abstract. The purpose of this study is to develop an innovative system for Coumarin-derived drug dosing, suitable for elderly patients. Recent research highlights that the pharmacological response of the patient is often affected by many exogenous factors other than the dosage prescribed and these factors could form a very complex relationship with the drug dosage. For this reason, new powerful computational tools are needed for approaching this problem. The system we propose is called Geometric Semantic Genetic Programming, and it is based on the use of recently defined geometric semantic genetic operators. In this paper, we present a new implementation of this Genetic Programming system, that allow us to use it for real-life applications in an efficient way, something that was impossible using the original definition. Experimental results show the suitability of the proposed system for managing anticoagulation therapy. In particular, results obtained with Geometric Semantic Genetic Programming are significantly better than the ones produced by standard Genetic Programming both on training and on out-of-sample test data.

1 Introduction

In the last few years researchers have dedicated several efforts to the definition of Genetic Programming (GP) [8] systems based on the semantics of the solutions, where by semantics we generally intend the behavior of a program once it is executed on a set of inputs, or more particularly the set of its output values on input training data [13]. In particular, very recently new genetic operators, called geometric semantic operators, have been proposed by Moraglio et al. [15]. As Moraglio et al. demonstrate in [15], these operators have the interesting property of inducing a unimodal fitness landscape on any problem consisting in finding the match between a set of input data and a set of known outputs (like for instance regression and classification), which facilitates GP evolvability. In this paper the objective is to evaluate the regression performance of this new GP system on a field of pharmacogenetics of oral anticoagulation therapy, comparing the results with the ones obtained by standard GP. The indication for the use of oral

L. Correia, L.P. Reis, and J. Cascalho (Eds.): EPIA 2013, LNAI 8154, pp. 78–89, 2013.
© Springer-Verlag Berlin Heidelberg 2013

anticoagulant in many patients is to reduce the embolic risk associated with diseases such as atrial fibrillation, left ventricular dysfunction, deep vein thrombosis and mechanical aortic valve replacement and could be useful for patients who had undergone orthopedic surgery. The trial-error basis of the methods currently in use to fine tune the dosage for a given patient along with the responses' variability due to genetic and behavioral factors can result in out of range periods and, therefore, in a non negligible risk of thromboembolic and bleeding events. Therefore, the problem addressed is the prediction of appropriate oral anticoagulant level of medical drugs. A difficulty with the use of oral anticoagulants is that prescription needs to be individually determined for each patient, usually by following a standard initial dosing protocol, measuring the coagulation rate regularly (using the international normalized ratio, INR, which is a measure of prothrombin time. A high INR value indicates overcoagulation) and then adjusting the dose until the required rate of coagulation is obtained. Relevant help could come from computer support. Mathematical models, able to predict the maintenance dose, were already elaborated more than 20 years ago [16]. These models have widely been applied only recently [7]. The use of computer-based techniques has been shown to have a favorable impact in this field [6] and computational techniques capable of producing reliable predictive models are needed.

Geometric Semantic GP could definitely be a promising approach for this issue, given its ability of inducing unimodal fitness landscapes on problems, independently of how complex they are. Nevertheless, as stated by Moraglio et al. [15], these operators have a serious limitation: by construction, they always produce offspring that are larger than their parents (expressed as the total number of tree nodes) and, as demonstrated in [15], this makes the size of the individuals in the population grow exponentially with generations. Thus, after a few generations the population is composed by individuals so big that the computational cost of evaluating their fitness is unmanageable. This limitation makes these operators impossible to use in practice, in particular on complex real-life applications.

The solution suggested in [15] to overcome this drawback is to integrate in the GP algorithm a "simplification" phase, aimed at transforming each individual in the population into an equivalent (i.e. with the same semantics) but possibly smaller one. Even though this is an interesting and challenging study, depending on the language used to code individuals simplification can be very difficult, and it is often a very time consuming task. For this reason, in this paper we propose a different strategy to solve the problem: we develop a GP system incorporating a new implementation of geometric semantic genetic operators that makes them usable in practice, and does so very efficiently, without requiring any simplification of the individuals during the GP run.

The paper is organized as follows: Section 2 describes the geometric semantic operators introduced by Moraglio et al., while Section 3 presents our new implementation that overcomes the current limitations of these operators, making them usable and efficient. Section 4 presents the medical problem addressed in this paper and highlights its importance for clinicians. Section 5 presents the experimental settings and discusses the obtained results. Finally, Section 6 concludes the paper and provides hints for future research.

2   Geometric Semantic Operators

While semantically aware methods [1,9,10] often produced superior performances with respect to traditional methods, many of them are indirect: search operators act on the syntax of the parents to produce offspring, which are successively accepted only if some semantic criterion is satisfied. As reported in [15], this has at least two drawbacks: (i) these implementations are very wasteful as heavily based on trial-and-error; (ii) they do not provide insights on how syntactic and semantic searches relate to each other. To overcome these drawbacks, new operators were introduced in [15] that directly search the semantic space.

To explain the idea behind these operators, let us first give an example using Genetic Algorithms (GAs). Let us consider a GA problem in which the target solution is known and the fitness of each individual corresponds to its distance to the target (our reasoning holds for any distance measure used). This problem is characterized by a very good evolvability and it is in general easy to solve for GAs. In fact, for instance, if we use point mutation, any possible individual different from the global optimum has at least one neighbor (individual resulting from its mutation) that is closer than itself to the target, and thus fitter. So, there are no local optima. In other words, the fitness landscape is unimodal, which usually indicates a good evolvability. Similar considerations hold for many types of crossover, including various kinds of geometric crossover as the ones defined in [10].

Now, let us consider the typical GP problem of finding a function that maps sets of input data into known target ones (regression and classification are particular cases). The fitness of an individual for this problem is typically a distance between its calculated values and the target ones (error measure). Now, let us assume that we are able to find a transformation on the syntax of the individuals, whose effect is just a random perturbation of one of their calculated values. In other words, let us assume that we are able to transform an individual G into an individual H whose output is like the output of G, except for one value, that is randomly perturbed. Under this hypothesis, we are able to map the considered GP problem into the GA problem discussed above, in which point mutation is used. So, this transformation, if known, would induce a unimodal fitness landscape on every problem like the considered one (e.g. regressions and classifications), making those problems easily evolvable by GP, at least on training data. The same also holds for transformations on pairs of solutions that correspond to GAs semantic crossovers.

This idea of looking for such operators is very ambitious and extremely challenging: finding those operators would allow us to directly search the space of semantics, at the same time working on unimodal fitness landscapes. Although not without limitations, contribution [15] accomplishes this task, defining new operators that have exactly these characteristics.

Here we report the definition of geometric semantic operators given in [15] for real functions domains, since these are the operators we will use in the experimental phase. For applications that consider other kinds of data, the reader is referred to [15].

Definition 1. (Geometric Semantic Crossover). *Given two parent functions* T_1, T_2 : $\mathbb{R}^n \to \mathbb{R}$, *the geometric semantic crossover returns the real function* $T_{XO} = (T_1 \cdot T_R) + ((1 - T_R) \cdot T_2)$, *where* T_R *is a random real function whose output values range in the interval* $[0, 1]$.

The interested reader is referred to [15] for a formal proof of the fact that this operator corresponds to a geometric crossover on the semantic space. Nevertheless, even without formally proving it, we can have an intuition of it by considering that the (unique) offspring generated by this crossover has a semantic vector that is a linear combination of the semantics of the parents with random coefficients included in $[0, 1]$ and whose sum is equal to 1. To constrain T_R in producing values in $[0, 1]$ we use the sigmoid function: $T_R = \frac{1}{1+e^{-T_{rand}}}$ where T_{rand} is a random tree with no constraints on the output values.

Definition 2. (Geometric Semantic Mutation). *Given a parent function* $T : \mathbb{R}^n \to \mathbb{R}$, *the geometric semantic mutation with mutation step* ms *returns the real function* $T_M = T + ms \cdot (T_{R1} - T_{R2})$, *where* T_{R1} *and* T_{R2} *are random real functions.*

Reference [15] formally proves that this operator corresponds to a box mutation on the semantic space, and induces a unimodal fitness landscape. However, even though without a formal proof, it is not difficult to have an intuition of it, considering that each element of the semantic vector of the offspring is a "weak" perturbation of the corresponding element in the parent's semantics. We informally define this perturbation as "weak" because it is given by a random expression centered in zero (the difference between two random trees). Nevertheless, by changing parameter ms, we are able to tune the "step" of the mutation, and thus the importance of this perturbation.

We also point out that at every step of one of these operators, offspring contain the complete structure of the parents, plus one or more random trees as its subtrees and some arithmetic operators: the size of each offspring is thus clearly much larger than the one of their parents. The exponential growth of the individuals in the population (demonstrated in [15]) makes these operators unusable in practice: after a few generations the population becomes unmanageable because the fitness evaluation process becomes unbearably slow. The solution that is suggested in [15] consists in performing an automatic simplification step after every generation in which the programs are replaced by (hopefully smaller) semantically equivalent ones. However, this additional step adds to the computational cost of GP and is only a partial solution to the progressive program size growth. Last but not least, according to the particular language used to code individuals and the used primitives, automatic simplification can be a very hard task.

In the next section, we present a new implementation of these operators that overcomes this limitation, making them efficient without performing any simplification step and without imposing any particular representation for the individuals (for example the traditional tree-based representation of GP individuals can be used, as well as a linear representation, or any other one).

For simplicity, from now on, our implementation of GP using the geometric semantic crossover and mutation presented in [15] will be indicated as GS-GP (Geometric-Semantic GP).

3 Implementation of Geometric Semantic GP

The implementation we propose can be described as follows. Note that, although we describe the algorithm assuming the representation of the individuals is tree based, the implementation fits any other type of representation.

In a first step, we create an initial population of (typically random) individuals, exactly as in standard GP. We store these individuals in a table (that we call P from now on) as shown in Figure 1(a), and we evaluate them. To store the evaluations we create a table (that we call V from now on) containing, for each individual in P, the values resulting from its evaluation on each fitness case (in other words, it contains the semantics of that individual). Hence, with a population of n individuals and a training set of k fitness cases, table V will be made of n rows and k columns.

Then, for every generation, a new empty table V' is created. Whenever a new individual T must be generated by crossover between selected parents T_1 and T_2, T is represented by a triplet $T = \langle \&(T_1), \&(T_2), \&(R) \rangle$, where R is a random tree and for any tree τ, $\&(\tau)$ is a *reference* (or memory pointer) to τ (using a C-like notation). This triplet is stored in an appropriate structure (that we call $\mathcal{M}$ from now on) that also contains the name of the operator used, as shown in Figure 1c. The random tree R is created, stored in P, and evaluated in each fitness case to reveal its semantics. The values of the semantics of T are also easily obtained, by calculating $(T_1 \cdot R) + ((1 - R) \cdot T_2)$ for each fitness case, according to the definition of geometric semantic crossover, and stored in V'. Analogously, whenever a new individual T must be obtained by applying mutation to an individual T_1, T is represented by a triplet $T = \langle \&(T_1), \&(R_1), \&(R_2) \rangle$ (stored in $\mathcal{M}$), where R_1 and R_2 are two random trees (newly created, stored in P and evaluated for their semantics). The semantics of T is calculated as $T_1 + ms \cdot (R_1 - R_2)$ for each fitness case, according to the definition of geometric semantic mutation, and stored in V'. In the end of each generation, table V' is copied into V and erased. At this point, all the rows of P and $\mathcal{M}$ referring to individuals that are not ancestors[1] of the new population can also be erased, because they will not be used anymore.

In synthesis, this algorithm is based on the idea that, when semantic operators are used, an individual can be fully described by its semantics (which makes the syntactic component much less important than in standard GP), a concept discussed in depth in [15]. Therefore, at every generation we update table V with the semantics of the new individuals, and save the information needed to build their syntactic structures without explicitly building them. In terms of computational time, we emphasize that the process of updating table V is very efficient as it does not require the evaluation of the entire trees. Indeed, evaluating each individual requires (except for the initial generation) a constant time, which is independent from the size of the individual itself. In terms of memory, tables P and $\mathcal{M}$ grow during the run. However, table P adds a maximum of $2 \times n$ rows per generation (if all new individuals are created by mutation) and table $\mathcal{M}$ (which contains only memory pointers) adds a maximum of n rows per generation. Even if we never erase the "ex-ancestors" from these tables (and never reuse random trees, which is also possible), we can manage them efficiently for several thousands of generations.

---

[1] We abuse the term "ancestors" to designate not only the parents but also the random trees used to build an individual by crossover or mutation.

The final step of the algorithm is performed after the end of the last generation: in order to reconstruct the individuals, we need to "unwind" our compact representation and make the syntax of the individuals explicit. Therefore, despite performing the evolutionary search very efficiently, in the end we cannot avoid dealing with the large trees that characterize the standard implementation of geometric semantic operators. However, most probably we will only be interested in the best individual found, so this unwinding (and recommended simplification) process may be required only once, and it is done offline after the run is finished. This greatly contrasts with the solution proposed by Moraglio et al. of building and simplifying every tree in the population at each generation online with the search process.

Let us briefly consider the computational cost of evolving a population of n individuals for g generations. At every generation, we need $O(n)$ space to store the new individuals. Thus, we need $O(ng)$ space in total. Since we need to do only $O(1)$ operations for any new individual (since the fitness can be computed directly from the semantics, which can immediately be obtained from the semantics of the parents), the time complexity is also $O(ng)$. Thus, we have a linear space and time complexity with respect to population size and number of generations.

Excluding the time needed to build and simplify the best individual, the proposed implementation allowed us to evolve populations for thousands of generations with a considerable speed up with respect to standard GP. Next we provide a simple example that illustrates the functioning of the proposed algorithm.

Example. Let us consider the simple initial population P shown in table (a) of Figure 1 and the simple pool of random trees that are added to P as needed, shown in table (b). For simplicity, we will generate all the individuals in the new population (that we call P' from now on) using only crossover, which will require only this small amount of random trees. Besides the representation of the individuals in infix notation, these tables contain an identifier (Id) for each individual (T_1, ..., T_5 and R_1, ..., R_5). These identifiers will be used to represent the different individuals, and the individuals created for the new population will be represented by the identifiers T_6, ..., T_{10}.

We now describe the generation of a new population P'. Let us assume that the (non-deterministic) selection process imposes that T_6 is generated by crossover between T_1 and T_4. Analogously, we assume that T_7 is generated by crossover between T_4 and T_5, T_8 is generated by crossover between T_3 and T_5, T_9 is generated by crossover between T_1 and T_5, and T_{10} is generated by crossover between T_3 and T_4. Furthermore, we assume that to perform these five crossovers the random trees R_1, R_2, R_3, R_4 and R_5 are used, respectively. The individuals in P' are simply represented by the set of entries exhibited in table (c) of Figure 1 (structure $\mathcal{M}$). This table contains, for each new individual, a *reference* to the ancestors that have been used to generate it and the name of the operator used to generate it (either "crossover" or "mutation").

Let us assume that now we want to reconstruct the genotype of one of the individuals in P', for example T_{10}. The tables in Figure 1 contain all the information needed to do that. In particular, from table (c) we learn that T_{10} is obtained by crossover between T_3 and T_4, using random tree R_5. Thus, from the definition of geometric semantic crossover, we know that it will have the following structure: $(T_3 \cdot R_5) + ((1 - R_5) \cdot T_4)$. The remaining tables (a) and (b), that contain the syntactic structure of T_3, T_4, and

Id	Individual
T_1	$x_1 + x_2 x_3$
T_2	$x_3 - x_2 x_4$
T_3	$x_3 + x_4 - 2x_1$
T_4	$x_1 x_3$
T_5	$x_1 - x_3$

(a)

Id	Individual
R_1	$x_1 + x_2 - 2x_4$
R_2	$x_2 - x_1$
R_3	$x_1 + x_4 - 3x_3$
R_4	$x_2 - x_3 - x_4$
R_5	$2x_1$

(b)

Id	Operator	Entry
T_6	crossover	$\langle \&(T_1), \&(T_4), \&(R_1) \rangle$
T_7	crossover	$\langle \&(T_4), \&(T_5), \&(R_2) \rangle$
T_8	crossover	$\langle \&(T_3), \&(T_5), \&(R_3) \rangle$
T_9	crossover	$\langle \&(T_1), \&(T_5), \&(R_4) \rangle$
T_{10}	crossover	$\langle \&(T_3), \&(T_4), \&(R_5) \rangle$

(c)

Fig. 1. Illustration of the example described in Section 3. (a) The initial population P; (b) The random trees used by crossover; (c) The representation in memory of the new population P'.

R_5, provide us with the rest of the information we need to completely reconstruct the syntactic structure of T_{10}, which is $((x_3 + x_4 - 2x_1) \cdot (2x_1)) + ((1 - (2x_1)) \cdot (x_1 x_3))$ and upon simplification can become $-x_1(4x_1 - 3x_3 - 2x_4 + 2x_1 x_3)$.

4   Oral Anticoagulant Therapy

Coumarins-derived Oral Anticoagulant therapy (OAT), prescribed to more than 2.5 million new patients per year, is commonly used as life-long therapy in the prevention of systemic embolism in patients with atrial fibrillation, valvular heart disease, and in the primary and secondary prevention of venous and pulmonary thromboembolism. It is also used for the prevention of thromboembolic events in patients with acute myocardial infarction and with angina pectoris, in patients with heart valves, and after some types of orthopedic surgery.

Due to the increased prevalence of atrial fibrillation and thromboembolic disorders in elderly people [3] oral anticoagulation is one of the most frequently prescribed therapy in elderly patients.

Aging is a complex process which is accompanied by a potential multitude of issues that include numerous health problems associated to a multiple administration of medications, often coupled with reduced mobility and greater frailty, with a tendency to fall. Despite its validated efficiency, all these conditions are often cited as reasons to preclude the elderly from being anticoagulated [4].

In all subjects a combination of personal, genetic and non-genetic factors are responsible for about 20-fold variation in the coumarins dose required to achieve the therapeutic range of drug action, evaluated by the prothrombin international normalized ratio (INR) measurement. In case of elderly patients, this variability is highlight by clinically significant interaction due to coadministration of different drugs [14], and by liver and renal impairment which can further emphasize this interaction or directly modify the anticoagulant action [2]. For this reasons, oral anticoagulant therapy Initiation in elderly is more challenging than other patients.

The safety and efficacy of warfarin therapy are dependent on maintaining the INR within the target range for the indication. Due to above-cited inter patient variability in drug dose requirements, empiric dosing results in frequent dose changes as the therapeutic international normalized ratio (INR) gets too high or low, leaving patients at risk for bleeding (over-coagulation) and thromboembolism (under-coagulation). This means that there is a need to carry on the research to develop predictive models that are able to account for strong intraindividual variability in elderly patients' response to coumarins treatment.

Most of the computational approaches in the literature for the definition of mathematical models to support management decisions for OAT, provide the use of regression methods. The widely applied technique is Multiple Linear Regression, especially used to predict the value of the maintenance dose [7]. Other linear and non linear approaches enclose Generalized Linear Models [5] and polynomial regression [11]. More complex machine learning techniques were also employed to support clinical decisions on therapy management. A review of these methods is proposed in [12] for a review).

5   Experimental Study

In this section the experimental phase is outlined. In particular, section 5.1 briefly described the data used in the experimental phase; section 5.2 presents the experimental settings for the considered systems, while section 5.3 contains a detailed analysis of the obtained results.

5.1   Data Description

We collect data from clinical computerized databases based on 950 genotyped over 65 years old patients in anticoagulant therapy. A data preprocessing approach returned 748 *cleaned* patients (i.e. with complete data, not missing any information) useful for analysis. The features of this dataset can be summarized in four main entities: personal characteristics, anamnestic features, genetic data and therapy's characteristics . Demographic information includes body mass index and smoke habit; the anamnestic data are related to medical evidence leading to OAT (Atrial Fibrillation, Deep Venous Thrombosis/Pulmunary Embolism, other), a comorbidity (yes or not) and polipharmacotherapy evaluations (digitalis, amiodarone, furosemide, nitrates, beta blockers, calcium channel blockers, ACE inhibitors, diuretic tiazidic, sartanic, statins and other) and a renal function parameter (creatinine clearance); genetic data include the information related to the genetic polymorphysms involved in the metabolism of anticoagulant drug (CYP2C9, VKORC1 and CYP4F2); therapy's features describe patient's INR range, the INR range assigned within which patient should remain during therapy (2-3, 2.5-3.5 , 2.5-3), target INR (represented by the average of the values of INR range, respectively 2.5, 3 and 2.75), vitamin k antagonist anticoagulant drug (warfarin 5mg and acenocumarol 4 or 1mg) and their dosage, which is the independent variable of the study. All data used in the study were checked by clinicians of the anticoagulation clinical center. Dataset includes two subsets of patients: 403 stable patients which reached a stable response to therapy (stay in assigned INR range without significant modification of drug dose) and 345 unstable patients which did not reach stability. Descriptive statistic table relative to all features of the dataset is available as supplementary material on the authors' website (<anonymized>).

5.2 Experimental Settings

We have tested our implementation of GP with geometric semantic operators (GS-GP) against a standard GP system (STD-GP). A total of 30 runs were performed with each technique using different randomly generated partitions of the dataset into training (70%) and test (30%) sets. All the runs used populations of 200 individuals allowed to evolve for 500 generations. Tree initialization was performed with the Ramped Half-and-Half method [8] with a maximum initial depth equal to 6. The function set contained the four binary arithmetic operators $+$, $-$, $*$, and $/$ protected as in [8]. Fitness was calculated as the root mean squared error (RMSE) between predicted and expected outputs. The terminal set contained the number of variables corresponding to the number of features in each dataset. Tournaments of size 4 were used to select the parents of the new generation. To create new individuals, STD-GP used standard (subtree swapping) crossover [8] and (subtree) mutation [8] with probabilities 0.9 and 0.1, respectively. In our system this means that crossover is applied 90% of the times (while 10% of the times a parent is copied into the new generation) and 10% of the offspring are mutated. For GS-GP, crossover rate was 0.7, while mutation rate was 0.3, since preliminary tests have shown that the geometric semantic operators require a relatively high mutation rate in order to be able to effectively explore the search space. Survival was elitist as it always copied the best individual into the next generation. No maximum tree depth limit has been imposed during the evolution.

5.3 Experimental Results

The experimental results are reported using curves of the fitness (RMSE) on the training and test sets and boxplots obtained in the following way. For each generation the training fitness of the best individual, as well as its fitness in the test set (that we call test fitness) were recorded. The curves in the plots report the median of these values for the 30 runs. The median was preferred over the mean because of its higher robustness to outliers. The boxplots refer to the fitness values in generation 500. In the following text we may use the terms fitness, error and RMSE interchangeably.

Figure 2(a) and Figure 2(b) report training and test error for STD-GP and GS-GP while generations elapse. These figures clearly show that GS-GP outperforms STD-GP on both training and test sets. Figure 2(c) and Figure 2(d) report a statistical study of the training and test fitness of the best individual, both for GS-GP and STD-GP, for each of the 30 performed runs. Denoting by IQR the interquartile range, the ends of the whiskers represent the lowest datum still within 1.5 IQR of the lower quartile, and the highest datum still within 1.5 IQR of the upper quartile. As it is possible to see, GS-GP produces solutions with a lower dispersion with respect to the ones produced by STD-GP. To analyze the statistical significance of these results, a set of tests has been performed on the median errors. As a first step, the Kolmogorov-Smirnov test has shown that the data are not normally distributed and hence a rank-based statistic has been used. Successively, the Wilcoxon rank-sum test for pairwise data comparison has been used under the alternative hypothesis that the samples do not have equal medians. The p-values obtained are 3.4783×10^{-4} when test fitness of STD-GP is compared to test fitness of GS-GP and 4.6890×10^{-7} when training fitness of STD-GP is compared to training fitness of GS-GP. Therefore, when employing the usual significance level $\alpha = 0.05$ (or even a smaller

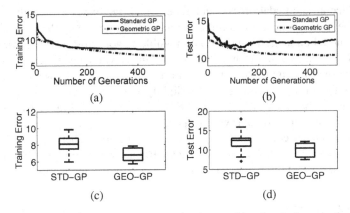

Fig. 2. Median of train (a) and test (b) fitness for the considered techniques at each generation calculated over 30 independent runs. Train (c) and test (c) fitness of the best individual produced in each of the 30 runs at the last performed generation.

one), we can clearly state that GS-GP produces fitness values that are significantly lower (i.e., better) than the STD-GP both on training and test data.

5.4 Generalization Ability

Given the very promising results obtained by GS-GP, we have performed a further experimental analysis to investigate the generalization ability of the models produced by the new technique.

A first indication about the behavior of GS-GP and STD-GP on unseen data comes from the Figure 2(b) of the previous section. From this figure, it seems that, differently from $ST - GP$, GS-GP is able to produce a model that does not overfit the unseen data.

To confirm this hypothesis, in this section we report the results obtained running GS-GP for 10000 generations. Given the fact that geometric semantic genetic operators induce a unimodal fitness landscape, we expected that the fitness on the training set will improve for GS-GP, but the main concern regards its generalization ability when the number of generations increases. In particular, in this section we want to answer the following question: do the good performances of GS-GP on training set result in an overfitted model for unseen data?

Figure 3(a) and Figure 3(b) allow to answer to this question. In particular, the good results that GS-GP has obtained on training data were expected: the geometric semantic

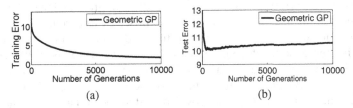

Fig. 3. Median of train (left) and test (right) fitness for GS-GP at each generation calculated over 30 independent runs

operators induce an unimodal fitness landscape, which facilitates evolvability. On the other hand, these excellent results on the training set do not affect the generalization ability of the model on unseen data. As it is possible to note from Figure 3(b), GS-GP produces a model that is able to handle unseen instances with a test fitness comparable to the one obtained in Figure 2(b). Furthermore, from Figure 3(b) we see that after generation 500 the error on the training set is slightly increasing, but not in a comparable way to the irregular behavior of the curve of STD-GP reported in Figure 2(b). This is a very promising result, in particular with respect to the considered applicative domain. Moreover, this results seems to indicate an important difference between $STD - GP$ and GS-GP: while $STD - GP$ overfits the data after a few generations, GS-GP seems to be much more robust to unseen data, at least for the studied application.

6   Conclusions

New genetic operators, called geometric semantic operators, have been proposed for genetic programming. They have the extremely interesting property of inducing a unimodal fitness landscape for any problem consisting in matching input data into known target outputs. This should make all the problems of this kind easily evolvable by genetic programming.

Nevertheless, in their first definition these new operators have a strong limitation that makes them unusable in practice: they produce offspring that are larger than their parents, and this results in an exponential growth of the size of the individuals in the population. This paper proposed an implementation of GP that uses the geometric semantic operators efficiently. This new GP system evolves the semantics of the individuals without explicitly building their syntax. It does so by keeping a set of trees (of the initial population and the random ones used by geometric semantic crossover and mutation) in memory and a set of pointers to them, representing the "instructions" on how to build the new individuals. Thanks to this compact representation, it is possible to make use of the great potential of geometric semantic GP to solve complex real-life problems.

The proposed GP system was used to address an important application in the field of pharmacogenetics. In particular, the problem addressed is the prediction of appropriate oral anticoagulant level of medical drugs. A difficulty with oral anticoagulants use is that prescription needs to be individually determined for each patient, usually by following a standard initial dosing protocol, measuring the coagulation rate regularly and then adjusting the dose until the required rate of coagulation is obtained.

The experimental results demonstrate that the new system outperforms standard genetic programming. Besides the fact that the new genetic programming system has excellent results on training data, we were positively surprised by its excellent generalization ability on the studied applications, testified by the good results obtained on test data.

Considering the good result achieved in this study, we will pursue it: beside new experimental validations on new data and different applications, we plan to orient our future activity towards more theoretical studies of the generalization ability of geometric semantic genetic programming. Moreover, regarding the oral anticoagulant therapy problem, we plan to start a work of interpretation of the models generated by GP, in strict collaboration with a team of expert clinicians.

Acknowledgments. This work was supported by national funds through FCT under contract PEst-OE/EEI/LA0021/2013 and projects EnviGP (PTDC/EIA-CCO/103363/2008) and MassGP (PTDC/EEI-CTP/2975/2012). Portugal.

References

1. Beadle, L., Johnson, C.: Semantically driven crossover in genetic programming. In: Proc. of the IEEE World Congress on Comput. Intelligence, pp. 111–116. IEEE Press (2008)
2. Capodanno, D., Angiolillo, D.J.: Antithrombotic therapy in the elderly. Journal of the American College of Cardiology 56(21), 1683–1692 (2010)
3. Anderson Jr., F.A., Wheeler, H.B., Goldberg, R.J., et al.: A population-based perspective of the hospital incidence and case-fatality rates of deep vein thrombosis and pulmonary embolism: The worcester dvt study. Archives of Internal Medicine 151(5), 933–938 (1991)
4. Fang, M.C., Chen, J., Rich, M.W.: Atrial fibrillation in the elderly. Am. J. Med. 120(6), 481–487 (2007)
5. Fang, M.C., Machtinger, E.L., Wang, F., Schillinger, D.: Health literacy and anticoagulation-related outcomes among patients taking warfarin. J. Gen. Intern. Med. 21(8), 841–846 (2006)
6. Jowett, S., Bryan, S., Poller, L., Van Den Besselaar, A.M.H.P., Van Der Meer, F.J.M., Palareti, G., Shiach, C., Tripodi, A., Keown, M., Ibrahim, S., Lowe, G., Moia, M., Turpie, A.G., Jespersen, J.: The cost-effectiveness of computer-assisted anticoagulant dosage: results from the European action on anticoagulation (eaa) multicentre study. J. Thromb. Haemost. 7(9), 1482–1490 (2009)
7. Klein, T.E., Altman, R.B., Eriksson, N., Gage, B.F., Kimmel, S.E., Lee, M.-T.M., Limdi, N.A., Page, D., Roden, D.M., Wagner, M.J., Caldwell, M.D., Johnson, J.A.: Estimation of the dose with clinical and pharmacogenetic data. New England Journal of Medicine 360(8), 753–764 (2009)
8. Koza, J.R.: Genetic Programming: On the Programming of Computers by Means of Natural Selection. MIT Press, Cambridge (1992)
9. Krawiec, K.: Medial crossovers for genetic programming. In: Moraglio, A., Silva, S., Krawiec, K., Machado, P., Cotta, C. (eds.) EuroGP 2012. LNCS, vol. 7244, pp. 61–72. Springer, Heidelberg (2012)
10. Krawiec, K., Lichocki, P.: Approximating geometric crossover in semantic space. In: GECCO 2009, July 8-12, pp. 987–994. ACM (2009)
11. Leichsenring, I., Plesch, W., Unkrig, V., Kitchen, S., Kitchen, D.P., Maclean, R., Dikkeschei, B., van den Besselaar, A.M.H.P.: Multicentre isi assignment and calibration of the inr measuring range of a new point-of-care system designed for home monitoring of oral anticoagulation therapy. Thromb. Haemost. 97(5), 856–861 (2007)
12. Martin, B., Filipovic, M., Rennie, L., Shaw, D.: Using machine learning to prescribe warfarin. In: Dicheva, D., Dochev, D. (eds.) AIMSA 2010. LNCS, vol. 6304, pp. 151–160. Springer, Heidelberg (2010)
13. McPhee, N.F., Ohs, B., Hutchison, T.: Semantic building blocks in genetic programming. In: O'Neill, M., Vanneschi, L., Gustafson, S., Esparcia Alcázar, A.I., De Falco, I., Della Cioppa, A., Tarantino, E. (eds.) EuroGP 2008. LNCS, vol. 4971, pp. 134–145. Springer, Heidelberg (2008)
14. Miners, J.O., Birkett, D.J.: Cytochrome p4502c9: an enzyme of major importance in human drug metabolism. British Journal of Clinical Pharmacology 45(6), 525–538 (1998)
15. Moraglio, A., Krawiec, K., Johnson, C.G.: Geometric semantic genetic programming. In: Coello, C.A.C., Cutello, V., Deb, K., Forrest, S., Nicosia, G., Pavone, M. (eds.) PPSN 2012, Part I. LNCS, vol. 7491, pp. 21–31. Springer, Heidelberg (2012)
16. Ryan, P.J., Gilbert, M., Rose, P.E.: Computer control of anticoagulant dose for therapeutic management. BMJ 299(6709), 1207–1209 (1989)

Dynamics of Neuronal Models
in Online Neuroevolution of Robotic Controllers

Fernando Silva[1,3], Luís Correia[3], and Anders Lyhne Christensen[1,2]

[1] Instituto de Telecomunicações, Lisboa, Portugal
[2] Instituto Universitário de Lisboa (ISCTE-IUL), Lisboa, Portugal
[3] LabMAg, Faculdade de Ciências, Universidade de Lisboa, Portugal
{fsilva,luis.correia}@di.fc.ul.pt, anders.christensen@iscte.pt

Abstract. In this paper, we investigate the dynamics of different neuronal models on online neuroevolution of robotic controllers in multirobot systems. We compare the performance and robustness of neural network-based controllers using summing neurons, multiplicative neurons, and a combination of the two. We perform a series of simulation-based experiments in which a group of e-puck-like robots must perform an integrated navigation and obstacle avoidance task in environments of different complexity. We show that: (i) multiplicative controllers and hybrid controllers maintain stable performance levels across tasks of different complexity, (ii) summing controllers evolve diverse behaviours that vary qualitatively during task execution, and (iii) multiplicative controllers lead to less diverse and more static behaviours that are maintained despite environmental changes. Complementary, hybrid controllers exhibit both behavioural characteristics, and display superior generalisation capabilities in simple and complex tasks.

Keywords: Evolutionary robotics, artificial neural network, evolutionary algorithm, online neuroevolution.

1 Introduction

Evolutionary computation has been widely studied in the field of robotics as a means to automate the design of robotic systems [1]. In evolutionary robotics (ER), robot controllers are typically based on artificial neural networks (ANNs) due to their capacity to tolerate noise in sensors. The parameters of the ANN, such as the connection weights, and occasionally the topology, are optimised by an evolutionary algorithm (EA), a process termed *neuroevolution* [2].

Online neuroevolution of controllers is a process of continuous adaptation that potentially gives robots the capacity to respond to changes or unforeseen circumstances by modifying their behaviour. An EA is executed on the robots themselves while they perform their task. The main components of the EA (evaluation, selection, and reproduction) are carried out autonomously by the robots without any external supervision. This way, robots may be capable of long-term self-adaptation in a completely autonomous manner.

L. Correia, L.P. Reis, and J. Cascalho (Eds.): EPIA 2013, LNAI 8154, pp. 90–101, 2013.

In a contribution by Watson *et al.* [3], the use of multirobot systems in online neuroevolution was motivated by the speed-up of evolution due to the inherent parallelism in groups of robots that evolve together in the task environment. Over the last decade, different approaches to online neuroevolution in multi-robot systems have been proposed [4]. Notwithstanding, properties at the level of individual neurons have largely been left unstudied. Online neuroevolution studies have been almost exclusively based on ANNs composed of a variation of the neuronal model introduced in the 1940s by McCulloch and Pitts [5], that is, summing neurons. With advances in biology, multiplicative-like operations have been found in neurons as a means to process non-linear interactions between sensory inputs [6,7]. In machine learning, multiplicative neurons have shown to increase the computational power and storage capacities of ANNs [8,9].

In this paper, we investigate the dynamics and potential benefits of multiplicative neurons and of summing neurons, separately and combined, in online neuroevolution in multirobot systems. We conduct a simulated experiment in which a group of robots modelled after the e-puck [10] must perform an integrated navigation and obstacle avoidance task in environments of distinct complexity. The task implies an integrated set of actions, and consequently a trade-off between avoiding obstacles, maintaining speed, and forward movement. The three types of controllers are compared with respect to the speed of convergence, task performance, generalisation capabilities, complexity of solutions, and diversity of behaviours evolved. The main conclusion is that the combination of summing neurons and multiplicative neurons leads to superior generalisation capabilities, and allows robots to exhibit different behaviours that vary qualitatively or that remain static, depending on task complexity.

2   Background and Related Work

In this section, we describe the neuronal models considered, and we introduce odNEAT, the online neuroevolution algorithm used in this study. We exclusively consider discrete-time ANNs with summing or multiplicative neuronal models. Models that have more complex or explicit time-dependent dynamics, such as spiking neurons [11], are left for posterior investigation.

2.1   Neuronal Models

In discrete-time ANNs, the classic summing unit is the most commonly used neuronal model. The summing unit model performs the computation as follows:

$$a_i = f\left(\sum_{j=1}^{N} w_{ji} \cdot x_j + w_0 \right).\tag{1}$$

where a_i is the activation level of a given neuron i, and f is the activation function applied on the weighted sum of inputs from incoming neurons x_j, plus the bias value w_0. The activation function f can take the form of, for instance, a threshold function or a sigmoid function. However, relying exclusively on sums of weighted inputs potentially limits performance and learning capabilities when complex or non-linear interactions are considered [6,9], as in the case of robotics.

In machine learning, multiplicative neuronal models were introduced more than 20 years ago. Examples include the pi-sigma and the sigma-pi units [9], and the more general product unit neuronal model [8], which we use in this study. In the product unit model, the activation of a neuron i is computed as follows:

$$a_i = f\Big(\prod_{j=1}^{N} x_j^{w_{ji}} \Big) . \tag{2}$$

with notations similar to Eq. 1. The number of exponents j gives the *order* of the neuron, thus denoting the ANN as a *higher-order neural network*. The exponents are real values, in which negative exponents enable division operations.

The potential benefits of the product unit model have been widely discussed in the literature. Using gradient descent methods, Durbin and Rumelhart [8] concluded that product units have superior information and learning capabilities compared to summing units. Complementary, Schmitt [9] analysed the gains in information processing and learning capabilities of product unit neurons in terms of solution complexity and computational power. Despite the positive results, only recently the potential benefits of product units were investigated in an evolutionary robotics context. Cazenille *et al.* [7] performed the offline evolution of ANN controllers for the *coupled inverted pendulums* problem, a benchmark in modular robotics. The authors investigated the interplay between microscopic properties such as the neuronal model, and macroscopic properties such as modularity and repeating motifs in ANNs. Surprisingly, their results suggested that product units may be counter-productive when used alone. If ANNs display regularities such as modularity, then product units may lead to better fitness scores, while requiring fewer evaluations for evolving a solution.

2.2   odNEAT: An Online Neuroevolution Algorithm

odNEAT [4] is a decentralised online neuroevolution algorithm for multirobot systems. odNEAT optimises both the parameters and the topology of the ANN controllers. The algorithm starts with networks with no hidden neurons, and with each input neuron connected to every output neuron. Topologies are gradually *complexified* by adding new neurons and new connections through mutation, thus allowing odNEAT to find an appropriate degree of complexity for the task.

odNEAT implements the online evolutionary process according to a physically distributed island model. Each robot optimises an internal population of genomes through intra-island variation, and genetic information between two or more robots is exchanged through inter-island migration. In this way, each robot is potentially self-sufficient and the evolutionary process capitalises on the exchange of genetic information between multiple robots for faster adaptation.

During task execution, each robot is controlled by an ANN that represents a candidate solution to a given task. Agents maintain a virtual energy level reflecting their individual performance. The fitness value is defined as the average of the energy level, sampled at regular time intervals. When the virtual energy level of a robot reaches zero, the current controller is considered unfit for the task. A new controller is then created by selecting two parents from the repository, each one via a tournament selection of size 2. Offspring is created through crossover of the parents' genomes and mutation of the new genome. Mutation is both structural and parametric, as it adds new neurons and new connections, and optimises parameters such as connection weights and neuron bias values.

Note that odNEAT is used in this study as a representative efficient online neuroevolution algorithm that optimises ANN weights and topologies [4]. As all experiments are based on odNEAT, the main distinctions among them will be the use of summing neurons, multiplicative neurons, or a combination of the two, rather than the online neuroevolution algorithm or its particular details.

3   Methods

In this section, we define our experimental methodology, including the robot model, the two variants of the navigation and obstacle avoidance task, the experimental parameters, and how we characterise the behaviours evolved.

3.1   Robot Model and Behavioural Control

To conduct our simulated experiments, we use JBotEvolver[1], an open-source, multirobot simulation platform, and neuroevolution framework. The simulator implements 2D differential drive kinematics. In our experimental setup, the simulated robots are modelled after the e-puck [10], a small (75 mm in diameter) differential drive robot capable of moving at a maximum speed of 13 cm/s. Each robot is equipped with eight infrared sensors, for obstacle detection, and communication of, for instance, genomes at a range of up to 25 cm between sender and receiver. Each infrared sensor and each actuator are subject to noise, which is simulated by adding a random Gaussian component within ±5% of the sensor saturation value or actuation value. Each robot also has an internal sensor that allows it to perceive its current virtual energy level.

Each robot is controlled by an ANN synthesised by odNEAT. The ANN's connection weights ∈ [-10,10]. The input layer consists of 17 neurons: (i) eight for robot detection, (ii) eight for wall detection, and (iii) one neuron for the virtual energy level sensor. The output layer contains two neurons, one for each wheel.

3.2   Integrated Navigation and Obstacle Avoidance

Navigation and obstacle avoidance is a classic task in evolutionary robotics, and an essential feature for autonomous robots operating in real-world environments.

---

[1] https://code.google.com/p/jbotevolver/

Robots have to simultaneously move as straight as possible, maximise wheel speed, and avoid obstacles. The task implies an integrated set of actions, and consequently a trade-off between avoiding obstacles in sensor range and maintaining speed and forward movement. Navigation and obstacle avoidance is typically conducted in single robot experiments. In multirobot experiments, each robot poses as an additional, moving obstacle for the remaining group.

In our experiments, a group of five robots operates in a square arena surrounded by walls. The size of the arena was chosen to be 3 x 3 meters. Initially, robots are placed in random positions. During simulation, the virtual energy level E is updated every 100 ms according to the following equation:

$$\frac{\Delta E}{\Delta t} = f_{norm}(V \cdot (1 - \sqrt{\Delta v}) \cdot (1 - d_r) \cdot (1 - d_w)) \,. \tag{3}$$

where V is the sum of rotation speeds of the two wheels, with $0 \leq V \leq 1$. $\Delta v \in [0, 1]$ is the normalised algebraic difference between the signed speed values of the wheels (positive in one direction, negative in the other). d_r and d_w are the highest activation values of the infrared sensors for robot detection and for wall detection, respectively. d_r and d_w are normalised to a value between 0 (no robot/wall in sight) and 1 (collision with a robot or wall). f_{norm} maps from the domain $[0, 1]$ into $[-1, 1]$.

Experimental Setup. We conducted experiments in two different environments to assess performance. The first environment is a plain arena, in which the only obstacles are the robots and the walls that confine the arena. The second environment is an arena with five additional obstacles with dimensions of 0.5 x 0.5 meters. The additional obstacles are of the same material as the walls, and intuitively increase the difficulty of the task by reducing the area for navigation. In each evolutionary run conducted in the second environment, obstacles are placed at random locations. The environments are illustrated in Fig. 1.

We performed three sets of evolutionary experiments in each environment, characterised by different neuronal models. In one set of experiments, neurons

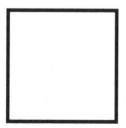      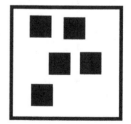

(a) Plain environment   (b) Complex environment — example 1   (c) Complex environment — example 2

Fig. 1. The two types of environments used to evolve controllers. Each of the arenas measures 3 x 3 meters. The dark areas denote physical obstacles, while the white areas denote the arena surface on which robots can navigate.

added through structural mutation are multiplicative product units. In the second set of experiments, neurons introduced are summing units. In the third set of experiments, each new neuron has an equal probability of 0.5, sampled from a uniform distribution, of being either a multiplicative or summing neuron. Multiplicative product units may therefore be combined with summing units, and introduce the ability for computing weighted sums of products, and vice-versa.

For each experimental configuration, we performed 30 independent evolutionary runs. Each run lasts for 100 hours of simulated time. The virtual energy level of robots is limited to the range [0,100] energy units. When the energy level reaches zero, a new controller is generated and assigned the default energy value of 50 units. Other parameters are the same as in [4].

3.3 Characterisation of Behavioural Diversity

Ultimately, online neuroevolution of controllers synthesises the *behavioural* control of robots. To characterise and compare behaviours evolved, we use a generic Hamming distance-based behavioural metric based on the mapping between sensors and actuators. This measure has shown to be, at least, as efficient as domain-dependent behavioural metrics [12]. The behaviour metric is based on the set of sensor readings and actuation values ϑ normalised into [0,1], as follows:

$$\vartheta = \Big[\{\mathbf{s}(t), \mathbf{a}(t)\}, t \in [0, T]\Big] . \tag{4}$$

where $\mathbf{s}(t)$ and $\mathbf{a}(t)$ are the sensor readings and actuation values at time t, respectively, and T is the number of observations. The binary version ϑ_{bin} of ϑ is computed as follows:

$$\vartheta_{bin} = \Big[\vartheta_{bin}(t), t \in [0, T]\Big] = \Big[\{\mathbf{s}_{bin}(t), \mathbf{a}_{bin}(t)\}, t \in [0, T]\Big] . \tag{5}$$

where each $s_{bin,i}(t) \in s_{bin}(t)$ is defined as 1 if $s_i(t) > 0.5$ and 0 otherwise, and each $a_{bin,i}(t) \in a_{bin}(t)$ is defined as 1 if $a_i(t) > 0.5$ and 0 otherwise. The Hamming distance between two behaviours is then computed as follows:

$$\sigma(\vartheta_1, \vartheta_2) = \sum_{t=0}^{T} h(\vartheta_{1,bin}(t), \vartheta_{2,bin}(t)) . \tag{6}$$

$$h(\vartheta_1, \vartheta_2) = \sum_{i=1}^{len(\vartheta_1)} 1 - \delta(\vartheta_1[i], \vartheta_2[i]) . \tag{7}$$

where $len(\vartheta_1) = len(\vartheta_2)$ denotes the length of the binary sequences ϑ_1 and ϑ_2, and $\delta(i,j)$ is the Kronecker delta defined as $\delta(i,j) = 1$ if $i = j$, and 0 otherwise. We further extend the generic Hamming distance between sensor readings and actuation values to capture the *intra-behaviour* distance as follows:

$$\sigma(\vartheta_{bin}) = \sum_{t=1}^{T} h(\vartheta_{bin}(t-1), \vartheta_{bin}(t)) . \tag{8}$$

$\sigma(\vartheta_{bin})$ captures the differences between consecutive observations of the relation between sensor readings and actuation values, thereby approximating to what extent the behaviour of a robot varies during task execution.

4   Experimental Results

In this section, we present and discuss the experimental results. We use the Mann-Whitney test to compute statistical significance of differences between sets of results because it is a non-parametric test, and therefore no strong assumptions need to be made about the underlying distributions. Success rates are compared using Fisher's exact test, a non-parametric test suitable for this purpose [13]. Statistical dependence between two variables is computed using the non-parametric Spearman's rank correlation coefficient.

4.1   Comparison of Performance

We start by comparing the performance of the three controller models. We focus on three aspects: (i) the number of evaluations, i.e., the number of controllers tested by each robot before a solution to the task is found, (ii) the complexity of solutions evolved, and (iii) their generalisation capabilities.

The number of evaluations for the three types of controllers is listed in Table 1. In the plain environment, neuroevolution of hybrid controllers required fewer evaluations to synthesise solutions for the task. Differences between hybrid controllers and multiplicative controllers are statistically significant ($\rho < 0.001$, Mann-Whitney). Differences between other types of controllers are not significant. In the complex environment, solutions are synthesised faster when summing controllers are evolved. Multiplicative controllers, once again, require more evaluations to evolve solutions. Differences are significant with respect to summing controllers ($\rho < 0.001$, Mann-Whitney), and to hybrid controllers ($\rho < 0.01$). Differences between summing controllers and hybrid controllers are not significant, although summing controllers converged to a solution in fewer evaluations on average. Interestingly, both hybrid controllers and multiplicative controllers are affected by the increase in task complexity. Summing controllers, on the other hand, require an approximate number of evaluations in the two tasks.

Table 1. Comparison of the number of evaluations between the three types of controllers considered in the two tasks (average ± std. dev.)

Controller	Plain environment	Complex environment
Summing	23.63 ± 19.16	22.61 ± 18.68
Multiplicative	26.08 ± 21.10	32.02 ± 24.93
Hybrid	21.54 ± 20.53	27.81 ± 29.16

By analysing the complexity of solutions evolved[2], we observed simple neural topologies for solving each task, as listed in Table 2. Overall, multiplicative controllers present the least complex topologies, in equality with summing controllers in the complex environment. Despite solving the tasks with less structure, the number of evaluations to synthesise suitable multiplicative neurons is higher, as discussed previously and shown in Table 1. This is due to multiplicative neurons requiring a finer-grain adjustment of parameters. Compared to summing neurons, multiplicative controllers required more adjustments of connection weights through mutation and, therefore, a higher number of evaluations.

Complementary, hybrid controllers present the most complex topologies. The crossover operator manipulates topological structure involving different neural dynamics, i.e., summing and multiplicative neurons. We analysed the effects of evolutionary operators and found a more accentuated decrease in fitness scores when hybrid controllers are recombined. This effect is progressively eliminated as new neurons and new connections are added to the network. Nonetheless, despite differences in terms of neural complexity and number of evaluations, each experimental configuration lead to the evolution of high-scoring controllers. The average fitness score of solutions varies from 91.31 to 94.59 in the plain environment, and from 91.24 to 95.49 in the complex environment. Differences in fitness scores are not statistically significant across all comparisons.

Table 2. Neural complexity of solutions evolved. Neurons and connections added through evolution (average ± std. dev.).

| Controller | Plain environment | | Complex environment | |
	Neurons	Connections	Neurons	Connections
Summing	3.14 ± 0.35	6.59 ± 0.90	1.17 ± 0.41	2.57 ± 1.03
Multiplicative	2.16 ± 0.40	4.62 ± 0.94	1.21 ± 0.42	2.84 ± 1.09
Hybrid	3.16 ± 0.39	6.56 ± 1.13	3.11 ± 0.32	6.47 ± 0.79

Testing for Generalisation. To analyse the generalisation capabilities of the different types of controllers, we conducted a series of generalisation tests. For each evolutionary run conducted previously, we restarted the task 100 times using the controllers evolved and not allowing further evolution. Each task restart is a generalisation test serving to assess if robots can continuously operate after redeployment, and for evaluating the ability to operate in conditions not experienced during the evolution phase. A group of robots passes a generalisation test if it continues to solve the task, i.e., if the virtual energy level of none of the robots reaches zero. Each generalisation test has the same duration as the evolutionary phase, 100 hours of simulated time. In the complex environment, the five obstacles are placed in random locations in each test.

---

[2] The complete set of networks evolved is available at http://dx.doi.org/10.6084/m9.figshare.705842

Table 3. Generalisation performance of each controller model. The table lists the average generalisation capabilities of each group of five robots, and the total of generalisation tests solved successfully.

	Plain environment		Complex environment	
Controller	Generalisation (%)	Succ. tests	Generalisation (%)	Succ. tests
Summing	73.40 ± 23.33	2202/3000	45.30 ± 30.36	1359/3000
Multiplicative	85.87 ± 13.07	2576/3000	57.03 ± 26.09	1711/3000
Hybrid	84.10 ± 24.24	2523/3000	80.60 ± 14.77	2418/3000

In Table 3, we show the generalisation performance of each controller model. In the plain environment, multiplicative controllers outperform summing controllers by 12% as they solve 374 tests more. Differences in successful and unsuccessful generalisation tests are statistically significant ($\rho < 1 \cdot 10^{-4}$, Fisher's exact test). The hybrid controllers display high generalisation performance, similar to multiplicative controllers. Differences between these two controllers are not statistically significant. In the complex task, multiplicative controllers also outperform summing controllers by approximately 12% due to solving 352 tests more ($\rho < 1 \cdot 10^{-4}$, Fisher's exact test). More importantly, hybrid controllers significantly outperform both summing controllers and multiplicative controllers ($\rho < 1 \cdot 10^{-4}$, Fisher's exact test), and maintain approximate generalisation levels across the two tasks. In this way, the higher number of evaluations to synthesise hybrid controllers for complex tasks is compensated for by the increase in generalisation performance caused by the interplay between summing neurons and multiplicative neurons in the same neural architecture.

The generalisation performance of hybrid controllers is a factor particularly important in the case of online evolution. Results obtained indicate that hybrid controllers can adapt more efficiently than multiplicative and summing controllers alone to contexts not explicitly encountered during evolution, thus avoiding the continuation of the evolutionary process. In this way, hybrid controllers revealed to be advantageous as they are high-scoring controllers with superior generalisation capabilities that require a competitive number of evaluations.

4.2 Analysis of Genotypic and Behavioural Search Space

To unveil the neuroevolution dynamics of summing neurons and of multiplicative neurons, we compared how the evolutionary search proceeds through the high-dimensional genotypic search space. To visualise the intermediate genotypes produced when using the two neuronal models, and how they traverse the search space *with respect to each other*, we use *Sammon's nonlinear mapping* [14]. Sammon's mapping performs a point mapping of high-dimensional data to two-dimensional spaces, such that the structure of the data is approximately preserved. The distance in the high-dimensional space δ_{ij} between two

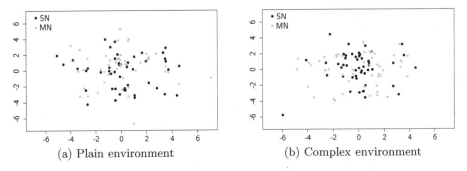

(a) Plain environment   (b) Complex environment

Fig. 2. Combined Sammon's mapping of intermediate genotypes. Genotypes representing networks using summing neurons are marked in black, and genotypes representing networks with multiplicative neurons are marked in gray.

genotypes i and j is based on genomic distance as used in odNEAT. The distance between two points in the two-dimensional space is their Euclidean distance.

In Fig 2, we show the Sammon's mapping for the plain environment and for the complex environment. In order to obtain a clearer visualisation, and a consistent and representative selection of genotypes, we map the 50 most genetically different genotypes, which in turn represent the 50 most diverse topologies with respect to each experimental configuration. Generally speaking, there is a balanced exploration of the search space, as different controller models discover similar regions of the genotypic search space. Genotypes representing hybrid controllers also explore similar regions of the search space (data not shown). The progress throughout the evolutionary process is therefore similar in the three controller models, indicating that the fitness landscape is not significantly altered by the use of different neuronal models. In this way, the dynamics of neuronal models may lead to the evolution of different *behaviours*, which in turn would account for differences in performance. To verify this hypothesis, we analysed the behaviour space explored by solutions evolved. We used the two generic Hamming distance-based behaviour metrics described in Sect. 3.3. For each solution, we analysed the behaviour during 10,000 seconds of simulated time with a sampling rate of 10 seconds, resulting in a behaviour vector of length 1000.

In Fig. 3, we show the distribution of inter-behaviour distances and the distribution of intra-behaviour distances. In all experimental configurations, there is a very strong monotonic correlation between the novelty of the behaviour in terms of average inter-behaviour distance and the degree of intra-behaviour variance ($\rho > 0.98$, Spearman's correlation). The more different the behaviour is compared to remaining behaviours, the more it varies during task execution. Complementary, more common behaviours present a lower intra-behaviour variance during task execution. Despite behavioural differences, there is no clear correlation between the inter- and intra-behaviour distances and fitness scores ($-0.09 < \rho < 0.34$), as both types of behaviours lead to high performance.

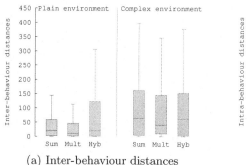 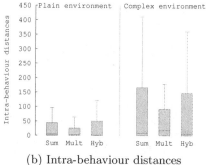

(a) Inter-behaviour distances   (b) Intra-behaviour distances

Fig. 3. Distribution of: (a) inter-behaviour distances between behaviours evolved for each task, by each type of controllers, and (b) intra-behaviour distances

Compared to multiplicative controllers, robots using summing controllers synthesise more diverse behaviours in the two environments ($\rho < 0.001$, Mann-Whitney) with higher intra-behaviour variance. Multiplicative controllers lead to less diverse behaviours with lower intra-behaviour variance, which maintain the same qualitative behaviour despite environmental changes. Summing controllers, on the other hand, adopt qualitatively different behaviours, especially in more complex tasks. Hybrid controllers appear to exhibit both behavioural characteristics. In the plain environment, robots evolved a greater diversity of robust behaviours with the higher intra-behaviour variance. With the increase of task complexity, robots exhibit intermediate behaviours in terms of inter- and intra-behaviour distances. In other words, hybrid controllers appear to capitalise on the tendency exhibited by multiplicative controllers to maintain the same qualitative behaviour, and on the ability of summing controllers to modify the behaviour of robots during task execution.

5   Conclusions and Future Work

In this paper, we investigated the effects of multiplicative neurons in online neuroevolution of robotic controllers. We compared the dynamics of multiplicative neurons, separately and when combined with summing neurons, in terms of speed of convergence, complexity of solutions evolved, task performance, generalisation capabilities, and diversity of behaviours evolved.

We showed that solutions are synthesised faster when summing neurons are used alone. We also showed that neural controllers using a combination of summing neurons and multiplications provide competitive results in terms of speed of convergence. In complex environments, the hybrid controllers rely on larger topologies, with one or two hidden neurons more. Nonetheless, these controllers exhibit significantly superior generalisation capabilities, and appear to balance between maintaining the same behaviour and modifying the behaviour depending on task requirements and complexity. An analysis of neural activation and structural patterns could potentially unveil differences in neural dynamics and the decision making mechanisms underlying the robot's behaviour.

The immediate follow-up work includes the study of macroscopic properties such as modularity, structural regularity and hierarchy, and investigating if these properties can facilitate online neuroevolution for real-world complex tasks.

Acknowledgments. This work was partly supported by the Fundação para a Ciência e a Tecnologia (FCT) under the grants PEst-OE/EEI/LA0008/2013 and SFRH/BD/89573/2012.

References

1. Floreano, D., Keller, L.: Evolution of adaptive behaviour by means of Darwinian selection. PLoS Biology 8(1), e1000292 (2010)
2. Floreano, D., Dürr, P., Mattiussi, C.: Neuroevolution: from architectures to learning. Evolutionary Intelligence 1(1), 47–62 (2008)
3. Watson, R., Ficici, S., Pollack, J.: Embodied evolution: Distributing an evolutionary algorithm in a population of robots. Robotics and Autonomous Systems 39(1), 1–18 (2002)
4. Silva, F., Urbano, P., Oliveira, S., Christensen, A.L.: odNEAT: An algorithm for distributed online, onboard evolution of robot behaviours. In: 13th International Conference on Simulation & Synthesis of Living Systems, pp. 251–258. MIT Press, Cambridge (2012)
5. McCulloch, W., Pitts, W.: A logical calculus of the ideas immanent in nervous activity. Bulletin of Mathematical Biology 5(4), 115–133 (1943)
6. Koch, C.: Biophysics of computation: information processing in single neurons. Oxford Univ. Press, Oxford (2004)
7. Cazenille, L., Bredeche, N., Hamann, H., Stradner, J.: Impact of neuron models and network structure on evolving modular robot neural network controllers. In: 14th Genetic and Evolutionary Computation Conference, pp. 89–96. ACM Press, New York (2012)
8. Durbin, R., Rumelhart, D.E.: Product units: A computationally powerful and biologically plausible extension to backpropagation networks. Neural Computation 1(1), 133–142 (1989)
9. Schmitt, M.: On the complexity of computing and learning with multiplicative neural networks. Neural Computation 14(2), 241–301 (2002)
10. Mondada, F., Bonani, M., Raemy, X., Pugh, J., Cianci, C., Klaptocz, A., Magnenat, S., Zufferey, J., Floreano, D., Martinoli, A.: The e-puck, a robot designed for education in engineering. In: 9th Conference on Autonomous Robot Systems and Competitions, IPCB, Castelo Branco, Portugal, pp. 59–65 (2009)
11. Floreano, D., Schoeni, N., Caprari, G., Blynel, J.: Evolutionary bits 'n' spikes. In: 8th International Conference on Simulation & Synthesis of Living Systems, pp. 335–344. MIT Press, Cambridge (2003)
12. Mouret, J., Doncieux, S.: Encouraging behavioral diversity in evolutionary robotics: An empirical study. Evolutionary Computation 20(1), 91–133 (2012)
13. Fisher, R.: Statistical Methods for Research Workers. Oliver & Boyd, Edinburgh (1925)
14. Sammon Jr., J.: A nonlinear mapping for data structure analysis. IEEE Transactions on Computers C-18(5), 401–409 (1969)

Evolving an Harmonic Number Generator with ReNCoDe

Rui L. Lopes and Ernesto Costa

Center for Informatics and Systems of the University of Coimbra
Polo II - Pinhal de Marrocos 3030-290 Coimbra, Portugal
{rmlopes,ernesto}@dei.uc.pt

Abstract. Evolutionary Algorithms (EA) are loosely inspired in the ideas of natural selection and genetics. Over the years some researchers have advocated the need of incorporating more ideas from biology into EAs, in particular with respect to the individuals' representation and the mapping from the genotype to the phenotype. One of the first successful proposals in that direction was the Artificial Regulatory Network (ARN) model. Soon after some variants of the ARN with increased capabilities were proposed, namely the Regulatory Network Computational Device (ReNCoDe). In this paper we further study ReNCoDe, testing the implications of some design choices of the underlying ARN model. A Genetic Programming community-approved symbolic regression benchmark (the harmonic number) is used to compare the performance of the different alternatives.

Keywords: genetic regulatory networks, genotype-phenotype mapping, genetic programming, regression, harmonic numbers.

1 Introduction

Nature-inspired algorithms are used today extensively to solve a multitude of learning, design, and optimisation problems, giving rise to a new research area called Evolutionary Computation (EC) [1]. Typically, the objects manipulated by the evolutionary algorithms are represented at two different levels. At a low level (the genotype) the representations are manipulated by the variation operators; at a high level (the phenotype) the objects are evaluated to determine their fitness and are selected accordingly. Because of that, we need a mapping between these two levels. Several proposals were made over time, yet many researchers complain about the simplistic approach adopted in most solutions, and have proposed more complex ones, like Grammatical Evolution [2], Self-Modifying Cartesian Genetic Programming [3], Gene Expression Programming [4], or Enzyme Genetic Programming [5].

In particular, in [6,7] the authors suggested that one should enrich the artificial model of evolution with the inclusion of feedback regulatory mechanisms, and showed how an Artificial Regulatory Network (ARN) could be used computationally in different settings [8]. This model simulates protein interaction

L. Correia, L.P. Reis, and J. Cascalho (Eds.): EPIA 2013, LNAI 8154, pp. 102–113, 2013.

and expression in an artificial cell. Some extensions to the initial model were made with the inclusion of extra input proteins and dividing the gene's products in transcription factors (regulatory proteins) and non-regulatory proteins [9], or by transforming the regulatory gene network into a computable graph, with or without feedback edges, similarly to what is done in GP [10].

The main goal of this paper is to study the best parameterisation of ReN-CoDe in a "community-approved" benchmark problem, with respect to: i) how to initialise the genome, ii)the influence of the size of the genome, and iii) the representational issue of allowing or not the existence of overlapping genes. This study will be done in the context of the discovery of a program capable of computing the harmonic number H_n, for different values of n.

The remaining sections are organised as follows. Section 2 describes the original ARN model. Then, Section 3 describes the Regulatory Network Computational Device, the target of study of this work. In Section 4 we describe the problem used and the experimental setup is presented in Section 5. The results are presented and analysed in Section 6. Finally, in Section 7 we draw some conclusions and present ideas for future work.

2   Artificial Regulatory Network

The Artificial Regulatory Network (ARN) [6] is an attempt to incorporate regulatory mechanisms between the genotype and the phenotype. There are no other products, i.e., DNA, and processes in between these two levels. The genome has fixed length and it is constructed by simple duplication with mutation events, called DM events. Regulation between genes is a mediated process, achieved by means of a binding process between proteins (i.e., transcription factors) and special zones in the genome that appear upstream the promoter of a gene (for an in depth description of ARNs see [10]).

Representation. The model presented in this paper uses a binary genome that includes genes, each one divided into a coding region and a regulatory region. The regulatory part is composed by a promoter (P), that is used to indicate the presence of a gene, and two binding sites, an enhancer (E) and an inhibitor (H). Each P, E, and H are 32-bit long sequences. The coding part is made of five 32-bit sequences, G1 to G5 (see Figure 1).

Gene Expression. The genotype - phenotype mapping is defined by expressing each 5*32=160-bit long gene, resulting in a 32-bit protein. A gene expresses a protein by a majority rule: if we consider a gene, for example G^m, divided into 5 parts of size 32 each, G_1^m to G_5^m, at position i, say, the protein's bit will have a value corresponding to the most frequent value in each of these 5 parts, at the same position, i.e.,

$$P_i^m = majority(G_{ki}^m, \quad \forall k = 1, \ldots, 5), \quad \forall i = 1, \ldots, 32$$

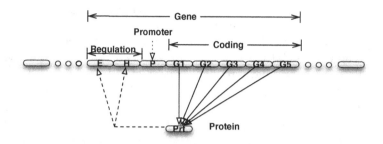

Fig. 1. A genome element

Regulation. Genes interact mediated by proteins, which bind to the regulatory region of each gene. In order for a link to exist between any two genes, the concentration of the corresponding protein must attain a certain level, and that depends on the strength of the binding. The strength of the binding is computed by calculating the degree of complementarity between the protein and each of the regulatory regions, according to formula 1:

$$x_i = \frac{1}{N} \sum_{\substack{0 < j \le N \\ j \ne i}} c_j \exp^{\beta(\mu_{ji} - \mu_{max})} \tag{1}$$

where x_i represents the binding strength of the enhancer (e_i) or the inhibitory (h_i) region, N is the number of proteins, c_j the concentration of protein j, μ_{ji} is the number of bits that are different in the protein j and in the regulation sites of protein i, μ_{max} is the maximum match achievable, and β is a scaling factor.

The production of a protein over time depends on its concentration, which in turn is a function of the way each protein binds to that gene's regulatory regions. It is defined by the differential equation

$$\frac{dc_i}{dt} = \delta(e_i - h_i)c_i \tag{2}$$

where e_i and h_i are defined by equation 1, and δ is a scaling factor.

Computational Device. Using this representation and processes of gene expression and regulation, we can build for each genome the corresponding artificial gene regulatory network (ARN): each node is a gene and the arcs express regulatory dependencies. From a problem-solving perspective we want to transform an ARN into a computational problem-solver. This can be achieved in different ways. For example, [8] proposed the introduction of extra enhancer and inhibitor sites, where all proteins bind. The variation of the binding strength of all proteins over time is defined as the time dependent output. In a different way, in [9] the authors defined some extra proteins, not produced by genes but contributing to regulation, that assume the role of inputs. Moreover, two types of promoters were introduced splitting the genes into those that contribute to regulation and those that do not and act as outputs. The computed function is defined by the input-output relationship.

3   The Regulatory Network Computational Device (ReNCoDe)

A different approach to the use of ARNs as a computational device was presented in [10], called ReNCoDe. It employs the original ARN architecture but without using the proteins' concentrations, and including new biologically inspired genetic operators. At the core of the approach a new method for extracting a program from an ARN.

Extracting Circuits from ARNs. The networks resultant from ARN genomes can be very complex, composed of multiple links (inhibition and excitation) between different nodes (genes). In order to extract a circuit from these networks they must first be reduced, the input and the output nodes must de identified, and we must ascribe a semantic to its nodes. In the simplest cases, the final product will be an executable feed-forward circuit, although it is also possible to use feedback connections creating cyclic graphs that can be iterated and allow to keep state information. Using feedback links or not is a design issue that depends on the type of the problem. The algorithm starts by transforming every pair of connections, excitation (e) and inhibition (h), into one single connection with strength equal to the difference between the originals (e-h). Every connection with negative or null strength is discarded. Moreover, if there are reciprocal connections between a pair of nodes, the weakest is also discarded. The node with the highest connectivity is chosen as the output. After this, the circuit is built backwards from the output node until the terminals are reached (nodes without input edges). If, at any point of this process, there is a deadlock (every node is input to some other), again the gene with highest connectivity is chosen.

To complete the process, a mapping is needed linking nodes (i.e., genes) to functions and terminals. To that end we use the gene-protein correspondence using the protein's signature to obtain the function/terminal by a majority vote process. As an example, to code the function set { +, -, *, / } only two bits are necessary. These are obtained by splitting the 32-bit protein into sixteen 2-bit chunks and applying the majority rule in each position of the chunks. This will provide us with an index to the function set. If the number of functions is not a power of two, it is rounded up and the set is used circularly. If some determined problem has more than one input, for instance four, then the majority rule is applied over the terminal node signature (its binary stream) to define to which input it corresponds.

Finally, in order to improve the evolvability of the genomes, in [10] the authors proposed also variation operators inspired by the concepts of transposons and non-coding DNA, which can copy a part of the genome (*transposon-like*), or introduce non-coding (*junk*) genetic material (streams with 0s). Also, a *delete* operator is used to allow the genomes to be shrunk if beneficial. Every iteration one of these operators may be applied with a predetermined probability. The results reported show an efficiency increase when using any of the operators with small lengths (versus fixed size genomes with only the mutation operator). The average results also show that the *transposon-like* operator is more *stable*.

For this problem the basic ReNCoDe will be used, without feedback connections nor the operators described above.

4   The Problem

Nowadays there is a GP community-driven effort to find agreement on appropriate benchmarks for the field. The latest progress includes a complete survey of the problems used in the field [11] and an inquiry to the community researchers [12]. This effort resulted so far in a list of problems that should not be used, and proposes some alternatives in the corresponding problem domains. One of the blacklisted problems was Koza's polynomial regression, which was used in the past to test ReNCoDe.

One of the alternatives proposed by the authors of the survey, was the harmonic number (defined in Equation 3). This series can be approximated using the asymptotic expansion presented in Equation 4, where γ is the Euler's constant ($\gamma \approx 0.57722$).

$$H_n = \sum_{i=1}^{n} \frac{1}{i} \tag{3}$$

$$H_n = \gamma + ln(n) + \frac{1}{2n} - \frac{1}{12n^2} + \frac{1}{120n^4} - \ldots \tag{4}$$

This problem is particularly interesting because there is not only an interpolation task (during the evolutionary process), but also an extrapolation task with the evolved solution. The goal is to evolve a program that approximates the harmonic number, given an input belonging to the set of natural integers from 1 to 50. Upon the evolutionary phase the best solution is tested for generalisation over the set of natural integers from 1 to 120. This problem was studied in the context of GP by [13], and the proposed approach was able to rediscover the asymptotic approximations, with absolute error sum in the order of 10^{-2} .

The function set used is composed of {+, *, reciprocalsum, ln, sqrt, cos}. The reciprocalsum function returns the sum of the reciprocals of each input, and is used to replace protected division. All non-terminal nodes have variable arity in ReNCoDe, thus the unary functions have to be adapted. A node mapped to either ln, sqrt, or cos with more than one input, will first sum the corresponding inputs and then apply the function. The terminal set is composed only of n. Every function is protected: in case of overflow or impossible values provided as inputs the return value is 10^6.

The fitness function used in this work is the Mean Squared Error (MSE) between the output of the individual and the target function. A linear-scaling approach to fitness is used [14]. Given $y = gp(x)$ as the output of an expression evolved by GP on the input data x, a linear regression on the target values t

can be performed using Equations 5 and 6, where $\overline{y}$ and $\overline{t}$ denote respectively the average output and the average target value. These equations calculate the *slope* and *intercept* of the set of outputs y, minimising the sum of squared errors between t and $a + by$ (with a different from 0 and b different from 1). With the constants a and b calculated by this simple regression, all that is left to the evolutionary run is to find a function with the correct shape, using as fitness the modified MSE presented in Equation 7.

$$b = \frac{\sum[(t - \overline{t})(y - \overline{y})]}{\sum[(y - \overline{y})^2]} \tag{5}$$

$$a = \overline{t} - b\overline{y} \tag{6}$$

$$MSE(t, a + by) = \frac{1}{N} \sum_i^N (a + by - t)^2 \tag{7}$$

It is worth mentioning that individuals who generate a constant output for every input are considered invalid, since the denominator in Equation 5 would result in 0, and thus the fitness returned is 10^6. In previous experiments with various symbolic regression problems it was noticed that individuals generating a constant output correspond many times to local minima and many evaluations were needed to step out. Hence avoiding these individuals constitutes an advantage of the linear scaling.

During the generalisation testing the constants a and b determined during evaluation are used to compute the output of the individual and compare with the target result.

5 Experimental Setup

The experiments described in this section have a double goal. First, they are aimed at testing the different design choices to find in which scenario the best performance is achieved and how relevant is each parameter. Second, a modern regression problem is tackled in which the evolved solution is also tested for generalisation performance.

Three design choices of the genotypic model of ReNCoDe - the ARN - were tested. Duplication and divergence (or mutation) is known to generate scale-free and small-world networks, similar to those found in bacteria [8]. Nonetheless, this is not necessarily beneficial in the context of optimisation as shown in [9] by comparing against randomly generated genomes. Here we perform the same study in the context of symbolic regression with ReNCoDe. Another aspect is the genome size, which in turn influences the number of coded proteins. We test different lengths to find out how it affects the performance of the approach. Finally, genes may overlap or not, changing the number of coded proteins and the organisation of the genome.

The summary of the values for each design detail is presented in Table 1. This sums a total of 20 experiments. However, the experiments with genome size 1024 and without overlapping genes were discarded, resulting in a total of 18 experiments. The reason behind is that the total coding area of a gene with promoter and binding sites is 256 bits, making it hard for a genome with 1024 bits to code for any proteins.

The remaining parameters are fixed, as presented in Table 2. These were determined in past experiments with the model. Every experiment was run 100 times, using a standard evolution strategy: (10+100)-ES. Crossover was never applied, nor the operators described in Section 3, using only bit-flip mutation with a 0.01% rate. A run terminates when a solution is found which error is less than 10^{-3} over the train set, or when 10^6 evaluations are reached (in which case it is considered a failure).

Table 1. Values of the different design alternatives. Initialisation may be random (rnd) or use DM events (dm). Genes may overlap or not (True/False). The size of the genome varies from 1024 to 16364 bits, with the corresponding number of DM events.

Initialisation	$\{dm, rnd\}$
Overlapping genes	$\{True, False\}$
Size	$\{1024, 2048, 4096, 8192, 16364\}$
#DM	$\{5, 6, 7, 8, 9\}$

Table 2. Evolutionary parameters values

Parameter	Value
Number of Runs	100
Initial Population	100
Max. Evaluations	10^6
DM Mutation Rate	0.02
Mutation Operator Rate	0.01
Protein Bind Threshold	16

Statistical Analysis. In order to compare the different experiments it is necessary to perform a statistical analysis of the results. Since the samples are independent and do not follow a normal distribution the analysis was performed by first running the Kruskal-Wallis test, followed by pairwise comparisons using the Mann-Whitney-Wilcoxon test [15]. Particularly, one-tailed tests were performed in both directions. Since we are dealing with multiple pairwise comparisons, Bonferroni correction was applied [16]. In the results discussion when the term *significant* is written the authors mean *statistically significant*.

6   Results and Analysis

Most of ReNCoDe's configurations were able to find a solution within a few hundred evaluations. Many circuits were found that approximate the target function with an error in the order of 10^{-5}. As an example two different circuits are shown Figure 2.

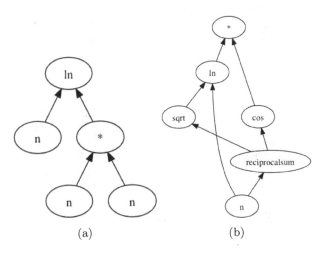

(a)   (b)

Fig. 2. Two circuits that approximates the harmonic number. a) $0.607878226352 + 0.495419818124 * ln(n + n^2)$; b) $0.64077307942 + 0.98526810735 * (ln(sqrt(^{1.0}/n) + n) * cos(^{1.0}/n))$.

The success rate and mean effort are shown in Table 3 (rounded to the unit). Note that the results for $dm5$ and $dm6F$ are influenced by the presence of outliers, raising drastically the mean effort to find a solution. A plot of the distribution of the number of evaluations is presented in Figure 3a, without showing the outliers. Analysing this data we can see that the most important of the design aspects is the generation method. There is a clear performance increase when using the random generation instead of a sequence of DM events. However, it appears that the genome size does not influence much the performance for this problem. The experiments where there were not genes overlapping show a decrease in performance for most of the combinations, although not significant.

The statistical analysis of the results corroborates these conclusions by showing that most of the random generation configurations are significantly better than the DM-generated. Running the Kruskal-Wallis rank sum test returns a p-value inferior to 0.05. The multiple pairwise comparisons are summarised in Table 4. It displays "+" if the scenario in the row performs significantly better than the one in the column, or "−" in the opposite case. A "∼" is presented in case the difference is not statistically significant.

Table 3. Summary of the results for each experiment. The labels indicate the initialisation method and gene size, with 'F' corresponding to no overlap (False).

Experiment	Success (%)	Avg. #Evaluations	std.dev.
dm5	100	3422	25154
dm6	100	340	203
dm6F	100	3449	9053
dm7	100	256	128
dm7F	100	385	457
dm8	100	183	59
dm8F	100	226	202
dm9	100	149	52
dm9F	100	165	50
rnd5	100	139	62
rnd6	100	109	29
rnd6F	100	121	43
rnd7	100	105	22
rnd7F	100	105	26
rnd8	100	115	39
rnd8F	100	103	17
rnd9	98	212	210
rnd9F	100	147	69

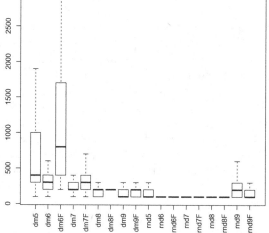

Fig. 3. Distribution of the number of evaluations to reach a solution, for each experiment. The labels indicate the initialisation method and gene size, with 'F' corresponding to no overlap (False). Outliers are omitted.

Table 4. Pairwise comparisons results. The experiments in the header line have the same ordering as in Table 3 (excluding *rnd9F*). The vertical line separates the *dm* from the *rnd* columns. "+" is displayed if the scenario in the row performs significantly better than the one in the column, or "−" in the opposite case. "~" is used where the difference is not statistically significant.

	dm5	dm6	dm6F	dm7	dm7F	dm8	dm8F	dm9	dm9F	rnd5	rnd6	rnd6F	rnd7	rnd7F	rnd8	rnd8F	rnd9
dm6	+																
dm6F	−	−															
dm7	+	+	+														
dm7F	+	~	+	−													
dm8	+	+	+	+	+												
dm8F	+	+	+	+	+	~											
dm9	+	+	+	+	+	+	+										
dm9F	+	+	+	+	+	~	+	~									
rnd5	+	+	+	+	+	+	+	~	+								
rnd6	+	+	+	+	+	+	+	+	+	+							
rnd6F	+	+	+	+	+	+	+	+	+	~	~						
rnd7	+	+	+	+	+	+	+	+	+	+	~	~					
rnd7F	+	+	+	+	+	+	+	+	+	+	~	+	~				
rnd8	+	+	+	+	+	+	+	+	+	~	~	~	~	~			
rnd8F	+	+	+	+	+	+	+	+	+	+	~	+	~	~	~		
rnd9	+	+	+	+	+	~	~	~	~	−	−	−	−	−	−	−	
rnd9F	+	+	+	+	+	+	~	~	~	−	~	−	−	−	−	+	

6.1 Generalisation

As mentioned in Section 5, after the evolutionary run, the solutions were tested for generalisation over the interval $[1 : 120 : 1]$, i.e. the set of natural integers from 1 to 120. Table 5 shows how many of the runs obtained a generalisation MSE lower than 10^{-3} (Suc.), as well as the respective means, deviations and best result.

The data presented shows that the small random genomes perform better in the extrapolation task. Running the statistical tests we found that the only significant differences are between the small random genomes and the DM generated. There are not significant differences amongst each group. As for the gene overlapping, despite the small differences presented before in the number of evaluations these results shows that there is not any advantage in extrapolation performance.

Table 5. Summary of the extrapolation results for each experiment

Experiment	Suc.(%)	Mean	std.dev.	Min
dm5	69	0.00036	0.00029	0.000005
dm6	76	0.00032	0.00023	0.000019
dm6F	76	0.00032	0.00023	0.000019
dm7	72	0.00034	0.00026	0.000043
dm7F	72	0.00034	0.00026	0.000043
dm8	76	0.00032	0.00023	0.000027
dm8F	76	0.00032	0.00023	0.000027
dm9	78	0.00037	0.00027	0.000021
dm9F	78	0.00037	0.00027	0.000021
rnd5	98	0.00021	0.00016	0.000005
rnd6	98	0.00021	0.00021	0.000006
rnd6F	98	0.00021	0.00021	0.000006
rnd7	94	0.00024	0.00020	0.000023
rnd7F	94	0.00024	0.00020	0.000023
rnd8	87	0.00032	0.00026	0.000001
rnd8F	87	0.00032	0.00026	0.000001
rnd9	77	0.00034	0.00027	0.000008
rnd9F	78	0.00034	0.00027	0.000008

7   Conclusion

The Regulatory Network Computational Device is a recent GP representation that uses the ARN model as genotype, extracting a executable graph from the network (the expressed phenotype). It has been applied with success to different benchmark problems of various classes. Here we tackled a symbolic regression problem that has been proposed by the latest community effort to standardise the benchmarks used in GP. Moreover, we studied how three design aspects of the genotypic model influence the performance of the system.

Different genome sizes were tested from 1024 bit to 16364. This aspect does not influence drastically the performance, but in general small genomes perform better. Allowing genes to overlap increased performance, particularly in DM genomes. However, the difference is not significant. Finally, we found that the most relevant aspect is the generation method: random or through DM events. The results clearly indicate that genomes that were randomly generated perform significantly better than those created by a series of duplication with divergence events.

One of the interesting points of the problem tackled in this work is the extrapolation task after the evolutionary process. Most of the evolved solutions perform well in this task. We found that random initialised genomes generalise better than the DM generated. Moreover, in this task nor the genome size nor the gene overlapping present significant differences in the results.

The evidence that randomly initialised genomes perform better than duplicated ones is clear. The authors have also analysed the results relatively ot the number of proteins coded by the genomes, although a correlation was not found.

In order to understand the reasons behind these results, future work should focus on the networks' properties. For instance, the degree of connectivity and the topologies generated by both methods could provide some insight into the differences found in the experiments reported here.

References

1. Eiben, A.E., Smith, J.E.: Introduction to Evolutionary Computing. Springer, Heidelberg (2003)
2. O'Neill, M., Ryan, C.: Grammatical Evolution: Evolutionary Automatic Programming in a Arbitrary Language. In: Genetic Programming. Kluwer Academic Publishers (2003)
3. Miller, J.F.: Cartesian Genetic Programming. Natural Computing Series. Springer (2011)
4. Ferreira, C.: Gene Expression Programming, 2nd edn. Springer (2006)
5. Lones, M.A., Tyrrell, A.M.: Biomimetic representation with genetic programming enzyme. Genetic Programming and Evolvable Machines 3(1), 193–217 (2002)
6. Banzhaf, W.: Artificial regulatory networks and genetic programming. In: Riolo, R.L., Worzel, B. (eds.) Genetic Programming Theory and Practice, ch. 4, pp. 43–62. Kluwer (2003)
7. Banzhaf, W., Beslon, G., Christensen, S., Foster, J., Képès, F., Lefort, V., Miller, J., Radman, M., Ramsden, J.: From artificial evolution to computational evolution: a research agenda. Nature Reviews Genetics 7(9), 729–735 (2006)
8. Dwight Kuo, P., Banzhaf, W., Leier, A.: Network topology and the evolution of dynamics in an artificial genetic regulatory network model created by whole genome duplication and divergence. Bio Systems 85(3), 177–200 (2006)
9. Nicolau, M., Schoenauer, M., Banzhaf, W.: Evolving Genes to Balance a Pole. In: Esparcia-Alcázar, A.I., Ekárt, A., Silva, S., Dignum, S., Uyar, A.Ş. (eds.) EuroGP 2010. LNCS, vol. 6021, pp. 196–207. Springer, Heidelberg (2010)
10. Lopes, R.L., Costa, E.: The Regulatory Network Computational Device. Genetic Programming and Evolvable Machines 13(3), 339–375 (2012)
11. McDermott, J., White, D.R., Luke, S., Manzoni, L., Castelli, M., Vanneschi, L., Jaśkowski, W., Krawiec, K., Harper, R., De Jong, K.: Genetic programming needs better benchmarks. In: Proceedings of the Fourteenth International Conference on Genetic and Evolutionary Computation Conference, pp. 791–798 (2012)
12. White, D.R., McDermott, J., Castelli, M., Manzoni, L., Goldman, B.W., Kronberger, G., Jaśkowski, W., O'Reilly, U.M., Luke, S.: Better GP benchmarks: community survey results and proposals. Genetic Programming and Evolvable Machines (December 2012)
13. Streeter, M.J.: Automated Discovery of Numerical Approximation Formulae via Genetic Programming. Ph.D. thesis, Worcester Polytechnic Institute, Worcester Polytechnic Institute (April 2001)
14. Keijzer, M.: Improving symbolic regression with interval arithmetic and linear scaling. In: Ryan, C., Soule, T., Keijzer, M., Tsang, E.P.K., Poli, R., Costa, E. (eds.) EuroGP 2003. LNCS, vol. 2610, pp. 275–299. Springer, Heidelberg (2003)
15. R Core Team: R: A Language and Environment for Statistical Computing. R Foundation for Statistical Computing, Vienna, Austria (2012), http://www.R-project.org
16. Field, A., Hole, G.: How to design and report experiments. Sage Publications London (2003)

GPU-Based Automatic Configuration
of Differential Evolution: A Case Study

Roberto Ugolotti, Pablo Mesejo, Youssef S.G. Nashed, and Stefano Cagnoni

Department of Information Engineering, University of Parma, Italy
{rob_ugo,pmesejo,nashed,cagnoni}@ce.unipr.it

Abstract. The performance of an evolutionary algorithm strongly depends on the choice of the parameters which regulate its behavior. In this paper, two evolutionary algorithms (Particle Swarm Optimization and Differential Evolution) are used to find the optimal configuration of parameters for Differential Evolution. We tested our approach on four benchmark functions, and the comparison with an exhaustive search demonstrated its effectiveness. Then, the same method was used to tune the parameters of Differential Evolution in solving a real-world problem: the automatic localization of the hippocampus in histological brain images. The results obtained consistently outperformed the ones achieved using manually-tuned parameters. Thanks to a GPU-based implementation, our tuner is up to 8 times faster than the corresponding sequential version.

Keywords: Automatic Algorithm Configuration, Global Continuous Optimization, Particle Swarm Optimization, Differential Evolution, GPGPU.

1 Introduction

In this paper we consider the problem of automatizing the configuration of an Evolutionary Algorithm (EA) [7]. This problem is particularly important because EAs are usually strongly influenced by the choice of the parameters which regulate their behavior [6].

Regarding automatic algorithm configuration, it is important to distinguish between parameter tuning and parameter control. In the former, parameter values are fixed in the initialization stage and do not change while the EA is running; in the latter, parameter values are adapted as evolution proceeds. This work is focused only on choosing parameter values, either numerical (integer or real) or nominal (representing different design options for the algorithm), which will then be kept constant during evolution.

Virtually, all EAs have free parameters that are partly set by the programmer and/or by the user. In this scenario, automatic parameter configuration is desirable, since *manual* parameter tuning is time consuming, error-prone, and may introduce a bias when comparing a new algorithm with a reference, caused by better knowledge of one with respect to the other or to possible differences in the amount of time spent tuning each of them. Nevertheless, it is important to take the computational burden of these tuners into account, since they usually need

L. Correia, L.P. Reis, and J. Cascalho (Eds.): EPIA 2013, LNAI 8154, pp. 114–125, 2013.

to perform many tests to find the best parameter configuration for the algorithm under consideration.

In this regard, GPU implementations have arisen as a very promising way to speed-up EAs. Modern graphics hardware has gained an important role in parallel computing, since it has been used to accelerate general computations (General Purpose Graphics Processing Unit programming), as well as in computer graphics. In particular, CUDA^{TM} (Compute Unified Distributed Architecture) is a parallel computing environment by nVIDIA^{TM}, which exploits the massively parallel computation capabilities of its GPUs. CUDA-C [17] is an extension of the C language that allows development of GPU routines (termed *kernels*), that can be executed in parallel by several different CUDA^{TM} threads. We use GPU implementations of real-valued population-based optimization techniques (Particle Swarm Optimization (PSO) [9] and Differential Evolution (DE) [21]) to automatically configure DE parameters. The implementations of both metaheuristics on GPU are publicly available in a general-purpose library[1]. DE has been chosen as the method to be tuned mainly because it is the optimizer initially proposed and employed in the real-world problem under consideration [13], while PSO was chosen as tuner because of its easy parallelization and its ability to reach good results using a limited number of fitness evaluations.

The choice of using Evolutionary Algorithms or Swarm Intelligence [2] algorithms as tuners derives from the large number of attractive features they offer: robust and reliable performance, global search capability, virtually no need of specific information about the problem to solve, easy implementation, and, above all, implicit parallelism, which allows one to obtain a parallel implementation starting from a design which is also conceptually parallel.

We first demonstrate that our method works properly by optimizing DE parameters for some well-known benchmark functions and compare these results with the ones obtained by a systematic search of the parameter space. Then, we apply the same approach to the real-world problem of localizing the hippocampus in histological brain images, obtaining a significant improvement with respect to the manually-tuned algorithm. Finally, we prove that our GPU-based implementation significantly reduces computation time, compared to a sequential implementation.

The remainder of this paper is structured as follows: in Section 2 we provide the theoretical foundations necessary to understand our work, including the metaheuristics and the problem used to test the system. In Section 3, a general overview of the method is presented, providing details about its GPU implementation. Finally, Section 4 presents the results of our approach followed, in Section 5, by some final remarks and a discussion about future developments.

---

[1] The code can be downloaded from `http://sourceforge.net/p/libcudaoptimize` [16].

2   Theoretical Background

In this section, we will briefly describe the metaheuristics used in this work (DE and PSO) and the hippocampus localization task used as test, and make a short review of the approaches in which metaheuristics have been used to tune EAs.

2.1   Particle Swarm Optimization

Particle Swarm Optimization is a well-established stochastic optimization algorithm which simulates the behavior of bird flocks. A set of N particles moves through a "fitness" function domain seeking the function optimum. Each particle's motion is described by two simple discrete-time equations which regulate the particles' velocity and position:

$$\begin{aligned}
\boldsymbol{v_n}(t) &= w \cdot \boldsymbol{v_n}(t-1) \\
&+ c_1 \cdot \boldsymbol{rand}() \cdot (\boldsymbol{BLP_n} - \boldsymbol{P_n}(t-1)) \\
&+ c_2 \cdot \boldsymbol{rand}() \cdot (\boldsymbol{BNP_n} - \boldsymbol{P_n}(t-1)), \\
\boldsymbol{P_n}(t) &= \boldsymbol{P_n}(t-1) + \boldsymbol{v_n}(t),
\end{aligned}$$

where $\boldsymbol{P_n}(t)$ and $\boldsymbol{v_n}(t)$ represent the nth particle's position and velocity at time t; c_1, c_2 are positive constants which represent how strongly cognitive and social information affects the behavior of the particle; w is a positive inertia factor (possibly depending on time t) used to balance PSO's global and local search abilities; $\boldsymbol{rand}()$ returns a vector of random values uniformly distributed in $[0,1]$; $\boldsymbol{BLP_n}$ and $\boldsymbol{BNP_n}$ are, respectively, the best-fitness location visited so far by particle n and by any particle in its neighborhood; this may include a limited set of particles (lbest PSO) or coincide with the whole population (gbest PSO). One of the most commonly used PSO versions is based on a "ring topology", i.e. each particle's neighborhood includes the particle itself, the K particles preceding it, and the K particles which immediately follow it.

2.2   Differential Evolution

Differential Evolution is one of the most successful EAs for global continuous optimization [3]. DE perturbs the current population members with the scaled difference of distinct individuals. At every generation, every individual $\boldsymbol{X_i}$ of the current population acts as a parent vector, for which a donor vector $\boldsymbol{D_i}$ is generated. In the original version of DE, the donor vector for the ith parent $(\boldsymbol{X_i})$ is created by combining three distinct randomly-selected elements $\boldsymbol{X_{r1}}$, $\boldsymbol{X_{r2}}$ and $\boldsymbol{X_{r3}}$. The donor vector is then computed as:

$$\boldsymbol{D_i} = \boldsymbol{X_{r1}} + F \cdot (\boldsymbol{X_{r2}} - \boldsymbol{X_{r3}}),$$

where the scale factor F (usually $F \in [0, 1.5]$) is a parameter that strongly influences DE's performances. This mutation strategy is called *random*, but other mutation strategies have been proposed. Two of the most successful ones are *best* (equation 1) and *target-to-best* (TTB) (equation 2):

$$D_i = X_{best} + F \cdot (X_{r1} - X_{r2}), \tag{1}$$
$$D_i = X_i + F \cdot (X_{best} - X_i) + F \cdot (X_{r1} - X_{r2}), \tag{2}$$

where X_{best} represents the best-fitness individual in the current population.

After mutation, every parent-donor pair generates one child (called trial vector, T_i) by means of a crossover operation. Two kinds of crossover are typically used: *binomial* (uniform) and *exponential*; both are regulated by a parameter called *crossover rate* ($Cr \in [0,1]$). In the former, a real number $r \in [0,1]$ and an integer $M \in [1, S]$, where S is the size of the search space, are randomly generated for each element of the vector. The trial vector is then defined as follows:

$$T_{i,j} = \begin{cases} D_{i,j} \text{ if } r \leq Cr \text{ or } j = M \\ X_{i,j} \qquad\qquad\quad \text{otherwise} \end{cases}, j = 1 \dots S$$

The exponential crossover, instead, generates an integer L according to this pseudo-code:

```
L = 0
do
L = L + 1
r = rand()
while ((r ≤ Cr) and (L ≤ S))
```

This value is then used to generate the trial vector:

$$T_{i,j} = \begin{cases} D_{i,j} \text{ for } j = \langle M \rangle_S \dots \langle M + L - 1 \rangle_S \\ X_{i,j} \qquad\qquad\qquad\qquad\quad \text{otherwise} \end{cases}, j = 1 \dots S$$

where $\langle \cdot \rangle_S$ represents the *modulo S* function.

After the crossover operation, the trial vector is evaluated and its fitness is compared to its parent's: only the one having the best fitness survives in the next generation, while the other is discarded from the population.

2.3   Hippocampus Localization in Histological Images

The hippocampus is a structure located in the mammalian brain that has long been known for its crucial role in learning and memory processes, as well as an early biomarker for Alzheimer disease and epilepsy. Thus, its automatic, robust and fast localization is of great interest for the scientific community.

As a real-valued benchmark for our method, we consider the problem of localizing the hippocampus in histological images (see Figure 1) extracted from the Allen Brain Atlas (ABA), a publicly available image database [1] which contains a genome-scale set of histological images (cellular-resolution gene-expression profiles) obtained by In Situ Hybridization of serial sections of mouse brains.

In particular, a localization method based on Active Shape Models has been recently presented to tackle this kind of images [13]. It uses a medial shape-based representation of the hippocampus in polar coordinates, with the objective of creating simple models that can be managed easily and efficiently. Two parametric models (see Figure 1b) representing the two parts that compose the hippocampus (called SP and SG, see Figure 1a) are moved and deformed by DE according to an intensity-based similarity function between the model and the object itself.

Each model comprises two sets of points: the goal is to overlap the first one (Inner Set, I) to the part of the hippocampus to be located, while placing the other one (Outer Set, O), obtained by rigidly shifting the first one, immediately outside it. This model is subjected to external forces (driven by the image features) and internal forces (driven by the model itself). The target function H to be maximized (see [13] for a more detailed description) has three components: the external energy EE, the internal energy IE, and the contraction factor C.

$$H = EE - (IE + C)$$

In turn, EE can be divided in two components: PE, that depends on the control points of the model (denoted by black dots in Figure 1b) and CE, that is computed considering the points which belong to the segments connecting them:

$$EE = \gamma_P \cdot PE + \gamma_C \cdot CE$$

$$PE = \sum_{i=1}^{n} [T(N_3(I_i)) - T(N_3(O_i))]$$

$$CE = \sum_{i=2}^{n} \sum_{j=1}^{p} T(I_{i-1} + \frac{j}{p+1}(I_i - I_{i-1})) - \sum_{i=2}^{n} \sum_{j=1}^{p} T(O_{i-1} + \frac{j}{p+1}(O_i - O_{i-1}))$$

where I_i and O_i represent the points of the two sets, γ_P and γ_C are positive values that weigh the two components, $N_3(P)$ is a 3×3 neighborhood centered in P, $T(P)$ is the intensity of P if P is a point, or the average intensity if P is a neighborhood, and p is the number of points sampled in each segment.

The internal energy IE is computed as:

$$IE = \xi_\rho \cdot \sqrt{\sum_{i=2}^{n} (\rho_i - \rho_{mi})^2} + \xi_\vartheta \cdot \sqrt{\sum_{i=2}^{n} (\vartheta_i - \vartheta_{mi})^2}$$

where ξ_ρ and ξ_ϑ are two positive weights that regulate the deformability of the model. (ρ_i, θ_i) represents a point of the model in polar coordinates, while (ρ_{mi}, θ_{mi}) represents a point of a reference model derived empirically from a set of "training" images.

Finally, the contraction factor C also regulates the model's deformability to avoid unfeasible situations that are not allowed by the nature of the hippocampus and is defined as follows:

$$C = \xi_c \cdot \|I_n - I_1\|$$

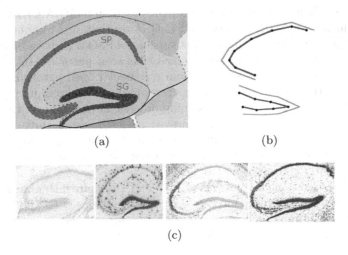

(a)   (b)

(c)

Fig. 1. In the upper row, an anatomical atlas representation of the hippocampal region that highlights the two parts to be recognized (SP and SG) and an example of the parametric models used to localize the hippocampus; below, four examples of hippocampus images taken from the ABA

2.4 Automatic Configuration of EAs

Parameter tuning for configuring and analyzing EAs is a topic frequently dealt with in the last years [6,19]. In fact, the idea of a meta-EA was already introduced in 1978 by Mercer and Sampson [12], while Grefenstette, in 1986 [8], conducted more extensive experiments with a meta-GA and showed its effectiveness. However, the problem of tuning EAs by means of meta-EAs implies tackling two challenges: the noise in (meta)-fitness values and the very expensive (meta)-evaluations. Our GPU-based implementation tries to reduce the impact of the latter problem. Recently, REVAC, a specific type of meta-EA where the population approximates the probability density function of the most promising areas of the utility-landscape (similar to Iterative F-RACE), was shown to be able to find much more robust parameter values than one could set by hand, outperforming the winner of the competition on the CEC 2005 test-suite [20]. An attempt to find the optimal parameters for DE using a metaheuristic called Local Unimodal Sampling is described in [18]. As well, PSO parameter tuning has already been proposed based on other overlaying optimizers, like in [11], where PSO itself is used to this end.

3 Parameter Configuration Using Metaheuristics

3.1 Parallel Implementation

The GPU-based implementation of the metaheuristics employed in this paper has been presented in [16].

The first implementations of PSO and DE based on nVIDIA CUDA™ were developed in 2009 and 2010, respectively [4,5]. After that, other implementations of DE have been developed [10], and fast versions of PSO have been implemented by removing the constraint of synchronicity between particles [15]. The early GPU PSO implementations suffered from a coarse-grained parallelization (one thread per particle), that did not offer the opportunity to compute the fitness function, usually the most time-consuming process, in parallel over the problem dimensions. This aspect has been improved by letting each thread manage a single dimension of each particle, adding a further level of parallelism [14]. Similar inefficiencies characterized the early implementations of DE. These problems were subsequently addressed by [10] and [16].

Our GPU-based implementations of PSO and DE share a similar kernel configuration. Both are implemented as three distinct kernels: (i) the first kernel generates the solutions to be evaluated, (ii) the second kernel implements the fitness function, and (iii) the last kernel updates the population.

3.2   General Design

A tuner (or meta-optimizer) is implemented exactly in the same way as any optimizer used to solve any other problem. The only significant difference consists in the evaluation of the fitness function F. In this case, each particle $X_i = \{x_1, \ldots, x_n\}$ of a tuner represents an optimizer (or better, a set of $m \leq n$ parameters that describes an optimizer). An optimizer $O_i(X_i)$ is instantiated using the parameters X_i and the entire optimization process is repeated on the test function T times to compute fitness f_i as the average fitness over the T runs. This optimizer, used to tune DE, has been introduced in [22].

PSO and DE are real-valued optimization methods, so the problem of representing nominal parameters needs to be addressed. We chose to represent each nominal value by a vector which associates a number to each option available; the option associated with the highest number is then selected. Figure 2 shows the encoding of a particle representing a DE instance. Table 1 shows the range and possible values of the DE parameters in the representation we have used.

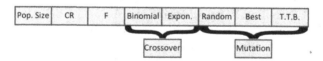

Fig. 2. Representation of a set of parameters used in the tuner. The first three elements represent numerical parameters (integer and real-valued numbers), while the last five encode the nominal parameters, i.e. crossover type and mutation strategy. In this case $n = 8$ values are needed to represent $m = 5$ parameters.

Table 1. Range of DE parameters allowed by the tuning algorithms. The last column shows the sampling step used in the systematic search used as reference.

Parameter	Values	Step
Crossover Rate	$[0.0, 1.0]$	0.1
Scale Factor (F)	$[0.0, 1.5]$	0.1
Population Size	$[30, 150]$	10
Crossover	{binomial, exponential}	-
Mutation	{random, best, target-to-best}	-

Table 2. Benchmark functions. For every function, the table displays name, search space, formula, modality (multimodal, unimodal) and separability (separable, non separable). Optimum fitness values are 0 for all functions.

Name	Range	Formula	Modality	Separability
Zakharov	$[-10, 10]^n$	$\left(\sum_{i=1}^{n} x_i{}^2\right) + \left(\sum_{i=1}^{n} 0.5 \cdot i \cdot x_i{}^2\right)^2 + \left(\sum_{i=1}^{n} 0.5 \cdot i \cdot x_i{}^2\right)^4$	U	S
Schwefel 1.2	$[-100, 100]^n$	$\sum_{i=1}^{n}\left(\sum_{j=1}^{i} x_j\right)^2$	U	NS
Rastrigin	$[-5.12, 5.12]^n$	$\sum_{i=1}^{n}\left\{x_i^2 - 10 \cdot \cos(2\pi x_i) + 10\right\}$	M	S
Rosenbrock	$[-100, 100]^n$	$\sum_{i=2}^{n} 100(x_i - x_{i-1}^2)^2 + (1 - x_{i-1})^2$	M	NS

4   Experiments

In this section, we will describe the two sets of tests performed on benchmark functions and on the hippocampus localization problem. Tests were run on a 64-bit Intel Core i7 CPU running at 3.40GHz using CUDA v. 4.2 on an nVidia GeForce GTX680 graphics card with compute capability 3.0.

4.1   Benchmark Functions

Four classical benchmark functions (see Table 2) having different nature (unimodal/multimodal and separable/non separable) have been selected to test our method. For each function, we systematically sampled (with the sampling steps shown in Table 1) the parameter search space and ran 10 independent repetitions for each of the 13728 parameter combinations generated. This way, we created a *ground truth* used to assess the results of the meta-optimization. The termination criteria was set after 1000 generations and the search space size was set to 16.

Parameter tuning has been repeated 10 times (5 using DE as meta-optimizater, 5 using PSO) on each function. Table 3 shows the parameters used by the tuners based on DE and PSO. In order to choose them, we performed the tuning procedure described here over 8 benchmark functions using "standard" parameters for PSO and DE (see [22]). The best parameter configuration found, based on

Table 3. Parameters of DE/PSO tuners

DE	PSO
TTB Mutation	$C_1 = 1.525$
Exponential Crossover	$C_2 = 1.881$
F = 0.52	$w = 0.443$
Cr = 0.91	Ring Topology ($K = 1$)
Population Size = 24, Generation = 64	

Table 4. Range of the parameters found by systematic search and meta-optimization. P is population, Mut is mutation strategy, $Cross$ is Crossover Type.

Zakharov	Schewfel 1.2	Rastrigin	Rosenbrock
Systematic Search			
$Cr \in [0.8, 0.9]$	$Cr \in [0.9, 1.0]$	$Cr \in [0.0, 0.4]$	$Cr \in [0.8, 0.9]$
$F \in [0.5, 0.6]$	$F \in [0.5, 0.7]$	$F \in [0.1, 0.6]$	$F \in [0.6, 0.7]$
$P \in [100, 150]$	$P \in [120, 150]$	$P \in [90, 150]$	$P \in [90, 150]$
$Mut \in \{Best, TTB\}$	$Mut \in \{Best, TTB\}$	$Mut \in \{Rand, Best\}$	$Mut \in \{Best, TTB\}$
$Cross \in \{Bin, Exp\}$	$Cross \in \{Bin, Exp\}$	$Cross \in \{Bin, Exp\}$	$Cross \in \{Bin, Exp\}$
Meta-Optimization			
$Cr \in [0.799, 1.0]$	$Cr \in [0.935, 1.0]$	$Cr \in [0.0, 0.344]$	$Cr \in [0.740, 0.939]$
$F \in [0.433, 0.714]$	$F \in [0.535, 0.667]$	$F \in [0.207, 0.596]$	$F \in [0.646, 0.686]$
$P \in [111, 146]$	$P \in [80, 150]$	$P \in [79, 149]$	$P \in [83, 150]$
$Mut \in \{Best, TTB\}$	$Mut \in \{Best, TTB\}$	$Mut \in \{Rand, Best\}$	$Mut \in \{Best, TTB\}$
$Cross \in \{Bin, Exp\}$	$Cross \in \{Bin, Exp\}$	$Cross \in \{Bin, Exp\}$	$Cross \in \{Bin, Exp\}$

the number of times in which it obtained the best fitness value, was used in our tests. Notice that, with this parameter configuration, a maximum of 1536 sets of parameters (64 generations × 24 individuals) are evaluated in each run.

Table 4 shows, for each function, the ranges of the parameters for the best 10 settings found in the systematic search and for the 10 combinations obtained by the meta-optimization process. As can be observed, all parameter ranges virtually overlap, which proves that the meta-optimizer operates correctly. For every set of parameters, we performed 100 independent runs on the corresponding function. All optimizers were able to obtain a median value of 0 over the corresponding function. The Wilcoxon signed-rank test confirmed the absence of statistically significant differences between the results of the systematic search and the tuning, since the p-values obtained were always larger than 0.01.

To assess the speedup of the GPU implementation, we performed the same experiments adapting a sequential C++ DE implementation[2]. We compared the execution time of our method and of a sequential version of the tuner, using DE as meta-optimizer on Rastrigin function. Using the same parameters previously shown, the GPU version takes an average of 164.4 s for the entire

---

[2] http://www1.icsi.berkeley.edu/~storn/code.html

Table 5. The ten sets of parameters generated by the meta-optimization processes. The three columns for SP and SG (see Figure 1a) indicate the average fitnesses and standard deviations and if there is a significant improvement against the original parameters (+), no difference (=) or a worse result (−). This is a maximization problem: a higher value represents a better solution.

	Parameters					SP			SG		
	Cr	F	P	Mut	$Cross$	Avg	Std	Cmp	Avg	Std	Cmp
Original	0.9	0.7	150	TTB	Exp	142.6	14.7		136.2	17.8	
DE 1	0.859	0.41	121	$Rand$	Bin	144.5	12.1	+	141.0	13.9	+
DE 2	0.952	0.427	150	$Rand$	Exp	145.0	11.3	+	141.9	13.2	+
DE 3	0.952	0.419	150	$Rand$	Exp	145.0	11.5	+	141.8	13.1	+
DE 4	0.9	0.431	144	$Rand$	Bin	144.8	11.7	+	141.3	13.8	+
DE 5	0.954	0.44	115	$Rand$	Exp	144.9	11.7	+	141.4	13.7	+
PSO 1	0.949	0.448	149	$Rand$	Exp	145.1	11.2	+	141.8	13.1	+
PSO 2	0.953	0.471	143	$Rand$	Exp	145.0	11.3	+	141.7	13.2	+
PSO 3	0.974	0.473	150	$Rand$	Exp	145.0	11.5	+	142.0	13.0	+
PSO 4	0.783	0.347	150	$Rand$	Bin	144.6	11.8	+	141.0	13.7	+
PSO 5	0.922	0.437	139	$Rand$	Exp	145.0	11.2	+	141.4	13.2	+

meta-optimization process, while the sequential version takes 1007.5 s (6.1 times slower). Moreover, if we increase the problem dimension from 16 to 64, the two versions take 458.2 s and 3683.7 s, respectively and, consequently, the speedup increases to 8. Please notice that these speedups are only indicative of the problem, since they depend on a very high number of parameters, like population size, number of evaluations, degree of parallelism and complexity of the fitness function and, obviously, on the hardware on which the optimizer is run.

4.2 Hippocampus Localization

After checking the correctness of our approach, we repeated the same procedure to solve the hippocampus localization task described in Section 2.3. The parameters of the meta-optimizer are the same used in the previous sections.

Table 5 shows the parameters obtained by our meta-optimizer and by the manually-tuned ones used in [13]. To exclude any possible bias in favor of the former, we used a slightly different (and better working) setting for the reference, increasing its population to the maximum allowed, 150. This choice has been made because the implementation in [13] was sequential and this led us to use a lower number of particles for speed issues, while in our GPU implementation population size has no impact on computation time.

Each row in the table presents the method used in the meta-optimization process and the parameters found. Each set of parameters have been tested 10 times over 320 different hippocampus images.

In order to check the statistical significance of the results obtained, a Kruskal-Wallis test was performed with a level of confidence of 0.01. Since the normality and homoscedasticity assumptions were not met, as checked through the application of Kolmogorov-Smirnov and Bartlett's tests, non-parametric tests were used. The p-value was close to zero, suggesting that at least one sample median is significantly different from the others. The columns termed "Cmp" indicate the results of the Wilcoxon signed-rank test used to assess the statistical significance of the difference between the generated optimizer and the reference manually-tuned one (Original) for the two parts that compose the model (SP and SG). This value is a "+" if the results of the automatically tuned DE are significantly better than the original ones (significance of $p = 0.01$, using the Bonferroni-Holm correction), a "−" when the original method is better, and a "=" when there are no statistically significant differences.

As can be seen, meta-optimization always leads to similar results, and the automatically selected parameters always lead to higher mean and lower standard deviation than the ones set after a time-consuming manual tuning.

5   Discussion

In this paper we proposed a method to find the best parameters for an Evolutionary Algorithm (in this case, Differential Evolution) to solve a real-world problem. This method uses an evolutionary process to search the space of the algorithm parameters to find the best possible configuration for it.

We tested this method on four benchmark functions and the comparison with a systematic search proved that our method is actual effective. Afterwards, we repeated the same procedure optimizing the DE algorithm parameters in a real-world application, where Differential Evolution is used to localize hippocampi in brain histological images. The results obtained by the automatically-tuned optimizers outperform the manually-tuned ones almost systematically.

As future work, we will try to increase the degree of parallelization in order to take even more advantage of the parallel nature of the algorithms.

We evaluate the goodness of our tuning procedure by the quality of the solution obtained by the tuned optimizer. In the same way, we could add another step to empirically find a good set of parameters for the meta-optimizer. Of course, this could lead to an infinite repetition of the same procedure and to an exponential increase in optimization time. However, our practical aim is just to improve the performance of manually-tuned optimizers without increasing complexity too much, when dealing with problems in which the optimizer's performance is particularly sensitive to the parameters.

Acknowledgments. Roberto Ugolotti is funded by Fondazione Cariparma. Pablo Mesejo and Youssef S. G. Nashed are funded by the European Commission (Marie Curie ITN MIBISOC, FP7 PEOPLE-ITN-2008, GA n. 238819).

References

1. Allen Institute for Brain Science: Allen Reference Atlases (2004-2006), http://mouse.brain-map.org
2. Bonabeau, E., Dorigo, M., Theraulaz, G.: Swarm Intelligence: From Natural to Artificial Systems, Oxford (1999)

3. Das, S., Suganthan, P.: Differential Evolution: A Survey of the State-of-the-Art. IEEE Transactions on Evolutionary Computation 15(1), 4–31 (2011)
4. de Veronese, L., Krohling, R.: Swarm's flight: Accelerating the particles using C-CUDA. In: Proc. IEEE Congress on Evolutionary Computation, pp. 3264–3270 (2009)
5. de Veronese, L., Krohling, R.: Differential evolution algorithm on the GPU with C-CUDA. In: Proc. IEEE Congress on Evolutionary Computation, pp. 1–7 (2010)
6. Eiben, A.E., Smit, S.K.: Parameter tuning for configuring and analyzing evolutionary algorithms. Swarm and Evolutionary Computation 1(1), 19–31 (2011)
7. Eiben, A.E., Smith, J.E.: Introduction to Evolutionary Computing. Springer (2003)
8. Grefenstette, J.: Optimization of control parameters for genetic algorithms. IEEE Trans. Syst. Man Cybern. 16(1), 122–128 (1986)
9. Kennedy, J., Eberhart, R.: Particle Swarm Optimization. In: Proc. IEEE International Conference on Neural Networks, vol. 4, pp. 1942–1948 (1995)
10. Krömer, P., Snášel, V., Platoš, J., Abraham, A.: Many-threaded implementation of differential evolution for the CUDA platform. In: Proc. of Genetic and Evolutionary Computation Conference (GECCO), pp. 1595–1602. ACM (2011)
11. Meissner, M., Schmuker, M., Schneider, G.: Optimized Particle Swarm Optimization (OPSO) and its application to artificial neural network training. BMC Bioinformatics 7 (2006)
12. Mercer, R., Sampson, J.: Adaptive search using a reproductive metaplan. Kybernetes 7, 215–228 (1978)
13. Mesejo, P., Ugolotti, R., Di Cunto, F., Giacobini, M., Cagnoni, S.: Automatic hippocampus localization in histological images using differential evolution-based deformable models. Pattern Recognition Letters 34(3), 299–307 (2013)
14. Mussi, L., Daolio, F., Cagnoni, S.: Evaluation of parallel Particle Swarm Optimization algorithms within the CUDA architecture. Information Sciences 181(20), 4642–4657 (2011)
15. Mussi, L., Nashed, Y.S.G., Cagnoni, S.: GPU-based asynchronous particle swarm optimization. In: Proc. of the Genetic and Evolutionary Computation Conference (GECCO), pp. 1555–1562. ACM (2011)
16. Nashed, Y.S.G., Ugolotti, R., Mesejo, P., Cagnoni, S.: libCudaOptimize: an open source library of GPU-based metaheuristics. In: Proc. of the Genetic and Evolutionary Computation Conference (GECCO) Companion, pp. 117–124. ACM (2012)
17. nVIDIA Corporation: nVIDIA CUDA Programming Guide v. 4.0 (2011)
18. Pedersen, M.E.H.: Tuning and Simplifying Heuristical Optimization. Master's thesis, University of Southampton (2010)
19. Smit, S.K., Eiben, A.E.: Comparing parameter tuning methods for evolutionary algorithms. In: Proc. of the IEEE Congress on Evolutionary Computation, pp. 399–406 (2009)
20. Smit, S.K., Eiben, A.E.: Beating the 'world champion' evolutionary algorithm via REVAC tuning. In: Proc. of IEEE Congress on Evolutionary Computation, pp. 1–8 (2010)
21. Storn, R., Price, K.: Differential Evolution - a simple and efficient adaptive scheme for global optimization over continuous spaces. Technical report, International Computer Science Institute (1995)
22. Ugolotti, R., Nashed, Y.S.G., Mesejo, P., Cagnoni, S.: Algorithm Configuration using GPU-based Metaheuristics. In: Proc. of the Genetic and Evolutionary Computation Conference (GECCO) Companion (2013)

Structure-Based Constants
in Genetic Programming

Christian B. Veenhuis

Berlin University of Technology, Berlin, Germany
veenhuis@googlemail.com

Abstract. Evolving constants in Genetic Programming is still an open issue. As real values they cannot be integrated in GP trees in a direct manner, because the nodes represent discrete symbols. Present solutions are the concept of ephemeral random constants or hybrid approaches, which have additional computational costs. Furthermore, one has to change the GP algorithm for them. This paper proposes a concept, which does not change the GP algorithm or its components. Instead, it introduces structure-based constants realized as functions, which can be simply added to each function set while keeping the original GP approach. These constant functions derive their constant values from the tree structures of their child-trees (subtrees). That is, a constant is represented by a tree structure being this way under the influence of the typical genetic operators like subtree crossover or mutation. These structure-based constants were applied to symbolic regression problems. They outperformed the standard approach of ephemeral random constants. Their results together with their better properties make the structure-based constant concept a possible candidate for the replacement of the ephemeral random constants.

Keywords: Genetic Programming, Constant, Structure-based Constant, Constant Function, Subtree Relationship, Full Tree Normalization, Generic Benchmark, Polynomial Benchmark, Sum-of-Gaussians Benchmark.

1 Introduction

If one applies Genetic Programming to problem domains, whose solutions are represented by mathematical expressions, as for instance in symbolic regression, one possibly needs to enable the usage of constants. Since constants are typically real values and the tree nodes represent discrete symbols like SIN or ADD, the question arises how to integrate these real values into the trees, a problem which is still considered as an open issue [8].

In order to overcome GP's weakness in discovering numerical constants, Koza has introduced the concept of Ephemeral Random Constants (ERC) [6,7]. These constants are represented by the terminal symbol $\Re$, whereby each terminal node additionally keeps a real value with the numerical value of this constant. For this concept it is necessary to change two aspects of GP: Firstly, the initialization

L. Correia, L.P. Reis, and J. Cascalho (Eds.): EPIA 2013, LNAI 8154, pp. 126–137, 2013.

component of GP needs to be changed. Each time, the terminal symbol of constants shall be added to an initial tree, a random number uniformly drawn from a given interval $[c_{min}, c_{max}]$ is added to its node, too. It is hoped that GP is able to produce all needed constants later on based on these pre-created ones by combining them with the (mathematical) operations contained in the set of functions. Secondly, the data structure of a node needs to be extended to be able to held this additional real value. Thus, this concept can not be used with every GP implementation without changing it. Beside these disadvantages, the ERC concept has the advantage that the number of created constants is not restricted, since a constant here is just a terminal symbol being randomly selected by the initializer.

Although the ephemeral random constants can still be considered as standard, other methods have been developed as well to improve the numeric quality of constants. In [1,2] the authors introduced an operator they called "Numeric Mutation", which extends the ERC concept. This operator is applied to a subset of the population at each generation and replaces all constants in existence by new ones uniformly drawn from a given constant interval, which is defined around the current constant value to be changed. The bounds of this interval are controlled by a "temperature factor" adopted from Kirkpatrick's simulated annealing method [5]. This "temperature factor" is set in dependence to the objective value of the best individual. If the run converges to good solutions, the interval becomes narrower, which leads to smaller changes in constant values. Otherwise, the interval becomes wider to allow bigger changes. Since this concept uses ephemeral random constants, it inherits their disadvantages: the need to change the node structure as well as to change the initialization component. Furthermore, a GP can not be run with standard configurations, because the new mutation operator needs to be considered.

Another method that changes constant values is introduced in [9]. There, the authors borrow the idea from [4] who use a table (array) of constants, whereby the constants in the trees use indices into this table. This concept was extended by using a sorted table so a smaller index also means a smaller constant value. Before creating the initial population, this sorted table is filled up with random values. Then, while creating the initial trees, each time the terminal symbol of constants shall be added, an index into the sorted table is added to its node, too. Based on these indices, the authors introduced two mutation operators. The first one is called "Uniform" and replaces an index of a constant by a new index, which is uniformly drawn from the whole range of indices. The second operator is called "Creep" and chooses the new index directly above or below the old one. This way also the change of the constant value is relatively small. In order to use this concept, one needs to change the node structure to be able to store an index as well as to change the initialization component. Furthermore, the sorted table mechanism needs to be integrated into the GP approach. This sorted table is of fixed size. Thus, the number of constants needs to be pre-specified. Finally, GP can not be run with standard configurations, because the new mutation operators need to be considered, too.

The above methods work by adding operators to GP or changing its initialization component. Another category of methods hybridize GP with other optimizers, whereby the other optimizers are used to adjust the constant values. These approaches are not GP methods as such, but hybrid concepts. In [10] GP is combined with a local optimizer based on gradient descent. There, at each generation all constant values of all trees are optimized by three iterations of a gradient descent approach. Since this concept uses ephemeral random constants, the node structure as well as the initialization component of a GP need to be changed, if one wants to use this concept. Furthermore, the local optimizer needs to be embedded as well.

In [3] another GP hybrid is presented, which uses a genetic algorithm as local optimizer for constants. Quite similar to the work of [4], the constants use indices into a table of constants. The difference is that each individual has its own local constant table. All these constant tables of the population are optimized by the employed genetic algorithm. If one wants to use this concept, a lot of GP aspects need to be changed: the node structure (for the index), the initialization component and the individual, which needs to keep its table of constants. Since this table is of fixed size, the number of constants per individual also needs to be pre-specified. Last but not least, the genetic algorithm needs to be integrated as well.

All in all, one can state that the methods developed so far have one "problem" in common: they all change either the GP algorithm as such by introducing new mechanisms like tables or by hybridizing GP with other algorithms, or they change components of GP like the initialization procedure to randomly produce constants or indices. If one wants to use constants, one can be certain to have to change his GP library for this. Furthermore, a hybrid could be considered to be not a GP anymore, but merely GP-like! A hybrid solves the constant problem just for the hybrid itself and not for GP in general. Thus, these approaches are not suitable to solve the constant problem in a sufficient manner. They are all "pragmatic" solutions. This raises the question, whether it is possible to replace the current concepts by another one, which *does not need to change GP or one of its components*. Further desirable properties would be that *as few parameters as possible* need to be specified and that the concept is *problem-independent*.

These are strong wishes, but can they be fulfilled? A main reason for the nature of current solutions to the constant problem is that intuitively everyone associates a constant with a value. And a value needs to be hold in a variable, it needs to be created and adjusted. This automatically leads to changes of data structures and of the procedures that deal with them. In order to get rid of this effect, one has to change the overall nature of solutions to the constant problem. But, if one wants to change the nature of current constant concepts, one needs to change the nature of constants, too.

Since the business of GP and its operators is to create and rearrange (sub)trees, a constant should also be a subtree to be under the influence of the typical genetic operators like subtree crossover or mutation. Therefore, in the proposed concept, a constant is not anymore a terminal, but a function. As a function it has subtrees as operands. The value of this constant function is derived from the tree structures of its operands. Thereby, the constant functions do not consider the content of the nodes. They are completely based on properties of the tree structures. Thus, they can be used in and added to each GP application being this way problem-independent.

The further paper is organized as follows. Section 2 introduces the proposed concept called *Structure-based Constants*. Two generic benchmark functions, which allow to specify the number of constants, are introduced in section 3. In section 4 the conducted experiments with their results are presented. Finally, in section 5 some conclusions are drawn.

2   Structure-Based Constants

Since the business of GP operators is to create and rearrange trees, a constant should also be represented in some way by a tree. This way, constants would also be under the influence of the subtree crossover and mutation operators. Therefore, in this section, a constant concept called *Structure-based Constants* (SC) is introduced, which replaces constant terminals by constant functions denoted by *SC*. A constant function has subtrees as operands, like all other functions, too. The value of this constant function is deterministically derived from the tree structures of its children (subtrees). This way, the constants are totally integrated into the GP trees without the need to pre-generate randomized ones or to determine them by additional algorithms in parallel or afterwards. This saves computation time and one can keep the *original GP algorithm*.

Let $\mathbb{T}$ denote the set of all trees

$$\mathbb{T} = \{T_1, \cdots, T_{|\mathbb{T}|}\}$$

and $|T|$ the number of nodes (or cardinality) of the given tree $T \in \mathbb{T}$. Each tree has at least a root node and is not empty.

Furthermore, let $[c_{min}, c_{max}]$ be a pre-specified interval of constant values as it is also used for ephemeral random constants. Then, the structure-based constants (SC), as proposed and used in this paper, can be defined as in the following subsections.

2.1   Subtree Relationships

This category of structure-based constants uses the child-trees of the constant function and sets their tree properties into relationship. A simple property of a tree is the number of nodes it is composed of. These numbers of nodes of the child-trees are combined to deterministically compute constants:

Definition 1 (*SC_{quot}*). *The structure-based constant*

$$SC_{quot} : \mathbb{T} \times \mathbb{T} \to \mathbb{R}$$

is the quotient of the number of nodes of its left and right child-trees T_{left} and T_{right}:

$$SC_{quot}(T_{left}, T_{right}) := \frac{|T_{left}|}{|T_{right}|}$$

Definition 2 (SC_{mmquot}). *The structure-based constant*

$$SC_{mmquot} : \mathbb{T} \times \mathbb{T} \to \mathbb{R}$$

is the minimum-maximum quotient of the number of nodes of both child-trees T_{left} and T_{right} mapped into the constant interval $[c_{min}, c_{max}]$:

$$SC_{mmquot}(T_{left}, T_{right}) := c_{min} + \frac{\min(|T_{left}|, |T_{right}|)}{\max(|T_{left}|, |T_{right}|)}(c_{max} - c_{min})$$

2.2   Full Tree Normalization

This category of structure-based constants uses a child-tree of the constant function and normalizes one of its tree properties with respect to a full tree structure. A full tree is a structurally complete tree (and some authors call it also complete tree, perfect tree or perfect A-ary tree):

Definition 3 (Full Tree). *A full tree denoted by $T_{L,A}$ is a rooted A-ary tree with L levels, i.e., a tree with exactly one root node and every internal node has exactly A children. All nodes of the last level are in existence and no internal node is missing as depicted in the following for a full tree $T_{3,2}$ with 3 levels and an arity of 2:*

A simple property of a tree is the number of nodes it is composed of. Thus, the number of nodes of a child-tree is normalized by the number of nodes of a full tree to deterministically compute constants. The total number of nodes of a full tree $T_{L,A}$ can be computed by Eq. (1).

$$N_{nodes}(L, A) = \sum_{i=0}^{L-1} A^i \qquad (L \geq 1, A \geq 1) \tag{1}$$

Definition 4 (SC_{full}). *The structure-based constant*

$$SC_{full} : \mathbb{T} \to \mathbb{R}$$

normalizes a child-tree T by its corresponding full tree based on the number of nodes and maps it into the constant interval [c_{min}, c_{max}]:

$$SC_{full}(T) := c_{min} + \frac{|T|}{N_{nodes}(level(T), arity(T))}(c_{max} - c_{min})$$

The function level(T) delivers the maximum depth of subtree T and arity(T) the maximum arity occurring in T.

The full tree represents the maximum structure of a tree. Thus, a subtree T can be either this full tree, which leads to a quotient of 1, or a structurally smaller version of the full tree, which produces a quotient < 1. In this sense, a tree is divided by a maximum tree performing this way a normalization.

The corresponding full tree used by SC_{full} is based on the arity and depth of the child-tree. Thus, a child-tree [sin[x]] as depicted in

is a full tree, because the maximum arity occurring is 1 and the depth is 2, which leads to $N_{nodes}(L, A) = N_{nodes}(2, 1) = 2 = |$ [sin[x]] $|$. But typically mathematical expressions also allow binary operations so the question arises whether it would not be better to use the global maximum arity from the whole function set. In this case the former example would not be anymore a full tree, because the corresponding full tree would have $N_{nodes}(L, A) = N_{nodes}(2, 2) = 3$ nodes. A child-tree that is a full tree represents c_{max}. Maybe using the global maximum arity produces more different constants, because the number of possible full trees is reduced. In order to examine this, a variation to SC_{full} is defined in the following, which uses the global maximum arity out of the function set denoted by A_{max}.

Definition 5 *(SC_{full-g}).* *The structure-based constant*

$$SC_{full-g} : \mathbb{T} \to \mathbb{R}$$

normalizes a child-tree T by a corresponding full tree based on the number of nodes and maps it into the constant interval [c_{min}, c_{max}]:

$$SC_{full-g}(T) := c_{min} + \frac{|T|}{N_{nodes}(level(T), A_{max})}(c_{max} - c_{min})$$

The function level(T) delivers the maximum depth of subtree T and A_{max} is the global maximum arity out of the function set.

3   Benchmark Functions

In order to evaluate the capabilities of the introduced structure-based constants concept, two generic benchmark functions (subsections 3.1 and 3.2) were specifically created for the work at hand and allow to specify the number of constants c_{num} as a sort of parameter of model complexity.

3.1 Polynomial Benchmark

This generic benchmark function allows to specify the number of constants c_{num} as a benchmark parameter. It is designed for the standard set of functions (see Table 1). Since the standard set is particularly able to build polynomials, this benchmark function is just a reduced polynomial, whose degree is used as the number of constants:

$$P_{c_{num}}(x) = \sum_{i=1}^{c_{num}} (c_{min} + \frac{i}{c_{num}}(c_{max} - c_{min})) \cdot x^i \tag{2}$$

The interval $[c_{min}, c_{max}]$ is the allowed range of constant values. Note that the x^0 term is omitted. This way, the degree of the polynomial is the number of wished constants c_{num}. The coefficients, which are the searched constants, increase from the lowest to the highest exponent.

3.2 Sum-of-Gaussians Benchmark

Like the previous one, also this benchmark function allows to specify the number of constants c_{num} as a benchmark parameter. But this one is designed for the extended set of functions (see Table 1). The idea is to use a Gaussian for each constant, whereby the constant shifts the Gaussian's position and works this way as an offset. They are shifted in a way that all Gaussians are distributed over the given range $[X_{min}, X_{max}]$. All shifted Gaussians are summed up to build the Sum-of-Gaussians Benchmark:

$$G_{c_{num}}(x) = \sum_{i=1}^{c_{num}} e^{-\left(x + X_{min} + (i-0.5)\frac{X_{max}-X_{min}}{c_{num}}\right)^2} \tag{3}$$

4 Experiments

The aim of the conducted experiments was to find out, whether the structure-based constants perform at least comparably to Koza's ERC concept. That they have better properties, because they do not change anything in the GP algorithm, is not enough. Thus, all four SCs ($SC_{quot}, SC_{mmquot}, SC_{full}, SC_{full-g}$) as well as Koza's ERC concept were applied to all benchmark functions as introduced in section 3. Each of the two generic benchmarks was used with $c_{num} = 1, \cdots ,$ 10 constants leading to 10 different benchmark functions per generic benchmark. The used $[c_{min}, c_{max}]$ interval for constants was set to $[-5, +5]$ for all structure-based constants as well as for ERC. According to the used function sets (see below), the global maximum arity A_{max} for SC_{full-g} was set to 2.

For each benchmark function and constant concept, 50 independent runs were performed. For all benchmark functions $P_{c_{num}}$ and $G_{c_{num}}$, samples of 100 points were used with all $x^{(k)}$ being uniformly distributed in $[X_{min}, X_{max}] = [-10, 10]$ and $y^{(k)} = P_{c_{num}}(x^{(k)})$ or $y^{(k)} = G_{c_{num}}(x^{(k)})$, respectively. For all experiments, a standard GP was used with the settings as given in Table 1.

Table 1. The settings of the used GP approach

Objective:	Symbolic regression with constants evaluated by sum of deviations over all points $F(T) = \sum_k |\mathcal{I}(T, x^{(k)}) - y^{(k)}|$ with tree $T \in \mathbb{T}$ and interpreter $\mathcal{I}$
Standard Function Set:	NEG, ADD, SUB, MUL, DIV and X as terminal plus $\mathfrak{R}$ or the appropriate $SC_{...}$ constant
Extended Function Set:	NEG, ADD, SUB, MUL, DIV, POW, ABS, SQRT, SIN, COS, TAN, EXP, LN, LOG and X as terminal plus $\mathfrak{R}$ or the appropriate $SC_{...}$ constant
Initialization:	Ramped half-and-half, min = 2, max = 6
Crossover:	Standard crossover, probability = 0.9
Mutation:	Subtree mutation, probability = 0.1
Selection:	Tournament selection, 3 competitors
Fitness:	Raw (standard) fitness
Replacement:	Generational replacement scheme
Parameters:	Population size = 500, generations = 40, no elitists

4.1   Polynomial Benchmark

For this benchmark, the experiments were conducted for the first ten numbers of constants ($c_{num} \in \{1, \cdots, 10\}$). In Table 2 (page 134), the results are given as numerical values. The winners are printed in boldface. Inspecting this table reveals that the Polynomial benchmarks were best solved by all SC approaches. But among the SCs, there is no clear winner. Three of them win in four and five of the ten cases and one only in two cases (SC_{full-g}). Among the best three SCs, one (SC_{quot}) performed only well for the lower numbers of constants, whereas the other two (SC_{mmquot}, SC_{full}) won cases over the whole range.

4.2   Sum-of-Gaussians Benchmark

Also for this benchmark, the experiments were conducted for the first ten numbers of constants ($c_{num} \in \{1, \cdots, 10\}$). According to Table 3 (page 135), the Sum-of-Gaussians benchmarks were best solved by SC_{mmquot}. It wins in 6 of the ten cases, outperforming this way all other methods (with respect to the number of won cases). The other three SCs win only in zero, one and two cases. Although very tight, the ERC method wins one case and outperformed SC_{quot}. The reached fitness values of all methods are close for all benchmarks.

4.3   Summary

In Table 4 (page 136) the final results over all benchmark functions are given in terms of the numbers of cases in which a constant concept performs as best. The last row gives the total numbers of won cases. From there it can be seen that the winner over all benchmarks is clearly SC_{mmquot}. The second bests are SC_{full}

Table 2. The obtained results for the **POLYNOMIAL** benchmark averaged over 50 independent runs. The columns 'Avg. Fitness' are the best fitness values reached on average and 'sd' are the appropriate standard deviations.

	ERC	SC_{quot}	SC_{mmquot}	SC_{full}	SC_{full-g}
c_{num}	Avg. Fitness	Avg. Fitness	Avg. Fitness	Avg. Fitness	Avg. Fitness
1	5.45922 (sd: 15.6871)	**0** (sd: 0)	**0** (sd: 0)	**0** (sd: 0)	**0** (sd: 0)
2	50.945 (sd: 92.6602)	0.000106878 (sd: 0.000748143)	**0** (sd: 0)	**0** (sd: 0)	**0** (sd: 0)
3	959.358 (sd: 1178.68)	**411.188** (sd: 365.07)	542.772 (sd: 399.006)	531.82 (sd: 467.849)	450.682 (sd: 307.877)
4	11401.3 (sd: 9706.61)	**2207.93** (sd: 4602.83)	4128.93 (sd: 3238.93)	3492.59 (sd: 3076.52)	4441.4 (sd: 3156.95)
5	102873 (sd: 113909)	**19747.8** (sd: 65657.6)	30664.9 (sd: 91940.9)	64953.9 (sd: 336065)	47246.9 (sd: 116539)
6	$1.12797e+006$ (sd: 1.14452e+006)	569027 (sd: 895749)	296845 (sd: 562651)	**284145** (sd: 314191)	436775 (sd: 640290)
7	$2.63987e+007$ (sd: 3.32783e+007)	$1.05336e+007$ (sd: 2.0743e+007)	**$8.3599e+006$** (sd: 1.38128e+007)	$8.57059e+006$ (sd: 9.45898e+006)	$1.11821e+007$ (sd: 2.49654e+007)
8	$1.14364e+008$ (sd: 1.77641e+008)	$5.03722e+007$ (sd: 5.97271e+007)	$3.87227e+007$ (sd: 6.56279e+007)	**$2.35991e+007$** (sd: 2.33994e+007)	$2.61194e+007$ (sd: 3.1674e+007)
9	$3.12286e+009$ (sd: 3.11849e+009)	$1.08845e+009$ (sd: 1.94807e+009)	**$7.09581e+008$** (sd: 8.15087e+008)	$1.48731e+009$ (sd: 2.2571e+009)	$1.24917e+009$ (sd: 2.32232e+009)
10	$1.63328e+010$ (sd: 1.74503e+010)	$7.36193e+009$ (sd: 9.44926e+009)	$3.70365e+009$ (sd: 4.755e+009)	**$3.0351e+009$** (sd: 4.06186e+009)	$4.18589e+009$ (sd: 9.61241e+009)

Table 3. The obtained results for the **SUM-OF-GAUSSIANS** benchmark averaged over 50 independent runs. The columns 'Avg. Fitness' are the best fitness values reached on average and 'sd' are the appropriate standard deviations.

	ERC	SC_{quot}	SC_{mmquot}	SC_{full}	SC_{full-g}
c_{num}	Avg. Fitness	Avg. Fitness	Avg. Fitness	Avg. Fitness	Avg. Fitness
1	7.92484 (sd: 2.45881)	7.06104 (sd: 3.36676)	**6.86777** (sd: 3.5924)	7.81503 (sd: 2.66586)	7.6386 (sd: 2.8158)
2	**17.5234** (sd: 0.0690329)	17.5466 (sd: 0.002927)	17.5473 (sd: 1.06581e−014)	17.5458 (sd: 0.00873155)	17.5473 (sd: 1.06581e−014)
3	22.7068 (sd: 4.75156)	24.0421 (sd: 3.74773)	**21.9589** (sd: 5.61472)	24.3213 (sd: 4.34347)	24.4268 (sd: 3.72864)
4	29.8323 (sd: 0.344268)	29.6072 (sd: 1.40193)	**27.1038** (sd: 4.65437)	27.5833 (sd: 3.84845)	27.7314 (sd: 2.86172)
5	30.2103 (sd: 0.707123)	30.04 (sd: 1.22604)	24.1434 (sd: 6.01809)	25.2328 (sd: 6.67785)	**23.6527** (sd: 6.88129)
6	20.7481 (sd: 5.78303)	19.5825 (sd: 6.47734)	16.0718 (sd: 2.42268)	**15.1753** (sd: 3.70877)	15.7694 (sd: 3.41595)
7	23.6415 (sd: 1.02372)	19.2354 (sd: 5.1414)	**17.5072** (sd: 4.74239)	18.3589 (sd: 4.79823)	17.8244 (sd: 5.47548)
8	19.4204 (sd: 0.208013)	14.377 (sd: 4.89492)	**11.6531** (sd: 5.85684)	13.708 (sd: 6.13544)	12.6103 (sd: 6.13404)
9	14.7881 (sd: 0.177806)	12.4597 (sd: 3.8964)	**8.36133** (sd: 5.28631)	10.0999 (sd: 5.31754)	9.5832 (sd: 4.95699)
10	10.8935 (sd: 0.0634859)	10.5945 (sd: 0.853805)	10.4401 (sd: 1.31411)	10.2638 (sd: 1.32116)	**9.99404** (sd: 1.56859)

and SC_{full-g}. The worst of the structure-based constants is SC_{quot}, which only performs well for the Polynomial benchmark $P_{c_{num}}$ with fewer constants. Note that SC_{quot} is the only SC, which neither specifies a constant interval nor represents negative constants. All other methods (including ERC) allow for negative constants by using appropriate c_{min} and c_{max} bounds. It seems that forcing negative constants by an interval is easier for GP than to build negative constants by applying the NEG function to a positive constant.

Considered over all 20 benchmark functions, the ERC concept only won 1 case. In all other 19 cases it was outperformed by most of the structure-based constants.

Table 4. The numbers of won cases over all benchmarks

Benchmark	ERC	SC_{quot}	SC_{mmquot}	SC_{full}	SC_{full-g}
# Polynomial	0	4	4	5	2
# Sum-of-Gaussians	1	0	6	1	2
Σ	1	4	10	6	4

5   Conclusions

In this paper a new constant concept called Structure-based Constants (SC) was introduced. It represents a constant by a tree so it is under the influence of the subtree crossover and mutation operators. Opposed to the common procedure, such a structure-based constant is not a terminal, but a function. The value of this constant function is derived from the tree structures of its child-trees. This new concept has a number of advantages compared to ERC and other solutions to the constant problem:

- Since a constant is a tree itself, it totally integrates into the GP trees. Thus, it is evolved with subtree crossover and mutation and **no additional optimizer is needed** anymore. This saves computation time.
- An SC only needs to be added to the function set the same way as all other functions, too.
- Neither the GP algorithm, nor one of its components need to be changed or extended. Thus, **one can keep the original GP** and use standard configurations.
- **Only few parameters** need to be specified. In fact, it is the interval of constant values as already used by the ERC approach. (For SC_{full-g} also A_{max} needs to be set. But this parameter cannot be freely adjusted by the user – it must be set to the global maximum arity of the used function set. Thus, it can not be considered to be a real parameter in the same sense as the interval bounds.)
- The SCs do not consider the content of tree nodes, because they are based on properties of tree structures. Thus, they are **problem-independent** and can be used in each GP application.

– Since the original GP is not changed, the SCs can be used with each GP implementation already in existence.

In 19 of 20 cases the SC concept outperformed ERC. With this result the demand that it must be at least comparable to ERC is fulfilled. It seems that the structure-based constants could be a suitable approach to replace Koza's ERC concept.

References

1. Evett, M., Fernandez, T.: Numeric Mutation Improves the Discovery of Numeric Constants in Genetic Programming. In: Proc. 3rd Annual Conference on Genetic Programming, pp. 66–71. Morgan Kaufmann (1998)
2. Fernandez, T., Evett, M.: Numeric Mutation as an Improvement to Symbolic Regression in Genetic Programming. In: Porto, V.W., Waagen, D. (eds.) EP 1998. LNCS, vol. 1447, pp. 251–260. Springer, Heidelberg (1998)
3. Howard, L.M., D'Angelo, D.J.: The GA-P: A Genetic Algorithm and Genetic Programming Hybrid. IEEE Intelligent Systems 10(3), 11–15 (1995), doi:10.1109/64.393137
4. Keith, M.J., Martin, M.C.: Genetic Programming in C++: Implementation Issues. In: Kinnear Jr., K.E. (ed.) Advances in Genetic Programming (1994)
5. Kirkpatrick, S., Gelatt, C.D., Vecchi, M.P.: Optimization by simulated annealing. Science 220, 671–680 (1983)
6. Koza, J.R.: Genetic Programming: A Paradigm for Genetically Breeding Populations of Computer Programs to Solve Problems, Stanford University, Computer Science Department. Technical Report STAN-CS-90-1314 (June 1990)
7. Koza, J.R.: Genetic Programming: On the Programming of Computers by Means of Natural Selection. MIT Press, Cambridge (1992)
8. O'Neill, M., Vanneschi, L., Gustafson, S., Banzhaf, W.: Open issues in genetic programming. Genetic Programming and Evolvable Machines 11(3-4), 339–363 (2010), doi:10.1007/s10710-010-9113-2
9. Ryan, C., Keijzer, M.: An Analysis of Diversity of Constants of Genetic Programming. In: Ryan, C., Soule, T., Keijzer, M., Tsang, E.P.K., Poli, R., Costa, E. (eds.) EuroGP 2003. LNCS, vol. 2610, pp. 404–413. Springer, Heidelberg (2003)
10. Topchy, A., Punch, W.F.: Faster Genetic Programming based on Local Gradient Search of Numeric Leaf Values. In: Proc. of the Genetic and Evolutionary Computation Conference (GECCO 2001), pp. 155–162. Morgan Kaufmann (2001)

Class Imbalance in the Prediction of Dementia from Neuropsychological Data

Cecília Nunes[1], Dina Silva[2], Manuela Guerreiro[2], Alexandre de Mendonça[2], Alexandra M. Carvalho[3], and Sara C. Madeira[1]

[1] Knowledge Discovery and Bioinformatics (KDBio) Group, INESC-ID and Instituto Superior Técnico (IST), Technical University of Lisbon, Lisbon, Portugal
cnunes@kdbio.inesc-id.pt, sara.madeira@ist.utl.pt
[2] Dementia Clinics, Institute of Molecular Medicine and Faculty of Medicine, University of Lisbon, Lisbon, Portugal
{dinasilva,mmgguerreiro,mendonca}@fm.ul.pt
[3] Instituto de Telecomunicações (IT) and Instituto Superior Técnico (IST), Technical University of Lisbon, Lisbon, Portugal
alexandra.carvalho@lx.it.pt

Abstract. Class imbalance affects medical diagnosis, as the number of disease cases is often outnumbered. When it is severe, learning algorithms fail to retrieve the rarer classes and common assessment metrics become uninformative. In this work, class imbalance is approached using neuropsychological data, with the aim of differentiating Alzheimer's Disease (AD) from Mild Cognitive Impairment (MCI) and predicting the conversion from MCI to AD. The effect of the imbalance on four learning algorithms is examined through the application of bagging, Bayes risk minimization and MetaCost. Plain decision trees were always outperformed, indicating susceptibility to the imbalance. The naïve Bayes classifier was robust but suffered a bias that was adjusted through risk minimization. This strategy outperformed all other combinations of classifiers and meta-learning/ensemble methods. The tree-augmented naïve Bayes classifier also benefited from an adjustment of the decision threshold. In the nearly balanced datasets, it was improved by bagging, suggesting that the tree structure was too strong for the attribute dependencies. Support vector machines were robust, as their plain version achieved good results and was never outperformed.

Keywords: Alzheimer's disease, dementia, classification, class imbalance, prognostic prediction, prediction of dementia, neuropsychological data.

1 Introduction

Alzheimer's Disease (AD) is the most common form of dementia among the elderly, affecting 26 million worldwide in 2006 [1]. It remains incurable and its prevalence is estimated to increase given the aging of the world population.

AD progression is categorized in (1) preclinical AD, (2) mild cognitive impairment (MCI) due to AD and (3) dementia due to AD [2]. Characterizing the

L. Correia, L.P. Reis, and J. Cascalho (Eds.): EPIA 2013, LNAI 8154, pp. 138–151, 2013.
© Springer-Verlag Berlin Heidelberg 2013

stage is of the utmost importance for managing the disease, as small delays in AD onset and progression would lead to significant reductions in its global burden [1]. However, the boundaries between the stages are unclear, making this an extremely challenging task [3]. As advanced diagnosis techniques are expensive, invasive and often unavailable, medical doctors rely on neuropsychological assessments. Maximizing the information provided by neuropsychological tests has thus been subject to attention [4, 5]. In this context, given that each MCI patient is subject to cognitive tests several times before progressing to dementia, datasets used for diagnosis contain more MCI evaluations than dementia-labeled evaluations. Furthermore, datasets used for prognosis assimilate the fact that 10-15% of the patients with cognitive complaints progress to dementia each year [2]. Neuropsychological data is hence prone to class imbalance.

Class imbalance is the disparity in the proportions of different classes in datasets used for classification. It affects classification in two ways. First, predictive models neglect the accuracy over the minority. Overcoming this problem involves understanding how classifiers are affected and proposing solutions. Second, the imbalance makes assessment metrics uninformative. For example, if we consider a majority class that corresponds to 90% of the data, an all-majority classifier has no predictive power and yet leads to a 90% accuracy. In this context, the aim of this work is to study the effect of the imbalance of neuropsychological data on four state-of-the-art algorithms. The classification tasks are differentiating MCI from AD (diagnosis) and predicting the conversion from MCI to AD (prognosis) in patients with cognitive complaints. The algorithms are decision trees (DT), the naïve Bayes (NBayes) classifier, the tree-augmented naïve (TAN) Bayes classifier, and Support Vector Machines (SVMs). They are used as base classifiers for the other strategies. Bagging, MetaCost and the minimization of Bayes risk are applied to those classifiers in order to understand their behavior and improve their performance. In addition, assessment metrics for imbalanced data are discussed. Each classifier revealed different behaviors. In particular, DT were unstable and plain SVMs were robust to the imbalance. The best results were achieved when the NBayes was used with risk minimization.

This paper is organized as follows: Section 2 describes the problem of class imbalance and the solutions that have been proposed. Assessment metrics for imbalanced data are briefly discussed. In Section 3, we report the experiment setup, including a description of the data, preprocessing, training and evaluation steps. The results are presented and discussed in Section 4.

2   Learning from Imbalanced Data

Class imbalance can be defined as the proportion of minority instances over the total number of instances. The majority class is hereafter considered the negative class, and the minority the positive. This type of imbalance is in fact between-class imbalance. It can bias the learners towards the overrepresented class, while the minority may go unlearned. As the imbalance grows, we not longer aim at maximizing accuracy. Instead, we want classifiers to pay more attention to the minority class, without jeopardizing the performance over the majority [6, 7]. The specific consequences of the imbalance depend on the algorithm [8–13].

2.1 Solutions to Class Imbalance

Proposed solutions for class imbalance can be divided into data-level strategies and algorithm-level or cost-sensitive(CS) strategies [14]. Data-level solutions re-sample the dataset to obtain optimal class proportions [8, 15]. They involve undersampling the majority class or oversampling the minority. Random re-sampling has some disadvantages [16]. To overcome the overfitting caused by random oversampling, Synthetic Minority Over-sampling Technique (SMOTE) has been developed [17]. This method may lead to overgeneralization, which can be avoided by adaptive synthetic sampling [18, 19].

Instead of manipulating the data, CS solutions draw the attention of the classifiers to the minority by means of a cost. Costs are scores attributed to correct or incorrect classifications, for instance according to the class. The existence of non-uniform unknown misclassification costs is closely related to class imbalance. The relation between both problems is task-specific and method-specific. Nonetheless, while cost-*in*sensitive solutions lead to sub-optimal performances in both cases [20], CS approaches offer a solution to both problems [21]. In effect, there is an equivalence between varying the class proportion, the class prior probabilities or the misclassification costs [9, 22]. Zadrozny *et al.* [23] divided CS approaches in three categories:

1. *CS model inference*: costs are incorporated directly in the classifier induction algorithm [24, 25]. These techniques are out of the scope of this work since they focus on a single learning algorithm.
2. *CS decision making*: as opposed to minimizing the misclassification rate, class predictions are made according to the minimization of the expected loss [9, 26]. This requires class-membership probabilities to be inferred by the classifiers and knowledge of the costs.
3. *CS ensembles*: cost-sensitivity is introduced into ensemble methods [14, 27]. Examples include CS Boosting [14] and MetaCost [27].

In the following paragraphs, we introduce the minimization of Bayes risk, as a way to minimize the expected loss, together with MetaCost.

Minimizing Bayes Risk. The scores attributed to different predictions are represented in a cost matrix, as depicted in Table 1. $C(k, j)$ is the cost of classifying a class-j instance as k.

The optimal solution is the one that minimizes the loss function for all instances, over all class hypotheses [26]. Since the predictions made for each instance are in-

Table 1. Cost matrix

predicted negative	predicted positive	
$C(0,0)$	$C(1,0)$	real negative
$C(0,1)$	$C(1,1)$	real positive

dependent, this is equivalent to classifying each instance $\mathbf{X}_i$ with the class k that minimizes a quantity known as *conditional risk* [28]:

$$L(\mathbf{X}_i|k) = \sum_i P(j|\mathbf{X}_i)C(k,j), \tag{1}$$

where $P(j|\mathbf{X}_i)$ is the probability of class j given $\mathbf{X}_i$, with $1 \leq i \leq |D|$, and D is the dataset. To use this strategy, class-membership probability estimates are required. It follows that accurate classifications depend on accurate estimates. As such, deciding upon a class can be viewed as estimating a score and comparing it to a decision threshold. In binary classification, most algorithms consider a 0.5 threshold. Changing the cost matrix is a way of adjusting the threshold [9].

But how to tackle the problem of unknown costs? Choosing a cost ratio equivalent to inverting the class proportions would not account for the different severity of the errors [29]. The problems related to class proportion and unknown costs can be tackled by searching for the best cost setup [21]. In this work, the optimal cost setup is sought by means of fixed class cost ratios (see Section 3).

MetaCost. This approach led to significant cost reductions in many datasets, while dealing with poor or non-existing probability estimates. It uses bagging [6] to train several weak models. Then, the class-membership probabilities of each instance are estimated by averaging the estimates of the weak models, or through voting in case they are not provided. Using those probabilities, each training instance is relabeled with the class that minimizes the total expected cost (1). Finally, the classifier is trained on the relabeled data to build the final model.

2.2 Assessment Metrics for Imbalanced Learning

The accuracy and the error rate involve ratios between sums of instances of different classes. As such, they are uninformative in imbalanced data [30]. A high accuracy can correspond to a correctly classified majority, and hide a misclassified minority. Alternatively, authors have turned their attention to other metrics. Besides the area under the Receiver Operating Characteristic (ROC) curve, two other single assessment metrics are frequently used in imbalanced learning: the geometric mean (G-mean) [8, 11, 29] and the F-measure [14]. These metrics are defined from the ratios computed from the confusion matrix, such as the True Positive Rate (TPR), also known as recall, the True Negative Rate (TNR), the False Positive Rate(FPR), and the Precision. The G-mean is defined as [31]:

$$G - mean = \sqrt{TPR \times TNR}.$$

The shortcoming of this metric is that it can be optimistic in large imbalances. Given the learning bias, the TNR can be very high regardless of the actual learning ability, and the effect spreads to the G-mean[1]. This shortcoming is largely avoided by the F-measure [32]:

---

[1] This also occurs in ROC space metrics. The learning bias can cause the FPR to be low even if a large number of false positives occurs.

$$F - measure = \frac{(1 + \beta^2) \cdot Recall \cdot Precision}{\beta^2 Recall + Precision},$$

where typically $\beta = 1$, leading to the harmonic mean of recall and precision. Unlike the FPR, which compares the false positives with the total number of negatives, precision compares the false positives with the true positives [32]. In an imbalanced set, the number of true positives is smaller than the total number of negatives and thus negative misclassifications are better captured. In addition, the harmonic mean of two values is closer the lowest of them than the arithmetic mean. As such, a high F-measure assures both a high precision and recall [10]. For these reasons, it is our metric of choice. A limitation of the F-measure is the fact that it disregards the performance of the negative class.

3   Methods

3.1   Data Description

The Cognitive Complaints Cohort (CCC) [4] is a study conducted at Instituto de Medicina Molecular (IMM) to investigate dementia on subjects with cognitive complaints. The CCC database comprises the results of neuropsychological tests applied to subjects respecting the inclusion criteria specified by Silva et al. [33]. The tests correspond to the Battery of Lisbon for the Assessment of Dementia (BLAD), proposed by Garcia [34]. The battery is validated for the Portuguese population and comprises tests targeting different cognitive domains. The test results are mapped to the stage of dementia provided by medical doctors in the categories: normal, pre-MCI, MCI and dementia due to AD. The latter is simply denoted as AD. The database contains 1642 evaluations of 950 subjects and 162 attributes. Each evaluation is an instance and the attributes are the neuropsychological tests. The original classes are the aforementioned stages.

3.2   Data Preprocessing

The first step was the removal of normal and pre-MCI instances. This was followed by the elimination of non-informative attributes, as well as instances of patients evaluated only once, given their uselessness in prognosis. At last, removing instances with more than 90% missing values yielded 677 instances of 336 patients and 157 attributes, with yet nearly 50% of missing values.

The diagnosis of dementia was done considering each evaluation as an independent instance. Dementia prognosis required relabeling the MCI evaluations according to the progression to dementia of the corresponding patient, withing a given time frame. The prognosis classes are evolution (Evol) and non-evolution (noEvol) to dementia, and 2, 3 and 4-year time frames are considered. The datasets were divided into training and validation data. The training data was used to build and test the models through cross-validation (CV). The validation data was used in a final assessment of the models built with the CV data, as in a hold-out (HO) test. The HO data contains 25% of the original data, sampled in a stratified way. No different evaluations of the same patient are contained in both CV and HO datasets. The final datasets are summarized in Table 2.

Preprocessing involved two final procedures: attribute selection and missing value imputation. Correlation-based feature subset selection [35] was performed on the training data and extrapolated to the validation data. Mean-mode missing value imputation was used on the training data. A more sophisticated technique was also tested [36] but it introduced a learning bias. Mean-mode imputation avoided this bias, providing a straightforward solution to the problem.

Table 2. Summary of the datasets, with their learning aim and imbalances

Learning task	Abbreviation	Imbalance	Minority class
distinguish MCI from AD	CV_Diag	15.7%	AD
predict MCI-to-AD conversion in 2 years	CV_2Y	36.5%	Evol
predict MCI-to-AD conversion in 3 years	CV_3Y	47.6%	noEvol
predict MCI-to-AD conversion in 4 years	CV_4Y	32.4%	noEvol
distinguish MCI from AD	HO_Diag	8.3%	AD
predict MCI-to-AD conversion in 2 years	HO_2Y	22.2%	Evol
predict MCI-to-AD conversion in 3 years	HO_3Y	41.1%	Evol
predict MCI-to-AD conversion in 4 years	HO_4Y	41.9%	noEvol

3.3   Classification

Bagging, risk minimization, and MetaCost were applied to the NBayes classifier, TAN Bayes classifier, DT and SVMs. The strategies were chosen since they can be applied to any classifier and give insight about its behavior. The NBayes classifier was used with kernel density estimation since it showed superior results compared to using a normal distribution, in all datasets. The TAN Bayes classifier was used due to its efficiency and efficacy [37].

When dealing with imbalanced datasets and DT, either no pruning or pruning preceded by Laplace smoothing is advised [38, 39]. Hence, both methods were tested and the best was used. The pruning confidence factor was subject to grid-search in order to maximize F-measure for each dataset. SVMs were used with the polynomial kernel and the Gaussian kernel function. Grid-search was also performed for the SVM parameters. Regarding bagging and MetaCost, results for 10 iterations were considered. Attempts with less iterations led to worse results, while using more than 10 iterations does not provide significant bagging improvements [6]. Given the moderate dataset size, a 100% bag size was used. Implementations were provided by WEKA 3.6 [40].

Regarding the cost-setup, correct predictions were defined to have zero cost, that is $C(0,0) = 0$ and $C(1,1) = 0$. Since the cost matrix is invariant to multiplication by a positive factor, the majority class error cost was kept equal to one, and the minority error cost was varied. The goal was then to find the optimal misclassification cost ratio (MCR):

$$MCR = \frac{C(0,1)}{C(1,0)} = C(0,1),$$

which corresponds to the value that maximizes F-measure. Empirical tests showed that the optimal MCRs were never superior to 14 (obtained applying risk minimization to DTs). The second highest MCR was 8 (obtained using MetaCost with the NBayes). Since the difference in F-measure between the two cases was only 0.02, the MCR search interval was restricted to $[1, 8]$ with a step of 0.25. The models were built using 10-fold CV, performed 10 times with different random seeds. The CV partitions were the same for all methods. A Friedman rank test and its post-hoc Nemenyi pairwise comparisons were applied, as advised when testing more than two algorithms over multiple datasets [41]. Rejecting the null hypothesis of the Friedman test means that at least two of the results of applying the base classifier, bagging, risk minimization and MetaCost come from populations with different medians [42], that is significant differences in the performances were found. A significance level of 0.05 was considered.

4   Results and Discussion

In this section, we first present and discuss the results obtained for each classifier. The average values of F-measure can be observed in Table 3 and the results of the Friedman test and the pairwise comparisons are depicted in Table 4. For the CS methods, the MCR that maximized the F-measure was selected (Table 6 in Appendix A). Finally, we compare the best strategies for each classifier (Table 5).

Table 3. F-measure averaged over 10 CV experiments for each dataset, base classifier and meta-learning/ensemble method. The base classifier is the classifier without any method. For the CS methods, the MCR that maximized F-measure was selected.

classifier	dataset	Base classifier	Bagging	Risk min	MetaCost
DT	CV_Diag	0.464	0.492	0.504	**0.531**
	CV_2Y	0.514	0.551	0.578	**0.621**
	CV_3Y	0.679	**0.729**	0.692	0.695
	CV_4Y	0.579	0.607	0.604	**0.630**
	Average	0.559	0.595	0.594	**0.619**
NBayes	CV_Diag	**0.598**	0.596	0.597	0.583
	CV_2Y	0.636	0.640	**0.693**	0.655
	CV_3Y	0.745	0.746	**0.784**	0.737
	CV_4Y	0.675	0.679	**0.693**	0.661
	Average	0.664	0.665	**0.692**	0.659
TAN	CV_Diag	0.548	0.557	0.551	**0.583**
	CV_2Y	0.623	0.631	0.653	**0.655**
	CV_3Y	0.744	**0.769**	0.748	0.737
	CV_4Y	0.625	**0.668**	0.660	0.661
	Average	0.635	0.653	0.656	**0.659**
Polynomial-kernel SVM	CV_Diag	0.550	0.561	**0.569**	0.563
	CV_2Y	0.649	0.643	**0.681**	0.659
	CV_3Y	**0.74**	0.740	0.763	0.770
	CV_4Y	**0.722**	0.713	0.714	0.715
	Average	0.667	0.664	**0.682**	0.677
Gaussian-kernel SVM	CV_Diag	0.556	**0.579**	0.579	0.576
	CV_2Y	0.651	0.644	**0.685**	0.655
	CV_3Y	0.733	0.727	0.760	**0.771**
	CV_4Y	**0.712**	0.703	0.695	0.697
	Average	0.663	0.663	**0.680**	0.675

4.1   Decision Trees

DT usually lack robustness to the imbalance. They tend to grow mixed leaves with few minority instances that get disregarded. In addition, minority instances may end up isolated in single leaves, leading to overfitting [8]. Accordingly, in Table 4 it is possible to observe that plain DT were outperformed by all other methods. Both bagging and MetaCost improved the performance of DT, which meets the expectations given their instability [6]. Although DTs can suffer from a learning bias, they provide inaccurate class-membership probabilities, and are therefore bad candidates for risk minimization. Indeed, the bias was preferably tackled through MetaCost, since it led to the highest values of F-measure in most datasets. In the most balanced dataset, the CV_3Y dataset, MetaCost was outperformed by bagging indicating the absence of the bias.

Table 4. Results of the Friedman tests and post-hoc pairwise comparisons over all datasets. For each classifier, rejecting the null hypothesis corresponds to finding significantly different performances among all methods. Each entry indicates if the F-measure obtained with the method in the corresponding column was significantly greater or smaller than the F-measure obtained with the method in the row. The entry is blank in case the comparison revealed no statistically significant difference.

	Base classifier	Bagging	Risk min	MetaCost	Best strategy
Decision Trees	p-value=1.37E-10				
Base classifier	-	greater	greater	greater	Bagging,
Bagging	smaller	-			Risk min and
Risk min	smaller		-		MetaCost
MetaCost	smaller			-	
NBayes	p-value=3.34E-12				
Base classifier	-		greater		
Bagging		-	greater	smaller	Risk min
Risk min	smaller	smaller	-	smaller	
MetaCost		greater	greater	-	
TAN	p-value=5.96E-5				
Base classifier	-		greater	greater	Risk min
Bagging		-			and
Risk min	smaller		-		MetaCost
MetaCost	smaller			-	
Polynomial-kernel SVM	p-value=1.96E-3				
Base classifier	-				
Bagging		-	greater	greater	Base
Risk min		smaller	-		classifier
MetaCost		smaller		-	
Gaussian-kernel SVM	p-value=3.64E-3				
Base classifier	-				
Bagging		-			Base
Risk min			-		classifier
MetaCost				-	

4.2   Naïve Bayes Classifier

In the NBayes classifier, computing the *Maximum a Posteriori Hypothesis* for the class involves estimating the class prior probabilities and the conditional probabilities. The imbalance mainly affects the prior probabilities, while the conditional probabilities are calculated for each class [9]. Therefore, although naïve Bayesian class-membership probability estimates are inaccurate [43], when test instances are ranked according to them, they tend to be ordered according to the class [9]. The decision threshold may thus benefit from an adjustment, which seems to have been the case in our results. Applying risk minimization to the NBayes classifier was statistically superior to all the approaches (Table 4). On the other hand, the fact that conditional probabilities are skew-independent makes the NBayes robust to the imbalance. Indeed, the plain NBayes led to the greatest F-measure in the most imbalanced dataset (Table 3).

Since the NBayes is a stable algorithm, strategies involving bagging are typically not suitable [6]. This is clear in the results, given that bagging had no benefit compared to the other methods, including the plain classifier. Moreover, MetaCost was outperformed by risk minimization.

4.3   Tree-Augmented Naïve Bayes classifier

Table 4 shows that the best methods for the TAN Bayes classifier were risk minimization and MetaCost, while bagging provided no benefit. As the NBayes, the TAN Bays classifier may benefit from an adjustment of the decision threshold, given the potential bias in prior probabilities. The effect of using risk minimization on this classifier has not been extensively studied in the literature. One study showed that the TAN Bayes was improved compared to the NBayes, when risk minimization or resampling followed by F-measure threshold optimization were employed [44]. However, if minority class dependencies are incorrectly modeled by the TAN Bayes network [10], shifting the threshold does not seem to be adequate. The effect of risk minimization on this learner is thus unpredictable. In our results, although risk minimization outperformed the plain classifier, it did not maximize F-measure in any dataset.

TAN Bayes classifiers are not good candidates for bagging [45]. However, imposing a tree structure to rare data can be too strong and lead to overfitting. In this case, bagging-based methods can be useful. This seems to have been the case, as MetaCost maximized the F-measure in the most imbalanced datasets. In the nearly-balanced datasets, CV_3Y and CV_4Y, the best results were obtained with bagging, possibly given the absence of a bias in the prior probabilities.

4.4   Support Vector Machines

Two behaviors were described for SVMs in imbalanced data [11]. If the imbalance is moderate, they perform well, while in severe imbalances, SVMs are likely to classify everything as majority. Plain SVMs were never outperformed by the other methods, indicating their robustness (Table 4). Indeed, the greatest CV imbalance is 15.7%, which is a moderate imbalance. In Gaussian-kernel SVMs, different datasets were better modeled by different methods. No method was statistically superior to the others.

SVMs do not predict class-membership probabilities with high accuracy. Furthermore, in case the data is non-separable, biasing the output of the model does not provide any benefit. Nonetheless, the highest values of F-measure for the most imbalanced datasets were obtained through risk minimization with polynomial-kernel SVMs, revealing that a learning bias was present. In spite of SVM stability, if the dataset is small or the minority is rare, SVMs can overfit the data and thus benefit from bagging [13]. A fact that may corrupt the success of bagging is that SVM parameters are optimized for one of the 10 rounds of CV. This optimization is lost when the training data is changed. This appears to have been the case of polynomial-kernel SVMs, as bagging did not improve the base classifier and was outperformed by the CS methods.

A final note goes to the validation results. They were consistent to the CV results for the DT and for the NBayes classifier. DT were always outperformed and the NBayes was preferentially improved by risk minimization. The results obtained for the TAN Bayes classifier and SVMs were not very consistent with the CV results. The TAN Bayes not improved by MetaCost, as was the case in CV. Possibly, the increase in the size of the training data reduced th instability that rendered this classifier suitable for ensembling. The SVMs benefited from risk minimization in the CV test and from MetaCost in the HO test.

4.5   Comparison of the Best Strategies for Each Classifier

The best strategies for each classifier were also compared through a Friedman test and Nemenyi pairwise comparisons (Table 5). All strategies outperformed strategies with DTs as base learner, which led to the lowest values of F-measure. Combining risk minimization with the NBayes classifier achieved greater values of F-measure than all other strategies except polynomial-kernel SVMs. Therefore, and given the efficiency and simplicity of the combination the NBayes with risk minimization, it is the preferred strategy.

Table 5. Results of the Friedman tests and post-hoc pairwise comparisons between the best strategies for each classifier. The p-value is 2.41E-41.

	DT+ Bag	DT+ Risk	DT+ MetaCost	NBayes+ Risk	TAN+ Risk	TAN+ MetaCost	Poly SVM	Gaussian SVM
DT+Bag	-			greater	greater	greater	greater	greater
DT+Risk		-		greater	greater	greater	greater	greater
DT+MetaCost			-	greater	greater	greater	greater	greater
NBayes+Risk	smaller	smaller	smaller	-	smaller	smaller		smaller
TAN+Risk	smaller	smaller	smaller	greater	-			
TAN+MetaCost	smaller	smaller	smaller	greater		-		
Polynomial SVM	smaller	smaller	smaller				-	
Gaussian SVM	smaller	smaller	smaller	greater				-

Dataset	F-measure							
CV_Diag	0.49	0.50	0.53	0.60	0.55	0.58	0.55	0.56
CV_2Y	0.55	0.58	0.62	0.69	0.65	0.65	0.65	0.65
CV_3Y	0.73	0.69	0.70	0.78	0.75	0.74	0.75	0.73
CV_4Y	0.61	0.60	0.63	0.69	0.66	0.66	0.71	0.71
Average	**0.59**	**0.59**	**0.62**	**0.69**	**0.65**	**0.66**	**0.67**	**0.66**

5 Conclusions

In this work, we examined the effect of the imbalance of neuropsychological data on DT, NBayes, TAN Bayes and SVMs, in the diagnosis and prognosis of dementia in patients with cognitive complaints. The most consistent results were obtained for DT and NBayes. The first learner benefited from any meta-learning/ensemble strategy, namely MetaCost, while the second is clearly improved by the risk minimization. As a consequence and given the good performances obtained by the NBayes classifier combined with risk minimization, this is our method of choice for reliable and predictable results in neuropsychological data. SVMs were robust to the imbalances, but it was not possible to conclude which method is the best match for it.

Directions for future work include the study of an assessment strategy that can avoid the optimism of the G-mean and complement the F-measure by focusing on the majority class. Moreover, it would be relevant to compare the presented results with a resampling method, such as SMOTE, and observe its effect on each classifier. A final comment goes to the other challenges posed by the data, such as the high attribute dimensionality, and the high percentage of missing values. A deeper study of these problems could reduce the effect of the class skew and make the neuropsychological tests more informative.

Acknowledgments: This work was partially supported by FCT (Fundação para a Ciência e a Tecnologia) under projects: PTDC/EIA-EIA/111239/2009 (NEURO-CLINOMICS - Understanding NEUROdegenerative diseases through CLINical and OMICS data integration) and PEst-OE/EEI/LA0021/2011 (INESC-ID multiannual funding). The work of AMC was partially supported by FCT under grant PEst-OE/EEI/ LA0008/2011 (IT multiannual funding).

References

1. Brookmeyer, R., Johnson, E., Ziegler-Graham, K., Arrighi, H.M.: Forecasting the global burden of Alzheimer's disease. Alzheimers Dementia the Journal of the Alzheimers Association 3(3), 186–191 (2007)
2. Alzheimer's Association: Alzheimer's Disease Facts and Figures. Technical report, Alzheimer's Association (2012)
3. Yesavage, J.A., O'Hara, R., Kraemer, H., Noda, A., Taylor, J.L., Rosen, A., Friedman, L., Sheikh, J., Derouesné, C.: Modeling the prevalence and incidence of Alzheimers disease and mild cognitive impairment. Journal of Psychiatric Research 36, 281–286 (2002)
4. Maroco, J., Silva, D., Rodrigues, A., Guerreiro, M., Santana, I., Mendonça, A.D.: Data mining methods in the prediction of Dementia: A real-data comparison of the accuracy, sensitivity and specificity of linear discriminant analysis, logistic regression, neural networks, support vector machines, classification trees and random forests. BMC Research Notes 4:299 (2011)
5. Lemos, L.: A data mining approach to predict conversion from mild cognitive impairment to Alzheimers Disease. Master's thesis, IST (2012)
6. Breiman, L.E.O.: Bagging Predictors. Machine Learning 24(2), 123–140 (1996)

7. Kearns, M., Valiant, L.: Cryptographic limitations on learning Boolean formulae and finite automata. Journal of the Association for Computing Machinery 41(1), 67–95 (1994)

8. Kubat, M., Matwin, S.: Addressing the curse of imbalanced training sets: one-sided selection. Training, 179–186 (1997)

9. Elkan, C.: The Foundations of Cost-Sensitive Learning. In: Int. Joint Conf. on Artificial Intelligence, vol. 17(1), pp. 973–978 (2001)

10. Sun, Y., Wong, A.K.C., Kamel, M.S.: Classification of Imbalanced Data: a Review. Int. Journ. of Pattern Recognition and Artificial Intelligence 23(04), 687–719 (2009)

11. Akbani, R., Kwek, S.S., Japkowicz, N.: Applying support vector machines to imbalanced datasets. In: Boulicaut, J.-F., Esposito, F., Giannotti, F., Pedreschi, D. (eds.) ECML 2004. LNCS (LNAI), vol. 3201, pp. 39–50. Springer, Heidelberg (2004)

12. Wu, G., Chang, E.Y.: Class-Boundary Alignment for Imbalanced Dataset Learning. In: ICML 2003 Workshop on Learning from Imbalanced Data Sets (2003)

13. Tao, D., Tang, X.: Assymmetric bagging and random subspace for support vector machines-based relevance feedback in image retrieval. IEEE Transactions on Pattern Analysis and Machine Intelligence 28(7), 1088–1099 (2006)

14. Sun, Y., Kamel, M.S., Wong, A.K.C., Wang, Y.: Cost-sensitive boosting for classification of imbalanced data. Pattern Recognition 40, 3358–3378 (2007)

15. Japkowicz, N.: The Class Imbalance Problem: Significance and Strategies. Complexity 1, 111–117 (2000)

16. McCarthy, K., Zabar, B., Weiss, G.: Does cost-sensitive learning beat sampling for classifying rare classes? In: Proceedings of the 1st Int. Work. on Utilitybased Data Mining, pp. 69–77. ACM Press, New York (2005)

17. Chawla, N.V., Bowyer, K.W., Hall, L.O., Kegelmeyer, W.P.: SMOTE: Synthetic Minority Over-sampling Technique. Journal of Artificial Intelligence Research 16(1), 321–357 (2002)

18. Han, H., Wang, W.-Y., Mao, B.-H.: Borderline-SMOTE: A New Over-Sampling Method in Imbalanced Data Sets Learning. In: Huang, D.-S., Zhang, X.-P., Huang, G.-B. (eds.) ICIC 2005. LNCS, vol. 3644, pp. 878–887. Springer, Heidelberg (2005)

19. Garcia, E.A.: ADASYN: Adaptive synthetic sampling approach for imbalanced learning. In: 2008 IEEE Int. Joint Conf. on Neural Networks (IEEE World Congress on Computational Intelligence), vol. (3), pp. 1322–1328 (June 2008)

20. Jo, T., Japkowicz, N.: Class imbalances versus small disjuncts. ACM SIGKDD Explorations Newsletter 6(1), 40–49 (2004)

21. Maloof, M.A.: Learning When Data Sets are Imbalanced and When Costs are Unequal and Unknown. Analysis 21(9), 1263–1284 (2003)

22. Breiman, L., Friedman, J.H., Stone, C.J., Olshen, R.A.: Classification and Regression Trees (1984)

23. Zadrozny, B., Langford, J., Abe, N.: Cost-Sensitive Learning by Cost-Proportionate Example Weighting. In: Third IEEE Int. Conf. on Data Mining, pp. 435–442 (2003)

24. Ting, K.M.: An instance-weighting method to induce cost-sensitive trees (2002)

25. Veropoulos, K., Campbell, C., Cristianini, N.: Controlling the Sensitivity of Support Vector Machines. Heart Disease, 55–60 (1999)

26. Bishop, C.M.: Pattern Recognition and Machine Learning. Information science and statistics, vol. 4. Springer (2006)

27. Domingos, P.: MetaCost: A General Method for Making Classifiers Cost-Sensitive. In: Proceedings of the Fifth Int. Conf. on Knowledge Discovery, pp. 155–164 (1999)

28. Duda, R.O., Hart, P.E., Stork, D.G.: Pattern Classification, 2nd edn. Wiley (2001)

29. Thai-nghe, N., Gantner, Z., Schmidt-thieme, L.: Cost-Sensitive Learning Methods for Imbalanced Data. In: The 2010 Int. Joint Conf. on Neural Networks, pp. 1–8 (2010)

30. Lawrence, S., Burns, I., Back, A., Tsoi, A.C., Giles, C.L.: Neural network classification and prior class probabilities. In: Orr, G.B., Müller, K.-R. (eds.) NIPS-WS 1996. LNCS, vol. 1524, pp. 299–314. Springer, Heidelberg (1998)

31. Kubat, M., Holte, R.C., Matwin, S.: Machine Learning for the Detection of Oil Spills in Satellite Radar Images. Machine Learning 30, 195–215 (1998)

32. Davis, J., Goadrich, M.: The relationship between Precision-Recall and ROC curves. In: Proceedings of the 23rd International Conference on Machine Learning, ICML 2006, pp. 233–240 (2006)

33. Silva, D., Guerreiro, M., Maroco, J.A., Santana, I., Rodrigues, A., Bravo Marques, J., de Mendonça, A.: Comparison of Four Verbal Memory Tests for the Diagnosis and Predictive Value of Mild Cognitive Impairment. Dementia and Geriatric Cognitive Disorders Extra 2(1), 120–131 (2012)

34. Garcia, C.: A Doença de Alzheimer, problemas do diagnóstico clínico. Phd, Universidade de Medicina de Lisboa (1984)

35. Hall, M.A.: Correlation-based Feature Selection for Machine Learning. Methodology 21i195-i20, 1–5 (1999)

36. Honghai, F., Guoshun, C., Cheng, Y., Bingru, Y., Yumei, C.: A SVM Regression Based Approach to Filling in Missing Values. In: Khosla, R., Howlett, R.J., Jain, L.C. (eds.) KES 2005. LNCS (LNAI), vol. 3683, pp. 581–587. Springer, Heidelberg (2005)

37. Friedman, N., Geiger, D., Goldszmidt, M.: Bayesian Network Classifiers. Machine Learning 29(1), 131–163 (1997)

38. Bradford, J., Kunz, C., Kohavi, R., Brunk, C.: Pruning Decision Trees with Misclassification Costs. In: Nédellec, C., Rouveirol, C. (eds.) ECML 1998. LNCS, vol. 1398, pp. 131–136. Springer, Heidelberg (1998)

39. Provost, F., Domingos, P.: Well-Trained PETs: Improving Probability Estimation Trees (2000)

40. Hall, M., Frank, E., Holmes, G., Pfahringer, B., Reutemann, P., Witten, I.H.: The WEKA data mining software: an update. SIGKDD Explorations 11(1), 10–18 (2009)

41. Demsar, J.: Statistical Comparison of Classifiers over Multiple Data Sets. Journal of Machine Learning Research 7(7), 1–30 (2006)

42. Sheskin, D.J.: Handbook of Parametric and Nonparametric Statistical Procedures, vol. 51. CRC Press (1997)

43. Domingos, P., Pazzani, M.: Beyond independence: Conditions for the optimality of the simple Bayesian classifier. Machine Learning 29(2/3), 105–112 (1997)

44. Thai-nghe, N., Schmidt-thieme, L., Techniques, A.M.: Learning Optimal Threshold on Resampling Data to Deal with Class Imbalance. In: 8th IEEE Int. Conf. on Computing and Communication Technologies: Research, Innovation, and Vision for the Future (2010)

45. Quinn, C.J., Coleman, T.P., Kiyavash, N.: Approximating discrete probability distributions with causal dependence trees (2010)

Appendix A. Misclassification Cost Ratios

Table 6. MCRs that maximized the value of F-measure in the CS methods

classifier	dataset	Risk min	MetaCost
DT	CV_Diag	7.75	5
	CV_2Y	6.	4
	CV_3Y	5.25	5
	CV_4Y	4.5	3.5
	Average	5.88	4.38
NBayes	CV_Diag	1.75	1
	CV_2Y	7.5	4
	CV_3Y	7.75	6.75
	CV_4Y	1.5	4.75
	Average	4.63	4.13
TAN	CV_Diag	3.5	1
	CV_2Y	2.75	4
	CV_3Y	1.25	6.75
	CV_4Y	3	4.75
	Average	2.63	4.13
Polynomial-kernel SVM	CV_Diag	1.5	1.25
	CV_2Y	2.5	2.5
	CV_3Y	2	1.5
	CV_4Y	1.25	2.5
	Average	1.82	1.94
Gaussian-kernel SVM	CV_Diag	4.5	2.5
	CV_2Y	2.75	2.5
	CV_3Y	2.5	2.25
	CV_4Y	1.25	1.5
	Average	2.75	2.19

Appendix B. Validation Results Using Optimal MCRs

Table 7. Values of F-measure obtained by training the models on the CV datasets and testing them on the corresponding HO datasets. MCRs in Table 6 were used.

classifier	dataset	Base classifier	Bagging	Risk min	MetaCost
DT	HO_Diag	0.444	**0.5**	0.367	0.391
	HO_2Y	0.267	0.364	**0.4**	0.36
	HO_3Y	0.7	0.776	0.795	**0.815**
	HO_4Y	0.595	**0.629**	0.618	0.6
NBayes	HO_Diag	**0.516**	0.5	**0.516**	0.457
	HO_2Y	0.429	0.429	**0.556**	0.529
	HO_3Y	0.818	0.818	**0.824**	**0.824**
	HO_4Y	0.647	0.686	**0.722**	0.718
TAN	HO_Diag	**0.6**	0.522	0.48	0.5
	HO_2Y	0.581	0.417	**0.6**	0.565
	HO_3Y	0.818	0.769	**0.836**	0.753
	HO_4Y	0.667	**0.686**	0.651	0.619
Polynomial-kernel SVM	HO_Diag	0.5	0.5	0.467	**0.552**
	HO_2Y	0.435	0.381	**0.632**	0.615
	HO_3Y	0.721	0.781	**0.806**	**0.806**
	HO_4Y	0.667	0.667	0.667	**0.684**
Gaussian-kernel SVM	HO_Diag	0.48	0.48	0.488	**0.533**
	HO_2Y	0.435	0.455	0.6	**0.615**
	HO_3Y	0.733	0.762	0.806	**0.841**
	HO_4Y	0.647	0.667	0.647	**0.706**

A Noise Removal Algorithm
for Time Series Microarray Data

Naresh Doni Jayavelu and Nadav Bar

Systems Biology Group, Department of Chemical Engineering,
Norwegian University of Science and Technology (NTNU), Trondheim, NO-7491,
Norway
nareshd@chemeng.ntnu.no

Abstract. High-throughput technologies such as microarray data are a
great resource for studying and understanding biological systems at a low
cost. However noise present in the data makes it less reliable, and thus
many computational methods and algorithms have been developed for
removing the noise. We propose a novel noise removal algorithm based on
Fourier transform functions. The algorithm optimizes the coefficients of
the first and second order Fourier functions and selects the function which
maximizes the Spearman correlation to the original data. To demonstrate
the performance of this algorithm we compare the prediction accuracy of
well known modelling tools, such as network component analysis (NCA),
principal component analysis (PCA) and k-means clustering. We compared the performance of these tools on the original noisy data and the
data treated with the algorithm. We performed the comparison analysis
using three independent real biological data sets (each data set with two
replicates). In all cases the proposed algorithm removes the noise in the
data and substantially improves the predictions of modelling tools.

Keywords: Microarray time series data, Noise, Smoothing, Fourier,
Network component analysis, Principal component analysis, clustering.

1 Introduction

High-throughput technologies such as microarray have emerged as promising
tools for studying, modelling and understanding complex biological systems at
a low cost. Most often these data sets are prone to noise. This noise arises from
stochastic variations in the experiments and changes in the biological processes,
for example during sample preparation and hybridization processes [14]. The
major challenge in this microarray analysis is to separate the true biological signals from the noisy measurements. Prediction abilities (knowledge discovery) of
modelling tools as network component analysis (NCA)[6], principal components
analysis [8] and clustering [15] based on these data sets depend heavily on the
amount of noise present [11].

There are attempts in the literature to quantify and remove the noise in the
microarray [14]. One approach is to replicate the measurements several times,

L. Correia, L.P. Reis, and J. Cascalho (Eds.): EPIA 2013, LNAI 8154, pp. 152–162, 2013.
© Springer-Verlag Berlin Heidelberg 2013

an expensive approach that requires manpower, time and resources. There are many mathematical models developed for noise removal and smoothing of data [7]. Most of these models are based on the assumption that data is approximately Gaussian distributed [5]. However, Hardin et al [4] reported that microarray data does not necessarily need to satisfy this assumption. There are several other models available. For example Tang et al [13] described a singular value decomposition combined with spectral analysis for noise filtering.

In this article we developed an algorithm based on the Fourier transform function for removing noise and smoothing of the gene expression data. The expression values of the genes in time series experiments are known to follow a specific trends depending on the type of treatment given to the cells. For example, stimuli-response experiments are characterized by transient responses and in cell cycle experiments cyclic patterns are observed. In general, gene expressions follow two simple response shapes: either short impulses or long sustained responses [1]. We fitted individual time series gene expression data using an optimization algorithm that estimate the parameters of first and second order Fourier series functions, and constrains the functions to these two shapes or their combinations. The fitted functions represent the smooth approximations to the original expression values and thus remove the noise. We have applied our algorithm on three real biological microarray data sets. The first data set is gastrin responsive transcriptome data measured at 11 time points. The other two are epidermal growth factor (EGF) and heregulin (HRG) stimuli-response experiments with 17 time points. The data sets can be downloaded from the Array Express Website with accession numbers: GSE32869 and GSE13009 respectively for gastrin and epidermal growth factor and heregulin. We showed that predictions of NCA and PCA on data with noise removed are more consistent and accurate, and the clusters from k-means are tighter and distinct from each other.

2 Methods

Prior to the expression data fitting, we performed the selection of the differentially expressed genes based on fold change and p-value including normalization. The noise reduction algorithm fits the data (expression values of each gene) with Fourier functions and smoothing splines. The function that maximizes the Spearman correlation between the fitted curve and the original noisy data at the given time points is chosen.

2.1 Fourier Series

The Fourier series is a sum of trigonometric functions that describes a periodic signal. The fitting procedure uses a nonlinear least squares regression to estimate the parameters of a Fourier function. This optimization algorithm is a robust Bi-Square fitting method that creates the curve by minimizing the summed square of the residuals, and reduces the weight of outliers using Bi-square weights.

The optimization is performed using a Trust-Region method with coefficient constraints. The Fourier function is represented here in its trigonometric form:

$$f(x) = a_0 + \sum_{i=1}^{n} a_i \cos(n\omega x) + b_i \sin(n\omega x) \qquad (1)$$

Where a_0, a_i, and b_i are the parameters to be estimated, a_0 is the constant term of the data and is associated with the i = 0 cosine term, ω is the fundamental frequency of the signal and $1 \leq n \leq \infty$ is the number of harmonics in the series. Since the temporal pattern of gene expression is known to be wave like or impulse-like shapes, the only harmonics we expect during a time span of less than 72 hours are $n = 1$ or $n = 2$, corresponding to the superposition of one or two harmonics, respectively. The fundamental frequency of the function is not expected to be larger than the time span of the micro array experiment, so the fitted function is most likely to consist of one or two peaks. The reason behind this assumption is that not many genes experience cyclic behavior of more than two cycles during a short, several hours time span.

For practical reasons, we use discrete Fourier transform (DFT) to estimate the above parameters, since the time points are discrete and finite. To compute the DFT, we used the fast Fourier transform algorithm in Matlab, Mathworks

2.2   Smoothing Spline

When the Fourier series function is poorly correlated (Spearman correlation) to the data, we fitted the expression data with smoothing spline function in Matlab

2.3   Network Component Analysis

The NCA [6]is a computational method for reconstructing the hidden regulatory signals (activity profiles of transcription factors) from gene expression data with known connectivity information in terms of matrix decomposition. The NCA method can be represented in matrix form as follows:

$$[E] = [C][T] \qquad (2)$$

where the matrix E represents the expression values of genes over time points, the matrix C is the control strength of each transcription factor on gene and matrix T represents the activity profiles of transcription factors. The dimensions of E, C and T are $N \times M$ (N is the number of genes and M is the number of time points or measurement conditions), $N \times L$ (L is the number of transcription factors), $L \times M$, respectively.

Based on above formulation, the decomposition of $[E]$ into $[C]$ and $[T]$ can be achieved by minimizing the following objective function:

$$min\|[E] - [C][T]\|^2, s.t. C \in Z_0$$

where Z_0 is the initial connectivity pattern. The estimation of $[C]$ and $[T]$ is performed by using a two-step least-squares algorithm. With NCA as reconstruction method, we predicted significant TFs and their activity profiles.

2.4 Principal Component Analysis

The PCA [9] is a model reduction method that reduces the dimensionality (number of variables) of a data set by preserving as much variance as possible. PCA rotates the original data space such that the axes of the new coordinate system point into the directions of highest variance of the data. The axes or new variables are termed principal components (PCs) and are ordered by variance. We start with the expression matrix, E, where each row corresponds to genes and each column corresponds to different measurement conditions or time points at which cells were treated with a stimuli. The E_{it} entry of the matrix contains the expression value of gene i at condition t. The principal components are computed as:

$$e'_{ij} = \sum_{t=1}^{n} e_{it} v_{tj} \tag{3}$$

Where E_{tj} is the ith coefficient for the jth principal component. e_{it} is the expression value for gene i under t^{th} condition. E' is the data in terms of principal components.

2.5 K-means Clustering

K-means clustering is a powerful technique often employed in gene expression analysis for elucidating a variety of biological inferences. We selected the Pearson correlation as similarity measure and repeated the algorithm with 10 different initial randomizations (initial cluster centroid positions) to avoid local minima and chose the one with smallest sum of point to centroid distances. Selecting the correct number of clusters for a given data is always critical in k-means clustering so we considered a range of 3 to 15 clusters. We chose sum of point to centroid distances within the cluster, 'sumd' (tightness within cluster) and silhouette measure (separation in-between clusters) as clustering evaluation criteria [10, 2, 15].

The efficiency of the noise reduction (NR) algorithm is evaluated by the ability of prediction tools to reproduce the same temporal behaviors, whether it is activity profile of transcription factors (for NCA) or gain in cumulative variance (for PCA) and tightness and separation (for k-means clustering) from original noisy data (N) and data treated with algorithm (NR) . All computations were performed using Matlab, Mathworks.

3   Results and Discussion

3.1 NR Algorithm

The performance of the proposed algorithm (we term it as NR algorithm) in terms of removing the noise from the time series expression data are presented in Figure 1. This algorithm fits the gene expression data with Fourier 1, (impulse response), Fourier 2 (sustained response) or smoothing spline (complex patterns) and chooses the best one based on high correlation between the original and fitted data. In the majority of the genes, noise is repressed by smoothing out strong fluctuations. For instance, the gene expression of ANKZF1 is noisy at late time samples and this noise is removed with NR algorithm (Figure 1). Most of the genes temporal patterns are approximated with either Fourier 1 (%14 of total) or Fourier 2 (%58 of total) in gastrin data set.

3.2   Application to Modelling Tools

In this section, we illustrate the importance of noise removal in the gene expression data and performance of the noise reduction (NR) algorithm on predictive abilities of several modelling tools such as network component analysis (NCA), principal component analysis (PCA) and k-means clustering.

3.3   Network Component Analysis

NCA is a computational method applied on gene expression data to reconstruct the activity profiles of important transcription factors involved in the gene regulatory network. We demonstrated the performance of the NR algorithm on prediction ability of NCA by comparing the results of NCA applied on original noisy data (N) and data treated with NR algorithm (NR). The NCA predicted AP1 activity from NR treated data (in gastrin regulated system) showed peak activation at 4 hours (6th time sample) and is in accordance with the previous experiments [3, 12](Figure 2). In contrast, AP1 activity from original noisy data displayed peak activation at 2 (5th time sample) hours. Another measure to assess the performance of the NR algorithm is the similar reconstructed activity profiles from two independent measurements (replicates, see Figure 2). The similarity measure considered here is the Pearson correlation coefficient between profiles from two replicates. We computed the Pearson correlation for all the transcription factor activity profiles from the noisy replicate data sets and the replicate data sets treated with NR algorithm (Table 1). NCA predicted similar activity profiles (from two replicates) for the noise treated data sets. In contrast NCA predicted dissimilar profiles for the original noisy replicate data sets. The low correlation (0.07) between the AP1 activities predicted from two noisy EGF replicates is increased by a factor of 7.5 (to 0.55) after applying the NR algorithm. Similar results are found for CEBPG, TFAP2A, USF1, NKX21 and PAX6. This correlation analysis depicted that at least 80% of all the predicted activity profiles of transcription factors displayed improved correlations (between replicates) after treating the data with our NR algorithm in all three data sets gastrin, EGF and HRG (Figure 3).

3.4   Principal Component Analysis

PCA is a multi variate data analysis method to reduce the dimensionality of data (i.e. number of variables) while maximizing the variance. We applied PCA on

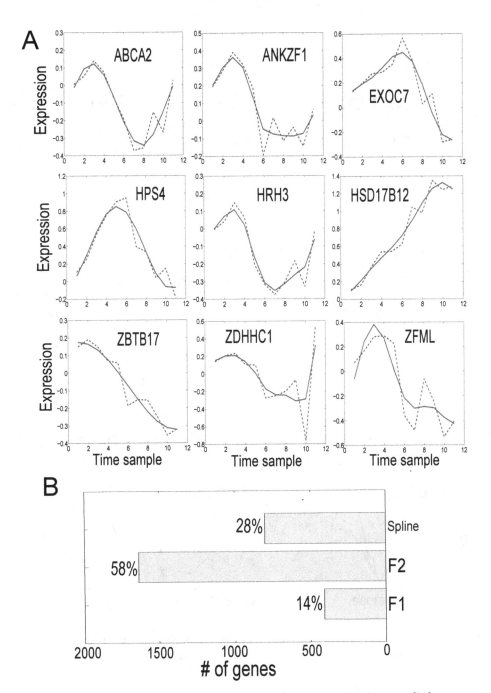

Fig. 1. (A) Fourier functions and smoothing spline approximations applied to gene expression data. Dotted blue lines denote original noisy data and solid red lines denote the fitted approximation. (B) Distribution of number of genes approximated by Fourier 1 (F1), Fourier 2 (F2) and smoothing spline.

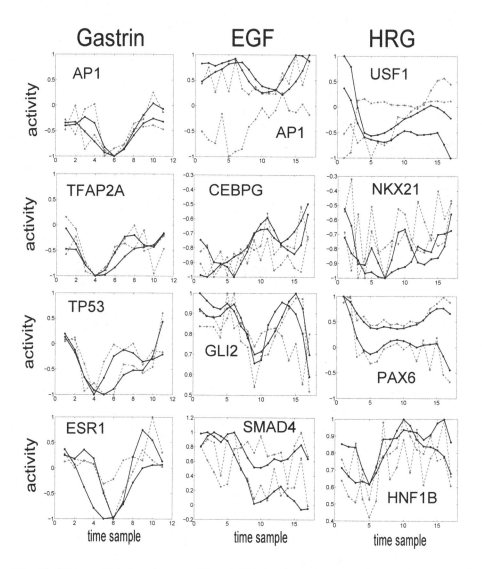

Fig. 2. NCA predicted activity profiles of TFs in three data sets (Gastrin, EGF and HRG) from two replicates. Black solid lines represent the noise reduced replicates and the dotted red lines represent the original noisy replicates.

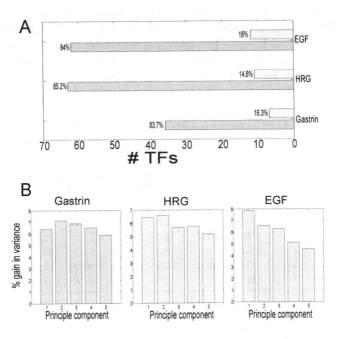

Fig. 3. (A) NCA results: Distribution of number of TFs with improved predictions (long bar) and unimproved (short bar) in three data sets. (B) PCA results: Gain in cumulative variance (difference in variance between noisy (N) and noise reduced data (NR)) associated with principal components in percent.

Table 1. Pearson correlation coefficient of the NCA predicted activity profiles of TFs between two replicates of the noise (N) and the noise reduced (NR) data sets

Data set	Transcription factor	Noise data (N)	Treated data NR	Improvement (fold)
Gastrin	AP1	0.395	0.9	2.27
	TFAP2A	0.473	0.738	1.56
	TP53	0.441	0.648	1.47
	ESR1	0.528	0.742	1.4
EGF	AP1	0.073	0.552	7.52
	CEBPG	0.03	0.645	21.37
	GLI2	0.125	0.795	6.32
	SMAD4	0.143	0.827	5.77
HRG	USF1	0.163	0.863	5.29
	NKX21	0.108	0.451	4.15
	PAX6	0.253	0.54	2.13
	HNF1B	0.361	0.69	1.9

the original noisy data (N) and NR algorithm treated data (NR) and compared the results. PCA reduced the NR treated data with at least 5% more cumulative variance associated with up to five principal components (PCs) than the noisy data (Figure 3 and Table 2) and this gain in variance is even more with fewer PCs. Similar results are obtained with all the three data sets (each with two replicates), which demonstrates the improvement of PCA analysis.

Table 2. Results of PCA applied on noisy (N) and noise reduced (NR) data sets on three systems. The values in the table represent the cumulative variance associated with principal components (PCs).

Principal component	Gastrin N	NR	EGF N	NR	HRG N	NR
1	60.9	67.3	72.3	80.1	61.1	67.5
2	74.6	81.7	82.6	89.2	75.6	82.2
3	86.9	93.7	88.3	94.6	84.3	90.0
4	90.1	96.6	91.5	96.6	89.9	95.6
5	92.1	98.0	93.3	97.9	92.6	97.8

3.5   K-means Clustering

K-means clustering is a powerful technique often employed in gene expression analysis for elucidating a variety of biological inferences such as shared regulation

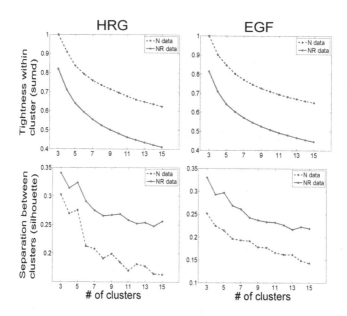

Fig. 4. K-means clustering: Statistical measures of sumd and silhouette are compared between noisy (N) and noise reduced(NR) data sets

or particular biological processes [2]. The k-means clustering applied in this analysis is generally sensitive to noisy data [11]. The Tightness within the cluster measure (sumd) in NR treated data is low compared to the noisy data for all cases of 3 to 15 clusters (Figure 4) in both EGF and HRG data sets. Smaller values of sumd corresponds to tighter clusters. This measure diminishes gradually from 3 clusters to 15 clusters (Figure 4) in both EGF and HRG systems. Another evaluation measure of silhouette is found to be higher in NR treated data sets (NR) than noise data (N). This indicates that NR treated data provides a clear separation of clusters in all 3 to 15 cluster cases, thus improving the performance of k-means clustering.

4   Conclusion

Noise is common in microarray gene expression data and reducing it is essential prior to the inference of knowledge from modelling which utilizes this data. The proposed NR algorithm fits the expression data with either Fourier 1, 2 or smoothing spline based on the temporal pattern and thus minimizes the noise. The performance of the algorithm is demonstrated based on improved prediction capabilities of several modelling tools. NCA reconstructed the very similar activity profiles of transcription factors from two replicate data sets after treating with the NR algorithm. PCA reduced the data with more cumulative variance in the NR treated data than noisy data. K-means clustering also performed better (in terms of statistical measures) with NR treated data.

References

[1] Bar-Joseph, Z., Gitter, A., Simon, I.: Studying and modelling dynamic biological processes using time-series gene expression data. Nat. Rev. Genet. 13(8), 552–564 (2012)

[2] D'haeseleer, P.: How does gene expression clustering work. Nat. Biotech. 23(12), 1499–1501 (2005), http://dx.doi.org/10.1038/nbt1205-1499

[3] Guo, Y.S., Cheng, J.Z., Jin, G.F., Gutkind, J.S., Hellmich, M.R., Townsend, C.M.: Gastrin stimulates cyclooxygenase-2 expression in intestinal epithelial cells through multiple signaling pathways: evidence for involvement of erk5 kinase transactivation of the epidermal growth factor. Journal of Biological Chemistry 277(50), 48755–48763 (2002)

[4] Hardin, J., Wilson, J.: A note on oligonucleotide expression values not being normally distributed. Biostatistics 10(3), 446–450 (2009), http://biostatistics.oxfordjournals.org/content/10/3/446.abstract

[5] Lewin, A., Bochkina, N., Richardson, S.: Fully bayesian mixture model for differential gene expression: simulations and model checks. Statistical Applications in Genetics and Molecular Biology 6 (2007)

[6] Liao, J.C., Boscolo, R., Yang, Y.L., Tran, L.M., Sabatti, C., Roychowdhury, V.P.: Network component analysis: reconstruction of regulatory signals in biological systems. Proc. Natl. Acad. Sci. U S A 100(26), 15522–15527 (2003)

[7] Posekany, A., Felsenstein, K., Sykacek, P.: Biological assessment of robust noise models in microarray data analysis. Bioinformatics (2011), http://bioinformatics.oxfordjournals.org/content/early/2011/01/19/bioinformatics.btr018.abstract

[8] Raychaudhuri, S., Stuart, J.M., Altman, R.B.: Principal components analysis to summarize microarray experiments: Application to sporulation time series. In: Pac. Symp. Biocomput., pp. 452–463 (2000)

[9] Ringner, M.: What is principal component analysis. Nat. Biotech. 26(3), 303–304 (2008), http://dx.doi.org/10.1038/nbt0308-303

[10] Rousseeuw, P.J.: Silhouettes: A graphical aid to the interpretation and validation of cluster analysis. Journal of Computational and Applied Mathematics 20, 53–65 (1987), http://www.sciencedirect.com/science/article/pii/0377042787901257

[11] Sloutsky, R., Jimenez, N., Swamidass, S.J., Naegle, K.M.: Accounting for noise when clustering biological data. Brief Bioinform. (October 2012), http://dx.doi.org/10.1093/bib/bbs057

[12] Subramaniam, D., Ramalingam, S., May, R., Dieckgraefe, B.K., Berg, D.E., Pothoulakis, C., Houchen, C.W., Wang, T.C., Anant, S.: Gastrin-mediated interleukin-8 and cyclooxygenase-2 gene expression: Differential transcriptional and posttranscriptional mechanisms. Gastroenterology 134(4), 1070–1082 (2008)

[13] Tang, V., Yan, H.: Noise reduction in microarray gene expression data based on spectral analysis. International Journal of Machine Learning and Cybernetics 3, 51–57 (2012), http://dx.doi.org/10.1007/s13042-011-0039-7

[14] Tu, Y., Stolovitzky, G., Klein, U.: Quantitative noise analysis for gene expression microarray experiments. Proceedings of the National Academy of Sciences 99(22), 14031–14036 (2002), http://www.pnas.org/content/99/22/14031.abstract

[15] Warren Liao, T.: Clustering of time series data-a survey. Pattern Recogn. 38(11), 1857–1874 (2005), http://dx.doi.org/10.1016/j.patcog.2005.01.025

An Associative State-Space Metric for Learning in Factored MDPs

Pedro Sequeira, Francisco S. Melo, and Ana Paiva

INESC-ID and Instituto Superior Técnico, Technical University of Lisbon
Av. Prof. Dr. Cavaco Silva, 2744-016 Porto Salvo, Portugal
pedro.sequeira@gaips.inesc-id.pt, {fmelo,ana.paiva}@inesc-id.pt

Abstract. In this paper we propose a novel *associative metric* based on the classical conditioning paradigm that, much like what happens in nature, identifies associations between stimuli perceived by a learning agent while interacting with the environment. We use an associative tree structure to identify associations between the perceived stimuli and use this structure to measure the degree of similarity between states in factored Markov decision problems. Our approach provides a *state-space metric* that requires no prior knowledge on the structure of the underlying decision problem and is designed to be learned online, *i.e.*, as the agent interacts with its environment. Our metric is thus amenable to application in reinforcement learning (RL) settings, allowing the learning agent to generalize its experience to unvisited states and improving the overall learning performance. We illustrate the application of our method in several problems of varying complexity and show that our metric leads to a performance comparable to that obtained with other well-studied metrics that require full knowledge of the decision problem.

1 Introduction

Associative learning is a paradigm from the field of behaviorism that posits that learning occurs whenever a change in behavior is observed [1]. Classical conditioning is one of the best-known associative learning paradigms. It is one of the most basic survival tools found in nature that allows organisms to expand the range of contexts where some of their already-known behaviors can be applied. By associating co-occurring stimuli from the environment, the organism can activate innate phylogenetic responses (*e.g.*, fight or flight responses) to new and previously unknown situations.

In this paper, we leverage this idea to reinforcement learning (RL). RL agents explore their environment and gather information that allows them to learn the best actions to take in different situations. Many classical RL methods, such as Q-learning, allow the agent to successively estimate how good each action is in every state, eventually conveying to the agent the information necessary to select only the best actions in all states. This typically requires the agent to experience every action in every state a sufficient number of times [2]. This need for "sufficient" visits to every state-action pair is often impractical, particularly in large environments, and several general approaches have been proposed to

L. Correia, L.P. Reis, and J. Cascalho (Eds.): EPIA 2013, LNAI 8154, pp. 163–174, 2013.
© Springer-Verlag Berlin Heidelberg 2013

mitigate this need, relying mostly on some form of function approximation (we refer to [3] for references and discussion).

However, certain scenarios present some particular structure that can be leveraged by the learning algorithm to improve the learning performance—namely, by alleviating the requirement of sufficient visits to every state-action pair. For example, in scenarios where the state is described by a finite set of state-variables (*i.e.*, where the state is *factored*), it is often possible to use this structure to improve the efficiency of RL [4]. This is particularly true if many of the state-variables are irrelevant for the task that to be learned, and it is possible to improve the learning performance by identifying such irrelevant state-variables, allowing the learning agent to focus only on those that are relevant [5,6].

Our approach builds on all aforementioned ideas. We introduce a method that allows the learning agent to identify *associations* between perceived stimuli during its interaction with the environment. Specifically, given a learning scenario with a factored state space, we use a pattern mining technique to build an *association tree* that identifies the occurrence of frequent *patterns* of state-variables (henceforth referred as *stimuli*) [7]. These associations are similar in spirit to those that natural organisms identify in their interaction with the environment, and are then used by the agent to build a metric that identifies two states as being "close" if they share multiple/frequent stimuli. This metric is learned online and combined with Q-learning, as proposed in [8] to improve the learning performance of our agents and use current information to update the value of states that are considered *similar* according to the associative metric.

The main contribution of our approach is to provide a general-purpose state-space metric that requires *no prior knowledge* on the structure of the underlying decision problem. The associative tree and the similarity metric are both learned online, *i.e.*, while the agent is interacting with its environment, making it particularly amenable to use in a reinforcement learning setting. We illustrate the application of our method in several factored Markov decision processes (MDPs) of varying complexity and show that our metric leads to a performance comparable to that obtained when using well-studied metrics from the literature [9].

2   Background

In this section we introduce the necessary background on both the biological and computational concepts that will be used throughout the paper.

2.1   Reinforcement Learning

The field of *reinforcement learning* (RL) addresses the general problem of an agent faced with a sequential decision problem [2]. By a process of trial-and-error, the agent must learn a "good" mapping that assigns states to actions. Such mapping determines how the agent acts in each possible situation and is commonly known as a *policy*. In a sense, reinforcement learning is the computational counterpart to the notion of *reinforcement* used in operant conditioning and behavior analysis [2,10].

RL agents can be modeled using *Markov decision processes* (MDPs). At every step t, the agent/environment is in state $X(t) = x$, with $x \in \mathcal{X}$ and chooses an action $A(t) = a$, with $a \in \mathcal{A}$. Both $\mathcal{X}$ and $\mathcal{A}$ are assumed finite. Given that $X(t) = x$ and $A(t) = a$, the agent/environment transitions to state $y \in \mathcal{X}$ with probability given by

$$\mathsf{P}(y \mid x, a) \triangleq \mathbb{P}\left[X(t+1) = y \mid X(t) = x, A(t) = a\right]$$

and receives a reward $r(x, a)$, and the process repeats. The agent must choose its actions so as to gather as much reward as possible, discounted by a positive discount factor $\gamma < 1$. Formally, this corresponds to maximizing the value

$$v = \mathbb{E}\left[\sum_t \gamma^t r(X(t), A(t))\right], \tag{1}$$

where, as before, $X(t)$ and $A(t)$ denote the state and action at time-step t, respectively. The reward function r implicitly encodes the *task* that the agent must accomplish. It is a well-known fact that in (finite) MDPs it is possible to find a *policy* $\pi^* : \mathcal{X} \to \mathcal{A}$ maximizing the value in (1). Associated with the optimal policy π^* is the *optimal Q-function*,

$$Q^*(x, a) = \mathbb{E}\left[\sum_t \gamma^t r(X(t), A(t)) \mid X(0) = x, A(0) = a\right],$$

from which the optimal policy can easily be computed [2].

In many MDPs the state $X(t)$ can be described using a finite set of *state features* $X_i(t), i = 1, \ldots, n$, each taking values in some feature space $\mathcal{X}_i$. The state-space thus corresponds to the cartesian product $\mathcal{X} = \mathcal{X}_1 \times \ldots \times \mathcal{X}_n$. The structure exhibited by such *factored MDPs*, both in terms of transition probabilities and reward function, can often be exploited, leading to more efficient solution methods [4,11]. The computational gains can be particularly noteworthy if many of the state-features are *irrelevant* for the underlying task to be solved by the agent. In fact, it is possible to greatly improve the performance of solution methods by identifying such irrelevant state-features and focusing only on those that are relevant [5,6]. In this paper, we refer to an element $x = (x_1, \ldots, x_n)$ as a state and to an element $x_i \in \mathcal{X}_i$ as a *stimulus*. We consider stimuli as *categorical nominal data*, *i.e.*, variables that describe discrete values.

If the MDP model is known, the function Q^* can easily be computed using, for example, dynamic programming. However, in RL settings, the dynamics P and reward r of the MDP model are typically unknown. The agent must thus *learn* Q^* through interactions with its environment. This can be achieved using, for example, the *Q-learning algorithm* [10], that updates the estimate for Q^* as

$$\hat{Q}(x(t), a(t)) \leftarrow (1 - \alpha_t)\hat{Q}(x(t), a(t)) + \alpha_t(r(t) + \gamma \max_b \hat{Q}(x(t+1), b)), \tag{2}$$

where $x(t)$ and $a(t)$ are the state and actions experienced (sampled) at time t, $r(t)$ is the received reward and $x(t+1)$ is the subsequent state. *Q*-learning is

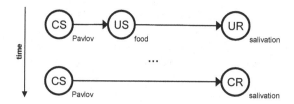

Fig. 1. Example of classical conditioning in a dog, inspired in Pavlov's experiments [12]

guaranteed to converge with probability 1 as long as every state-action pair is visited infinitely often and the step-size sequence, $\{\alpha_t\}$, verifies standard stochastic approximation conditions.

The need for infinite visits to every state-action pair is unpractical in many situations, and several general approaches have been proposed to mitigate this need. In this paper we adopt a simple technique proposed in [8], where Q-learning is combined with a *spreading function* that "spreads" the estimates of the Q-function in a given state to neighboring states. Formally, given a similarity function $\sigma_t(x, y)$ that measures how close two states x and y are, the Q-learning with spreading update is given by

$$\hat{Q}(x, a(t)) \leftarrow (1 - \alpha_t)\hat{Q}(x, a(t)) + \alpha_t \sigma_t(x, x(t))\big(r(t) + \gamma \max_b \hat{Q}(x(t+1), b)\big). \quad (3)$$

As discussed in [8], convergence of Q-learning with spreading to the optimal Q-function can be guaranteed as long as the spreading function σ_t converges to the Kronecker delta-function at a suitable rate.[1]

2.2   Classical Conditioning

Figure 1 illustrates a typical setting for a classical conditioning experimental procedure. In a first phase, known as *initial pairing* or *training*, an organism's biologically significant *unconditioned stimulus* (US) is paired with a neutral, biologically meaningless stimulus, called the *conditioned stimulus* (CS) [1, 12]. The US—for example food or an electrical shock,—reflexively evokes innate, automatic unconditioned responses (UR)—for example, salivating or freezing. The neutral CS can be any event that does not result in an overt behavioral response from the organism under investigation (*e.g.*, the sound of a bell, a light or even the presence of a person). In a second phase (*testing*), and after a few pairings between the US and CS, have occurred, the experimenter measures the level of response from the organism when exposed to the CS alone, with no US being presented. The experimenter typically observes a change in response from the organism in the presence of the CS, which now evokes a conditioned response (CR) similar to the UR evoked by the US.

Following the example in Fig. 1, the presence of Pavlov alone made the dogs start salivating in anticipation of food delivery. This change in response is due to the development of an *association* between a representation of the CS and one of the US, arising from the *co-occurrence of both stimuli*. This is the main idea

---

[1] Actually, the algorithm described in [8] also considers spreading across actions. In this paper we address only spreading across states.

behind Pavlov's *stimulus substitution theory* [12], where the CS "substitutes" the US in evoking the reflexive response.

The evolutionary advantages behind such associative mechanism could be the ability of organisms to broaden the contexts where they apply some advantageous response, and to anticipate the biological significance of co-occurring events [13]. By determining associations between stimuli in the environment, animals are able to: (i) recognize contexts (states) of the environment and thus anticipate rewards or punishments and consequences of behavior that are similar to those observed in previous interactions; (ii) integrate information from previous observations with new, never before experienced stimuli.

Inspired in these ideas from classical conditioning, our learning approach: (i) spreads action and reward information (the Q-values) between similar states; (ii) integrates information in new, unknown states, from the Q-values of previously experienced similar states.

3   Associative Metric for RL

In this section we introduce a new associative metric to be used in factored MDPs. We take inspiration in the classical conditioning paradigm introduced in the previous section, and port some of the underlying principles into an RL context, effectively improving the performance of RL agents.

To better explain our learning procedure let us consider a behavior phenomenon associated with the classical conditioning paradigm known as *secondary conditioning* or *sensory preconditioning* as an example to follow throughout this section. Secondary conditioning takes place whenever a CS (CS1) that is trained to predict some US is paired with a different CS (CS2), either before or after CS1 and US are paired. By means of this secondary association, CS2 also becomes associated with the US value through its association with CS1, and ends up evoking the same kind of CR [14]. Figure 2 illustrates an example of the secondary conditioning phenomenon, where, for explanatory purposes, we consider that the stimuli come from different perceptual modalities.

Biologically speaking, after being trained with sound-shock pairings followed by sound-light pairings, the agent should be able to predict the presence of the shock whenever it perceives the light, even if the two stimuli never co-ocurred. From a more computational perspective, the learning procedure should discover that environmental states involving light and shock are somehow *associated*. In this manner, whatever value is associated with the stimulus "shock" should, to some extent, also be associated with "light", and the outcome of executing similar actions in associated states should, to some extent, be similar.

We can therefore decompose the learning problem into two sub-problems: *identifying associated states* and *using the information* about some experienced states in other (associated) states. Sections 3.1 and 3.2 describe our approach in addressing the first sub-problem, where we propose the combination of a sensory pattern tree and a new associative metric to measure the distance between associated states. In Section 3.2 we discuss how this metric can then be combined with Q-learning with spreading to improve the performance of an RL agent.

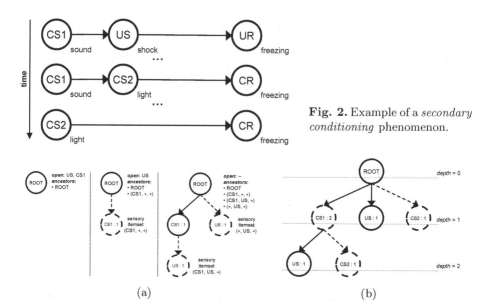

Fig. 2. Example of a *secondary conditioning* phenomenon.

(a)   (b)

Fig. 3. The construction of an associative sensory tree. The updated and inserted nodes at each step are marked with a dashed line. (a) The steps involving the construction of the tree after an initial observation of state $x(1) = (\mathsf{CS1}, \mathsf{US}, \emptyset)$, where the sensory itemsets associated with each new node are explicitly indicated; (b) Updated tree after observing state $x(2) = (\mathsf{CS1}, \emptyset, \mathsf{CS2})$. The depth of each node is explicitly indicated.

3.1 Sensory Pattern Mining

As we have seen in Section 2.2, one of the fundamental aspects in the classical conditioning paradigm is the ability of individuals to establish associations between the stimuli they perceive. Stimuli that are frequently perceived together are more likely to lead to similar *value* and *outcome* than stimuli that seldom co-occur. Inspired by this idea, our approach aims at endowing learning agents with a mechanism allowing them to determine how "similar" two states are based on how many associated stimuli they share. To determine such associations we follow a method introduced in [7], where an online sensory pattern mining technique is proposed to identify associations between stimuli occurring in the agent's perceptions, while the agent interacts with its environment. This method identifies such associations by incrementally constructing an *associative sensory tree*, using a variation of the FP-growth algorithm [15] commonly used for transactional pattern mining.

We denote a collection of state-elements $\mathbf{s} = \langle x_{i_1}, \ldots, x_{i_k} \rangle$, with $k \leq n$, as a *sensory itemset*. We assume without loss of generality that each feature-space $\mathcal{X}_i, i = 1 \ldots, n$, is an ordered set[2]. The general sensory pattern mining algorithm in [7] dynamically builds a sensory tree as follows:

---

[2] We note that the specific order of the elements $\mathcal{X}_i \in \mathcal{X}$ is not important, as long as it remains fixed throughout learning. This is a requirement of the tree construction algorithm that guarantees a minimal representation [7].

- At every time-step t, the agent observes state $X(t) = (x_1(t), \ldots, x_n(t))$;
- Given $X(t)$, the algorithm updates the tree by keeping two lists: an "open" list, initially containing all elements $x_i(t) \in X(t)$ to be inserted into the tree; an "ancestor" list containing the nodes in the tree updated so far, which at the beginning of each update contains only the ROOT node (see Fig. 3(a));
- The algorithm then picks one element $x_i(t)$ from the "open" list at a time, ignoring absent elements ($\emptyset$). For each element, a child node $\mathbf{s}$ is created for each node in the "ancestor" list, with counter $n(\mathbf{s}) = 1$. If the child node already exists, its counter is incremented by 1 (see Fig 3(b)). Each new *node* in the tree represents a *sensory itemset*, *i.e.*, a sub-combination of elements obtained from $X(t)$. The nodes' *counter* represents the number of times the corresponding itemset was observed by the agent so far.

Referring back to the example in the beginning of this section, let us consider that we have the state-space $\mathcal{X} = \mathcal{X}_1 \times \mathcal{X}_2 \times \mathcal{X}_3$, where $\mathcal{X}_1 = \{\mathsf{CS1}, \emptyset\}$, $\mathcal{X}_2 = \{\mathsf{US}, \emptyset\}$ and $\mathcal{X}_3 = \{\mathsf{CS2}, \emptyset\}$, where $\emptyset$ represents the absence of a particular stimulus. In other words, each state $x \in \mathcal{X}$ is described by the presence or absence of each of the three stimuli $X_i, i = 1, \ldots, 3$. Figure 3 shows the steps involving the construction of the tree when the agent perceives the state $X(1) = x = (\mathsf{CS1}, \mathsf{US}, \emptyset)$ (sound-shock pairing) from the environment followed by state $X(2) = y = (\mathsf{CS1}, \emptyset, \mathsf{CS2})$ (sound-light pairing).

Given the associative sensory tree, one can measure at each time-step the *degree of association* between stimuli in some sensory itemset $\mathbf{s}$, using the *Jaccard index* [16] which can be used to measure the similarity of sample sets. Given the itemset $\mathbf{s} = \langle x_{i_1}, \ldots, x_{i_k} \rangle$, let $\mathrm{d}(\mathbf{s})$ and $n(\mathbf{s})$ denote, respectively, the *depth* of and the *counter* associated with the corresponding node in the tree. For nodes not directly below the ROOT ($d > 1$), the Jaccard index of $\mathbf{s}$ is given by

$$J(\mathbf{s}) = \frac{n(\mathbf{s})}{\sum_{\mathbf{s}_d} (-1)^{\mathrm{d}(\mathbf{s}_d)+1} n(\mathbf{s}_d)}, \tag{4}$$

where the summation is taken over all nodes $\mathbf{s}_d$ in the *dependency tree* of $\mathbf{s}$, *i.e.*, the subtree containing all nodes in the "ancestor" list obtained after introducing itemset $\mathbf{s}$ in the tree.

Returning to our example, we can now calculate the Jaccard index of state x by solving (4):

$$J(\mathbf{s}) = \frac{n(\mathsf{CS1}, \mathsf{US}, *)}{n(\mathsf{CS1}, *, *) + n(*, \mathsf{US}, *) - n(\mathsf{CS1}, \mathsf{US}, *)} = \frac{1}{2}$$

As expected, the index is inferior to 1, as stimulus $\mathsf{CS1}$ also appears in y, where the US is absent.

We conclude by noting that associative sensory trees are variations of *FP-trees*, which are known to provide a compact representation of large transactional databases [15]. Associative sensory trees have an important advantage over FP-trees, since all information necessary to compute the degree of association between stimuli is trivially accessible from the tree (unlike in an FP-tree). We refer to [7] for further discussion.

3.2 Associative Metric for Factored MDPs

To define a metric using the associative sensory tree described in the previous section, we introduce some additional notation that facilitates the presentation. For any state $x \in \mathcal{X}$, let $\mathcal{S}(x)$ denote the set of all sensory itemsets associated with x. This corresponds to the set of all sub-combinations of stimuli in the dependency tree of the sensory itemset $\mathbf{s}$ associated with x.

We are now in position to introduce our state-space metric. Given the sensory tree at time-step t, we consider the distance between two states x and y as

$$d_A(x, y) = 1 - \frac{\sum_{\mathbf{s} \in \mathcal{S}(x) \cap \mathcal{S}(y)} J_t(\mathbf{s})}{\sum_{\mathbf{s} \in \mathcal{S}(x) \cup \mathcal{S}(y)} J_t(\mathbf{s})}. \tag{5}$$

The distance d_A is indeed a proper *metric*, as it can be reduced to the Tanimoto distance [17] between two vectors associated with x and y, each containing the Jaccard indices for the sensory patterns associated with x and y, respectively. Intuitively, the metric in (5) translates the rationale that two states x and y are "similar" if either they share many stimuli and/or many associated stimuli (stimuli that often co-exist).

Having defined the associative metric we can tackle the first problem defined in the beginning of the section and determine whether two states are similar or not. We can define $\mathcal{S}(x) = \{(\mathsf{CS1}, \mathsf{US}, *), (\mathsf{CS1}, *, *), (*, \mathsf{US}, *)\}$ and $\mathcal{S}(y) = \{(\mathsf{CS1}, *, \mathsf{CS2}), (\mathsf{CS1}, *, *), (*, *, \mathsf{CS2})\}$. The distance between x and y in the example can then be calculated from (5) as

$$d_A(x, y) = 1 - \frac{1}{0.5 + 0.5 + 1 + 0.5 + 0.5} = \frac{2}{3}$$

This means that the degree of similarity between the two states is $1/3$. It follows that our proposed model supports the secondary conditioning phenomenon: the light and foot shock stimuli have some degree of association by means of the sound stimulus, although CS2 and US were never observed together by the agent.

Now that we are able to identify similar states we describe how the metric in (5) can be combined online with Q-learning with spreading. In the experiments reported in this paper, we use the spreading function $\sigma_t(x, y) = e^{-\eta_t d_A(x,y)^2}$. The sequence $\{\eta_t\}$ is a slowly increasing value that ensures that σ_t approaches the Kronecker delta function at a suitable rate, and d_A is the metric defined in (5). As seen in Section 2.1, at each time step t the spreading function σ_t uses information from the current state $X(t)$ to update all other states $y \in \mathcal{X}$, depending on the similarity between $X(t)$ and y calculated according to the structure of the sensory tree at t.

3.3 MDP Metrics and Function Approximation in RL

The notion of "similarity" between states has recently been explored in the MDP literature as a means to render solution methods for MDPs more efficient [9,18]. In fact, by identifying "similar" states in an MDP $\mathcal{M}$, it may be possible to construct a smaller MDP $\mathcal{M}'$ that can more easily be solved.

As established in [19], "similarity" between MDP states is best captured by the notion of *bisimulation*. Bisimulation is an equivalence relation $\sim$ on $\mathcal{X}$ in which two states x and y are similar if $r(x, a) = r(y, a)$ for all $a \in \mathcal{A}$ and

$$\mathbb{P}\left[X(t+1) \in U \mid X(t) = x, A(t) = a\right] = \mathbb{P}\left[X(t+1) \in U \mid X(t) = y, A(t) = a\right],$$

where U is some set in the partition induced by $\sim$. Lax bisimulation is a generalization of bisimulation that also accounts for action relabeling. Both bisimulation and lax bisimulation led to the development of several *MDP metrics* in which, if the distance between two states x, y is zero, then $x \sim y$ [9].

While MDP metrics such as the one above were designed to improve efficiency in MDP solution methods, the best grounded MDP metrics—namely, those relying in the so-called Kantorovich metric—are computationally very demanding [9]. Additionally, they require complete knowledge of the MDP parameters, which renders them unsuitable for RL.

Nevertheless, many RL methods using function approximation implicitly or explicitly exploit some state-space metric [20, 21]. Metrics with well-established theoretical properties (*e.g.*, the bisimulation metric discussed above) could potentially bring significant improvements to RL with function approximation.

The metric proposed in this paper, being computed online, is suitable for RL. Besides, as our results show, in MDPs with a factored structure, our metric is able to attain a generalization performance that matches that obtained with more powerful metrics (such as the bisimulation metric).

4   Experimental Results

In this section we describe several simple experiments aiming at illustrating the applicability of our method. The experiments show the potential of combining the proposed associative metric with spreading in Q-learning, providing a boost in the agent's performance in several factored MDP problems. The main conclusions stemming from the experiments are analyzed and discussed.

To assess the applicability of our method, we applied Q-learning with spreading using σ_t defined earlier and our associative metric in several factored environments, with a state-space that could be factored into between 1 and 4 factors, with a number of states between 20 and 481, and 5 actions. The transition probabilities between states and the reward function were generated randomly. We present the results obtained in 4 of those environments having, respectively, 20 states (5×4), 60 states ($5 \times 4 \times 3$), 120 states ($5 \times 4 \times 3 \times 2$) and 481 states ($9 \times 7 \times 7$, where the dimension and number of factors was chosen randomly). In all scenarios we use $\gamma = 0.95$ and uniform exploration.

We compare the performance of standard Q-learning with that of Q-learning with spreading using several metrics. In particular, we compare 3 metrics:

- A *local metric*, d_ℓ, computed from the transition probabilities of the MDP. Given two states $x, y \in \mathcal{X}$, $d_\ell(x, y)$ corresponds to the average number of steps necessary to transition between the two states, which in grid-world scenarios roughly corresponds to the Manhattan distance. The distance between states that do not communicate was set to an arbitrary large constant.

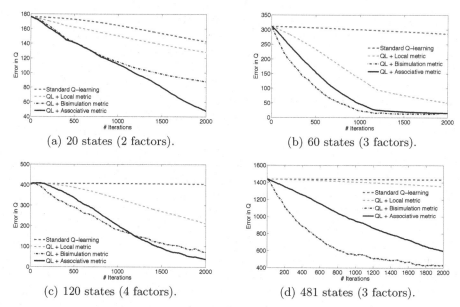

Fig. 4. Performance of Q-learning with spreading in different factored scenarios measuring the error in the Q-values. We compare different metrics with varying knowledge of the MDP parameters. Results are averages over 10 independent Monte-Carlo trials.

- A simplified *bisimulation metric*, d_b [9]. The distance d_b is a simplified version of the bisimulation metric that relies on the total variation norm discussed in Section 3.3 and originally proposed in [9, Section 4.2].[3] We note that this is a theoretically sound metric that, however, requires complete knowledge of both P and r.
- The associative metric d_A described in Section 3.2.

For each of the test scenarios, we ran 10 independent Monte-Carlo trials, and evaluated the learning performance of the different methods by comparing the speed of convergence to the optimal Q-function. The parameter η_t was optimized empirically for each metric in each environment so as to optimize the performance of the corresponding method. Figure 4 depicts the average results.

We note, first of all, that our method always outperforms both standard Q-learning and the local metric. The fact that our method learns faster than standard Q-learning indicates that, in these scenarios, the associations between stimuli provide a meaningful way to generalize the Q-values across states. It is also not surprising that our method generally outperforms the local metric, since it implicit assumes that there is some "spacial" regularity that can be used to generalize Q-values across neighboring states. However, this is generally not the case, meaning that in some scenarios the local metric does not provide a significant improvement in performance—see, for example, Figs. 4(a) and (d).

The bisimulation metric, although a simplified from [9], is a metric that takes into consideration both the transition structure and the reward function of the

---

[3] To simplify, we treat each state as an *equivalence class*. We refer to [9].

MDP. As such, it is not surprising that it allows for good generalization. The fact that our metric performs close to the bisimulation metric in several scenarios—see, for example, Figs. 4(a), (b) and (c)—is, on the other hand, a significant result, since our metric is *learned online*, while the agent interacts with the environment and so uses no prior knowledge on the MDP.

Finally, we note that our metric relies on the factorization of the state-space to build the sensory tree, since the latter is built by associating state-variables that co-occur frequently. In a non-factored MDP, our method would essentially reduce to standard Q-learning. The reliance of our metric on the factorization of the state-space justifies, to some extent, the result in Fig. 4(d). In fact, this corresponds to a large MDP where the "factors" of the state-space are also large. Therefore, not only is the problem larger and, thus, harder to learn, but also our method is able to generalize less than in other more factored scenarios.

5 Concluding Remarks

In this paper we proposed a new state-space associative metric for factored MDPs that draws inspiration from classical conditioning in nature. Our metric relies on identified associations between state-variables perceived by the learning agent during its interaction with the environment. These associations are learned using a sensory pattern-mining algorithm and determine the similarity between states, thus providing a state-space metric that requires no prior knowledge on the structure of the underlying decision problem. The sensory pattern-mining algorithm relies on the *associative sensory tree*, that captures the frequency of co-occurrence of stimuli in the agent's environment.

It is worth mentioning that the size of the associative sensory tree exhibits a worst-case exponential dependence in the number of *state-factors* (not states). However, aside from the memory requirements associated therewith, the structure of the tree is such that the computation of the distance is linear in the number of factors, which is extremely convenient for the online processing of distances. Moreover, as discussed in Section 3.1, the adopted tree representation can safely be replaced by other equivalent representations, such as the FP-tree [15] that, while more efficient in terms of memory requirements, may render the computation of the distance computationally more expensive.

Additionally, we note that the maximal size of the tree is only achieved when *all* the state space has been explored. However, it is in the early stages of the learning process—when little of the state space has been explored—that the use of associative metric may be more beneficial. Our results indicate that the combination of our metric with standard Q-learning does lead to an improved learning performance that is comparable to that obtained with other more powerful metrics that use information both from the transitions and rewards of the MDP. The specific strategy used to integrate the metric with Q-learning (*i.e.*, the decaying spreading function) enforces that when the size of the tree approaches its maximum size, the contribution of the associative metric to learning is generally small. Therefore, limiting the tree size to some pre-specified maximum or using tree-pruning techniques as those discussed in [7] should have little impact on the performance of our proposed method.

Acknowledgments. This work was partially supported by the Portuguese Fundação para a Ciência e a Tecnologia (FCT) under project PEst-OE/EEI/LA0021/2013 and the EU project SEMIRA through the grant ERA-Compl /0002/2009. The first author acknowledges grant SFRH/BD/38681/2007 from FCT.

References

1. Anderson, J.: Learning and Memory: An Integrated Approach. Wiley (2000)
2. Sutton, R., Barto, A.: Reinforcement Learning: An Introduction. MIT Press (1998)
3. Szepesvári, C.: Algorithms for Reinforcement Learning. Morgan & Claypool (2010)
4. Kearns, M., Koller, D.: Efficient reinforcement learning in factored MDPs. In: Proc. 1999 Int. Joint Conf. Artificial Intelligence, pp. 740–747 (1999)
5. Jong, N., Stone, P.: State abstraction discovery from irrelevant state variables. In: Proc. 19th Int. Joint Conf. Artificial Intelligence, pp. 752–757 (2005)
6. Kroon, M., Whiteson, S.: Automatic feature selection for model-based reinforcement learning in factored MDPs. In: Proc. 2009 Int. Conf. Machine Learning and Applications, pp. 324–330 (2009)
7. Sequeira, P., Antunes, C.: Real-time sensory pattern mining for autonomous agents. In: Cao, L., Bazzan, A.L.C., Gorodetsky, V., Mitkas, P.A., Weiss, G., Yu, P.S. (eds.) ADMI 2010. LNCS, vol. 5980, pp. 71–83. Springer, Heidelberg (2010)
8. Ribeiro, C., Szepesvári, C.: Q-learning combined with spreading: Convergence and results. In: Proc. Int. Conf. Intelligent and Cognitive Systems, pp. 32–36 (1996)
9. Ferns, N., Panangaden, P., Precup, D.: Metrics for finite Markov decision processes. In: Proc. 20th Conf. Uncertainty in Artificial Intelligence, pp. 162–169 (2004)
10. Watkins, C.: Learning from delayed rewards. PhD thesis, King's College, Cambridge Univ. (1989)
11. Guestrin, C., Koller, D., Parr, R., Venkataraman, S.: Efficient solution algorithms for factored MDPs. J. Artificial Intelligence Research 19, 399–468 (2003)
12. Pavlov, I.: Conditioned reflexes: An investigation of the physiological activity of the cerebral cortex. Oxford Univ. Press (1927)
13. Cardinal, R., Parkinson, J., Hall, J., Everitt, B.: Emotion and motivation: The role of the amygdala, ventral striatum, and prefrontal cortex. Neuroscience and Biobehavioral Reviews 26(3), 321–352 (2002)
14. Balkenius, C., Morén, J.: Computational models of classical conditioning: A comparative study. In: Proc. 5th Int. Conf. Simulation of Adaptive Behavior: From Animals to Animats, vol. 5, pp. 348–353 (1998)
15. Han, J., Pei, J., Yin, Y., Mao, R.: Mining frequent patterns without candidate generation. Data Mining and Knowledge Disc. 8, 53–87 (2004)
16. Jaccard, P.: The distribution of the flora in the alpine zone. New Phytologist 11(2), 37–50 (1912)
17. Lipkus, A.: A proof of the triangle inequality for the Tanimoto distance. J. Mathematical Chemistry 26(1), 263–265 (1999)
18. Ravindran, B., Barto, A.: Approximate homomorphisms: A framework for non-exact minimization in Markov decision processes. In: Proc. 5th Int. Conf. Knowledge-Based Computer Systems (2004)
19. Givan, R., Dean, T., Greig, M.: Equivalence notions and model minimization in Markov Decision Processes. Artificial Intelligence 147, 163–223 (2003)
20. Szepesvári, C., Smart, W.: Interpolation-based Q-learning. In: Proc. 21st Int. Conf. Machine Learning, pp. 100–107 (2004)
21. Ormoneit, D., Sen, S.: Kernel-based reinforcement learning. Machine Learning 49, 161–178 (2002)

A Distributed Approach to Diagnosis Candidate Generation

Nuno Cardoso and Rui Abreu

Department of Informatics Engineering
Faculty of Engineering, University of Porto
Porto, Portugal
nunopcardoso@gmail.com, rui@computer.org

Abstract. Generating diagnosis candidates for a set of failing transactions is an important challenge in the context of automatic fault localization of both software and hardware systems. Being an NP-Hard problem, exhaustive algorithms are usually prohibitive for real-world, often large, problems. In practice, the usage of heuristic-based approaches trade-off completeness for time efficiency. An example of such heuristic approaches is STACCATO, which was proposed in the context of reasoning-based fault localization. In this paper, we propose an efficient distributed algorithm, dubbed MHS^2, that renders the sequential search algorithm STACCATO suitable to distributed, Map-Reduce environments. The results show that MHS^2 scales to larger systems (when compared to STACCATO), while entailing either marginal or small runtime overhead.

Keywords: Fault Localization, Minimal Hitting Set, Map-Reduce.

1 Introduction

Detecting, localizing and correcting erroneous behavior in software, collectively known as debugging, is arguably one of the most expensive tasks in almost all software systems' life cycle [9]. While still being essentially a manual process, a wide set of techniques have been proposed in order to automate this process [1–3, 6]. With regard to automatic fault localization, which is the scope of this paper, two main problems exist:

- Finding sets of components (known as diagnostic candidates) that, by assuming their faultiness, would explain the observed erroneous behavior.
- Ranking the candidates according to their likelihood of being correct.

In this paper, we propose a Map-Reduce [5] approach, dubbed MHS^2[1], aimed at computing minimal diagnosis candidates in a parallel or even distributed fashion in order to broaden the search scope of current approaches.

This paper makes the following contributions:

---

[1] MHS^2 is an acronym for Map-reduce Heuristic-driven Search for Minimal Hitting Sets.

L. Correia, L.P. Reis, and J. Cascalho (Eds.): EPIA 2013, LNAI 8154, pp. 175–186, 2013.
© Springer-Verlag Berlin Heidelberg 2013

- We introduce an optimization to the sequential algorithm, which is able to prevent a large number of redundant calculations.
- We propose MHS^2, a novel Map-Reduce algorithm for dividing the minimal hitting set (MHS) problem across multiple CPUs.
- We provide an empirical evaluation of our approach, showing that MHS^2 efficiently scales with the number of processing units.

The remainder of this paper is organized as follows. In Section 2 we describe the problem as well as the sequential algorithm. In Section 3 we present our approach. Section 4 discusses the results obtained from synthetic experiments. Finally, in Section 5 we draw some conclusions about the paper.

2   Preliminaries

As previously stated, the first step when diagnosing a system is to generate valid diagnosis candidates, conceptually known as hitting sets. In scenarios where only one component is at fault this task is trivial. However, in the more usual case where multiple faults are responsible for erroneous behavior, the problem becomes exponentially more complex. In fact, for a particular system with M components there are 2^M possible (but not necessarily valid) candidates. To reduce the number of generated candidates only minimal candidates are considered.

Definition 1 (Minimal Valid Candidate/Minimal Hitting Set). *A candidate d is said to be both valid (i.e., a hitting set) and minimal if (1) every failed transaction involved a component from candidate d and (2) no valid candidate d' is contained in d.*

We use the hit spectra data structure to encode the involvement of components in pass/failed system transactions (a transaction can represent, for instance, a test case from a particular test suite).

Definition 2 (Hit Spectra). *Let S be a collection of sets s_i such that*

$$s_i = \{j \mid if\ component\ j\ participated\ in\ transaction\ i\}$$

Let $N = |S|$ denote the total number of observed transactions and $M = |COMPS|$ denote the number of system components. A hit spectra [10] is a pair (A, e), where A is a $N \times M$ activity matrix of the system and e the error vector, defined as

$$A_{ij} = \begin{cases} 1, & if\ j \in s_i \\ 0, & otherwise \end{cases} \qquad e_i = \begin{cases} 1, & if\ transaction\ i\ failed \\ 0, & otherwise \end{cases}$$

As an example, consider the hit spectra in Figure 1a for which all 2^M possible candidates (i.e., the power set $\mathcal{P}_{COMPS}$) are presented in Figure 1b. For this particular example, two valid minimal candidates exist: $\{3\}$ and $\{1, 2\}$. Even though the candidate $\{1, 2, 3\}$ is also valid, it is not minimal as it can be subsumed by either $\{3\}$ or $\{1, 2\}$.

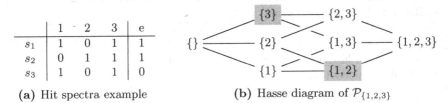

	1	2	3	e
s_1	1	0	1	1
s_2	0	1	1	1
s_3	1	0	1	0

(a) Hit spectra example

(b) Hasse diagram of $\mathcal{P}_{\{1,2,3\}}$

Fig. 1. Running example

Despite the advantage of only computing minimal candidates, the problem is known to be NP-hard [7]. Being an NP-hard problem, the usage of exhaustive search algorithms (e.g., [8, 12, 14]), is prohibitive for most real-world problems. In order to solve the candidate generation problem in a reasonable amount of time, approaches that relax the strict minimality[2] constraint have been proposed [2, 4, 6, 11, 13, 15]. In particular, STACCATO [2], the sequential algorithm which serves as foundation for our distributed algorithm, uses the Ochiai heuristic [3] in order to increase the likelihood of finding the actual fault explanation, yielding significant efficiency gains.

In Algorithm 1 (discarding, for now, the highlighted lines) a simplified version of STACCATO that captures its fundamental mechanics is presented[3]. The algorithm can perform two different tasks (lines 2–4 and 6–11), depending on whether or not the candidate d' is a hitting set. By definition, d' is a hitting set if all failing transaction in the spectra under analysis contain at least one component in d'. Due to the structure of the algorithm d' is a hitting set whenever e does not contain any errors (i.e., $\not\exists e_i \in e : e_i = 1$). In the first case, where d' is a hitting set (lines 2–4), the algorithm checks if d' is minimal (line 2) with regard to the already discovered minimal hitting set collection D'. If d' is minimal, all hitting sets in D' subsumable by d' are purged (line 3) and d' is added to D' (line 4). In the second case, where d' is not a hitting set (lines 6–11), the algorithm composes candidates in the form of $d' + \{j\}$. The order in which the components $j \in R \subset COMPS$ are selected is determined by some arbitrary heuristic $\mathcal{H}$ (line 6). This heuristic is responsible for both driving the algorithm towards high potential candidates and reducing the amount of candidates that need to be checked. Such tasks are application dependent and, in the particular case of model/reasoning-based fault diagnosis (MBD), the Ochiai heuristic [2, 3] was shown to exhibit good accuracy levels. Whenever a component j is selected, a temporary (A', e') is created where all transactions s_i such that $j \in s_i$ as well as column j are omitted (function STRIP, line 9). Finally, the algorithm makes a recursive call in order to explore (A', e') with candidate $d' + \{j\}$ (line 10).

---

[2] We use the term minimal in a more liberal way due to mentioned relaxation. A candidate d is said to be minimal if no other calculated candidate is contained in d.

[3] It is important to note that the cut-off conditions were removed for the sake of simplicity and any direct implementation will not be suitable to large problems. Both the details regarding search heuristics and cut-off parameters are outside the scope of this paper. Refer to [2] for more information.

Algorithm 1. STACCATO / MHS^2 map task

Inputs:
    Matrix (A, e)
    Partial minimal hitting set collection D' (default: $\emptyset$)
    Candidate d' (default: $\emptyset$)
Parameters:
    Ranking heuristic $\mathcal{H}$
    Branch level L
    Load division function SKIP
Output:
    Minimal hitting set collection D
    Partial minimal hitting set collection D'_k

```
1  if ∄e_i ∈ e : e_i = 1 then              # Minimality verification task
2     if MINIMAL(D', d') then
3         D' ← PURGE_SUBSUMED(D', d')
4         D' ← D' ∪ {d'}
5  else                                     # Candidate compositions task
6     R ← RANK(H, A, e)                     # Heuristic ranking
7     for j ∈ R do
8        if ¬(SIZE(d') + 1 = L ∧ SKIP()) then  # Load Division
9            (A', e') ← STRIP(A, e, j)
10           D' ← STACCATO(A', e', D', d' + {j})
11       A ← STRIP_COMPONENT(A, j)         # Optimization
12 return D'
```

	1	2	3	e
s_1	1	0	1	1
s_2	0	1	1	1
s_3	1	0	1	0

(a) After STRIP$(A, e, 3)$

	1	2	3	e
s_1	1	0	1	1
s_2	0	1	1	1
s_3	1	0	1	0

(b) After STRIP$(A, e, 2)$

	1	2	3	e
s_1	1	0	1	1
s_2	0	1	1	1
s_3	1	0	1	0

(c) After STRIP$(A, e, 1)$

Fig. 2. Evolution of (A, e)

To illustrate how STACCATO works, recall the example in Figure 1a. In the outer call to the algorithm as $\exists e_i \in e : e_i = 1$ (i.e., candidate d' is not a hitting set), candidates in the form of $\emptyset + \{j\}$ are composed. Consider, for instance, that the result of the heuristic over (A, e) entails the ranking $(3, 2, 1)$. After composing the candidate $\{3\}$ and making the recursive call, the algorithm adds it to D' as $\nexists e_i \in e : e_i = 1$ (Figure 2a) and $D' = \emptyset$, yielding $D' = \{\{3\}\}$. After composing candidate $\{2\}$, a temporary (A', e') is created (Figure 2b). Following the same logic, the hitting sets $\{2, 1\}$ and $\{2, 3\}$ can be found but only $\{2, 1\}$ is minimal as $\{2, 3\}$ can be subsumed by $\{3\}$[4]. Finally the same procedure is repeated for component 1 (Figure 2c), however no new minimal hitting set is found. The result for this example would be the collection $D = \{\{1, 2\}, \{3\}\}$.

---

[4] Actually, STACCATO would not compose candidate $\{2, 3\}$ due to an optimization that is a special case of the one proposed in this paper (see Section 3).

3   MHS^2

In this section, we propose MHS^2, our distributed MHS search algorithm. The proposed approach can be viewed as a Map-Reduce task [5]. The map task, presented in Algorithm 1 (now also including highlighted lines), consists of an adapted version of the sequential algorithm just outlined.

In contrast to the original algorithm, we added an optimization that prevents the calculation of the same candidates in different orders (line 11), as it would be the case of candidates $\{1, 2\}$ and $\{2, 1\}$ in the example of the previous section. Additionally, it generalizes over the optimization proposed in [2], which would be able to ignore the calculation of $\{2, 3\}$ but not the redundant reevaluation of $\{1, 2\}$. The fundamental idea behind this optimization is that after analyzing all candidates that can be generated from a particular candidate d' (i.e., the recursive call), it is guaranteed that no more minimal candidates subsumable by d' will be found[5]. A consequence of this optimization is that, as the number of components in (A, e) is different for all components $j \in R$, the time that the recursive call takes to complete may vary substantially.

To parallelize the algorithm across np processes, we added a parameter L that sets the *split-level*, i.e., the number of calls in the stack minus 1 or $|d'|$, at which the computation is divided among the processes. When a process of the distributed algorithm reaches the target level L, it uses a load division function (SKIP) in order to decide which elements of the ranking to skip or to analyze. The value of L implicitly controls the granularity of decision of the load division function at the cost of performing more redundant calculations. Implicitly, by setting a value $L > 0$, all processes redundantly calculate all candidates such that $|d'| <= L$.

With regard to the load division function SKIP, we propose two different approaches. The first, referred to as *stride*, consists in assigning elements of the ranking R to processes in a cyclical fashion. Formally, a process $p_{k \in [1..np]}$ is assigned to an element $R_l \in R$ if $(l \mod np) = (k - 1)$. The second approach, referred to as *random*, uses a pseudo-random generator in order to divide the computation. This random generator is then fed into an uniform distribution generator that assures that, over time, all p_k get assigned a similar number of elements in the ranking R although in random order (specially for large values of L). This method is aimed at obtaining a more even distribution of the problem across processes than *stride*. A particularity of this approach is that the seed of the pseudo random generator must be shared across process in order to assure that no further communication is needed.

Finally, the reduce task, responsible for merging all partial minimal hitting set collections $D'_{k \in [1..np]}$ originating from the map task (Algorithm 1), is presented in Algorithm 2. The reducer works by merging all hitting sets in a list

---

[5] Visually, using a Hasse diagram (Figure 1b), this optimization can be represented by removing all unexplored edges touching nodes subsumable by d'. As every link represents an evaluation that would be made without any optimizations (several links to same node means that the set is evaluated multiple times), it becomes obvious the potential of such optimization.

Algorithm 2. MHS^2 reduce task

Inputs:
    Partial minimal hitting set collections $D'_1, ..., D'_K$
Output:
    Minimal hitting set collection D

1 $D \leftarrow \emptyset$
2 $D' \leftarrow \text{SORT}(\bigcup_{k=1}^{K} D'_k)$ # Hitting sets sorted by cardinality
3 **for** $d \in D'$ **do**
4   **if** $\text{MINIMAL}(D, d)$ **then**
5          $D \leftarrow D \cup \{d\}$
6 **return** D

---

ordered by cardinality. The ordered list is then iterated, adding all minimal hitting sets to D. As the hitting sets are inserted in an increasing cardinality order, it is not necessary to look for subsumable hitting sets (PURGE_SUBSUMED in Algorithm 1) in D.

4   Results

In order to assess the performance of our algorithm we implemented it in C++ using OpenMPI as the parallelization framework. All the benchmarks were conducted in a single computer with 2× Intel Xeon CPU X5570 @ 2.93GHz (4 cores each). Additionally, we generated several (A, e) by means of a Bernoulli distribution, parameterized by r (i.e., the probability that a component is involved in a row of A equals r) for which solutions have been computed with different parameters. In order to ease the comparison of results, *all* transactions in *all* generated cases fail[6]. For each set of parameters, we generated 50 inputs for each $r \in \{0.25, 0.5, 0.75\}$, and the results represent the average of the observed metrics. Both due to space constraints and the fact that the diagnosis efficiency of the algorithm has already been studied in [2], we only analyze the performance gains obtained from the parallelization.

4.1   Benchmark 1

In the first benchmark, we aimed at observing the behavior of MHS^2 for small scenarios ($N = 40, M = 40, L = 2$) where all minimal candidates can be calculated. In Figure 3, we observe both the speedup (defined as $Sup(np) = \frac{T_1}{T_{np}}$,

---

[6] To illustrate the potential problems of having successful transactions in the test cases, consider the extreme case of a set of test cases with no failures versus a set of test cases with no nominal transactions. For the first scenario, all test cases only have one minimal hitting set (the empty set) whereas, for the second scenario, a potentially large number of minimal hitting sets may exist. As it is demonstrated in this section, the number of minimal hitting sets has a large impact in the algorithm's run-time.

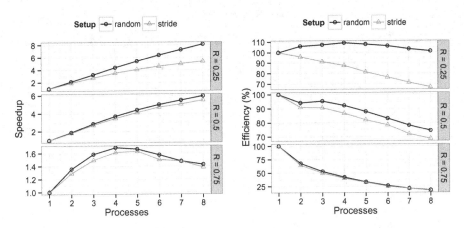

Fig. 3. Small scenarios speedup (left) and efficiency (right)

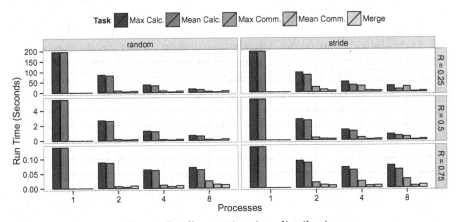

Fig. 4. Small scenarios time distribution

where T_{np} is the time needed to solve the problem for np processes) and the efficiency (defined as $Ef(np) = \frac{Sup(np)}{np}$) for both the *stride* and *random* load distribution functions.

The analysis of Figure 3 shows different speedup/efficiency patterns for r values. Despite, *random* consistently outperforms *stride*. Additionally, for $r = 0.75$ and in contrast to $r \in \{0.25, 0.5\}$, the speedup/efficiency is low (note the differences in y-axis scales). The observed speedup/efficiency patterns can be explained by analyzing Figure 4 where the total runtime is divided amongst the composing tasks (calculation, communication, and the merging of results). Additionally the maximum times for calculation and communication are shown to compare them with the respective mean values. First, it is important to note that the maximum runtime for different values of r varies substantially: ≈ 200 seconds for $r = 0.25$ vs. > 0.2 seconds for $r = 0.75$. This difference in runtimes

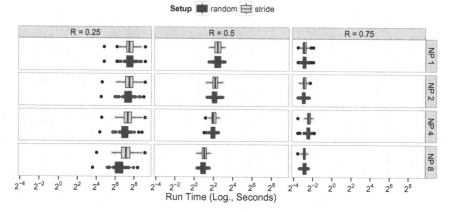

Fig. 5. Small scenarios runtime

exists due to the fact that for higher values of r, both the size and amount of minimal candidates tends to be smaller and, due to the optimization proposed in this paper, the number of candidates analyzed is also smaller. On the one hand, when the runtime is smaller, the parallelization overheads represent a higher relative impact in the performance (in extreme cases, the runtime can even increase with the number of processes). On the other hand, when both the cardinality and the amount of minimal hitting sets increase (small values of r) the parallelization overhead becomes almost insignificant. In such cases a larger efficiency difference between *stride* and *random* is observed due to a better load division by the *random* mechanism.

In Figure 4 this is visible when comparing the time bars for the maximum and mean calculation and communication times (actually, as the communication time includes waiting periods, it is dependent on the calculation time). For the *random* case, the maximum and mean calculation times are almost equal thus reducing waiting times. In contrast, in the *stride* case, the maximum and mean calculation times are uneven and, as a consequence, the communication overhead becomes higher: average ≈ 7 seconds for *random* vs. ≈ 28 seconds for *stride*. In scenarios where a large number of hitting sets exist and due to the fact of the function PURGE_SUBSUMED having a complexity greater than $O(n)$, the efficiency of the *random* load division can be greater than 100%. While good results are also observable for $r = 0.5$, the improvement is less significant.

Table 1. Small scenarios means and standard deviations

| | R=0.25 | | | | R=0.5 | | | | R=0.75 | | | |
| | Random | | Stride | | Random | | Stride | | Random | | Stride | |
NP	mean	sd	mean	sd	mean	sd	mean	sd	mean	sd	mean	sd
1	196.08	94.42	196.08	94.42	5.44	1.43	5.44	1.43	0.14	0.03	0.14	0.03
2	92.54	43.47	102.08	49.16	2.89	0.68	2.99	0.74	0.10	0.01	0.10	0.02
4	44.89	21.77	55.96	27.85	1.47	0.28	1.56	0.31	0.08	0.01	0.08	0.01
8	24.20	11.23	36.61	18.36	0.91	0.16	0.99	0.18	0.10	0.01	0.09	0.01

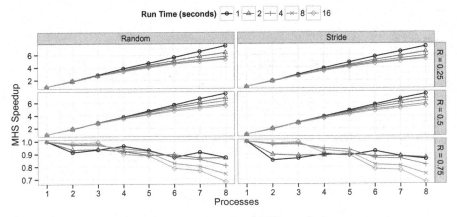

Fig. 6. Big scenarios MHS speedup

Finally, in Figure 5 the runtime distribution of the tests is plotted and in Table 1 both the means and standard deviations are presented. Both the figure and the table show that for the same generation parameters $(r, M, N$ and $L)$ the runtime exhibits a considerable variance (note that the x-axis has a logarithmic scale). It is important to note that, while the value of r has an important effect on the performance, the real key for efficiency is the problem complexity (i.e., the time needed to solve the problem). For complex enough (but still calculable) problems, the efficiency of the algorithm would increase asymptotically to 100% (or even past 100%) as the polynomial parallelization overhead would eventually be overwhelmed by the exponential complexity of the problem.

4.2 Benchmark 2

The second benchmark is aimed at observing the behavior of MHS^2 for realistic scenarios ($N = 100, M = 10000, L = 2$) where it is impractical to calculate all minimal candidates. In all the following scenarios a time based cut-off was implemented and we define the metric $CSup(np) = \frac{|D_1|}{|D_{np}|}$, where $|D_{np}|$ is the number of candidates generated using np processes for the same time, henceforward referred to as *MHS speedup*. While higher values of this metric do not necessarily mean higher diagnosis capabilities[7], in most cases they are positively correlated, due to the usage of the heuristic ranking function (see [2]).

Figures 6 and 7 show the results of computing candidates for big problems for runtimes $rt \in \{1, 2, 4, 8, 16\}$ and $np \in [1..8]$ (entailing a total runtime of $rt \times np$). It is clear that, for big problems, there is no significant difference in terms of the amount of generated candidates between *random* and *stride*. Figure 6 shows that for $r \in \{0.25, 0.5\}$ the MHS speedup scales well with the number of processes, however as the time increases, it becomes harder to find new minimal candidates

---

[7] As an example consider a set of failures for which the true explanation would be $d = \{1\}$ and two diagnostic collections $D^1 = \{\{1\}\}$ and $D^2 = \{\{1, 2\}, \{1, 3\}\}$. While $|D^2| > |D^1|$, D^1 has better diagnostic capabilities than D^2 as $d \in D^1 \wedge d \notin D^2$.

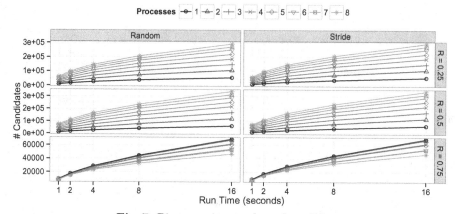

Fig. 7. Big scenarios number of candidates

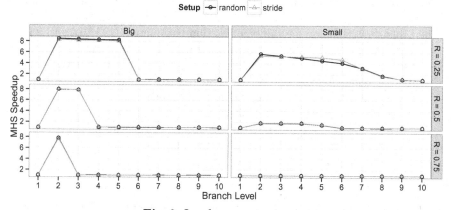

Fig. 8. Level parameter impact

(Figure 7). Regarding $r = 0.75$ we see that the MHS speedup pattern is not as smooth as for the former cases. Additionally it is important to note that, in contrast to the cases $r \in \{0.25, 0.5\}$, for $r = 0.75$, the number of candidates generated by $np = 8$ is smaller than for all other cases (Figure 7). This is due to the fact that for higher values of r both the cardinality and the number of minimal candidates becomes smaller, enabling the algorithm to explore a larger percentage of the search tree. As a consequence, and due to the limitations of an heuristic search, it happens that some of the candidates found first are subsumable by candidates found later, reducing the candidate collection size over time.

4.3   Benchmark 3

The final benchmark is aimed at observing the influence of parameter L in the number of generated candidates. In this benchmark we calculated candidates for both big and small problems using $np = 8$ and $rt = 10$ and $L \in [1..10]$. The analysis of Figure 8 reveals the great impact L has on the number of generated candidates. In the conducted experiments no MHS speedup lesser than 1 was observed, meaning that it should be a sensible choice to set L to be greater than 1. Optimal selection of values for L yielded an eightfold performance improvement for all the big scenarios. In the small scenarios, this improvement is still observable but with lesser magnitude. For the small scenarios with $r = 0.75$, L play no apparent role as the MHS speedup is always 1. This is due to the fact that all minimal candidates were always calculable within the defined time frame. A closer inspection of the data revealed an isolated threefold speedup peak at $L = 2$.

Another interesting pattern is the correlation between the Bernoulli distribution parameter r and the number of near-optimal values for L. This can be explained by using the argument relating r and the candidate size. The average candidate sizes (in optimal conditions) for $r \in \{0.25, 0.5, 0.75\}$ were $\{8, 6, 3\}$ for the small scenarios and $\{6, 4, 3\}$ for the large scenarios. If we observe, for each plot, the last value of L for which the performance is near optimal we see that it matches the average candidate size minus 1. Even though several levels may be near optimal, it is better to use the smaller still near optimal value for L ($L = 2$ for all the conducted experiments) as it implies less redundant calculation with an acceptable level of granularity for load division.

As a final note, although all benchmarks were conducted within a single host, implying low communication latencies, we expect that the algorithm is able to efficiently perform in distributed environments. In the conducted experiments, the communication sizes (using a non-compressed binary stream) were bounded by a 3.77 megabytes maximum.

5   Conclusions

In this paper, we proposed a distributed algorithm aimed at computing diagnosis candidates for a set of failing observations, dubbed MHS^2. This algorithm is not only more efficient in single CPU scenarios than the existent STACCATO algorithm but also is able to efficiently use the processing power of multiple CPUs to calculate minimal diagnosis candidates.

The results showed that, specially for large problems, the algorithm is able to scale with negligible overhead. The usage of parallel processing power enables the exploration of a larger number of potential candidates, increasing the likelihood of actually finding the "actual" set of failing components.

Future work would include the analysis of the algorithm's performance with a larger set of computation resources as also the analysis of its performance under a wider set of conditions. Additionally, it would be interesting to study the performance in massively parallel computing Hadoop-based infrastructures.

Acknowledgements. We would like to thank Lígia Massena, André Silva and José Carlos de Campos for the useful discussions during the development of our work. This material is based upon work supported by the National Science Foundation under Grant No. CNS 1116848, and by the scholarship number SFRH/BD/79368/2011 from Fundação para a Ciência e Tecnologia (FCT).

References

1. Abreu, R.: Spectrum-based Fault Localization in Embedded Software. PhD thesis, Delft University of Technology (November 2009)
2. Abreu, R., van Gemund, A.J.C.: A low-cost approximate minimal hitting set algorithm and its application to model-based diagnosis. In: Symposium on Abstraction, Reformulation, and Approximation, SARA 2009 (2009)
3. Abreu, R., Zoeteweij, P., van Gemund, A.J.C.: On the accuracy of spectrum-based fault localization. In: Testing: Academic and Industrial Conference Practice and Research Techniques, TAICPART 2007 (2007)
4. de Kleer, J., Williams, B.C.: Readings in model-based diagnosis (1992)
5. Dean, J., Ghemawat, S.: Mapreduce: simplified data processing on large clusters. In: Symposium on Opearting Systems Design & Implementation, OSDI 2004 (2004)
6. Feldman, A., Provan, G., Van Gemund, A.J.C.: Computing minimal diagnoses by greedy stochastic search. In: AAAI Conference on Artificial intelligence, AAAI 2008 (2008)
7. Garey, M.R., Johnsonp, D.S.: Computers and Intractability; A Guide to the Theory of NP-Completeness (1990)
8. Greiner, R., Smith, B.A., Wilkerson, R.W.: A correction to the algorithm in Reiter's theory of diagnosis. Artificial Intelligence 41(1) (1989)
9. Hailpern, B., Santhanam, P.: Software debugging, testing, and verification. IBM Syst. J. 41(1) (January 2002)
10. Harrold, M.J., Rothermel, G., Wu, R., Yi, L.: An empirical investigation of program spectra. In: Program Analysis for Software Tools and Engineering, PASTE 1998 (1998)
11. Pill, I., Quaritsch, T.: Optimizations for the boolean approach to computing minimal hitting sets. In: European Conference on Artificial Intelligence, ECAI 2012 (2012)
12. Reiter, R.: A theory of diagnosis from first principles. Artificial Intelligence 32(1) (1987)
13. Ruchkys, D.P., Song, S.W.: A parallel approximation hitting set algorithm for gene expression analysis. In: Symposium on Computer Architecture and High Performance Computing (2002)
14. Wotawa, F.: A variant of Reiter's hitting-set algorithm. Information Processing Letters 79(1) (2001)
15. Zhao, X., Ouyang, D.: Improved algorithms for deriving all minimal conflict sets in model-based diagnosis. In: Huang, D.-S., Heutte, L., Loog, M. (eds.) ICIC 2007. LNCS, vol. 4681, pp. 157–166. Springer, Heidelberg (2007)

Probabilistic Vector Machine: Scalability through Clustering Hybridization

Mihai Cimpoeşu, Andrei Sucilă, and Henri Luchian

Alexandru Ioan Cuza University, Faculty of Computer Science, Iasi, Romania
mihai.cimpoesu@info.uaic.ro

Abstract. In this paper, a hybrid clustering and classification algorithm is obtained by exploring the specific statistical model of a hyperplane classifier. We show how the seamless integration of the clustering component allows a substantial cost decrease in the training stage, without impairing the performance of the classifier. The algorithm is also robust to outliers and deals with training errors in a natural and efficient manner.

Keywords: algorithm, classification, clustering, hybrid, statistical model, probabilistic classifier.

1 Introduction

One side-effect of the need for accurate prediction in critical fields such as disease identification in biological data or identifying credit fraud in financial data has been a steady increase in the size of the datasets to be processed. Most prediction tasks require classification in order to sort and label dataset records. With the help of decision trees, a binary classifier often can be used to solve this type of problems. A binary classifier assigns either a positive or a negative label to each new unlabeled record, after having been developed using a training set of labeled records.

The introduction in [1], [2] and [3] of Support Vector Machine (SVM) and specifically, soft margin SVMs, has provided a good tool to deal with classification tasks. SVM is one of the most succesful algorithms in the field of classification, with numerous applications in text processing, bioinformatics and many other fields. However, the basic model of a soft margin SVM has some drawbacks. Some of these are: the inability to directly deal with outliers, defining the decision function based solely on border points and ignoring the distribution of the training set. These have been addressed by the introduction of error bounds in [4] and based on these bounds, a method for choosing parameters for SVMs has been presented in [5]. The bounds developed are not tight and overestimate the probability of error.

A serious problem with SVMs has been the handling of large databases, which becomes especially cumbersome as the training phase requires five fold cross-validation of parameters. The existing solvers, such as libSVM presented in [6] and [7], have a scalability problem, both in execution time and memory requirements which increase by an $O(n^2)$ factor in the database size. As a consequence, the authors of [8], [9] and [10] integrate clustering techniques in the training

L. Correia, L.P. Reis, and J. Cascalho (Eds.): EPIA 2013, LNAI 8154, pp. 187–198, 2013.
© Springer-Verlag Berlin Heidelberg 2013

phase in order to reduce the size of databases and [11] changes the formulation of SVMs to a second order cone programming problem, to allow for better scalability with the increase of database size.

We propose an alternative hyperplane classifier, based on the Probabilistic Vector Machine (PVM) classifier, introduced in [12], which has a built–in approach to dealing with outliers and classification errors. This leads to a smaller number of parameters, which decreases the cost of the five fold cross-validation in the search of the effective parameters.

Although the classifier only requires to solve linear feasibility problems, these are dense and can lead to memory problems in the linear programming algorithms. Therefore, the training phase can become very expensive –similar to the SVM– and a reduction in the problem size is in order. Clustering and the PVM algorithm have a natural way of being hybridized, allowing for a unified approach of the classification problem. The idea of bringing clustering and classification together, applied on SVM, were pursued in [13], [14] and [15]. Besides applying clustering before training the PVM, we also bring them closer together by readjusting the cluster architecture during the training of the classification algorithm.

The paper is structured as follows: a quick presentation of the background on PVM in Section 2; a first variant of the PVM model, followed by the combined implicit clustering variant in Section 3; experimental results and a discussion based on these results in Section 4 and conclusions in Section 5.

2   Probabilistic Vector Machine

The classifier upon which we base our hybrid approach is the Probabilistic Vector Machine, which has been recently introduced in [12]. It is a statistical binary classifier which learns a separating hyperplane, used to make the decision whether to label a point positively or negatively. The hyperplane is chosen such as to optimize a statistical measure of the training data.

Let $S = \{(x_i, y_i | x_i \in \mathbb{R}^n), y_i \in \{\pm 1\}\}$ be the training set and let $S_+ = \{x_i \in S | y_i = 1\}$ and $S_- = \{x_i \in S | y_i = -1\}$ be the training subsets comprized of positively and negatively labeled points. For a hyperplane determined by its normal, $w \in \mathbb{R}^n$, and its offset, or bias, $b \in \mathbb{R}$, we denote by E_+ and E_- the average signed distance to the hyperplane of the points in S_+ and S_- respectively. We also denote by σ_+, σ_- the average deviations from E_+ and E_- of the corresponding points. This can be written as:

$$E_+ = b + \frac{1}{|S_+|} \sum_{x_i \in S_+} <w, x_i>, \ E_- = -b - \frac{1}{|S_-|} \sum_{x_i \in S_-} <w, x_i>$$

$$\sigma_+ = \frac{1}{|S_+| - 1} \sum_{x_i \in S_+} |<w, x_i> +b - E_+|$$

$$\sigma_- = \frac{1}{|S_-| - 1} \sum_{x_i \in S_-} |<w, x_i> +b + E_-|$$

PVM searches for the hyperplane which optimizes the function:

$$\begin{cases} \min \max\{\frac{\sigma_+}{E_+}, -\frac{\sigma_-}{E_-}\} \\ E_+ \geq 1, E_- \leq -1 \end{cases} \tag{2.1}$$

This system is solvable through a series of linear feasibility problems which searches for the objective via bisection.

Nonlinear training may be achieved by using kernel functions. Let $\Phi : \mathbb{R}^n \to H$, where H is a Hilbert space, be the a projection function. Let $K : \mathbb{R}^n \times \mathbb{R}^n \to \mathbb{R}$ be the kernel function, defined as $K(u,v) =< \Phi(u), \Phi(v) >_H$. We search for w as a linear combination of the training points, $w = \sum_{i=1}^m \alpha_i \Phi(x_i)$.

Let $K_+^i = \frac{1}{|S_+|} \sum_{x_j \in S_+} K(x_i, x_j)$ and $K_-^i = \frac{1}{|S_-|} \sum_{x_j \in S_-} K(x_i, x_j)$. The positive and negative averages are, then:

$$E_+ = \sum_{i=1}^m \alpha_i K_+^i, \quad E_- = -\sum_{i=1}^m \alpha_i K_-^i$$

The linear feasibility systems formulated by PVM, parameterized by $t \in [0, \infty)$:

$$Feas(t) = \begin{cases} \left|\sum_{x_i \in S} \alpha_i(K(x_i, x_j) - K_+^i)\right| \leq \sigma_+^j \ , \ \forall x_j \in S_+ \\ (|S_+| - 1)t \cdot \left(b + \sum_{x_i \in S} \alpha_i K_+^i\right) - \sum_{x_i \in S_+} \sigma_+^i \geq 0 \\ b + \sum_{x_i \in S} \alpha_i K_+^i \geq 1 \\ \\ \left|\sum_{x_i \in S} \alpha_i(K(x_i, x_j) - K_-^i)\right| \leq \sigma_-^j \ , \ \forall x_j \in S_- \\ (|S_-| - 1)t \cdot \left(-b - \sum_{x_i \in S} \alpha_i K_-^i\right) - \sum_{x_i \in S_-} \sigma_-^i \geq 0 \\ -b - \sum_{x_i \in S} \alpha_i K_-^i \geq 1 \end{cases} \tag{2.2}$$

where σ_+^j is a majorant for the deviation of $x_j \in S_+$ and σ_-^j is a majorant for the deviation of $x_j \in S_-$. The optimal hyperplane for System 2.1 may be found by searching the minimal value of t, denoted as $t_{optimal}$, for which the linear system $Feas(t)$ is still feasible.

3   Integration of a Data Clustering Step

Although PVM can be solved efficiently using any of the readily available linear solvers, it turns out that solving times can be quite long. This is in part due to the fact that system (2.2) is a dense system, which, when we deal with m entries, has $2m + 4$ equations, with $2m$ unknowns and a ratio of non zero terms of approximately 0.5 of the total size of the system. Another factor is that the systems required to be solved will always turn into degenerate ones, as –whilst $t \to t_{optimal}$– the feasible area shrinks to just one point.

The strong constraints on the feasibility systems, as well as the memory complexity explosion of simplex solvers naturally lead to the idea of sampling the training database.

The first thing to consider when reducing the training set is to observe that PVM is based on a statistical model. One can obtain the same separating hyperplane by sampling the initial training set such that only a small deviation of the statistical measures is induced. Suppose that the signed distances from the hyperplane is normally distributed, with average μ. Let E be the average signed distance of a sample set. E has the property that $\frac{(E-\mu)\sqrt{N}}{s}$ follows a Student-T distribution with $N-1$ degrees of freedom, where s is the standard deviation of the sample set and N is the sample set size. s has the property that $\frac{(N-1)s^2}{\sigma^2}$ follows a χ^2 distribution, where σ is the standard deviation of the whole population. Therefore, for very large databases, we may compute the required sample set size for a desired level of confidence.

3.1   Using Weights

A first approach in reducing the database size is to combine the training with a clustering step. It is shown below that, by using weights in the training phase, one can obtain the same statistical model for the clustered set as that obtained for the original set.

Let $C_1, C_2, \ldots, C_k \subset S$ denote a partitioning of the training data with the property that a cluster contains only points of one label:

$$C_1 \cup C_2 \cup \ldots \cup C_k = S$$
$$C_i \neq \emptyset, \forall i \in \{1, \ldots, k\}$$
$$C_i \cap C_j = \emptyset, \forall i, j \in \{1, \ldots, k\}, i \neq j$$
$$y_j = y_l, \forall x_j, x_l \in C_i, \forall i \in \{1, \ldots, k\}$$

Let $P_i, i \in \{1, \ldots, k\}$ be the centers of these clusters:

$$P_i = \frac{1}{|C_i|} \sum_{x \in C_i} x$$

In order to obtain the same statistical model, a weighted average is used instead of the simple average when training on the cluster centers. Let C_+ be the set of clusters which contain only positively labeled points and with C_- the set of clusters which contain only negatively labeled points.

Note that $\cup_{C \in C_+} C = S_+$ and $\cup_{C \in C_-} C = S_-$. The positive and negative averages are, then :

$$E_{C_+} = b + \frac{1}{|S_+|} \sum_{C_i \in C_+} |C_i| <w, P_i>$$
$$E_{C_-} = -b - \frac{1}{|S_-|} \sum_{C_i \in C_-} |C_i| <w, P_i>$$

Given the way P_i are computed and since $\cup_{C \in C_+} C = S_+$, we have:

$$E_{C_+} = b + \frac{1}{|S_+|} \sum_{C_i \in C_+} |C_i| < w, P_i > \tag{3.1}$$

$$= b + \frac{1}{|S_+|} \sum_{C_i \in C_+} \left(|C_i| < w, \frac{1}{|C_i|} \sum_{x \in C_i} x > \right) \tag{3.2}$$

$$= b + \frac{1}{|S_+|} \sum_{C_i \in C_+} \left(\sum_{x \in C_i} < w, x > \right) \tag{3.3}$$

$$= b + \frac{1}{|S_+|} \sum_{x \in S_+} < w, x > \qquad = E_+ \tag{3.4}$$

and similarly, $E_{C_-} = S_-$. Hence, computing the average in a weighted manner leads to obtaining the same average as on the nonclustered set. The same weighing is used when the average deviations are computed:

$$\sigma_+ = \frac{1}{|S_+| - 1} \sum_{C_i \in C_+} |C_i| \sigma_+^i$$
$$\sigma_- = \frac{1}{|S_-| - 1} \sum_{C_i \in C_-} |C_i| \sigma_-^i$$

but in this case one no longer obtains the same average deviations as on the original set. This is due to the inequality:

$$\left| \sum_{x \in C_i} (< w, x > + b - E) \right| \leq \sum_{x \in C_i} | < w, x > + b - E| \tag{3.5}$$

which becomes a strict inequality when there are at least two members in the first sum with opposite signs.

The two sides in (3.5) are equal if all the terms $< w, x > + b - E$ have the same sign for all $x \in C_i$.

When using kernels, the clustering will take place in the projection space. It is important to reformulate the system such that the cluster centers themselves will no longer be explicitly used.

Let $\Phi : \mathbb{R}^n \to H$ be the implicit projection function and redefine P_i, the cluster centers as:

$$P_i = \frac{1}{|C_i|} \sum_{x \in C_i} \Phi(x)$$

The normal to the hyperplane then becomes:

$$w = \sum_{i=1}^{k} \alpha_i |C_i| P_i$$

$$= \sum_{i=1}^{k} \alpha_i \left(\sum_{x \in C_i} \Phi(x) \right)$$

The $< w, P_j >_H$ scalar products in H can be rewritten as:

$$< w, P_j >_H = \sum_{i=1}^{k} \alpha_i \left(\sum_{x \in C_i} < \Phi(x), P_j >_H \right)$$

$$= \sum_{i=1}^{k} \alpha_i \frac{1}{|C_j|} \sum_{x_i \in C_i, x_j \in C_j} < \Phi(x_i), \Phi(x_j) >_H$$

$$= \frac{1}{|C_j|} \sum_{i=1}^{k} \alpha_i \sum_{x_i \in C_i, x_j \in C_j} K(x_i, x_j)$$

Hence, the feasibility system (2.2) can be rewrittten without explicitly using the cluster centers. Instead, they can be replaced by the average of the projected points in the clusters, resulting in the following system for $Feas(t)$:

$$\begin{cases} E_+ = b + \sum_{i=1}^{k} \alpha_i \sum_{x_i \in C_i} K_+^i \\ \left| \frac{1}{|C_j|} \sum_{i=1}^{k} \alpha_i \sum_{x_i \in C_i, x_j \in C_j} K(x_i, x_j) + b - E_+ \right| \leq \sigma_+^j, \\ \forall C_j \in C_+ \\ \sigma_+ = \frac{1}{|S_+|-1} \sum_{C_i \in C_+} |C_i| \sigma_+^i \\ \sigma_+ \leq t E_+ \\ E_+ \geq 1 \\ E_- = -b - \sum_{i=1}^{k} \alpha_i \sum_{x_i \in C_i} K_-^i \\ \left| \frac{1}{|C_j|} \sum_{i=1}^{k} \alpha_i \sum_{x_i \in C_i, x_j \in C_j} K(x_i, x_j) + b + E_- \right| \leq \sigma_-^j, \\ \forall C_j \in C_- \\ \sigma_- = \frac{1}{|S_-|-1} \sum_{C_i \in C_-} |C_i| \sigma_-^i \\ \sigma_- \leq t E_- \\ E_- \geq 1 \end{cases} \quad (3.6)$$

In doing so, we eliminate the necessity of explicitly using the cluster centers, which leads to correct average computations even when using kernels. Furthermore, the advantages of using a clustered data set are preserved, as the actual problem size coincides with that of the clustered data set, due to the decrease of the number of unknowns in the linear feasibility problem.

3.2   Dividing Clusters

Due to the inequality (3.5), the average deviations used for the clustered set can substantially differ from the ones that would result on the unclustered data. Initial testing confirmed that this induces a negative effect marked by a severe degradation of the classification ability. Therefore, it is important to obtain a set of clusters which lead to relatively small differences between the members of inequality (3.5).

We define the relative induced distortion for cluster C_j, denoted $A(C_j)$, as:

$$1 - \frac{\left|\sum_{i=1}^{k} \alpha_i \sum_{x_i \in C_i, x_j \in C_j} K(x_i, x_j) + |C_j|b - |C_j|E\right|}{\sum_{x_j \in C_j} \left|b + \sum_{i=1}^{k} \alpha_i \sum_{x_i \in C_i} K(x_i, x_j) - E\right|} \tag{3.7}$$

where E is either E_+ or E_- depending on the labeling of the points in cluster C_j. This measure reflects the relative amount by which the average deviation differs in the clustered data from the nonclustered data.

For a cluster where all its points lie on the same side of E, the nominator and denominator in (3.7) will be equal and, hence, $A(C_j) = 0$. After training the classifier for a given cluster set, we will check which clusters have a relative induced distortion over a certain threshold and split each such cluster into two new clusters. For a cluster with high relative induced distortion, C_j, the two new clusters are:

$$C_{j+} = \{x_j \in C_j | b + \sum_{i=1}^{k} \alpha_i \sum_{x_i \in C_i} K(x_i, x_j) \geq E\}$$
$$C_{j-} = \{x_j \in C_j | b + \sum_{i=1}^{k} \alpha_i \sum_{x_i \in C_i} K(x_i, x_j) < E\}$$

Consequently:

$$C_{j+} \neq \emptyset, C_{j-} \neq \emptyset$$
$$C_{j+} \cap C_{j-} = \emptyset$$
$$C_{j+} \cup C_{j-} = C_j$$
$$A(C_{j+}) = A(C_{j-}) = 0$$

The idea is to cycle through training and splitting clusters until a sufficiently low error in the statistical model is reached. The only question that remains open is what would be a good halting criteria.

After training the classifier with a cluster set, the actual σ_+, σ_- for the non-clustered sets can be calculated. Let σ_+^c, σ_-^c be the average deviations obtained for the clustered set and let σ_+^n, σ_-^n be the average deviations for the nonclustered set. Let:

$$t_{opt}^c = \max\{\frac{\sigma_+^c}{E_+}, \frac{\sigma_-^c}{E_-}\}$$
$$t_{opt}^n = \max\{\frac{\sigma_+^n}{E_+}, \frac{\sigma_-^n}{E_-}\}$$

Note that, when the relative distortion of all clusters is 0, the only difference between $t_{optimal}$ and t_{opt}^n is the subspace of H where w is sought. When training over nonclustered data, the dimension of the subspace, W_{train}, is at most m, the number of training points. When training over clustered data, some points are enforced to have the same weight in the expression of w. The subspace in which we search for w, $W_{train_{clst}}$, is k-dimensional, where k is the number of clusters, and $W_{train_{clst}} \subset W_{train}$.

From (3.5) it follows that $t_{opt}^c \leq t_{opt}^n$. As the hyperplane obtained in the clustered form satisfies the conditions of system (2.1), but is not necessarily optimal for system (2.1), $t_{optimal} \leq t_{opt}^n$ will also hold.

Suppose that the linear subspace generated by the implicit cluster centers is identical to the linear subspace generated by the points themselves, $W_{train_{clst}} =$

W_{train}. As the statistical model obtained in the clustered form always has lower average deviations than the non clustered form, we will also have $t^c_{opt} \leq t_{optimal}$. Furthermore, under this condition we have:

$$t^c_{opt} \leq t_{optimal} \leq t^n_{opt}$$

Notice that, according to Section 2, t^n_{opt} is a good indicator for the probability of error. We will use the relative difference between t^c_{opt} and t^n_{opt} as a halting criteria for our algorithm. The complete procedure for training with implicit clusters is presented in Algorithm 1.

Algorithm 1. Algorithm for training with implicit clusters

Function bool SplitClusters(double maxDistortion)
$retValue \leftarrow FALSE$
for i = Clusters.count - 1; i $\geq$ 0; i−− **do**
  if Distortion(Clusters[i]) $\geq$ maxDistortion **then**
    Clusters.Append(Clusters[i].SplitPositive())
    Clusters.Append(Clusters[i].SplitNegative())
    Clusters.Erase(i)
    $retValue \leftarrow TRUE$
  end if
end for
return $retValue$
EndFunction
$maxDistortion \leftarrow 1.0$
$distortionDecreaseRate \leftarrow 0.9$
$\epsilon \leftarrow 0.05$
while true **do**
  TrainUsingCurrentClusters()
  RecomputeStatistics(t^c_{opt}, t^n_{opt})
  if $t^n_{opt} < t^c_{opt} \cdot (1 + \epsilon)$ **then**
    break
  end if
  while !SplitClusters(maxDistortion) **do**
    $maxDistortion* = distortionDecreaseRate$
  end while
end while

For all the trainings, we used an *distortionDecreaseRate* of 0.9. This parameter controls how fast the maximum relative induced distortion will decrease. Smaller values for this determine a faster termination of the algorithm, but increase the final number of clusters. The reason for this behaviour is that, when the *distortionDecreaseRate* is smaller the subsequent values obtained for *maxDistortion* are less coarse and, hence, may miss the value for which the number of clusters is minimal.

In Figure 1 we show how t^c_{opt} and t^n_{opt}, labeled as Clustered T and Real T, evolve over successive iterations on the tested databases.

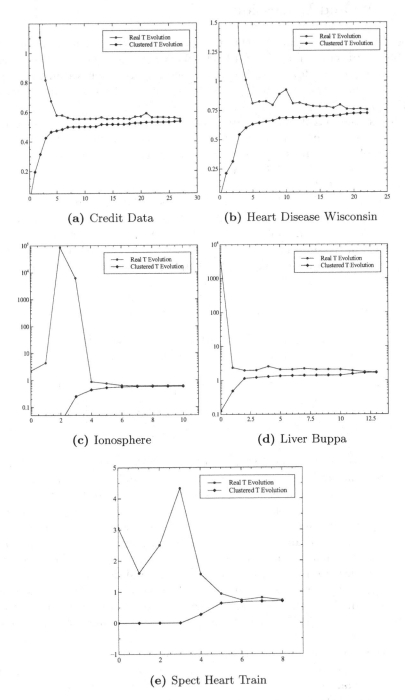

(a) Credit Data

(b) Heart Disease Wisconsin

(c) Ionosphere

(d) Liver Buppa

(e) Spect Heart Train

Fig. 1. Evolution of t_{opt}^c(Clustered T) and t_{opt}^n(Real T). The X axis represents the iteration count, the Y axis the values for t_{opt}^c and t_{opt}^n.

4   Results and Discussion

For the *TrainUsingCurrentClusters* procedure of Algorithm 1 we have used the
GNU Linear Programming Kit (GLPK), which is freely available. A version of
the algorithm, with a few extra features which are not discussed in this paper,
can be found at https://code.google.com/p/dpvm/. This code also functions
with CPLEX.

We tested the classifier on 5 databases taken from the UCI ML repository.
The datasets used can be found at http://archive.ics.uci.edu/ml/.

On each of the databases, 30 random splits were used separately, with each
split containing 80% of the data for training phase and 20% for the testing phase.
The results presented are averages over these 30 runs.

For reference, we also included in the comparison the soft margin SVM, which
we have trained using the same kernel. We have used five fold cross-validation
in order to determine the optimal value of the SVM parameter which controls
error tradeoff in the objective. This parameter is usually denoted as C. We then
used the obtained value for running the trainings on the 30 slices. The results

Table 1. Results for the Clustering PVM with cluster splitting

Problem Name	SVM	PVM	$t_{optimal}$	C-PVM	t_{opt}^n
Credit Screening	76.98	84.13	0.5397	84.76	0.5514
Heart Disease Cleaveland	54.23	82.11	0.7437	80.91	0.7543
Ionosphere	64.28	82.18	0.5922	82.13	0.6112
Liver	68.45	66.91	1.6927	66.31	1.715
Heart Spect Train	76.45	62.31	0.7018	66.90	0.7545

in Table 1 are obtained using an ϵ value of 0.05 in Algorithm 1. There is almost
no decrease in accuracy and the value of t_{opt}^n closely follows that of $t_{optimal}$.

For $\epsilon = 0$, the final number of clusters increases only slightly, but the overall
running time increases considerably as in the final iterations only a very small
clusters require splitting. However, for $\epsilon = 0$, we obtain a completely identical
model for the clustered and nonclustered training, giving perfectly identical results in the testing phase. This implies that the separating hyperplane is equally
well expressed using only the subspace of the clustered data.

Table 2 shows how the average number of clusters obtained for a problem over
the course of the 30 slices compares to the original size of the problem. These
were obtained for $\epsilon = 0.05$. The ratio of clustered versus unclustered data size
is, on average, 0.37.

The starting number of clusters has been set to 3% of the original dataset
size. We have experimented with different numbers of starting clusters, but have
found only a small variation in both the final number of clusters and the accuracy
of the resulting classifier.

Table 2. Original and clustered dataset sizes

Problem Name	Original Size	Average Clustered Size
Credit Screening	690	225.03
Heart Disease Cleaveland	303	104.83
Ionosphere	351	145.7
Liver	341	67.4
Heart Spect Train	80	48.17

5 Conclusions

We have shown how, through exploitation of the specific structure of the PVM classifier, hybridization with a clustering step is possible, designed to reduce the size of the training problem whilst maintaining the properties of the statistical classifier, namely its good generalization ability and its natural stability to outliers. One important observation is that the number of clusters does not increase linearly with the size of the original dataset. The reduction is significant as the simplex solvers used are sensible to the size of the problem, in terms of memory and running time.

Future development on the matter will search for a suitable way of unifying clusters after the split procedure in order to further reduce the size of the problem. We also plan to develop a distributed solver for the problem for tackling very large databases.

Future research will also target the clustering method used, as K-means is not necessarily optimal and was chosen in this research only for its ease of usage.

Acknowledgements. This work has been supported by the European Social Fund in Romania, under the responsibility of the Managing Authority for the Sectoral Operational Programme for Human Resources Development 2007-2013 [grant POSDRU/CPP 107/DMI 1.5/S/ 78342].

References

1. Vapnik, V.N., Boser, B.E., Guyon, I.: A training algorithm for optimal margin classifiers. In: COLT 1992 Proceedings of the Fifth Annual Workshop on Computational Learning, Pittsburgh, PA, USA, vol. 5, pp. 144–152. ACM, New York (1992)
2. Cortes, C., Vapnik, V.N.: Support-vector networks. Machine Learning 20(3), 273–297 (1995)
3. Vapnik, V.N.: Statistical learning theory. John Wiley and Sons Inc. (1998)
4. Chapelle, O., Vapnik, V.N.: Bounds on error expectation for support vector machines. Neural Computation 12(9), 2012–2036 (2000)
5. Chapelle, O., Vapnik, V.N.: Choosing multiple parameters for support vector machines. Machine Learning 46(1-3), 131–159 (2001)

6. Fan, R.-E., Chen, P.-H., Lin, C.-J.: Working set selection using second order information for training support vector machines. The Journal of Machine Learning Research 6, 1889–1918 (2005)

7. Chang, C.-C., Lin, C.-J.: Libsvm: a library for support vector machines. ACM Transactions on Intelligent Systems and Technology 2(3) (2011)

8. Yu, H., Yang, J., Han, J.: Classifying large data sets using svms with hierarchical clusters. In: Proceedings of the Ninth ACM SIGKDD International Conference on Knowledge Discovery and Data Mining, Washington, DC, USA, pp. 306–315. ACM, New York (2003)

9. Ligges, U., Krey, S.: Feature clustering for instrument classification. Computational Statistics 23, 279–291 (2011)

10. Awad, M., Khan, L., Bastani, F., Yen, I.-L.: An effective support vector machines (svm) performance using hierarchical clustering. In: Proceedings of the 16th IEEE International Conference on Tools with Artificial Intelligence, Boca Raton, FL, USA, pp. 663–667. IEEE Computer Society, Washington (2004)

11. Nath, J.S., Bhattacharyya, C., Murty, M.N.: Clustering based large margin classification: A scalable approach using socp formulation. In: Proceedings of the 12th ACM SIGKDD International Conference on Knowledge Discovery and Data Mining, Philadelphia, USA, pp. 374–379. ACM, New York (2006)

12. Sucilă, A.: A Distributed Statistical Binary Classifier. Probabilistic Vector Machine. Ph.D. Thesis, Alexandru Ioan Cuza University, Faculty of Computer Science (2012)

13. Cervantes, J., Li, X., Yu, W.: Support vector machine classification based on fuzzy clustering for large data sets. In: Gelbukh, A., Reyes-Garcia, C.A. (eds.) MICAI 2006. LNCS (LNAI), vol. 4293, pp. 572–582. Springer, Heidelberg (2006)

14. Cervantes, J., Li, X., Yu, W., Li, K.: Support vector machine classification for large data sets via minimum enclosing ball clustering. Neurocomputing 71(4-6), 611–619 (2008)

15. Nath, J.S., Bhattacharyya, C., Murty, M.N.: Clustering based large margin classification: a scalable approach using socp formulation. In: Proceedings of the 12th ACM SIGKDD International Conference on Knowledge Discovery and Data Mining, KDD 2006, pp. 674–679. ACM, New York (2006)

Resource Constrained Project Scheduling
with General Precedence Relations Optimized with SAT

Rui Alves[1], Filipe Alvelos[2], and Sérgio Dinis Sousa[2]

[1] Center for Computer Graphics
Campus de Azurém, 4800-058 Guimarães, Portugal
rmfalves@gmail.com
[2] Department of Production and Systems / Centro Algoritmi
University of Minho
Campus de Gualtar, 4710-057 Braga, Portugal
falvelos@dps.uminho.pt
sds@dps.uminho.pt

Abstract. This paper presents an approach, based on propositional satisfiability (SAT), for the resource constrained project scheduling problem with general precedence relations. This approach combines propositional satisfiability formulations with a bisection method, in order to achieve an optimal solution. The empirical results suggest that when the optimal schedule is significantly affected by the availability of resources, this strategy outperforms the typical integer linear programming approach.

1   Introduction

This paper addresses the Resource Constrained Project Scheduling Problem with General Precedence Relations (RCPSP-GPR) which is a generalization of the Resource Constrained Project Scheduling Problem (RCPSP) [1]. They both deal with the problem of determining a schedule for the activities of a project, given their durations, the precedence relations between them, and the resources demands and capacities, so that the schedule is consistent with those constraints and the project time span is minimal.

In the RCPSP, a precedence relation between two activities means that the first activity must finish for the second activity to start. In the RCPSP-GPR, a precedence relation between two activities has four different combined possibilities: the first activity must either start or finish, in order either to start or to finish the second activity. Thus, there are four different types of precedence relations: finish to start (FS), in which the first activity must finish for the second one to start; finish to finish (FF), in which the first activity must finish for the second one to finish as well; start to start (SS), in which the first activity must start for the second one to start either; start to finish (SF), in which the first activity must start for the second one to finish.

Resources play an important role in project scheduling and can be classified either as renewable or as non-renewable. Non-renewable resources, such as funding, energy, or material, are consumed by the activities. Renewable resources, such as storage

L. Correia, L.P. Reis, and J. Cascalho (Eds.): EPIA 2013, LNAI 8154, pp. 199–210, 2013.
© Springer-Verlag Berlin Heidelberg 2013

space, or manpower, are not consumed and keep the same stock along the entire lifetime of the project. This work deals only with renewable resources. For more details on precedence relations and resource constraints see Tavares [2].

Both problems (RCPSP and RCPSP-GPR) are NP-hard [3], hence no polynomial algorithms are currently known for them. The RCPSP is denoted as *PS | prec | C_{max}* as per the classification by Brucker et al [1].

The goal of this work is to develop an efficient approach to solve the RCPSP-GPR. Therefore an approach with that goal in mind is proposed, based on propositional satisfiability (SAT) [4] combined with a specific iterative algorithm.

The RCPSP has been a subject of study for several decades, and many different approaches have been proposed to solve it. The operational research community has provided several methods. Some of them are based on integer linear programming (ILP) models, such as the ones proposed by Talbot [5] or by Reyck and Herroelen [6]. Other operational research methods lay on heuristic processes, as is the case of the one proposed by Wiest [7]. In the constraint programming community, there are approaches such as the one proposed by Liess and Michelon [8], as well as others based on stochastic local search strategies [9][10]. There are also contributions from artificial intelligence domains that exploit the potential of genetic algorithms [11] and neural networks [12]. In what concerns to propositional satisfiability, henceforth mentioned as SAT, its potential for planning is well known [13], and two models have recently been proposed, one by Coelho and Vanhoucke [14], and other by Horbach [15].

The work by Coelho and Vanhoucke [14] addresses the Multi-Mode Resource Constrained Project Scheduling Problem (MMRCPSP). In this variant, each activity admits several modes, representing a relation between resource requirements and duration. Their approach combines SAT-k and an efficient meta-heuristic procedure from the literature. Thus, this present work differs from that one also because it uses pure SAT encoding, combined with a new search method to reach the optimal solution. Furthermore, the current work addresses general precedence relations.

Horbach [15] proposes a pure SAT model for the RCPSP problem, although without general precedence constraints and without targeting an optimal solution. A pure SAT formulation, by itself, is unable to return an optimal schedule. Instead, it requires that a project time span is given as a parameter, and then it simply finds a schedule, within that time span, consistent with the project constraints. That schedule, though, is not guaranteed to be optimal, since it may be possible to find another one within a lesser time span.

The present work takes into account with general precedence relations and introduces an algorithm to converge to an optimal schedule with minimum time span.

This paper is organized as follows: the next section presents a definition of the problem; the third section contains the model and the method that are the essence of this work; the fourth section has an integer linear programming model proposed by Talbot [5], against which the present approach is compared; finally, the fifth and sixth sections show some experimental evidence and conclusions.

2   Problem Definition

The RCPSP-GPR problem can be formally defined as follows:

Given a project P, such that:

a)   The project P contains a set A of activities, where $A=\{a_1,a_2,a_3,...,a_n\}$

b)   The project P contains a set R of resources, where $R=\{r_1,r_2,r_3,...,r_m\}$

c)   For any activity a, the starting instant of the activity a is represented by $s(a)$.

d)   For any activity a, the duration of the activity a is represented by $d(a)$.

e)   The project P contains a set S of general precedence relations, where a general precedence relation $prec(a,a')$ is defined as a functional application such that prec: AxA $\rightarrow$ {SS,SF,FS,FF}.

f)   For any resource r, the capacity of the resource r at any instant of the lifetime of the project P is represented by $c(r)$.

g)   For any activity a, and for any resource r, the demand (usage) of the resource r by the activity a, at any instant of the lifetime of the activity a, is represented by $u(a,r)$.

The goal is to minimize the duration of the project. Representing the start of an activity a by $s(a)$, the objective function is:

Minimize T such that $T = \max_{a\in A}[s(a) + d(a)]$

Under the following constraints:

1. $\sum_{a\in A} u(a,r) \leq c(r)$, for any resource $r \in R$, at any instant t of the project.

2.   If there is an activity a' such that $prec(a,a') \in S$, then:

prec(a,a')=SS $\Rightarrow$ s(a) < s(a')

prec(a,a')=FS $\Rightarrow$ s(a)+ d(a) - 1 < s(a')

prec(a,a')=FF $\Rightarrow$ s(a) + d(a) < s(a') + d(a')

prec(a,a')=SF $\Rightarrow$ s(a) < s(a') + d(a') - 1

3   A Propositional Satisfiability Approach

The propositional satisfiability approach presented here has two main components. Firstly, there is a satisfiability model that accounts for precedence relations and returns a consistent scheduling within a given project time span. Next, it is shown how this model can be iteratively invoked to converge towards a scheduling in which the project time span is the optimal one.

3.1   The Propositional Satisfiability Problem

Given a propositional formula, the problem of the propositional satisfiability consists of finding valid instantiations for its variables such that the formula becomes true, or otherwise determining that such values do not exist, in which case the formula is considered UNSAT. The satisfiability problem usually abbreviates to SAT in the literature. The SAT problem is proved to be NP-complete [3], meaning that unless an affirmative answer to the "P=NP" problem exist, any algorithm to solve it is worst case exponential. In many of the publicly available solvers, the basic solving process is based on the Davis-Putnam-Logemann-Loveland (DPLL) algorithm [16], [17]. However, the SAT algorithms have known further substantial improvements, with dynamic decision heuristics as VSIDS [18], VMTF [19] and Berkmin [20], efficient binary propagation techniques as the TWL [18], smart restart policies [21], or conflict driven clause learning [22]. As a result of these major breakthroughs, SAT algorithms became efficient enough to be suitable in various fields [23], [24], [25], [26], [27].

3.2   A Propositional Satisfiability Model

Next follows the SAT encoding for the RCPSP-GPR problem, as proposed on this paper.

Let us consider the following logical decision variables:

a) The set of variables $s_{i,t}$ where $s_{i,t}$ is true if and only if the activity a_i starts at the instant t, where i=1,2,3,...,n and t=0,1,2,...,T-1. The variable $\overline{s_{i,t}}$ is true otherwise.

b) The set of variables $x_{i,t}$ where $x_{i,t}$ is true if and only if the activity a_i is running at the instant t. The variable $\overline{x_{i,t}}$ is true otherwise.

There is a valid schedule for the project P if and only if there is an instantiation for the above variables so that all of the following conditions stand:

1.   Every activity a_i has a unique start instant. Given any activity a_i, and given a subset of variables S'={s_{it}:t=0,1,2,...,T-1}, then one and only one of the variables from S' is true.

2.   No activity starts too late to finish within the project time span. Thus for any activity a_i the logical proposition (1) is true

$$\bigwedge_{t=T-d(a_i)+1}^{T-1} \overline{s}_{it}$$

(1)

3. For every activity a_i, if a_i starts at an instant t, then a_i runs at the instants t, $t+1, t+2, \ldots, t+d(a_i)-1$, and does not run otherwise. Thus for any activity a_i the logical proposition (2) is true

$$s_{it} \implies \bigwedge_{j=0}^{t-1} \overline{x}_{ij} \quad \text{and} \quad \bigwedge_{j=t}^{t+d(a_i)-1} x_{ij} \quad \text{and} \quad \bigwedge_{j=t+d(a_i)}^{T-1} \overline{x}_{ij}$$

(2)

4. There are no time gaps, that is, before the project completion there is not any instant empty of running activities. If such an instant occurs, then all the activities must be finished, and so the next instant is also empty of ongoing activities. Thus for any activity a_i the logical proposition (3) is true

$$\bigwedge_{i=1}^{n} \overline{x}_{it} \implies \bigwedge_{i=1}^{n} \overline{x}_{it+1}$$

(3)

5. Given two activities a_i and a_j, if $(a_i,a_j,FS) \in G$, then the activity a_j does not start before the activity a_i has finished. Thus for every two activities a_i and a_j such that $(a_i,a_j,FS) \in G$, with $i=1,2,\ldots,n$, with $j=1,2,\ldots,n$, and for every $t=0,1,2,\ldots,T-1$, the logical proposition (4) is true

$$s_{it} \implies \bigwedge_{k=0}^{t+d(a_i)-1} \overline{s}_{jk}$$

(4)

6. Given two activities a_i and a_j, if $(a_i,a_j,SS) \in G$, then the activity a_j does not start before the activity a_i has started. Thus for every two activities a_i and a_j such that $(a_i,a_j,SS) \in G$, with $i=1,2,\ldots,n$, with $j=1,2,\ldots,n$, and for every $t=0,1,2,\ldots,T-1$, the logical proposition (5) is true

$$s_{it} \implies \bigwedge_{k=0}^{t-1} \overline{s}_{jk}$$

(5)

7. Given two activities a_i and a_j, if $(a_i,a_j,FF) \in G$, then the activity a_j does not finish before the activity a_i has finished. Thus for every two activities a_i and a_j such that $(a_i,a_j,FF) \in G$, with $i=1,2,\ldots,n$, with $j=1,2,\ldots,n$, and for every $t=0,1,2,\ldots,T-1$, the proposition (6) is true

$$s_{it} \implies \bigwedge_{k=0}^{t+d(a_i)-d(a_j)-1} \overline{s}_{jk} \tag{6}$$

8. Given two activities a_i and a_j, if $(a_i, a_j, SF) \in G$, then the activity a_j does not finish before the activity a_i has finished. Thus for every two activities a_i and a_j such that $(a_i, a_j, SF) \in G$, with i=1,2,...,n, with j=1,2,...,n, and for every t=0,1,2,...,T-1, the proposition (7) is true

$$s_{it} \implies \bigwedge_{k=0}^{t-d(a_j)+1} \overline{s}_{jk} \tag{7}$$

9. At any instant t, the total usage of a resource r by the ongoing activities cannot exceed $c(r)$. Hence, if there is a minimal set M of activities a_k that verifies the condition stated in (8)

$$\exists r \in R: \sum_{a_k \in M} u(a_k, r) > c(r) \tag{8}$$

then the logical proposition (9) is true for every instant t of the project, with t=0,1,2,...,T-1

$$\bigvee_{a_k \in M} \overline{x}_{kt} \tag{9}$$

This model, by itself, generates a consistent scheduling, but does not assure the optimality of that scheduling. In other words, the generated scheduling meets the precedence relation constraints, as well as the resource usage constraints, but does not necessarily correspond to a scheduling within the minimal possible time span for the project. The SAT method takes a project time frame as one of the inputs of the problem, and generates a propositional formula based upon it. The output is either a consistent scheduling, or the information that a consistent scheduling is not possible within the time frame passed as input (the formula is impossible, aka UNSAT).

3.3 A Bisectional Method for Accomplishing an Optimal Solution

The bisection method begins by setting an initial range, with sequential numbers, where the optimal time span is guaranteed to exist. The lower bound of that range is any point T_{min} at which $F(T_{min})$ is UNSAT, whereas the upper bound is any point T_{max}

such that there is a solution for $F(T_{max})$. For instance, T_{min} can be given by the total duration of the critical activities minus one, and T_{max} may be the total duration of every activities.

Once T_{min} and T_{max} are set, the middle point $T_{mid}=(T_{max} + T_{min}) / 2$ (truncated) is computed, and $F(T_{mid})$ is solved. If $F(T_{mid})$ results UNSAT, then the lower bound T_{min} rises to T_{mid}, otherwise, the upper bound T_{max} drops to T_{mid}. After this, there is a new sequential range, with about half the length of the previous range, where $F(T_{min})$ is UNSAT and $F(T_{max})$ has a solution. The procedure is then repeated for the most recent lower and upper bounds, so that this new search range is truncated again into half. This shrinking process continues forth, until only two sequential elements remain. At that point, as a consequence from the rule above, it can be deduced that T_{max} is optimal, since $F(T_{max})$ has solution but $F(T_{max}-1)$, actually $F(T_{min})$, is UNSAT.

With this method, the width of the search range halves at each iteration, hence it requires about $log_2(M)$ iterations to complete, where M is the width of the initial range. The Figure 1 formally defines the algorithm.

FUNCTION BISECTIONAL SEARCH

INPUT: Project P and its constraints

OUTPUT: Optimal scheduling

Tmin := Total duration of the critical activities of P, minus one

Tmax := Total duration of the activities of P (critical and non-critical)

WHILE Tmax – Tmin > 1

      Tmed := Trunc((Tmax + Tmin) / 2)

      IF Solution(F(T)) = $\varnothing$ **THEN** Tmin := Tmed **ELSE** Tmax := Tmed

END WHILE

RETURN Tmax, Scheduling F(T)

Fig. 1. The bisectional search algorithm

4   An Integer Linear Programming Model

This paper compares the proposed approach against a well known integer linear programming formulation for the RCPSP by Talbot [5].

Talbot formulates the RCPSP as an integer programming problem where x_{jt} is a binary variable, such that $x_{jt}=1$ if and only if the activity j finishes at the instant t, and $x_{jt}=0$ otherwise. The activities are indexed from 1 to N, where N is the index of a sink activity that must be unique. If that activity does not naturally exist in the project, then it must be added as an artificial activity, with zero duration and no resource consumption. As seen before, every a_j activity is associated an earlier finish and a latest finish time by the critical path method. In the specific case of the terminal activity N, its latest finish time L_N is assigned H, where H is computed through some heuristic to estimate an upper limit for the project time span, as for example, the total duration of the activities. Since the project time span is given by the finish time of the latest project activity (which is unique), the earlier this time is, the lower the project time span will be.

The objective function takes the variables x_{Nt} associated to the time range in which the activity N may finish, with $t=E_N,E_N+1,E_N+2,...,L_N$. All of these variables are set to zero, except one of them (as ensured by the constraints further on). The function is built in terms that its value equals the index of the variable set to one. That way, the function value decreases as that index is lower, and the function value directly returns the minimal time span found for the project. Hence the objective function is given by the expression (10):

$$\text{Minimize} \sum_{t=E_N}^{L_N} t x_{Nt}$$

(10)

That said about the objective function, the following constraints apply:

$$\sum_{t=E_j}^{L_j} x_{jt} = 1 \quad \text{for} \quad j = 1, \ldots, N$$

(11)

$$-\sum_{t=E_a}^{L_a} t x_{at} + \sum_{t=E_b}^{L_b} (t - d_b) x_{bt} \geq 0$$

(12)

$$-\sum_{t=E_a}^{L_a} t x_{at} + \sum_{t=E_b}^{L_b} t x_{bt} \geq 0$$

(13)

$$-\sum_{t=E_a}^{L_a} (t - d_a) x_{at} + \sum_{t=E_b}^{L_b} (t - d_b) x_{bt} \geq 0$$

(14)

$$-\sum_{t=E_a}^{L_a} (t - d_a) x_{at} + \sum_{t=E_b}^{L_b} t x_{bt} \geq 0$$

(15)

The resource constraints, as defined next, require that for any resource k, at any time instant t, its total demand r_{kt} by the ongoing activities does not exceed its capacity R_k:

$$\sum_{j=1}^{N} \sum_{q=t}^{t+d_j-1} r_{jk} \, x_{jq} \leqslant R_k, \qquad k=1,\ldots,K, \quad t=1,\ldots,H \tag{16}$$

Finally every variable x_{jt} is binary, such that $x_{jt} \in \{0,1\}$ for every j and t.

5   Computational Experiments

The proposed SAT based approach was implemented over the MiniSat framework [28]. The method was tested on 60 instances of 30 activities and on 60 instances of 60 activities from the PSPLIB site [29], and it was compared against the linear integer programming model proposed by Talbot [5]. The Talbot formulation was solved over the CPLEX framework [30]. The tests were conducted on an Intel Core2 Duo 3.00 GHz, in Linux CentOS 5.4 environment. The timeout was set to 7200 seconds.

From the experimental results (http://pessoais.dps.uminho.pt/sds/rcpsp_gp r_results.pdf), two trends arise. First, there is a set of instances in which the linear integer programming approach takes marginal time, and its performance over SAT is clearly superior. In a second group of instances, however, the instances take considerable time for the linear integer programming case, with this one being outperformed by the SAT approach. A statistical analysis suggests that those differences seem to correlate with the level of competition for resources among the activities.

The statistical analysis is based on the differences between the optimal time span and the minimal admissible time span returned by the critical path method, that is, the optimal time span regardless of resource availabilities and considering just the precedence relations constraints. This indicator is henceforth designated as *delta*. Thus, an instance with a zero delta means that the optimal schedule is absolutely coincident with the critical path and that the resources availabilities have null or no relevant effect on the schedule. In opposition, a large delta indicates a scheduling that is considerably affected by the resources constraints.

For each of both methods, and for every instance that was successfully solved with that method (without reaching time out), the delta was grouped by categories, with the first category ranging from zero to four, the second category from five to nine, and so forth, up to a last category for every delta equal or superior to 30. Finally, for each category, it was computed the average solution time of its instances. The results are depicted on the Table 1 and illustrated in the chart of the Figure 2.

As the chart shows, a lower delta category takes a lower average time, whereas a higher delta category corresponds to a higher average time. Since a higher delta is a consequence of a greater impact of the resources constraints, it seems likely that the instances with higher resources competition for resources among its activities are the ones that require higher resolution times.

The SAT approach, on the other hand, behaves in a more independent and stable pattern towards delta. The statistical analysis described above was also performed for the instances that were successfully solved by the SAT approach, and as shown in the Table 1, no correlation was detected.

Table 1. Average solution times by delta category

Range	Average Time with ILP (s)	Average Time with SAT (s)
delta < 5	128	175
5 ≤ delta < 10	882	60
10 ≤ delta < 15	2 414	71
15 ≤ delta < 20	4 161	48
20 ≤ delta < 25	4 127	174
25 ≤ delta < 30	4 675	164
delta ≥ 30	4 847	285

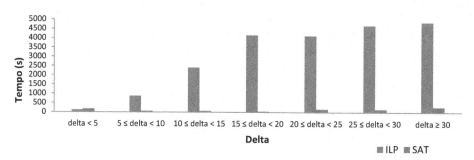

Fig. 2. Column chart of the solution times by delta category

This result suggests that the SAT approach is less vulnerable to the growth of delta. Being so, as the delta increases, this method tends to be more efficient comparatively to the integer linear programming approach.

Another relevant point of analysis is the effect on the performance of the SAT method when it is applied to instances with general precedence relations. For that purpose, there were 30 instances with 30 activities and 30 instances with 60 activities that were modified in order to check if, and to what extent, the performance of this method is affected under the existence of general precedence relations. The modification consisted on replacing each strict precedence relation from the original instance by a new precedence relation between the same activities, by the same order, but where the precedence type was randomly set from any of the four possible types (FS, FF, SS, SF). Since the results have shown that in 76% of the cases the new modified instances were solved in less time than their original counterpart, it seems reasonable to suppose that the general precedence relations has no negative impact on the performance of the proposed SAT method.

6   Conclusion and Further Research

On this paper it was proposed a SAT model for the RCPSP-GPR problem, complemented by a search method to converge to the optimal solution. The efficiency of this approach was compared against an alternative approach from the literature, based on integer linear programming, as proposed by Talbot [5].

The experimental data suggests that the performance of the integer programming method may differ dramatically from one instance to another, even between instances from the same group (instances with the same number of activities), while the performance of the proposed SAT approach tends to be more stable, without such huge differences as in the former method. The statistical analysis indicates that those differences seem to correlate with the level of competition for resources among the activities, so it seems reasonable to state the following:

In the case of the instances in which the optimal scheduling is unaffected or marginally affected by the availability of resources (the optimal time scheduling is close to the minimum possible time project given by the critical path), the integer programming has superior performance relatively to the SAT approach proposed in this work; otherwise, if the optimal scheduling is significantly affected by the availability (or scarcity) of resources, the approach proposed in this work has generally higher efficiency.

As an example of improvement, the search method can evolve in order to start from a closer search range, as a possible way to shorten the number of required iterations. One way by which this can possibly be achieved is to hybridize this method with an integer programming model under a linear relaxation, in order to set a shorter initial search range.

References

1. Brucker, P., Drexl, A., Möhring, R., Neumann, K.: Resource-constrained project scheduling: Notation, classification, models, and methods. European Journal of Operational Research 112(1), 3–41 (1999)
2. Tavares, L.V.: Advanced Models for Project Management. Kluwer Academic Publishers (1998)
3. Garey, M.R., Johnson, D.S.: Computers and Intractability: A Guide to the Theory of NP Completeness. W.H. Freeman and Company (1979)
4. Hammer, P.L., Rudeanu, S.: Boolean Methods in Operations Research and Related Areas. Springer (1968)
5. Talbot, B.: Resource-Constrained Project Scheduling with Time-Resource Tradeoffs: The Nonpreemptive Case. Management Science 28(10), 1197–1210 (1982)
6. Reyck, B., Herroelen, W.: A branch-and-bound procedure for the resource-constrained project scheduling problem with generalized precedence relations. Operations Research 45(2) (March-April 1997)
7. Wiest, J.D.: A heuristic model for scheduling large projects with limited resources. Management Science 13(6), 359–377 (1967)
8. Liess, O., Michelon, P.: A constraint programming approach for the resource-constrained project scheduling problem. Annals of Operation Research 157(1), 25–36 (2008)
9. Cho, J.H., Kim, Y.D.: A simulated annealing algorithm for resource constrained project scheduling problems. Journal of the Operational Research Society 48(7), 736–744 (1997)

10. Fu, N., Lau, H.C., Varakantham, P., Xiao, F.: Robust Local Search for Solving RCPSP/max with Durational Uncertainty. Journal of Artificial Intelligence Research 43, 43–86 (2012)
11. Hartmann, S.: A competitive genetic algorithm for resource constrained project scheduling. Naval Research Logistics 45(7), 733–750 (1998)
12. Colak, S., Agarwal, A., Erenguc, S.: Resource Constrained Scheduling Problem: A Hybrid Neural Approach. Perspectives in Modern Project Scheduling 112(1983), 3–41 (1999)
13. Huang, R., Chen, Y., Zhang, W.: SAS+ Planning as Satisfiability. Journal of Artificial Intelligence Research 43(1), 293–328 (2012)
14. Coelho, J., Vanhoucke, M.: Multi-mode resource-constrained project scheduling using RCPSP and SAT solvers. Journal of Operational Research 213(1), 73–82 (2011) ISSN 0377-2217
15. Horbach, A.: A Boolean satisfiability approach to the resource-constrained project scheduling problem. Annals of Operations Research 181(1), 89–107, doi:10.1007/s10479-010-0693-2
16. Davis, M., Putnam, H.: A Computing Procedure for Quantification Theory. Journal of the Association of Computing Machinery (1960)
17. Davis, M., Logemann, G., Loveland, D.: A Machine Program for Theorem Proving. Communications of the ACM 5, 394–397 (1962)
18. Moskewicz, M.W., Madigan, C.F., Zhao, Y., Zhang, L., Malik, S.: Chaff: Engineering an Efficient SAT Solver. In: Annual ACM IEEE Design Automation Conference (2001)
19. Ryan, L.: Efficient Algorithms for Clause-Learning SAT Solvers, M.S. Thesis, Simon Fraser (2004)
20. Goldberg, E., Novikov, Y.: BerkMin: A fast and robust Sat-solver. Discrete Applied Mathematics 155(12) (June 2007)
21. Biere, A.: Adaptive Restart Strategies for Conflict Driven SAT Solvers. In: Kleine Büning, H., Zhao, X. (eds.) SAT 2008. LNCS, vol. 4996, pp. 28–33. Springer, Heidelberg (2008)
22. Marques, J.P., Karem, S., Sakallah, A.: Conflict analysis in search algorithms for propositional satisfiability. In: Proc. of the IEEE Intl. Conf. on Tools with Artificial Intelligence (1996)
23. Chakradhar, S.T., Agrawal, V.D., Tothweiller, S.G.: A Transitive Closure Algorithm for Test Generation. IEEE Trans. Computer-Aided Design 12(7), 1015–1028 (1993)
24. Larrabee, T.: Efficient Generation of Test Patterns Using Boolean Satisfiability. PhD Dissertation, Dept. of Computer Science, Stanford Univ., STAN-CS-90-1302 (February 1990)
25. Kautz, H., Selman, B.: Planning as Satisfiability. In: Proceedings of the 10th European Conference on Artificial Intelligence, ECAI 1992 (August 1992)
26. Lynce, I., Marques-Silva, J.: Efficient Haplotype Inference with Boolean Satisfiability. In: AAAI 2006 (2006)
27. Ivancic, F., Yang, Z., Ganai, M.K., Gupta, A., Ashar, P.: Efficient SAT-based bounded model checking for software verification. Journal of Theoretical Computer Science 404(3) (September 2008)
28. Eén, N., Sörensson, N.: An Extensible SAT-solver. In: Giunchiglia, E., Tacchella, A. (eds.) SAT 2003. LNCS, vol. 2919, pp. 502–518. Springer, Heidelberg (2004)
29. Kolisch, R., Sprecher, A.: PSPLIB – A project scheduling problem library. European Journal of Operational Research, 205–216 (1996)
30. IBM   CPLEX,   http://www-01.ibm.com/software/integration/optimization/cplex-optimizer/ (January 10, 2013)

A Statistical Binary Classifier: Probabilistic Vector Machine

Mihai Cimpoeşu, Andrei Sucilă, and Henri Luchian

Alexandru Ioan Cuza University, Faculty of Computer Science, Iasi, Romania
mihai.cimpoesu@info.uaic.ro

Abstract. A binary classification algorithm, called Probabilistic Vector Machine – PVM, is proposed. It is based on statistical measurements of the training data, providing a robust and lightweight classification model with reliable performance. The proposed model is also shown to provide the optimal binary classifier, in terms of probability of error, under a set of loose conditions regarding the data distribution. We compare PVM against GEPSVM and PSVM and provide evidence of superior performance on a number of datasets in terms of average accuracy and standard deviation of accuracy.

Keywords: binary classification, hyperplane classifier, SVM, probabilistic, statistical model, kernel.

1 Introduction

In the field of supervised classification, linear binary discriminative classifiers have long been a useful decision tool. Two representatives of this class are the Perceptron, introduced in [7], and Support Vector Machines (SVM), introduced in [19], [5], [20], which have been highly successful in tackling problems from diverse fields, such as bioinformatics, malware detection and many others.

Binary classification problems have as input a training set, which we will denote as $S = \{(x_i, y_i) | x_i \in \mathbb{R}^n, y_i \in \{\pm 1\}, i \in \overline{1..m}\}$, consisting of points $x_i \in \mathbb{R}^n$ and their labels $y_i \in \{\pm 1\}$. We will also denote by $S_+ = \{x_i \in S | y_i = 1\}$ and by $S_- = \{x_i \in S | y_i = -1\}$ the positively and, respectively, negatively labeled training points.

For an input training set a decision function, $f : \mathbb{R}^n \to \{\pm 1\}$, is obtained, for which the parameters are chosen according to the training set. This decision function is later used to classify new examples, referred to as test examples.

Linear binary classifiers search for a hyperplane defined by its normal vector, $w \in \mathbb{R}^n$, and its offset $b \in \mathbb{R}$. This hyperplane is used to define the decision function as $label(x) = sgn(<w, x> + b)$, thus assigning positive labels to the points found in the positive semispace and negative labels to those found in the negative semispace. The hyperplane is refered to as a separating hyperplane.

The focus of this paper is to introduce a binary classifier - named Probabilistic Vector Machine (PVM) - which is based on statistical information derived from the training data. This will provide a strong link to the generalization ability

L. Correia, L.P. Reis, and J. Cascalho (Eds.): EPIA 2013, LNAI 8154, pp. 211–222, 2013.

and a robust model that would be resilient to outliers. The mathematical modeling should also allow for complete and efficient resolution of the optimization problem.

The paper is structured as follows: related work in Section 2; model and motivation in Section 3; kernel usage for nonlinear classification in Section 4; solution for the proposed optimization problem in Section 5; test results in Section 6 and conclusions in Section 7.

2   Related Work

Since the first linear binary classifier, the perceptron, was first introduced in [7] many other classifiers using the same type of decision function have been introduced. The differences among linear binary classifiers are given by the criteria for choosing the separating hyperplane. SVM, introduced in [19], [5], [20], searches for the separating hyperplane which maximizes the distance to the closest point from the S_+ and S_- sets. It was originally developed only for the separable case, in which $conv(S_+) \cap conv(S_-) = \emptyset$, also referred to as the hard margin SVM. However, this rather stringent constraint on the input was later relaxed by allowing training errors, for which the objective is altered. This introduces a tradeoff which is sometimes hard to balance and requires a grid search for the parameters, where every combination of parameters is evaluated via a fold procedure. This can be very time consuming.

SVMs are shown by Vapnik to be built upon the Structural Risk Minimization (SRM) principle. Shortly after its introduction, it quickly outperformed the existing classifiers as shown in domains such as pattern classification [15], text processing [12], bioinformatics [17] and many others. The choice of the hyperplane as done by SVM has been shown to be relatable to the probability of error in [3] and [4]. Some of the shortcomings of SVM stem from the fact that it bases the hyperplane construction on the border elements of S_+ and S_-.

To address some of the limitations of SVMs, [9] introduces a bound on the expected generalization error and subsequently uses this bound in [10] to develop a classifier which minimizes this measure. This allows the separating hyperplane to be built using information derived from the distribution of the entire training set.

The bounds approach, however, produces optimization problems with nonconvex objectives. It leads to models which are difficult to solve efficiently. Currently, the methods based on error bounds resort to simple hill climbing, as the optimization process can become very complicated if one aims at fully solving the underlying problems.

A different approach is taken by [8], where the proximal support vector machines (PSVM) are introduced. PSVM uses two parallel hyperplanes in order to label points. The label is attributed based on which of the two hyperplanes is closer. This, of course, is equivalent to having a single separating hyperplane. This approach eliminates the quadratic program and provides similar classification ability. [16] later introduces the generalized eigenvalue proximal SVM

(GEPSVM), a classification process based on two nonparallel hyperplanes. This idea is further built upon in [11] by introducing the twin SVM (TSVM), where two hyperplanes are used to represent the data. Later, [14] use the same idea to introduce the least squares TSVM (LSTSVM) which also searches for two nonparallel hyperplanes that minimize the distances introduced in [11] in a least squares sense. [18] present a noise resistant variant of TSVM entitled robust twin SVM (R-TWSVM) designed to correctly classify data that contains measurement errors. The various versions that have been introduced in previous years have also provided a incremental increase in classiffication ability.

A detailed overview of some of the most popular classification techniques can be found in [21] and [13].

3   Probabilistic Vector Machine

For a motivation of our model, let $h = (w, b) \in \mathbb{R}^n \times \mathbb{R}$ be a hyperplane in the problem space. Let $x \in \mathbb{R}^n$ be a point whose actual label is $y \in \{\pm 1\}$. The probability of error may then be expressed ás:

$$
\begin{aligned}
P(err) =& P(y \neq sgn(<w, x> +b)) \\
=& P((y = 1) \cap (<w, x> +b < 0)) \\
&+ P((y = -1) \cap (<w, x> +b > 0))
\end{aligned}
$$

So the probability of error is the sum of the probabilities of false negatives (FN) and false positives (FP). The objective proposed by PVM is the minimization of the maximum between these two components. Specifically, the hyperplane sought is:

$$
\begin{aligned}
(w, b) = \arg_{(w,b)} \min \max \{ & P((y = 1) \cap (<w, x> +b < 0)), \\
& P((y = -1) \cap (<w, x> +b > 0)) \}
\end{aligned}
$$

This choice is motivated by several arguments:

- In practical settings, the accuracy of a classifier is not the only measure. In order to be of use, both FN and FP probabilities have to be low.
- The reduction of the two also leads to the reduction of their sum, although, admittedly, sometimes not the lowest possible value. However, there is clearly a strong link between minimizing the maximum of the two and minimizing their sum.
- Minimizing the maximum of the two leads, under certain conditions, to a mathematical model which can be solved with convex optimizers.

During the training stage of the problem we only have the S_+ and S_- to base the choice of the hyperplane upon. Therefore, the probabilities for FN and FP have to be related to these sets. The objective expressed using only S_+ and S_- is expressed as:

$$
\begin{aligned}
(w, b) = \arg_{(w,b)} \min \max \{ & P((<w, x> +b < 0)|x \in S_+), \\
& P((<w, x> +b > 0)|x \in S_-) \}
\end{aligned} \tag{3.1}
$$

Assume that the signed distances to the hyperplane are normally distributed. Let E_+, E_- be the averages over the positive and, respectively, negative training set. Let σ_+, σ_- be the respective standard deviations. Note that these are in fact induced by the choice of the hyperplane. The probabilities can be expressed as a function of the distance between the hyperplane and the average divided by the corresponding $\sigma_\pm$. The hyperplane coresponds to a signed distance of 0, so if we denote by $\lambda_+ = \frac{E_+}{\sigma_+}$ and $\lambda_- = \frac{-E_-}{\sigma_-}$, then we get:

$$P((< w, x > +b < 0)|x \in S_+) = \int_{-\infty}^{-\lambda_+} f(s)ds$$
$$P((< w, x > +b > 0)|x \in S_-) = \int_{-\infty}^{-\lambda_-} f(s)ds$$

where $f : \mathbb{R} \to [0, 1]$ is the gaussian density of probability function. The objective can then be stated as:

$$(w, b) = \arg_{(w,b)} \min \max\{-\lambda_+, -\lambda_-\}$$
$$= \arg_{(w,b)} \max \min\{\lambda_+, \lambda_-\}$$

The condition that the distributions are normal can be replaced by a weaker one.

Definition 1. *Two random variables, D_+ and D_-, with means E_+, E_- and standard deviations σ_+, σ_-, are called similarly negatively distributed if:*

$$P(\frac{D_+ - E_+}{\sigma_+} \leq -\lambda) = P(\frac{D_- + E_-}{\sigma_-} \leq -\lambda), \forall \lambda \in [0, \infty) \quad (3.2)$$

This property obviously holds if D_+ and D_- are each normally distributed. It also holds for random variables which are distributed simetrically and identically, but with different means.

If the distances to the separating hyperplane have the property (3.2), then the optimal hyperplane in terms of (3.1) can be found by optimizing:

$$(w, b) = \arg_{(w,b)} \max \min\{\frac{E_+}{\sigma_+}, -\frac{E_-}{\sigma_-}\}$$

Making the choice of the separating hyperplane in this way would yield signed distance distributions as shown in Figure 1.

The standard deviation is then replaced by the average deviation, in order to obtain a model to which convex optimization can be applied. This will be the case for all σ definitions. The system that defines our problem is then be expressed as:

$$\begin{cases} minmax\{\frac{\sigma_+}{E_+}, \frac{\sigma_-}{E_-}\} \\ \frac{1}{|S_+|} \sum_{x_i \in S_+} d(x_i, h) = E_+ \\ -\frac{1}{|S_-|} \sum_{x_i \in S_-} d(x_i, h) = E_- \\ E_+ > 0, E_- > 0 \\ \sigma_+ = \frac{1}{|S_+|-1} \sum_{x_i \in S_+} |d(x_i, h) - E_+| \\ \sigma_- = \frac{1}{|S_-|-1} \sum_{x_i \in S_-} |d(x_i, h) + E_-| \end{cases}$$

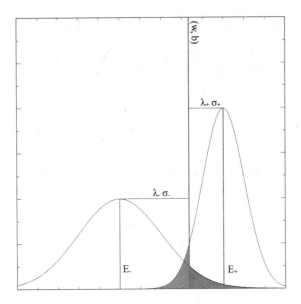

Fig. 1. Induced distance distributions. The hyperplane is sought such as to minimize the maximum between the red and blue areas, which correspond to the FN and FP probabilities. Equivalently, the hyperplane has to maximize the minimum between λ_+ and λ_-.

Note that $d(x_i, h) = \frac{<w,x>+b}{||w||}$, so:

$$\frac{1}{||w|| \cdot |S_+|} \sum_{x_i \in S_+} <w, x_i> +b = E_+$$
$$-\frac{1}{||w|| \cdot |S_-|} \sum_{x_i \in S_-} <w, x_i> +b = E_-$$
$$\sigma_+ = \frac{1}{||w|| \cdot |S_+|-1} \sum_{x_i \in S_+} |<w, x_i> +b - ||w|| \cdot E_+|$$
$$\sigma_- = \frac{1}{||w|| \cdot |S_-|-1} \sum_{x_i \in S_-} |<w, x_i> +b + ||w|| \cdot E_-|$$

Since the objective depends upon $\frac{\sigma_+}{E_+}$ and $\frac{\sigma_-}{E_-}$, the system may be rewritten as:

$$
\begin{cases}
minmax\{\frac{\sigma_+}{E_+}, \frac{\sigma_-}{E_-}\} \\
\frac{1}{|S_+|} \sum_{x_i \in S_+} <w, x_i> +b = E_+ \\
-\frac{1}{|S_-|} \sum_{x_i \in S_-} <w, x_i> +b = E_- \\
E_+ \geq 1, E_- \geq 1 \\
\sigma_+ = \frac{1}{|S_+|-1} \sum_{x_i \in S_+} |<w, x_i> +b - E_+| \\
\sigma_- = \frac{1}{|S_-|-1} \sum_{x_i \in S_-} |<w, x_i> +b + E_-|
\end{cases}
\tag{3.3}
$$

Note that, in order for the system to be feasible, all that is required is that $\frac{1}{|S_+|} \sum_{x_i \in S_+} x_i \neq \frac{1}{|S_-|} \sum_{x_i \in S_-} x_i$, because then one can choose $w_0 \in \mathbb{R}^n$ such that:

$$\frac{1}{|S_+|} \sum_{x_i \in S_+} w_0 \cdot x_i \neq \frac{1}{|S_-|} \sum_{x_i \in S_-} w_0 \cdot x_i$$

. If $\frac{1}{|S_+|} \sum_{x_i \in S_+} w_0 \cdot x_i < \frac{1}{|S_-|} \sum_{x_i \in S_-} w_0 \cdot x_i$, then $w_1 = -w_0$; else, $w_1 = w_0$.
We now have:

$$\frac{1}{|S_+|} \sum_{x_i \in S_+} w_0 \cdot x_i > \frac{1}{|S_-|} \sum_{x_i \in S_-} w_0 \cdot x_i$$

and may choose the offset $b \in \mathbb{R}$ such that:

$$b + \frac{1}{|S_+|} \sum_{x_i \in S_+} w_0 \cdot x_i > 0 > b + \frac{1}{|S_-|} \sum_{x_i \in S_-} w_0 \cdot x_i$$

.

which, after scaling, gives a feasible solution to System (3.3).

If the system is not feasible, then, from the viewpoint of a normal distribution of distances, it would be pointless to search for a separation, because the accuracy rate would be at most 50%.

Lemma 1. *The optimization problem* (3.3) *is equivalent to solving:*

$$\begin{cases} minmax\{\frac{\sigma_+}{E_+}, \frac{\sigma_-}{E_-}\} \\ b + \frac{1}{|S_+|} \sum_{x_i \in S_+} < w, x_i > = E_+ \\ -b - \frac{1}{|S_-|} \sum_{x_i \in S_-} < w, x_i > = E_- \\ E_+ \geq 1, E_- \geq 1 \\ |< w, x_i > + b - E_+| \leq \sigma_+^i, \forall x_i \in S_+ \\ |< w, x_i > + b + E_-| \leq \sigma_-^i, \forall x_i \in S_- \\ \sigma_+ = \frac{1}{|S_+|-1} \sum_{x_i \in S_+} \sigma_+^i \\ \sigma_- = \frac{1}{|S_-|-1} \sum_{x_i \in S_-} \sigma_-^i \end{cases} \quad (3.4)$$

Note that, apart from the objective function, system (3.4) uses only linear equations. This will lead to an easy solution, as detailed in Section 5.

The important properties of the model proposed thus far are that it has a direct connection to the generalization error embedded in the objective function and that it is likely to be resilient to outliers; the latter results from the fact that it is based on a statistical model of the training data which provides a built–in mechanism for dealing with outliers.

Also note that, because of the way the system is built, it does not require for S_+ and S_- to be linearly separable, as hard margin SVM would require, and does not require special treatment for classification errors, thus avoiding the introduction of a tradeoff term in the objective function.

4   Using Kernels

The system introduced thus far focuses on simply the linear aspect of the problem. However, many practical applications require nonlinear separation. The way to achieve this is by using kernel functions. In order to do so, we first project the points into a Hilbert space, H, via a projection function $\Phi : \mathbb{R}^n \to H$. In the H space we train our classifier in the linear manner described in Section 3. Since the constraint equations are linear and the separation is linear as well, the search for w can be restricted to the linear subspace generated by the training points.

Consider $w = \sum_{i=1}^{m} \alpha_i \Phi(x_i)$, where $\alpha_i \in \mathbb{R}$. The scalar products can be expressed as:

$$< w, \Phi(x) > = < \sum_{i=1}^{m} \alpha_i \Phi(x_i), \Phi(x) > = \sum_{i=1}^{m} \alpha_i < \Phi(x_i), \Phi(x) >$$

By defining $K : \mathbb{R}^n \times \mathbb{R}^n \to \mathbb{R}$ as $K(u, v) = < \Phi(u), \Phi(v) >$, the projection function does not require an explicit definition. Indeed, using Mercer's theorem, one only needs to define the kernel function, K, and have the projection function Φ implicitly defined.

Replacing the scalar product accordingly in system (3.4), we obtain the following system:

$$\begin{cases} \min \max\{\frac{\sigma_+}{E_+}, \frac{\sigma_-}{E_-}\} \\ b + \sum_{i=1}^{m} [\alpha_i \cdot \frac{1}{|S_+|} \sum_{x_j \in S_+} K(x_i, x_j)] = E_+ \\ -b - \sum_{i=1}^{m} [\alpha_i \cdot \frac{1}{|S_-|} \sum_{x_j \in S_-} K(x_i, x_j)] = E_- \\ |\sum_{x_i \in S} \alpha_i K(x_i, x_j) + b - E_+| \leq \sigma_+^j \ , \ \forall x_j \in S_+ \\ |\sum_{x_i \in S} \alpha_i K(x_i, x_j) + b + E_-| \leq \sigma_-^j \ , \ \forall x_j \in S_- \\ \frac{1}{|S_+|-1} \sum_{x_i \in S_+} \sigma_+^i = \sigma_+ \\ \frac{1}{|S_-|-1} \sum_{x_i \in S_-} \sigma_-^i = \sigma_- \\ \sigma_+ \leq t \cdot E_+ \\ \sigma_- \leq t \cdot E_- \\ E_+ \geq 1, \ E_- \geq 1 \end{cases} \quad (4.1)$$

5   Solving the PVM Problem

While thus far we have proposed a model for the problem, we have not discussed yet the way this system can be solved.

In the current formulation, since fractions do not preserve convexity, the objective $\min \max\{\frac{\sigma_+}{E_+}, \frac{\sigma_-}{E_-}\}$ is not a convex function. However, each of the fractions used has linear factors. Hence, one deals with a quasilinear function (see [1], Chapter 3, pg. 95 for details), meaning that each sublevel and superlevel set is a convex set. Moreover, the maximum of two quasiconvex functions is a quasiconvex function. To see this, let $t \in [0, +\infty)$. Restricting our domain to $E_+ > 0, E_- > 0$, we get:

$$\max\{\frac{\sigma_+}{E_+}, \frac{\sigma_-}{E_-}\} \leq t \Leftrightarrow$$

$$\Leftrightarrow \begin{cases} \frac{\sigma_+}{E_+} \leq t \\ \frac{\sigma_-}{E_-} \leq t \end{cases} \Leftrightarrow \begin{cases} \sigma_+ \leq t \cdot E_+ \\ \sigma_- \leq t \cdot E_- \end{cases}$$

which, for a fixed t, is a set of linear equations, the intersection of which is a convex domain.

A quasiconvex function has only one (strict) local optimum, which is also the global one. As a consequence, one can solve system (4.1) via a set of feasibility problems. To see this, let us denote by $Feas(t)$ the feasibility problem obtained by enforcing the condition $\max\{\frac{\sigma_+}{E_+}, \frac{\sigma_-}{E_-}\} \leq t$:

$$\begin{cases} \sigma_+^j - \sum_{x_i \in S} \alpha_i(K(x_i, x_j) - K_+^i) \geq 0 \ , \ \forall x_j \in S_+ \\ \sigma_+^j + \sum_{x_i \in S} \alpha_i(K(x_i, x_j) - K_+^i) \geq 0 \ , \ \forall x_j \in S_+ \\ \sigma_-^j - \sum_{x_i \in S} \alpha_i(K(x_i, x_j) - K_-^i) \geq 0 \ , \ \forall x_j \in S_- \\ \sigma_-^j + \sum_{x_i \in S} \alpha_i(K(x_i, x_j) - K_-^i) \geq 0 \ , \ \forall x_j \in S_- \\ (|S_+| - 1)t \cdot (b + \sum_{x_i \in S} \alpha_i K_+^i) - \sum_{x_i \in S_+} \sigma_+^i \geq 0 \\ (|S_-| - 1)t \cdot (-b - \sum_{x_i \in S} \alpha_i K_-^i) - \sum_{x_i \in S_-} \sigma_-^i \geq 0 \\ b + \sum_{x_i \in S} \alpha_i K_+^i \geq 1 \\ -b - \sum_{x_i \in S} \alpha_i K_-^i \geq 1 \end{cases} \quad (5.1)$$

The optimal solution to system (4.1) is, then:

$$t_{optimal} = \inf\{t \in \mathbb{R}_+ | Feas(t) \text{ is feasible}\}$$
$$= \sup\{t \in \mathbb{R}_+ | Feas(t) \text{ is infeasible}\}$$

The $t_{optimal}$ value can thus be easily found using a simple bisection procedure. Initialize $0 = t_{left} < t_{optimal} < t_{right} = \infty$. Each iteration, let $t = 0.5 \cdot (t_{left} + t_{right})$. If $Feas(t)$ is feasible, then $t_{right} = t$, otherwise $t_{left} = t$.

The feasibility problems formulated during this bisection procedure can be solved using one of the many linear programming solvers freely available. Note, however, that as the bounds on $t_{optimal}$ get closer together, the feasible region of $Feas(t)$ approaches a single point and this can lead to numerical problems in the linear solvers.

6   Results

For the linear solver required by PVM, we have used the GNU Linear Programming Kit (GLPK) which is freely available. The CPLEX library can also be used with the code. A version of the algorithm, with a few extra features which are not discussed in this paper, can be found at https://code.google.com/p/dpvm/.

6.1   Testing on Aritificial Data

In the figures shown in this section the stars have positive labels, the circles have negative labels and the obtained separating hyperplane is represented by a single black line.

In Figure 2 we compared our method with the soft margin SVM. For training the SVM we use a well known package, libSVM, presented in [6], [2]. The comparisons where done using artificial datasets. This is for illustrating the way in which PVM works. As can be seen, PVM takes into account the distribution

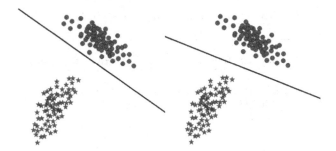

Fig. 2. PVM Separation on the left; SVM Separation on the right; PVM separation takes into account the distributions of the two sets, not just the closest members

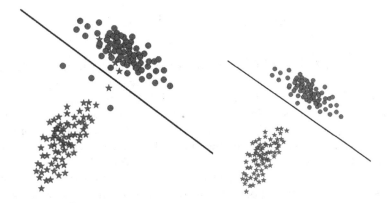

Fig. 3. Comparing the result of the training process on a set of points with the same training result on the same set to which some outliers have been added

of the entire training dataset, while SVM takes into account only the bordering elements of S_+ and S_-.

Figure 3 compares the training result when using the same data as in Figure 2 with the training result obtained when adding some outliers to this data. As is evident, the outliers bear little influence on the training process.

6.2   Testing on Public Datasets

We have conducted a set of tests on datasets from the University of California Irvine Machine Learning (UCI ML) repository. The datasets used can be found at http://archive.ics.uci.edu/ml/. Table 1 describes the datasets used. Preprocessing consisted in normalizing the features in each database. PVM has been compared to GEPSVM, presented in [16] and PSVM, presented in [8], which both proved similar or superior performance to SVMs. The parameters of each algorithm were obtained via grid searches and the evaluation of a combination of parameters was done using tenfold cross validation. The optimal parameters

Table 1. UCI ML datasets description

Dataset	Records	Features	Positive	Negative
Hepatitis	156	19	32	123
WPBC	199	34	47	151
Sonar	209	60	111	97
Heart-statlog	270	14	150	120
Heart-c	303	14	139	164
Bupa Liver	345	7	200	145
Ionosphere	351	34	126	225
Votes	434	16	267	167
Australian	690	14	307	383
Pima-Indian	768	8	500	268

where then used for a set of 100 tenfold cross validation tests, or equivalently 1000 tests. Table 2 shows the results in terms of average accuracy and standard deviation obtained for the RBF kernel for PVM, GEPSVM and PSVM. The average accuracy and its standard deviation where computed over the 1000 tests. As the results presented in Table 2 suggest, PVM outperforms the other classfiers in terms of average accuracy, winning on 5 of the datasets, with a large improvement on 3 of these, namely Hepatitis, Heart–c and Votes. It is important to note that PVM outperforms the other algorithms especially on the larger datasets. One possibility why this happens is that the statistical measurements used offer more relevant and stable information once the number of training points is large

Table 2. Comparison between PVM, GEPSVM and PSVM on the UCI ML datasets on the RBF kernel. The results indicate the average accuracy over 100 tenfold runs. The accuracy is measured in percentages. The best results for a dataset are shown in bold.

Dataset	PVM	GEPSVM	PSVM
Hepatitis	**87.151**±0.94	79.28±5.2	78.57±0.24
WPBC	78.853±0.64	80±5.97	**80.55**±3.92
Sonar	86.995±1.37	80±5.97	**90**±7.21
Heart–statlog	77.548±1.84	**86.52**±7.36	70.74±6.86
Heart–c	**77.119**±1.21	70.37±8.90	70.68±7.66
Bupa Liver	73.021±0.92	68.18±6.2	**74.84**±9.04
Ionosphere	92.903±1.36	84.41±6.2	**95**±4.17
Votes	**96.54**±0.39	94.5±3.37	95.95±2.25
Australian	**83.623**±0.94	69.55±5.37	73.97±6.16
Pima-Indian	**77.133**±0.31	75.33±4.91	76.8±3.83

enough. One other important observation is that, of the algorithms tested here, PVM has a distinctly lower standard deviation of the accuracy. This implies that the derived separating hyperplane is more stable than the one derived by PSVM or the two hyperplanes of GEPSVM. The datasets used in this comparison have not been chosen such that the condition (3.2) is satisfied. However, PVM proves to be competitive. This suggests that, although the setting in which PVM provides an optimal separation hyperplane requires a special condition, practical examples do not always stray from the proposed model more than they stray from the models proposed by GEPSVM or PSVM.

7   Conclusion

We have introduced a new linear binary classifier designed to use statistical measurements of the training datasets. The underlying model of PVM is robust to outliers and shows good generalization ability on the datasets tested. PVM can also use linear programming tools, which are well established in the literature.

Future work will include the study of dedicated linear programming tools and will focus on developing a stable distributed feasibility solver to tackle the optimization problem proposed by the theoretical foundation of the algorithm.

Acknowledgements. This research has been supported through financing offered by the POSDRU/ 88/1.5/S/47646 Project for PHDs. This work has also been supported by the European Social Fund in Romania, under the responsibility of the Managing Authority for the Sectoral Operational Programme for Human Resources Development 2007-2013 [grant POSDRU/CPP 107/DMI 1.5/S/ 78342].

References

1. Boyd, S., Vandenberghe, L.: Convex Optimization. Cambridge University Press (2004)
2. Chang, C.-C., Lin, C.-J.: Libsvm: a library for support vector machines. ACM Transactions on Intelligent Systems and Technology 2(3) (2011)
3. Chapelle, O., Vapnik, V.N.: Bounds on error expectation for support vector machines. Neural Computation 12(9), 2012–2036 (2000)
4. Chapelle, O., Vapnik, V.N.: Choosing multiple parameters for support vector machines. Machine Learning 46(1-3), 131–159 (2001)
5. Cortes, C., Vapnik, V.N.: Support-vector networks. Machine Learning 20(3), 273–297 (1995)
6. Fan, R.-E., Chen, P.-H., Lin, C.-J.: Working set selection using second order information for training support vector machines. The Journal of Machine Learning Research 6, 1889–1918 (2005)
7. Frank, R.: The perceptron a perceiving and recognizing automaton. Technical Report 85-460-1, Cornell Aeronautical Laboratory (1957)
8. Fung, G., Mangasarian, O.L.: Proximal support vector machine classifiers. In: Proceedings of the Seventh ACM SIGKDD International Conference on Knowledge Discovery and Data Mining, San Francisco, CA, USA, pp. 77–86. ACM, New York (2001)

9. Garg, A., Har-Peled, S., Roth, D.: On generalization bounds, projection profile, and margin distribution. In: ICML 2002 Proceedings of the Nineteenth International Conference on Machine Learning, Sydney, Australia, pp. 171–178. Morgan Kaufmann Publishers Inc. (2002)
10. Garg, A., Roth, D.: Margin distribution optimization. Computational Imaging and Vision 29, 119–128 (2005)
11. Jayadeva, Khemchandani, R., Chandra, S.: Twin support vector machines for pattern classification. IEEE Transactions on Pattern Analysis and Machine Intelligence 29(5), 905–910 (2007)
12. Joachims, T., Ndellec, C., Rouveriol, C.: Text categorization with support vector machines: learning with many relevant features. In: Nédellec, C., Rouveirol, C. (eds.) ECML 1998. LNCS, vol. 1398, pp. 137–142. Springer, Heidelberg (1998)
13. Kotsiantis, S.B.: Supervised machine learning: A review of classification techniques. Informatica 31, 3–24 (2007)
14. Kumar, A.M., Gopal, M.: Least squares twin support vector machines for pattern classification. Expert Systems with Applications: An International Journal 36(4), 7535–7543 (2009)
15. Lee, S., Verri, A.: SVM 2002. LNCS, vol. 2388. Springer, Heidelberg (2002)
16. Mangasarian, O.L., Wild, W.: Multisurface proximal support vector machine classification via generalized eigenvalues. IEEE Transaction on Pattern Analysis and Machine Intelligence 28(1), 69–74 (2005)
17. Noble, W.S.: Support vector machine applications in computational biology. In: Kernel Methods in Computational Biology, pp. 71–92 (2004)
18. Qi, Z., Tian, Y., Shi, Y.: Robust twin support vector machine for pattern classification. Pattern Recognition (June 27, 2012) (accepted)
19. Vapnik, V.N., Boser, B.E., Guyon, I.: A training algorithm for optimal margin classifiers. In: COLT 1992 Proceedings of the Fifth Annual Workshop on Computational Learning, Pittsburgh, PA, USA, vol. 5, pp. 144–152. ACM, New York (1992)
20. Vapnik, V.N.: Statistical learning theory. John Wiley and Sons Inc. (1998)
21. Witten, I.H., Frank, E., Hall, M.A.: Data Mining: Practical Machine Learning Tools and Techniques. Morgan Kaufmann (2011)

Towards Practical Tabled Abduction in Logic Programs

Ari Saptawijaya* and Luís Moniz Pereira

Centro de Inteligência Artificial (CENTRIA), Departamento de Informática
Faculdade de Ciências e Tecnologia, Univ. Nova de Lisboa, 2829-516 Caparica, Portugal
ar.saptawijaya@campus.fct.unl.pt, lmp@fct.unl.pt

Abstract. Despite its potential as a reasoning paradigm in AI applications, abduction has been on the back burner in logic programming, as abduction can be too difficult to implement, and costly to perform, in particular if abductive solutions are not tabled. If they become tabled, then abductive solutions can be reused, even from one abductive context to another. On the other hand, current Prolog systems, with their tabling mechanisms, are mature enough to facilitate the introduction of tabling abductive solutions (tabled abduction) into them. The concept of tabled abduction has been realized recently in an abductive logic programming system TABDUAL. Besides tabling abductive solutions, TABDUAL also relies on the dual transformation. In this paper, we emphasize two TABDUAL improvements: (1) the dual transformation *by need*, and (2) a new construct for accessing ongoing abductive solutions, that permits modular mixes between abductive and non-abductive program parts. We apply subsequently these improvements on two distinct problems, and evaluate the performance and the scalability of TABDUAL on several benchmarks on the basis of these problems, by examining four TABDUAL variants.

Keywords: tabled abduction, abductive logic programming, tabled logic programming, dual transformation.

1 Introduction

Abduction has already been well studied in the field of computational logic, and logic programming in particular, for a few decades by now [1, 3, 4, 8]. Abduction in logic programs offers a formalism to declaratively express problems in a variety of areas, e.g. in decision-making, diagnosis, planning, belief revision, and hypothetical reasoning (cf. [2, 5, 9–11]). On the other hand, many Prolog systems have become mature and practical, and thus it makes sense to facilitate the use of abduction into such systems.

In abduction, finding some best explanations (i.e. abductive solutions) to the observed evidence, or finding assumptions that can justify a goal, can be very costly. It is often the case that abductive solutions found within a context are relevant in a different context, and thus can be reused with little cost. In logic programming, absent of abduction, goal solution reuse is commonly addressed by employing a tabling mechanism. Therefore, tabling is conceptually suitable for abduction, to deal with the reuse of abductive solutions. In practice, abductive solutions reuse is not immediately amenable to

---

* Affiliated with Fakultas Ilmu Komputer at Universitas Indonesia, Depok, Indonesia.

L. Correia, L.P. Reis, and J. Cascalho (Eds.): EPIA 2013, LNAI 8154, pp. 223–234, 2013.

tabling, because such solutions go together with an ongoing abductive context. It also poses a new problem on how to reuse them in a different but compatible context, while catering as well to loops in logic programs, i.e. positive loops and loops over negation, now complicated by abduction.

A concept of tabled abduction in abductive normal logic programs, and its prototype TABDUAL, to address the above issues, was recently introduced [16]. It is realized via a program transformation, where abduction is subsequently enacted on the transformed program. The transformation relies on the theory of the dual transformation [1], which allows one to more efficiently handle the problem of abduction under negative goals, by introducing their positive dual counterparts. We review TABDUAL in Section 2.

While TABDUAL successfully addresses abductive solution reuse, even in different contexts, and also correctly deals with loops in programs, it may suffer from a heavy transformation load due to performing the *complete* dual transformation of an input program in advance. Such heavy transformation clearly may hinder its practical use in real world problems. In the current work, we contribute by enacting a *by-need* dual transformation, i.e. dual rules are only created as they are needed during abduction, either eagerly or lazily – the two approaches we shall examine. We furthermore enhance TABDUAL's flexibility by introducing a new system predicate to access ongoing abductive solutions, thereby permitting modular mixes of abductive and non-abductive program parts. These and other improvements are detailed in Section 3.

Until now there has been no evaluation of TABDUAL, both in terms of performance and scalability, as to gauge its suitability for likely applications. In order to understand better the influence of TABDUAL's features on its performance, we separately factor out its important features, resulting in four TABDUAL variants of the same underlying implementation. One evaluation uses the well-known N-queens problem, as the problem size can be easily scaled up and in that we can additionally study how tabling of conflicts or nogoods of subproblems influences the performance and the scalability of these variants. The other evaluation is based on an example from declarative debugging, previously characterized as belief revision [13, 14]. The latter evaluation reveals the relative worth of the newly introduced dual transformation by need. We discuss all evaluation results in Section 4, and conclude in Section 5.

2   Tabled Abduction

A *logic rule* has the form $H \leftarrow B_1, \ldots, B_m, not\ B_{m+1}, \ldots, not\ B_n$, where $n \geq m \geq 0$ and H, B_i with $1 \leq i \leq n$ are atoms. In a rule, H is called the head of the rule and $B_1, \ldots, B_m, not\ B_{m+1}, \ldots, not\ B_n$ its body. We use '*not*' to denote default negation. The atom B_i and its default negation $not\ B_i$ are named positive and negative *literals*, respectively. When $n = 0$, we say the rule is a *fact* and render it simply as H. The atoms *true* and *false* are, by definition, respectively true and false in every interpretation. A rule in the form of a denial, i.e. with empty head, or equivalently with *false* as head, is an *integrity constraint* (IC). A *logic program* (LP) is a set of logic rules, where non-ground rules (i.e. rules containing variables) stand for all their ground instances. In this work we focus on *normal logic programs*, i.e. those whose heads of rules are positive literals or empty. As usual, we write p/n to denote predicate p with arity n.

2.1   Abduction in Logic Programs

Let us recall that abduction, or inference to the best explanation (a designation common in the philosophy of science [7, 12]), is a reasoning method, whence one chooses hypotheses that would, if true, best explain observed evidence – whilst meeting any prescribed ICs – or that would satisfy some query. In LPs, abductive hypotheses (or *abducibles*) are named literals of the program having no rules, whose truth value is not assumed initially. Abducibles can have parameters, but must be ground when abduced. An *abductive normal logic program* is a normal logic program that allows for abducibles in the body of rules. Note that the negation '*not a*' of an abducible *a* refers not to its default negation, as abducibles by definition lack any rules, but rather to an assumed hypothetical negation of *a*.

The truth value of abucibles may be distinctly assumed *true* or *false*, through either their positive or negated form, as the case may be, as a means to produce an abductive solution to a goal query in the form of a consistent set of assumed hypotheses that lend support to it. An *abductive solution* to a query is thus a consistent set of abducible ground instances or their negations that, when replaced by their assigned truth value everywhere in the program P, provides a model of P (for the specific semantics used on P), satisfying both the query and the ICs – called an *abductive model*. Abduction in LPs can be accomplished naturally by a top-down query-oriented procedure to identify an (abductive) solution by need, i.e. as abducibles are encountered, where the abducibles in the solution are leaves in the procedural query-rooted call-graph, i.e. the graph recursively generated by the procedure calls from the literals in bodies of rules to the heads of rules, and subsequently to the literals in the rule's body.

2.2   Tabled Abduction in TABDUAL

Next, we recall the basics of tabled abduction in TABDUAL. Consider an abductive logic program, taken from [16]:

Example 1. Program P_1: $q \leftarrow a.$ $s \leftarrow b, q.$ $t \leftarrow s, q.$
where a and b are abducibles.

Suppose three queries: q, s, and t, are individually posed, in that order. The first query, q, is satisfied simply by taking $[a]$ as the abductive solution for q, and tabling it. Executing the second query, s, amounts to satisfying the two subgoals in its body, i.e. abducing b followed by invoking q. Since q has previously been invoked, we can benefit from reusing its solution, instead of recomputing, given that the solution was tabled. That is, query s can be solved by extending the current ongoing abductive context $[b]$ of subgoal q with the already tabled abductive solution $[a]$ of q, yielding $[a, b]$. The final query t can be solved similarly. Invoking the first subgoal s results in the priorly registered abductive solution $[a, b]$, which becomes the current abductive context of the second subgoal q. Since $[a, b]$ subsumes the previously obtained abductive solution $[a]$ of q, we can then safely take $[a, b]$ as the abductive solution to query t. This example shows how $[a]$, as the abductive solution of the first query q, can be reused from an abductive context of q (i.e. $[b]$ in the second query, s) to another context (i.e. $[a, b]$ in the third query, t). In practice the body of rule q may contain a huge number of subgoals, causing potentially expensive recomputation of its abductive solutions and thus such unnecessary recomputation should be avoided.

Tabled abduction is realized in a prototype TABDUAL, implemented in the most advanced LP tabling system XSB Prolog [19], which involves a program transformation of abductive logic programs. Abduction can then be enacted on the transformed program directly, without the need of a meta-interpreter. Example 1 already indicates two key ingredients of the transformation: (1) *abductive context*, which relays the ongoing abductive solution from one subgoal to subsequent subgoals, as well as from the head to the body of a rule, via *input* and *output* contexts, where abducibles can be envisaged as the terminals of parsing; and (2) *tabled predicates*, which table the abductive solutions for predicates defined in the input program.

Example 2. The rule $t \leftarrow s, q$ from Example 1 is transformed into two rules:
$$t_{ab}(E) \leftarrow s([\,], T), q(T, E). \qquad t(I, O) \leftarrow t_{ab}(E), produce(O, I, E).$$
Predicate $t_{ab}(E)$ is the tabled predicate which tables the abductive solution of t in its argument E. Its definition, in the left rule, follows from the original definition of t. Two extra arguments, that serve as input and output contexts, are added to the subgoals s and q in the rule's body. The left rule expresses that the tabled abductive solution E of t_{ab} is obtained by relaying the ongoing abductive solution in context T from subgoal s to subgoal q in the body, given the empty input abductive context of s (because there is no abducible in the body of the original rule of t). The rule on the right expresses that the output abductive solution O of t is obtained from the the solution entry E of t_{ab} and the given input context I of t, via TABDUAL system predicate $produce(O, I, E)$, that checks consistency. The other rules are transformed following the same idea.

An abducible is transformed into a rule that inserts it into the abductive context. For example, the abducible a of Example 1 is transformed into: $a(I, O) \leftarrow insert(a, I, O)$, where $insert(a, I, O)$ is a TABDUAL system predicate which inserts a into the input context I, resulting in the output context O, while also checking consistency. The negation $not\ a$ of the abducible a is transformed similarly, except that it is renamed into not_a in the head: $not\_a(I, O) \leftarrow insert(not\ a, I, O)$.

The TABDUAL program transformation employs the *dual transformation* [1], which makes negative goals 'positive', thus permitting to avoid the computation of all abductive solutions, and then negating them, under the otherwise regular negative goals. Instead, we are able to obtain one abductive solution at a time, as when we treat abduction under positive goals. The dual transformation defines for each atom A and its set of rules R in a program P, a set of dual rules whose head not_A is true if and only if A is false by R in the employed semantics of P. Note that, instead of having a negative goal $not\ A$ as the rules' head, we use its corresponding 'positive' one, not_A. Example 3 illustrates only the main idea of how the dual transformation is employed in TABDUAL and omits many details, e.g. checking loops in the input program, all of which are referred in [16].

Example 3. Consider the following program fragment, in which p is defined as:
$$p \leftarrow a. \qquad p \leftarrow q, not\ r.$$

where a is an abducible. The TABDUAL transformation will create a set of dual rules for p which falsify p with respect to its two rules, i.e. by falsifying both the first rule *and* the second rule, expressed below by predicate p^{*1} and p^{*2}, respectively:

$$not\_p(I, O) \leftarrow p^{*1}(I, T), p^{*2}(T, O).$$

In the TABDUAL transformation, this rule is known as the first layer of the dual transformation. Notice the addition of the input and output abductive context arguments, I and O, in the head, and similarly in each subgoal of the rule's body, where the intermediate context T is used to relay the abductive solution from p^{*1} to p^{*2}.

The second layer contains the definitions of p^{*1} and p^{*2}, where p^{*1} and p^{*2} are defined by falsifying the body of p's first rule and second rule, respectively. In case of p^{*1}, the first rule of p is falsified by abducing the negation of a:

$$p^{*1}(I, O) \leftarrow not\_a(I, O).$$

Notice that the negation of a, i.e. *not* a, is abduced by invoking the subgoal $not\_a(I, O)$. This subgoal is defined via the transformation of abducibles, as discussed earlier. In case of p^{*2}, the second rule of p is falsified by alternatively failing one subgoal in its body at a time, i.e. by negating q or, alternatively, by negating *not* r.

$$p^{*2}(I, O) \leftarrow not\_q(I, O). \qquad p^{*2}(I, O) \leftarrow r(I, O).$$

Finally, TABDUAL transforms integrity constraints like any other rules, and top-goal queries are always launched by also satisfying integrity constraints.

3 Improvements on TABDUAL

The number of dual rules for atom A, produced by a naive dual transformation, is generally exponential in the number of A's rules, because all combinations of body literals from the positive rules need to be generated. For complexity result of the core TABDUAL transformation, the reader is referred to [17]. We propound here, for the first time, two approaches of dual transformation by need, as a means to avoid a heavy TABDUAL transformation load due to superfluous dual transformation. We also extend TABDUAL features here with a new system predicate that allows accessing ongoing abductive solutions for dynamic manipulation.

3.1 By-need Dual Transformation

By its very previous specification, TABDUAL performs a *complete* dual transformation for every defined atom in the program in advance, i.e. as an integral part of the whole TABDUAL transformation. This certainly has a drawback, as potentially massive dual rules are created in the transformation, though only a few of them might be invoked during abduction. That is unpractical, as real-world problems typically consist of a huge number of rules, and such a complete dual transformation may hinder abduction to start taking place, not to mention the compile time, and space requirements, of the large thus produced transformed program.

One pragmatic and practical solution to this problem is to compute dual rules *by need*. That is, dual rules are created during abduction, based on the need of the on-going invoked goals. The transformed program still contains the first layer of the dual transformation, but its second layer is defined using a newly introduced TABDUAL system predicate, which will be interpreted by the TABDUAL system on-the-fly, during abduction, to produce the concrete definitions of the second layer. Recall Example 3. The by-need dual transformation contains the same first layer: $not\_p(I, O) \leftarrow p^{*1}(I, T), p^{*2}(T, O)$. The second layer now contains, for each $i \in \{1, 2\}$, rule $p^{*i}(I, O) \leftarrow dual(i, p, I, O)$. The newly introduced system predicate $dual/4$ facilitates the by-need construction of generic dual rules (i.e. without any context attached to them) from the i-th rule of $p/1$, during abduction. It will also instantiate the generic dual rules with the provided arguments and contexts, and subsequently invoke the instantiated dual rules.

Extra computation load that may occur during the abduction phase, due to the by-need construction of dual rules, can be reduced by memoizing the already constructed generic dual rules. Therefore, when such dual rules are later needed, they are available for reuse and thus their recomputation can be avoided. We discuss two approaches for memoizing generic dual rules; each approach influences how generic dual rules are constructed:

- *Tabling generic dual rules*, which results in an *eager* construction of dual rules, due to the local table scheduling employed by default in XSB. This scheduling strategy may not return any answers out of a strongly connected component (SCC) in the subgoal dependency graph, until that SCC is completely evaluated [19]. For instance, in Example 3, when $p^{*2}(I, O)$ is invoked, all two alternatives of generic dual rules from the second rule of p, i.e. $p^{*2}(I, O) \leftarrow not\_q(I, O)$ and $p^{*2}(I, O) \leftarrow r(I, O)$ are constructed before they are subsequently instantiated and invoked. XSB also provides other (in general less efficient) table scheduling alternative, i.e. batched scheduling, which allows constructing only one generic dual rule at a time before it is instantiated and invoked. But the choice between the two scheduling strategies is fixed in each XSB installation, i.e. at present one cannot switch from one to the other without a new installation. Hence, we propose the second approach.
- *Storing generic dual rules in a trie*, which results in a *lazy* construction of dual rules. The trie data structure in XSB allows facts to be stored in a tree representation, and is built from a prefix ordering of each fact; thus, in this case factors out the common prefix of the (stored) facts [20]. In our context, facts are used to represent generic dual rules, which can be directly stored and manipulated in a trie. This approach permits simulating batched scheduling within the local scheduling employed by default in XSB. It hence allows constructing one generic dual rule at a time, before it is instantiated and invoked, and memoizing it explicitly through storing it in the trie for later reuse. By constructing only one generic dual rule at a time, additional information to track through literal's position, used in constructing the latest dual rule, should then be maintained. All these steps are made possible through XSB's trie manipulation predicates.

3.2 Towards a More Flexible TABDUAL

TABDUAL encapsulates the ongoing abductive solution in an abductive context, which is relayed from one subgoal to another. In many problems, it is often the case that one needs to access the ongoing abductive solution in order to manipulate it dynamically, e.g. to filter abducibles using preferences. But since it is encapsulated in an abductive context, and such context is only introduced in the transformed program, the only way to accomplish it would be to modify directly the transformed program rather than the original problem representation. This is inconvenient and clearly unpractical when we deal with real world problems with a huge number of rules. We overcome this issue by introducing a new system predicate $abdQ(P)$ that allows to access the ongoing abductive solution and to manipulate it using the rules of P. This system predicate is transformed by unwrapping it and adding an extra argument to P for the ongoing abductive solution.

Example 4. Consider a program fragment: $q \leftarrow r, abdQ(s), t.$ $s(X) \leftarrow v(X).$ along with some other rules. Note that, though predicate s within system predicate wrapper $abdQ/1$ has no argument, its rule definition has one extra argument for the ongoing abductive solution. The tabled predicate q_{ab} in the transformed program would be $q_{ab}(E) \leftarrow r([\ ], T_1), s(T_1, T_1, T_2), t(T_2, E)$. That is, $s/3$ now gets access to the ongoing abductive solution T_1 from $r/2$, via its additional first argument. It still has the usual input and output contexts, T_1 and T_2, respectively, in its second and third arguments. Rule $s/1$ in P_3 is transformed like any other rules.

The new system predicate $abdQ/1$ permits modular mixes of abductive and non-abductive program parts. For instance, the rule of $s/1$ in P_3 may naturally be defined by some non-abductive program parts. Suppose that the argument of s represents the ongoing board configuration for the N-queens problem (i.e. the ongoing abductive solution is some board configuration). Then, the rule of $s/1$ can be instead, $s(X) \leftarrow prolog(safe(X))$, where the $prolog/1$ wrapper is the existing TABDUAL predicate to execute normal Prolog predicates, i.e. those not transformed by TABDUAL. Predicate $safe(X)$ can then be defined in the non-abductive program part, to check whether the ongoing board configuration of queens is momentarily safe.

The treatment of facts in programs may also benefit from modular mixes of abductive and non-abductive parts. As facts do not induce any abduction, a predicate comprised of just facts can be much more simply transformed: only a bridge transformed rule for invoking Prolog facts, is needed, which keeps the output abductive context equal to the input one. The facts are listed, untransformed, in the non-abductive part.

4   Evaluation of TABDUAL

As our benchmarks do not involve loops in their representation, we employ in the evaluation the version of TABDUAL which does not handle loops in programs, such as positive loops (e.g. program $P_1 = \{p \leftarrow p.\}$), or loops over negation (e.g. program $P_2 = \{p \leftarrow not\ q.\ ;\ q \leftarrow p.\}$). Loops handling in TABDUAL and the evaluation pertaining to that are fully discussed in [17]. For better understanding of how TABDUAL's features influence its performance, we consider four distinct TABDUAL variants (of the same underlying implementation), obtained by separately factoring out its important features:

1. TABDUAL-need, i.e. without dual transformation by need,
2. TABDUAL+eager, i.e. with eager dual transformation by need,
3. TABDUAL+lazy, i.e. with lazy dual transformation by need, and
4. TABDUAL+lazy-tab, i.e. TABDUAL+lazy without tabling abductive solutions.

The last variant is accomplished by removing the table declarations of abductive predicates $*^{ab}$, where $*$ is the predicate name (cf. Example 2), in the transformed program.

4.1 Benchmarks

The first benchmark is the well-known N-queens problem, where abduction is used to find safe board configurations of N queens. The problem is represented in TABDUAL as follows (for simplicity, we omit some syntactic details):

$q(0, N).$
$q(M, N) \leftarrow M > 0, q(M - 1, N), d(Y), pos(M, Y), not\ abdQ(conflict).$
$conflict(BoardConf) \leftarrow prolog(conflictual(BoardConf)).$

with the query $q(N, N)$ for N queens. Here, $pos/2$ is the abducible predicate representing the position of a queen, and $d/1$ is a column generator predicate, available as facts $d(i)$ for $1 \leq i \leq N$. Predicate $conflictual/1$ is defined in a non-abductive program module, to check whether the ongoing board configuration $BoardConf$ of queens is conflictual. By scaling up the problem, i.e. increasing the value of N, we aim at evaluating the scalability of TABDUAL, concentrating on tabling nogoods of subproblems, i.e. tabling conflictual configurations of queens (essentially, tabling the ongoing abductive solutions), influences scalability of the four variants.

The second benchmark concerns an example of declarative debugging. Our aim with this benchmark is to evaluate the relative worth of the by-need dual transformation in TABDUAL, with respect to both the transformation and the abduction time. Declarative debugging of normal logic programs has been characterized before as belief revision [13, 14], and it has been shown recently that they can also be viewed as abduction [18]. There are two problems of declarative debugging: incorrect solutions and missing solutions. The problem of debugging incorrect solution S is characterized by an IC of the form $\leftarrow S$, whereas debugging missing solution S by an IC of the form $\leftarrow not\ S$. Since with this benchmark we aim at evaluating the by-need dual transformation, we consider only the problem of debugging incorrect solutions (its form of IC perforce constructs and invokes dual rules). The benchmark takes the following program to debug, where the size $n > 1$ can easily be increased:

$$q_0(0, 1). \qquad q_0(X, 0).$$
$$q_1(1). \qquad q_1(X) \leftarrow q_0(X, X).$$
$$q_n(n). \qquad q_n(X) \leftarrow q_{n-1}(X).$$

with the IC: $\leftarrow q_m(0)$, for $0 \leq m \leq n$, to debug incorrect solution $q_m(0)$.

4.2 Results

The experiments were run under XSB-Prolog 3.3.6 on a 2.26 GHz Intel Core 2 Duo with 2 GB RAM. The time indicated in all results refers to the CPU time (as an average of several runs) to aggregate all abductive solutions.

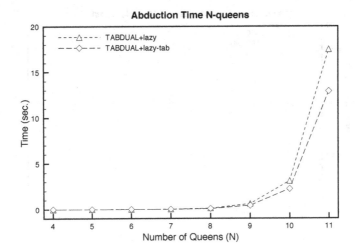

Fig. 1. The abduction time of different N queens.

N-queens. Since this benchmark is used to evaluate the benefit of *tabling* nogoods of subproblems (as abductive solutions), and *not* the benefit of the dual by-need improvement, we focus only on two TABDUAL variants: one with tabling feature, represented by TABDUAL+lazy, and the other without it, i.e. TABDUAL+lazy-tab. The transformation time of the problem representation is similar for both variants, i.e. around 0.003 seconds. Figure 1 shows abduction time for N queens, $4 \leq N \leq 11$. The reason that TABDUAL+lazy performs worse than TABDUAL+lazy-tab is that the conflict constraints in the N-queens problem are quite simple, i.e. consist of only column and diagonal checking. It turns out that tabling such simple conflicts does not pay off, that the cost of tabling overreaches the cost of Prolog recomputation. But what if we increase the complexity of the constraints, e.g. adding more queen's attributes (colors, shapes, etc.) to further constrain its safe positioning?

Figure 2 shows abduction time for 11 queens with increasing complexity of the conflict constraints. To simulate different complexity, the conflict constraints are repeated m number of times, where m varies from 1 to 400. It shows that TABDUAL+lazy's performance is remedied and, benefitting from tabling the ongoing conflict configurations, it consistently surpasses the performance of TABDUAL+lazy-tab (with increasing improvement as m increases, up to 15% for $m = 400$). That is, it is scale consistent with respect to the complexity of the constraints.

Declarative Debugging. Since our aim with this benchmark is to evaluate the relative worth of the by-need dual transformation, we focus on three variants: TABDUAL-need, TABDUAL+eager, and TABDUAL+lazy. We evaluate the benchmark for $n = 600$, i.e. debugging a program with 1202 rules. After applying the declarative debugging transformation (of incorrect solutions), which results in an abductive logic program, we apply the TABDUAL transformation. The TABDUAL transformation of variants employing the dual transformation by need takes 0.4691 seconds (compiled in 3.3 secs with 1.5 MB of memory in use), whereas TABDUAL-need takes 0.7388 seconds (compiled in 2.6

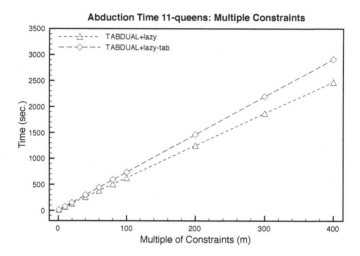

Fig. 2. The abduction time of 11 queens with increasing complexity of conflict constraints

secs with 2 MB of memory in use). Whereas TABDUAL-need creates 1802 second layer dual rules during the transformation, both TABDUAL+eager and TABDUAL+lazy creates only 601 second layer dual rules in the form similar to that of Example 4. And during abduction, the latter two variants construct only, by need, 60% of the complete second layer dual rules produced by the other variant.

Figure 3 shows how the by-need dual transformation influences the abduction time, where different values of m in the IC: $\leftarrow q_m(0)$ are evaluated consecutively, $100 \leq m \leq 600$; in this way, greater m may reuse generic dual rules constructed earlier by smaller m. We can observe that TABDUAL-need is faster than the two variants with the dual transformation by need. This is expected, due to the overhead incurred for computing dual rules on-the-fly, by need, during abduction. On the other hand, the overhead is compensated with the significantly less transformation time: the total (transformation plus abduction) time of TABDUAL-need is 0.8261, whereas TABDUAL+eager and TABDUAL+lazy need are 0.601 and 0.6511, respectively. That is, either dual by-need approach gives an overall better performance than TABDUAL-need.

In this scenario, where all abductive solutions are aggregated, TABDUAL+lazy is slower than TABDUAL+eager; the culprit could be the extra maintenance of the tracking information needed for the explicit memoization. But, TABDUAL+lazy returns the first abductive solution much faster than TABDUAL+eager, e.g. at $m = 600$ the lazy one needs 0.0003 seconds, whereas the eager one 0.0105 seconds. Aggregating all solutions may not be a realistic scenario in abduction as one cannot wait indefinitely for all solutions, whose number might even be infinite. Instead, one chooses a solution that satisfices so far, and may continue searching for more, if needed. In that case, it seems reasonable that the lazy dual rules computation may be competitive with the eager one. Nevertheless, the two approaches may become options for TABDUAL customization.

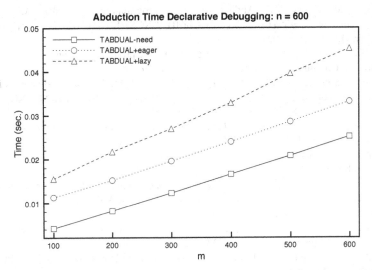

Fig. 3. The abduction time for debugging incorrect solution $q_m(0)$ (with $n = 600$)

5   Conclusion and Future Work

We introduced in TABDUAL two approaches for the dual transformation by need, and enhanced it with a system predicate to access ongoing abductive solutions for dynamic manipulation. Employing these improvements, we evaluated TABDUAL by factoring out their important features and studied its scalability and performance. An issue that we have touched upon the TABDUAL evaluation is tabling nogoods of subproblems (as ongoing abductive solutions) and how it may improve performance and scalability. With respect to our benchmarks, TABDUAL shows good scalability as complexity of constraints increases. The other evaluation result reveals that each approach of the dual transformation by need may be suitable for different situation, i.e. both approaches, lazy or eager, may become options for TABDUAL customization.

TABDUAL still has much room for improvement, which we shall explore in the future. We also look forward to applying TABDUAL, integrating it with other logic programming features (e.g. program updates, uncertainty), to moral reasoning [6, 15].

Acknowledgements. Ari Saptawijaya acknowledges the support of FCT/MEC Portugal, grant SFRH/BD/72795/2010. We thank Terrance Swift and David Warren for their expert advice in dealing with implementation issues in XSB.

References

1. Alferes, J.J., Pereira, L.M., Swift, T.: Abduction in well-founded semantics and generalized stable models via tabled dual programs. Theory and Practice of Logic Programming 4(4), 383–428 (2004)
2. de Castro, J.F., Pereira, L.M.: Abductive validation of a power-grid expert system diagnoser. In: Orchard, B., Yang, C., Ali, M. (eds.) IEA/AIE 2004. LNCS (LNAI), vol. 3029, pp. 838–847. Springer, Heidelberg (2004)

3. Denecker, M., Kakas, A.C.: Abduction in logic programming. In: Kakas, A.C., Sadri, F. (eds.) Computat. Logic (Kowalski Festschrift). LNCS (LNAI), vol. 2407, pp. 402–436. Springer, Heidelberg (2002)

4. Eiter, T., Gottlob, G., Leone, N.: Abduction from logic programs: semantics and complexity. Theoretical Computer Science 189(1-2), 129–177 (1997)

5. Gartner, J., Swift, T., Tien, A., Damásio, C.V., Pereira, L.M.: Psychiatric diagnosis from the viewpoint of computational logic. In: Palamidessi, C., et al. (eds.) CL 2000. LNCS (LNAI), vol. 1861, pp. 1362–1376. Springer, Heidelberg (2000)

6. Han, T.A., Saptawijaya, A., Pereira, L.M.: Moral reasoning under uncertainty. In: Bjørner, N., Voronkov, A. (eds.) LPAR-18. LNCS, vol. 7180, pp. 212–227. Springer, Heidelberg (2012)

7. Josephson, J.R., Josephson, S.G.: Abductive Inference: Computation, Philosophy, Technology. Cambridge U. P. (1995)

8. Kakas, A., Kowalski, R., Toni, F.: The role of abduction in logic programming. In: Gabbay, D., Hogger, C., Robinson, J. (eds.) Handbook of Logic in Artificial Intelligence and Logic Programming, vol. 5. Oxford U. P. (1998)

9. Kakas, A.C., Michael, A.: An abductive-based scheduler for air-crew assignment. J. of Applied Artificial Intelligence 15(1-3), 333–360 (2001)

10. Kowalski, R.: Computational Logic and Human Thinking: How to be Artificially Intelligent. Cambridge U. P. (2011)

11. Kowalski, R., Sadri, F.: Abductive logic programming agents with destructive databases. Annals of Mathematics and Artificial Intelligence 62(1), 129–158 (2011)

12. Lipton, P.: Inference to the Best Explanation. Routledge (2001)

13. Pereira, L.M., Damásio, C.V., Alferes, J.J.: Debugging by diagnosing assumptions. In: Fritzson, P.A. (ed.) AADEBUG 1993. LNCS, vol. 749, pp. 58–74. Springer, Heidelberg (1993)

14. Pereira, L.M., Damásio, C.V., Alferes, J.J.: Diagnosis and debugging as contradiction removal in logic programs. In: Damas, L.M.M., Filgueiras, M. (eds.) EPIA 1993. LNCS (LNAI), vol. 727, pp. 183–197. Springer, Heidelberg (1993)

15. Pereira, L.M., Saptawijaya, A.: Modelling Morality with Prospective Logic. In: Anderson, M., Anderson, S.L. (eds.) Machine Ethics, pp. 398–421. Cambridge U. P. (2011)

16. Pereira, L.M., Saptawijaya, A.: Abductive logic programming with tabled abduction. In: Procs. 7th Intl. Conf. on Software Engineering Advances (ICSEA), pp. 548–556. ThinkMind (2012)

17. Saptawijaya, A., Pereira, L.M.: Tabled abduction in logic programs. Accepted as Technical Communication at ICLP 2013 (2013), http://centria.di.fct.unl.pt/~lmp/publications/online-papers/tabdual_lp.pdf

18. Saptawijaya, A., Pereira, L.M.: Towards practical tabled abduction usable in decision making. In: Procs. 5th. KES Intl. Symposium on Intelligent Decision Technologies (KES-IDT). Frontiers of Artificial Intelligence and Applications (FAIA). IOS Press (2013)

19. Swift, T., Warren, D.S.: XSB: Extending Prolog with tabled logic programming. Theory and Practice of Logic Programming 12(1-2), 157–187 (2012)

20. Swift, T., Warren, D.S., Sagonas, K., Freire, J., Rao, P., Cui, B., Johnson, E., de Castro, L., Marques, R.F., Saha, D., Dawson, S., Kifer, M.: The XSB System Version 3.3.x vol.ume 1: Programmer's Manual (2012)

Aerial Ball Perception Based on the Use of a Single Perspective Camera

João Silva, Mário Antunes, Nuno Lau, António J. R. Neves,
and Luís Seabra Lopes

IEETA / Department of Electronics, Telecommunications and Informatics
University of Aveiro, Portugal
{joao.m.silva,mario.antunes,nunolau,an,lsl}@ua.pt

Abstract. The detection of the ball when it is not on the ground is an important research line within the Middle Size League of RoboCup. A correct detection of airborne balls is particularly important for goal keepers, since shots to goal are usually made that way. To tackle this problem on the CAMBADA team , we installed a perspective camera on the robot. This paper presents an analysis of the scenario and assumptions about the use of a single perspective camera for the purpose of 3D ball perception. The algorithm is based on physical properties of the perspective vision system and an heuristic that relates the size and position of the ball detected in the image and its position in the space relative to the camera. Regarding the ball detection, we attempt an approach based on a hybrid process of color segmentation to select regions of interest and statistical analysis of a global shape context histogram. This analysis attempts to classify the candidates as round or not round. Preliminary results are presented regarding the ball detection approach that confirms its effectiveness in uncontrolled environments. Moreover, experimental results are also presented for the ball position estimation and a sensor fusion proposal is described to merge the information of the ball into the worldstate of the robot.

1 Introduction

In the context of the Middle Size League (MSL) of RoboCup where the ball is shot through the air when the robots try to score goals, it is important to have some estimation of the ball path when it is in the air.

The CAMBADA team robots are equipped with an omnidirectional vision system which is capable of detecting the ball when it is on the ground, but fails to detect the ball as soon as it goes higher than themselves. In this paper, we present a new proposal to achieve a perception of the ball on the air using a single perspective camera, installed in the robots.

To achieve this objective, we explore the physical properties of the vision system and the correspondent geometric approximations to relate the position of the detected ball on the image and its position over the field. The ball detection is based on a hybrid approach. This approach is based on color segmentation for

L. Correia, L.P. Reis, and J. Cascalho (Eds.): EPIA 2013, LNAI 8154, pp. 235–246, 2013.
© Springer-Verlag Berlin Heidelberg 2013

Region Of Interest (ROI) selection and subsequent global shape context analysis for circularity estimation.

In section 2, the problem is briefly exposed and some related work overviewed. Section 3 describes the used vision system and section 4 presents the detection and estimation of ball candidates on the image. In section 5, the algorithm for estimating the position of the ball candidates on the space in front of the camera is presented and section 6 introduces some guidelines for the integration of information from both the cameras of the robots. Section 7 presents a brief comment on the work. Finally, in section 8, some final remarks are presented as future guidelines for this problem.

2   Problem Statement and Related Work

The work presented in this document is focused on the perception of a ball in a robotic soccer scenario. The soccer ball to detect is a size 5 FIFA ball, which has approximately 22 cm of diameter. According to the rules, it has a known predominant color for each tournament. Most teams take advantage of this restriction while the ball is on the ground, since in that case, the environment is very controlled (green floor with some white lines and black robots) and the ball candidates can be expected to be surrounded by this reduced set of colors. In the air, these restrictions are completely lost and thus the approach should be more shape based. The existing shape based approaches are mainly aiming at detecting a ball through shape on the omnidirectional camera.

Several teams already presented preliminary work on 3D ball detection using information from several sources, either two cameras or other robots information [1,2]. However, these approaches rely on the ball being visible by more than one source at the same time, either two cameras with overlapping visible areas or two robots, and then triangulate it. This is not possible if the ball is above the robot omnidirectional camera.

Regarding the detection of arbitrary balls, the MSL league is the most advanced one. Many of the algorithms proposed during previous research work showed promising results but, unfortunately, in some of them, the processing time do not allow its use during a game, being in some cases over one second per video frame [3].

Hanek et al. [4] proposed a Contracting Curve Density algorithm to recognize the ball without color labeling. This algorithm fits parametric curve models to the image data by using local criteria based on local image statistics to separate adjacent regions. The author claims that this method can extract the contour of the ball even in cluttered environments under different illumination, but the vague position of the ball should be known in advance. The global detection cannot be realized by this method.

Treptow et al. [5] proposed a method for detecting and tracking the ball in a RoboCup scenario without the need for color information. It uses Haar-like features trained by an adaboost algorithm to get a colorless representation of the ball. Tracking is performed by a particle filter. The author claims that the algorithm is able to track the ball with 25 fps using images with a resolution of 320×240 pixels. However, the training process is too complex and the algorithm cannot perform in real time for images with higher resolutions. Moreover, the results still show the detection of some false positives.

Mitri *et al.* [6] presented a scheme for color invariant ball detection, in which the edged filtered images serve as the input of an Adaboost learning procedure that constructs a cascade of classification and regression trees. This method can detect different soccer balls in different environments, but the false positive rate is high when there are other round objects in the environment.

Lu *et al.* [7] considered that the ball on the field can be approximated by an ellipse. They scan the color variation to search for the possible major and minor axes of the ellipse, using radial and rotary scanning, respectively. A ball is considered if the middle points of a possible major axis and a possible minor axis are very close to each other in the image. However, this method has a processing time that can achieve 150 ms if the tracking algorithm fails.

More recently, Neves *et al.* [8] proposed an algorithm based on the use of an edge detector, followed by the circular Hough transform and a validation algorithm. The average processing time of this approach was approximately 15 ms. However, to use this approach in real time it is necessary to know the size of the ball along the image, which is simple when considering the ground plane in a omnidirectional vision system. This is not the case when the ball is in the air, in a completely unknown environment without any defined plane.

3   The Perspective Camera

The used camera contains a CCD of $6.26 \times 5.01mm$ and pixel size $3.75 \mu m$. The maximum resolution is 1296×964 and the used lens has a $4mm$ focal length. The camera is fixed to the robot in such a way that the axis normal to the CCD plane is parallel to the ground (Fig. 1).

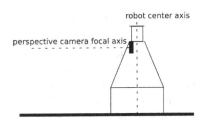

Fig. 1. Illustration of the perspective camera positioning on the robot. It is placed in such a way that the camera focal axis is parallel to the ground and the focal point is slightly displaced from the robot center axis.

Based on the CCD size and the focal length, we can derive the opening angle, both along the horizontal and vertical axis. We will use α to identify the horizontal opening angle and β to identify the vertical opening angle.

Given α and β, we can also estimate the theoretical relation of the field of view (FOV) of the camera at different distances (these calculations do not take into account any distortion that may be caused by the lens). For a given distance, we can then estimate the width of the plane parallel to the camera CCD (as illustrated in Fig. 2) by:

$$tan(\alpha) = \frac{1}{2}\frac{hFOV}{Y} \Rightarrow hFOV = 2 \times Y \times tan(\alpha) \qquad (1)$$

where $hFOV$ is the width of the FOV plane and Y is the distance of the FOV plane to the camera focal point.

For the height of the same plane, the analysis is similar in every way, now considering the ccd height and a β angle.

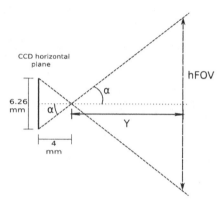

Fig. 2. Scheme of the relation between the CCD width and focal length with the opening angle. The geometric relation is used for estimating the FOV at a given distance.

Due to the position of the camera relative on the robot, the analysis of the FOV and associated geometric characteristics of the camera image have a more direct application. The idea of this application is to provide an approximation of the ball position mainly when it is in the air. The camera captures the images using format7, which allows to use the full resolution of the camera to capture the images but also allows to get and use only a specific ROI. For the objectives of the described work, we opted to use the full horizontal size of the image, while the vertical size was cropped to the first 500 lines. This value was defined to cope with the choice that the camera is used to detect aerial balls and thus it only needs the image above the horizon line. Since the maximum vertical resolution is 964, we use the top of the image, with a small margin.

4   Ball Visual Detection and Validation

Our proposal to detect ball candidates in the air is to use a hybrid approach of color segmentation and statistical analysis of a global shape context histogram.

On a first phase, and since the ball main color is known, a color segmentation of the image is made in order to obtain blobs of the ball color. This is achieved by a horizontal scan of the image rows. On each row, the ball color is detected and a blob is built row by row. The generated blobs have some properties which are immediately analyzed. Based on thresholds for minimum size and solidity of the blob convex hull, each solid blob is selected as a candidate for ball while blobs with very low values of solidity are discarded. An image of the ROI with the blob is created (Fig. 3b). The method to calibrate the colors and camera parameters is the same as the one used for the team omnidirectional camera and is described in [9].

These images are then analyzed by a modified global shape context classifier [10]. Each image is pre-processed with an edge detector and a polar histogram is created. This histogram is then statistically analyzed and returns a measure of circularity of the candidate. The edges image is divided in n layers and m angles, creating the polar histogram, as defined in [11]. The analysis of the histogram is made layer by layer, covering all the angles. An estimation of the average number of edge points on each slice and its standard deviation allows a rough discrimination between circular and non-circular contours, as exemplified in Fig. 3c. A ratio of edge points is also part of the statistics of the candidate.

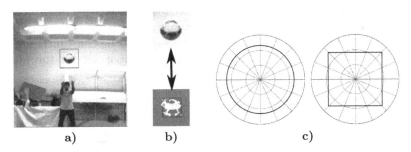

a)   b)   c)

Fig. 3. a): image from the camera with a ROI around the ball, which is the portion of the image used for the histogram analysis; **b)**: the ROI created based on the correspondent color blob; **c)**: rough representation of the polar histogram. The histogram creation fits its radius with the outer points of the edge image, which is not fully represented in these pictures. *Left*: A perfect circular edge on the polar histogram would look something like this. All the edge points are on the same layer and each of its slices have a similar number of points; *Right*: A square on the polar histogram. Due to the fitting properties of the histogram itself, the square edge points should be divided in more than one layer, which would not yield good statistics as circles.

The previously described step always returns the layer with the best statistics, which is currently the layer that has higher average value with minimum standard deviation (the maximum difference between average and standard deviation). This should represent the layer with the most consistent number of edge pixels and thus should be the rounder layer. The next step must then select which of these statistics make sense. Currently, three characteristics are analysed:

– The ratio of edge points on the layer must be within a given range. This range was empirically estimated through testing of several examples of ball and no ball candidates.
– The order of magnitude of the mean should be greater than or equal to the order of magnitude of the standard deviation.
– The candidate diameter estimated by color segmentation and the diameter estimated by the classifier must be coherent. Since the radius of the histogram is dependent on the number of layers, the coherence between the measures is dependent on an error margin based on the histogram radius.

4.1 Experimental Results

Some experiments were performed by throwing the ball through the air from a position approximately 6 meters away from the camera and in its direction. In the acquired videos the ball is always present and performs a single lob shot.

As expected, since the ball is moving almost directly to the camera, the variation of the ball center column on the image is very small (Fig. 4). The row of the ball center, however, was expected to vary. Initially the ball was being held low on the image (meaning the row of the image was a high value) and as it was thrown, it was expected that it went up on the image, then down again. Fig. 4 allows us to verify that this behavior was also observed as expected.

On the other hand, since the ball is coming closer to the camera every frame, it was also expectable that its size on the image would be constantly growing. The correct evaluation of the width of the ball is important for the position estimation described in the next section.

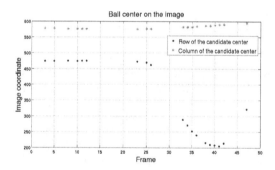

Fig. 4. Results from the image analysis of a ball being thrown in the camera direction: values of the row (Blue dots) and column (Red stars) where the ball center was detected on the image

The main contribution of this color/shape hybrid approach, however, is the reliability of the acquired data, respecting the real time constraints of the application. A test scenario was created, where the ball was thrown in such a way that it was visible in every frame of the videos, and several runs were made. Although the results are strongly affected by the environment around the ball, we obtained promising preliminary results with relatively high precision, even if the recall has shown lower results. A pure color approach yielded better results in a very controlled environment, but in a more uncontrolled environment we obtained a very low precision. We consider that having a higher precision is more advantageous, even when facing the loss of some recall. The processing time for the algorithm was 11.9 ± 2.8 ms which is still within our time restrictions.

5   Ball Position Estimation

After having the candidates selected as balls, there is the need to estimate their position. To accomplish that, we first analyze the candidate radius in pixels. The size that each pixel represents at each distance increases with distance to the camera focal point. This is due to the fact that the resolution is constant but the FOV is not. With the FOV width relation, we can estimate the size that each pixel represents at each given distance, by a relation of the estimated distance and the horizontal resolution:

$$pS = \frac{hFOV}{hR} \tag{2}$$

and thus, since we know that the ball has $0.22m$, we can estimate the number of pixel expected for the blob width at a given distance:

$$pW = \frac{0.22}{pS} \tag{3}$$

where pS is the pixel size in the plane with the defined horizontal FOV($hFOV$), hR is the CCD horizontal resolution and pW is the expected ball width, in pixels, for the plane with the given pixel size.

For the same setpoint distances as before the ball width, in pixels, was estimated. Table 1 presents those results.

Table 1. Table with the theoretical ball width at several distances. Distances are in meters, ball width are in pixels.

Distance to camera	1	2	3	4	5	6	7	8	9
Expected ball width	182	91	61	46	36	30	26	23	20

From this analysis, Equation 3 can be developed using Equations 2 and 1:

$$pW = \frac{0.22 \times hR}{hFOV} = \frac{0.22 \times hR}{2 \times Y \times tan(\alpha)} \tag{4}$$

from which we get an inverse relation function of pixel width pW and distance to camera focal point Y.

Given the known distance of the ball candidate, which is our YY coordinate, and the linear relation of the pixel size, we can estimate the XX coordinate. This is true due to the camera positioning on the robot that, besides having the focal axis parallel to the ground, it is also coincident with the robot YY axis. To accomplish the estimation of the XX coordinate we have to analyze the horizontal pixel coordinate of the ball center from the image center and apply the calculated pixel size.

We can thus obtain the XX and YY coordinates of the ball on the ground plane, relative to the perspective camera, from which we know the relative coordinates from the center of the robot.

5.1   Experimental Results

An experimental analysis was performed to verify the relation between the detected ball width in pixels and the distance it is from the camera.

Unfortunately, and like most practical scenario, it was verified that the expected theoretical values of the ball pixel width according to the distance was not verified in practice. To verify the ball width according to the distance from it to the camera, an experimental setup was mounted.

The experiment was performed by placing the camera with its axis along a field line and placing the ball in front of it. The ball was on top off a support, which maintained ball height, and was placed at the several setpoint distances from the camera (from one to nine meters). These distances were measured with tape and used as ground truth data. The ball pixel width was measured by manually stopping the video at frames corresponding to the setpoint distances and verifying the estimate width of the detected ball. The results are presented in Table 2.

Table 2. Table with the measured ball width at several distances. Distances are in meters, ball width are in pixels.

Distance to camera	1	2	3	4	5	6	7	8	9
Measured ball width	200	110	72	52	42	32	28	22	16

Based on the values for the relation between the ball pixel width and the distance to camera, a 3rd degree polynomial function is used to, given a ball blob's width, estimate its distance to camera (the YY coordinate). This relation is slightly different from the theoretical relation presented in Equation 4, due to factors like lens distortion that were not accounted in the previous analysis.

To make an approximation for this data, we can estimate a polynomial function that, given a pixel width of a given candidate, returns the distance at which that candidate is from the camera. To keep computational time affordable, we do not wish to have high degree polynomial functions, and thus we tested the fit of functions up to 4th degree, and verified that a 3rd degree function would fit the data acceptably. However, the function behavior at shorter distances is not proper (Fig. 5a).

For that reason, and given the short distances, a linear approximation of the data would fit the correspondent data in a better way. Fig. 5b represents the

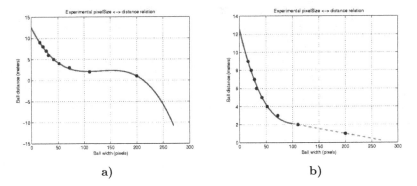

a)   b)

Fig. 5. Left **a)**: Third degree polynomial (red line) fitting the defined experimental setpoints (blue dots). Although the polynomial function fits acceptably for sizes corresponding to distances above 2 meters, closer distances sizes do not seem to evolve according to the same curve; Right **b)**: Third degree polynomial function (red line) fitting the defined experimental setpoints (blue dots) and linear polynomial function (green dashed) for the sizes corresponding to closer distances.

two polynomial functions considered. The used separation point was empirically estimated.

In the same experiment of throwing the ball from 6 meters away from the camera, described in Section 4.1, the results of the positions evaluated by the previously described algorithm were captured. Fig. 6 depicts these results. The path formed by the estimated XY positions approximates the path of the ball arc through the air, projected on the ground. This data allows a robot equipped with such camera to estimate the path of the incoming airborne ball and place itself in front of the ball, for defending in the case of the goal keeper. Given the nature of the task, there is no need for an excellent precision on the estimations, just a general direction which provides the target for the robot. The perspective vision process, from capture to the production of the XX and YY coordinates took an average time of around 12.5 ms to execute in a computer with an Intel Core 2 duo at 2.0 GHz. The tests were made for the perspective camera process running standalone at 30 frames per second.

6   Ball Integration

Being the ball the main element of a soccer game, its information is very important and needs to be as precise as possible. Failure on its detection can have very negative impact on the team performance. Probably even worse than failing to detect the ball (situation on which the robots can strategically move in search for it) is the identification of false positives on the ball. This can deviate the attention of a robot or even the entire team from the real ball, which can be catastrophic. To avoid false positives and keep coherence on the information of the ball, several contextual details are taken into account.

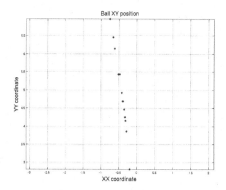

Fig. 6. Picture of a data capture of a ball kicking test. The ball was thrown by the air from a position around (-0.5, 6.0) in the approximate direction of the camera (which is the origin of the referential). The blue dots represent the estimated ball positions.

Given the several possible sources of information, the priority for ball position is the omnidirectional camera. Details about the visual ball detection and validation on the omnidirectional camera can be found in [9]. The next source to use is the shared information between team mates, because if they see the ball on the ground, there is no need to check the perspective camera information. Finally, the agent tries to fit the information from the perspective camera into the worldstate (Fig. 7). This is an improvement of the integration algorithm for the ball position information presented in [12].

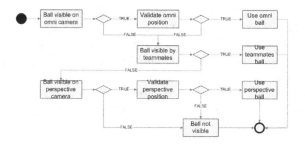

Fig. 7. Ball integration diagram

At this point, the visual information is a list of candidates that can be the ball.

The validation of a perspective ball candidate depends on some context of its detection, based on the analysis of information known at this point:

– The first analysis to make is whether the ball candidate is inside the field or not. If it is outside the field of play, there is no need to use it, since even if it is truly the ball, the game would be stopped.

- To maintain coherence between the perspective vision ball and the omni vision ball, an analysis of the last omni camera ball positions is made and a perspective candidate is considered only if the this candidate position is in a given vicinity of the omni camera ball position. Since we are interested in getting the general ball path, the angular difference is the measure considered for defining this vicinity. A candidate from the perspective camera is only accepted if the difference to the last omni camera ball position is below a given threshold. This validation is performed on the first detections by the frontal camera, when the ball has also just became or is becoming not visible for the omni directional camera.

- Another filtering that is done is an analysis of the number of cycles with the ball visible on the perspective camera. Again, the objective of the perspective camera is to detect aerial balls. During the game, the ball leaves the ground only on short periods. When a kick raises the ball, it will inevitably be in the air for only a few instants, which can be periodic if the ball bounces several times, but still the appearances are short. A ball constantly detected for more than a given amount of time is then discarded, since it is probably a false positive or, for instance, a stop game situation and the referee is holding the ball on his hands.

7 Conclusions

This paper presents a new approach for aerial ball perception based on the use of a single perspective camera. This approach is based on three main steps, the visual ball detection followed by an estimation of the ball position based on a geometric analysis of the vision system and finally a sensor fusion approach of this information with other sources of information.

The hybrid approach for visual detection of the ball uses a fast color segmentation based algorithm combined with the application of a polar histogram analysis. Although a pure shape based algorithm could provide more accurate results, the fact that this application has real-time restrictions, lead us to include the color segmentation based algorithm to reduce the shape analysis to limited small size ROIs.

8 Future Work

The initial analysis of the performance of this approach showed that there are some limitations to its use on a real game scenario, mainly due to the fact that the object of interest, the ball, moves at a very high speed. In many frames of a video capture, it is verified that the distortion blur is very high and thus, the shape analysis is compromised, forcing us to wide the range of detection, thus lowering the effectiveness of the algorithm.

As future approaches, we intent to explore two scenarios to try to deal with this problem:

- to export and use the detection approach on high speed cameras, which would probably provide us frames with a reduced blur effect (even if we could/should not process all the frames)
- to try a new approach based on 3D Kinect camera to detect aerial objects.

Aknowledgements. This work was developed in the Institute of Electronic and Telematic Engineering of University of Aveiro and was partially funded by FEDER through the Operational Program Competitiveness Factors - COMPETE, also by National Funds through FCT - Foundation for Science and Technology in the context of the project FCOMP-01-0124-FEDER-022682 (FCT reference PEst-C/EEI/UI0127/2011) and finally by project Cloud Thinking (funded by the QREN Mais Centro program, ref. CENTRO-07-ST24-FEDER-002031).

References

1. Voigtlrande, A., Lange, S., Lauer, M., Riedmiller, M.: Real-time 3D ball recognition using perspective and catadioptric cameras. In: Proc. of the 3rd European Conference on Mobile Robots, Freiburg, Germany (2007)
2. Burla, A.: 3D ball detection in robocup. Master's thesis, University of Stuttgart (2007)
3. Mitri, S., Frintrop, S., Pervolz, K., Surmann, H., Nuchter, A.: Robust object detection at regions of interest with an application in ball recognition. In: Proc. of the 2005 IEEE International Conference on Robotics and Automation, ICRA 2005, Barcelona, Spain, pp. 125–130 (2005)
4. Hanek, R., Schmitt, T., Buck, S.: Fast image-based object localization in natural scenes. In: Proc. of the 2002 IEEE/RSJ Int. Conference on Intelligent Robotics and Systems, Lausanne, Switzerland, pp. 116–122 (2002)
5. Treptow, A., Zell, A.: Real-time object tracking for soccer-robots without color information. Robotics and Autonomous Systems 48, 41–48 (2004)
6. Mitri, S., Pervolz, K., Surmann, H., Nuchter, A.: Fast color independent ball detection for mobile robots. In: Proc. of the 2004 IEEE Int. Conference on Mechatronics and Robotics, Aachen, Germany, pp. 900–905 (2004)
7. Lu, H., Zheng, Z., Liu, F., Wang, X.: A robust object recognition method for soccer robots. In: Proc. of the 7th World Congress on Intelligent Control and Automation, Chongqing, China (2008)
8. Neves, A.J.R., Azevedo, J.L., Cunha, B., Lau, N., Silva, J., Santos, F., Corrente, G., Martins, D.A., Figueiredo, N., Pereira, A., Almeida, L., Lopes, L.S., Pinho, A.J., Rodrigues, J., Pedreiras, P.: CAMBADA soccer team: from robot architecture to multiagent coordination. In: Papic, V. (ed.) Robot Soccer, pp. 19–45. INTECH (2010)
9. Neves, A.J., Pinho, A.J., Martins, D.A., Cunha, B.: An efficient omnidirectional vision system for soccer robots: From calibration to object detection. Mechatronics 21, 399–410 (2011)
10. Pereira, R., Lopes, L.S.: Learning visual object categories with global descriptors and local features. In: Lopes, L.S., Lau, N., Mariano, P., Rocha, L.M. (eds.) EPIA 2009. LNCS, vol. 5816, pp. 225–236. Springer, Heidelberg (2009)
11. Antunes, M.L.P.: Semantic vision agent for robotics. Master's thesis, University of Aveiro (2011)
12. Silva, J., Lau, N., Neves, A.J.R., Rodrigues, J., Azevedo, J.L.: World modeling on an MSL robotic soccer team. Mechatronics 21, 411–422 (2011)

Loop Closure Detection with a Holistic Image Feature

Francisco M. Campos[1], Luís Correia[2], and João M.F. Calado[3]

[1] LabMAg and the Mechanical Engineering Department, Instituto Superior de Engenharia de Lisboa, Lisbon, Portugal
`fcampos@dem.isel.pt`

[2] LabMAg, Computer Science Department, Universidade de Lisboa, Lisbon, Portugal
`Luis.correia@di.fc.ul.pt`

[3] IDMEC and the Mechanical Engineering Department, Instituto Superior de Engenharia de Lisboa, Lisbon, Portugal
`jcalado@dem.isel.pt`

Abstract. In this paper we introduce a novel image descriptor, LBP-gist, suitable for real time loop closure detection. As the name suggests, the proposed method builds on two popular image analysis techniques: the gist feature, which has been used in holistic scene description and the LBP operator, originally designed for texture classification. The combination of the two methods gives rise to a very fast computing feature which is shown to be competitive to the state-of-the-art loop closure detection. Fast image search is achieved via Winner Take All Hashing, a simple method for image retrieval that exploits the descriptive power of rank-correlation measures. Two modifications of this method are proposed, to improve its selectivity. The performance of LBP-gist and the hashing strategy is demonstrated on two outdoor datasets.

Keywords: loop closure detection, visual features, gist, Local Binary Patterns, image retrieval.

1 Introduction

This paper is concerned with vision based loop closure detection, i.e. recognizing when a robot has returned to a previously visited place by relying on visual data. The success in correctly establishing this correspondence strongly depends on the discriminative power of the appearance representation. In order to increase discriminativity, many studies have exploited local image features, which capture appearance in localized areas in the image, denoted Regions of Interest (ROI). The success of this model, known as Bag of Features (BoF), is due to the good descriptive power of local features and a compact representation that allows fast searching of similar images. The BoF model has been widely applied in robot localization and loop closure detection [1–3], however, the resulting systems are based on a fairly complex framework, involving i) the detection of ROIs, ii) feature extraction, iii) quantization and iv) retrieval from the inverted file. Often, the retrieved images are further checked for geometric consistency, before a loop closure is signaled. Adding to the complexity of the

L. Correia, L.P. Reis, and J. Cascalho (Eds.): EPIA 2013, LNAI 8154, pp. 247–258, 2013.
© Springer-Verlag Berlin Heidelberg 2013

approach is the requirement to build a vocabulary beforehand. An alternative to the BoF model is to use holistic features, which have been proven effective in describing scene categories [4]. These features contrast to the BoF model by removing the need to segment the image. Instead, statistics over the entire image are computed, and rough spatial information may be captured by imposing a fixed geometry sampling. Even though the simplicity of this approach and its low computation demands make holistic features very attractive from a robotics perspective, they have seldom been exploited. Motivated by the potential of this kind of representation, in this paper we introduce LBP-gist, a feature that combines the ideas behind gist [4], a popular holistic feature, and the LBP (Local Binary Pattern) method [5] for texture extraction.

A key component of an efficient loop closure detector is the image search algorithm. This component is responsible for lowering the computational cost of the approach, by selecting only the relevant database images to be further processed by a more expensive similarity test. Furthermore, the computational complexity of the search algorithm itself is a concern, hence two requirements should be fulfilled: it should use a compact image representation and fastly index similar images. Recently, hashing methods have been seen as an attractive solution to image retrieval that meets these requirements. In particular, Locality Sensitive Hashing (LSH) methods [6] are well suited when one aims at finding similar images, as in place recognition, rather than exact duplicates. Given an appropriate similarity measure, the basic idea in LSH is to hash points so that similar points are hashed to the same bucket and dissimilar points are hashed to different buckets, with high probability. A category of these algorithms is particularly interesting for guaranteeing a collision probability that is equal to a similarity measure. In this work we use Winner Take All (WTA) hashing [7], a method offering this similarity estimation capability. Building on this property, we extend WTA by i) enhancing the similarity estimation through bucket weighting and ii) setting a relative distance condition that is shown to significantly reduce the number of images in the relevant set.

This paper is organized as follows: after reviewing related work in section 2, we present an overview of the LBP method, in section 3; section 4 introduces the LBP-gist feature and the corresponding similarity measure; section 5 details our implementation of WTA hashing and section 6 presents the evaluation of both the hashing scheme and the image feature; finally, section 7 draws the conclusions of the paper.

2   Related Work

Along with numerous studies on the BoF model, a few examples of place recognition with holistic features can be found in the literature. Most of the recent systems using holistic features are inspired by the gist feature, introduced in [4] to model the contents of a scene. The gist feature is extracted by first applying steerable filters at different scales and orientations to the image and dividing it into a 4-by-4 grid. Then, mean responses of these filters are computed on each sub-window, and gathered in a gist descriptor. From this definition, the idea of applying a fixed geometry and computing statistics over the image has been adopted and developed in various ways in place recognition works [8–12]. In [8] a biologically inspired vision system is coupled with a Monte Carlo algorithm to localize a robot. In that system good localization performance is achieved by combining gist with saliency-based local features, for increased descriptive power. A specific representation for omni-directional imagery, denoted gist panorama, is introduced in [9]. Low computation cost is achieved in that work through gist descriptor quantization, while localization performance is improved by a panorama alignment technique. Gist has been applied together with the epitome concept by Ni *et al* [10], in a relatively more complex representation that aims at increasing scale and translation invariance. The method proposed in this paper contrasts to the aforementioned approaches for using the fast extracting LBP features for texture description.

An approach close to ours is presented in [11], where the authors rely on the census transform to extract textural information. Although similar to the LBP operator, the census transform may not provide the same flexibility, due to its small spatial support. Also, in that work only the localization problem was addressed, with a discriminative approach that is not suitable for loop closure detection. Another fast computing feature is BRIEF-Gist, recently proposed by Sunderhauf and Protzel [12]. The approach brings together the concept of capturing the spatial structure, as in gist, and BRIEF, a visual feature originally proposed for describing local patches. Low computation time is guaranteed by the good properties of BRIEF, which is based on simple pixel comparison and produces a binary descriptor. The resulting BRIEF-Gist descriptors can thus be compared extremely fast through Hamming distance. In spite of the efficiency of the method, the authors note that it may be sensitive to rotation and translation, which may lower its accuracy.

Several works have shown performance improvements in vocabulary based approaches via the so called term frequency-inverse document frequency (tf-idf) weighting scheme [13]. This is achieved by replacing the visual word histogram by a more discriminative descriptor, where each component is a product of two factors. The first factor, tf, is the frequency of a given word in the current document, indicating how well this word describes the document; the second factor, idf, captures the informativeness of the word by considering its occurrence throughout the whole database, being used to downweight the contribution of frequently occurring words. While the tf-idf is not directly applicable to the LBP-gist descriptor, it is conceivable to measure the importance of buckets in a hashtable after database descriptors have been inserted. Here we propose a modified similarity measure for the WTA hash that takes into account an idf factor computed on buckets.

3   Overview of the LBP Approach to Texture Analysis

The first step in the LBP method [5] amounts to generating a label image, where each pixel of the intensity image is assigned a code that summarizes the local structure around the pixel. The LBP operator is made very efficient by deriving the codes from simple binary comparisons. In the original formulation of LBP, the code for a pixel at (x_c, y_c) results from the binary comparison of its gray level, g_c, to those of eight neighborhood pixels, denoted g_0, g_1,..., g_7, by a threshold function $T(g_c, g_i)$. The eight bits obtained through the comparison are arranged in a bit string resulting in a label from of 2^8 possible codes. While in this formulation the operator support is limited to a block of 3x3 pixels, in a later development, denoted multiresolution LBP [14], the method was generalized to arbitrary neighborhood size and number of sample neighbours. In this generic version, the neighborhood is defined as a set of P sampling points evenly distributed on a circle centered on the pixel to be labeled, thus allowing for capturing structures at different scales, by varying the radius R of the circle. Sampling points that do not fall exactly on an image pixel are assigned a gray level obtained by bilinear interpolation. The choice of the number of sampling points determines the granularity of the texture being encoded. In order to limit the number of different patterns to a manageable size, the number of sampling points should be kept low, even as the radius increases. In this case, the authors of [14] advocate that the gray level of a sampling point should represent a larger area than just a pixel (or the immediate neighbours, when interpolation is needed) and propose filtering the intensity image through a low-pass gaussian filter, prior to applying the LBP operator.

In many applications, the codes obtained through thresholding are mapped onto a smaller subset, based on the concept of uniformity. As defined in [15], the uniformity measure U is the number of bitwise transitions from 0 to 1 and vice-versa, when the bit string is considered circular. According to this definition, codes of lower U are more uniform, thus a local binary pattern was called uniform if its U value is at most 2. Noticing that uniform patterns account for a large fraction of the codes occurring in natural images, in [15] non-uniform patterns are merged into a single label in the so-called uniform mapping.

In the second step of the LBP method, texture in the whole image or image sub-blocks is described by the distribution of patterns, captured as a histogram of label occurrence in the area being analyzed.

4 LBP-gist Method

4.1 The LBP-gist Descriptor

The LBP-gist descriptor combines the LBP operator with the holistic representation of gist to achieve a rich and fast computing visual feature. The method we propose makes use of the LBP operator with several modifications which enhance its descriptive power. In order to extend the LBP kernel support, we rely on the multiresolution method, which can be adjusted through the R and P parameters for a suitable size and granularity. Additionally, we resort to a modified threshold function for the reduction of noise in the labels. In fact, LBP is known to lack robustness in describing flat areas, like the sky and uniform walls, where negligible variations in pixel values give rise to different labels. Since such flat patterns often occur in a place recognition problem, this issue should be addressed in developing a robust image descriptor. In [16], flat patterns are handled by setting a constant offset at which the threshold function changes value, thus allowing for small variations in gray level to equally produce a binary comparison of 0. In [17] the authors point out that a constant offset thresholding is not invariant to gray scale transforms and propose a variable offset, proportional to the gray level of the center pixel. However, as this later approach results in small offsets in dark areas, the descriptor may still be affected by noise. In order to overcome this limitation, in the LBP-gist method we propose using an offset that results from the combination of a constant and a variable term. Accordingly, the threshold function we use is defined as:

$$T(g_c, g_i) = \begin{cases} 1, & g_i \geq (1+\tau_1)g_c + \tau_0 \\ 0, & g_i < (1+\tau_1)g_c + \tau_0 \end{cases} \tag{1}$$

where τ_0 is the constant offset term and τ_1 is a small value determining the gray level scaling offset.

In order to achieve better descriptive power, we use a mapping that represents codes with U measure higher than 2, instead of merging them. Specifically, our mapping considers codes with U measure immediately superior to the uniform patterns, i.e. U=4, and merges codes with a higher value of U. In this mapping a histogram will have a length of 58 uniform patterns + 141 non-uniform patterns = 199.

Once a label image is obtained, the second step in LBP-gist extraction amounts to dividing this image in N equally sized regions and computing LBP histograms over each block. Denoting by h^i the histogram of block i, the final descriptor is defined as the set of histograms, $D=\{h^1, h^2,\ldots, h^N\}$. In this work images are divided in horizontal blocks, a partition geometry that is more robust against variations arising from in plane motion of the robot. Concretely, we extract 3 non-overlapping horizontal blocks plus 2 overlapping blocks, from each image (see Fig. 1). Thus, descriptor D will contain 5 histograms of length 199.

Fig. 1. Horizontal image partitioning in 5 blocks. Left: non-overlapping blocks; right: overlapping blocks.

4.2 Similarity Computation

In order to compare LBP-gist descriptors, the issue of how information extracted from different image blocks is combined must be addressed. The combination of multiple LBP histograms in the computation of a similarity score has often been done by concatenating individual histograms into single vectors, which were then used as inputs to a histogram comparison measure. However, this approach fails in accounting for the uneven relevance different image blocks may bear. In a face recognition system

[18], this issue was tackled by aggregating the distances obtained for individual histograms through a weighted sum, with weights being set in an offline procedure. Contrasting to that system, a loop closure detector must cope with an unknown environment, which means that the relevance of image regions cannot be known beforehand and may change with location. Hence, we propose a new context dependent similarity measure that assigns relevance to an image pattern based on its occurrence in locations of similar appearance. The procedure amounts to first arranging the database images in ascending order of their distances to the test image, computed by the Chi square measure on the concatenated histograms. In a database containing nl locations/images, the ordered set of descriptors $\{D_1, D_2,..., D_{nl}\}$ will be obtained. In the next step, similarity scores for each image block are computed. Denoting by d^i_j the Chi square distance between block i of the test image and the corresponding block of the j database image, this distance is converted to a similarity value, s^i_j, through the exponential function as follows

$$s^i_j = \exp\left(d^i_j / \alpha\right) \tag{2}$$

with α being a constant scaling factor. The relevance assigned to block i, quantified by weight w_i, is derived from the score it obtained in a set of images, through

$$w^i = \left(\sum_{j=k_1}^{k_2} s^i_j\right)^{-1} \tag{3}$$

In this expression, the reason for selecting images starting at rank k_1 is that, when a loop closure actually occurs, the high similarity of the top ranked images would in practice reduce the relevance weights, even if the visual patterns in the image are not frequent. On the other hand, it is expected that a visual pattern which is frequent should occur both in top and lower rank images and therefore have its relevance reliably estimated by the proposed method. In our experiments, the values of k_1 and k_2 were respectively adjusted to 10 and 20, for best performance; however, the overall algorithm did not show to be significantly sensitive to changes in these parameters. The final similarity score is defined as the weighted sum of individual similarities:

$$s_j = \sum w^i s^i_j \tag{4}$$

5   Hashing Method

A hash method enables efficient retrieval by mapping points in the input space onto a hash table, through a so called hash function. At query time, a hash table bucket is indexed, by hashing the test point through the same function and database points found in that bucket are retrieved. This is also the concept underlying the WTA hashing algorithm, which is used in this work. In this section we review this hashing scheme and introduce two modifications to the original algorithm, aiming at improved performance.

5.1 WTA Hashing Overview

The WTA algorithm [7] relies on hash functions h_i defined as follows. First, a random permutation of the descriptor dimensions is generated and the first K elements of the permutation are saved, yielding a partial permutation θ_{Ki}. The hash value of point p is obtained by applying partial permutation θ_{Ki} and extracting the index of the maximum component of the resulting vector. If many h_i functions are used, the collision rate between points p_1 and p_2 is an estimate of a rank-order similarity measure $S(p_1,p_2)$ (see [7] for details).

As in other LSH schemes, efficiency of WTA hashing may be increased by building composite hash functions $g_j = (h_{j,1}, h_{j,2}, ..., h_{j,T})$, where $h_{j,i}$ are randomly generated hash functions. Assuming two dissimilar points have probability P of collision, through a function $h_{j,i}$, this probability becomes $P \cdot T$, when g_j is used instead. It follows that the number of dissimilar points returned will be lower. On the other hand, a number L of g_j functions are commonly used in parallel, so that the probability of retrieving a similar point is above a prescribed value (see [19] for the selection of parameters criteria).

5.2 Proposed Modifications to WTA Hashing

Usually a LSH algorithm is considered completed when collisions have been detected and the database items colliding at least once with the test point have been retrieved for further processing. Here we take a different approach in that we use the number of collisions as an estimate of similarity. Let $c_j(p_1,p_2)$, $c_j \in \{0,1\}$ be a function indicating a collision between points p_1 and p_2, through g_j, with $c_j=1$ when a collision is detected and $c_j =0$ in the opposite case. We define a similarity estimate by adding up the collisions detected by the set of functions g_j:

$$S(p_1, p_2) = \sum_{j=1}^{L} c_j(p_1, p_2) \tag{5}$$

In the following we introduce two methods, idf weighting (idf) and relative similarity threshold (RST) that respectively aim at improving the similarity estimate and reduce the number of irrelevant images returned by the hashing algorithm.

Idf Method -The tf-idf scheme [13], as defined for vocabulary based systems, downweights word contribution according to informativeness, which decreases with word frequency. This notion of informativeness can be integrated in our hashing algorithm by allowing hashtable buckets to replace the concept of visual words of the traditional tf-idf scheme. In this case, a idf weight is defined as $Widf_j = \log(N/nb_j)$, where N is the number of images inserted in the hashtable and nb_j is the number of images colliding in bucket *j*. The new similarity measure accumulates collisions between two points, now weighted by the *Widf* factor:

$$S(p_1, p_2) = \sum_{j=1}^{L} (c_j(p_1, p_2) \times Widf_j) \tag{6}$$

RST Method – This is a query adaptive method that aims to reduce the number of irrelevant images retrieved by exploiting the collision-based similarity measure. The idea is to use the similarity values for a given query to determine a variable threshold below which images are discarded. Such criterion may be beneficial in environments with many similar looking places, which produce high values of similarity. In this case, a threshold derived from the similarity ratio would be useful to reject those places that are less likely to be a match. Accordingly, in this work we define the RST method as the selection of only those images having a similarity (eq. 6) above half the maximum similarity measured for a given query.

6   Results

We tested the proposed model on two publicly available datasets, which will be referred as Malaga and City Centre (see table 1 for details on each dataset). Both datasets provide imagery captured by two perspective cameras installed on a mobile platform. In the case of the Malaga dataset, both cameras are forward-pointing while the City Centre cameras are oriented to the left and right of the robot movement. As in the first case the two cameras provide similar information, only the images from one camera (left) are used in this work. The two datasets also feature different sampling strategies. Since high frame rates imply a large overlapping between consecutive images, usually the detector is applied on a subsampled set of images. This is the case of the Malaga image sequence, which was subsampled to a lower frame rate. Differently, images from the City Centre dataset were captured at approximately 1.5m intervals, which inherently corresponds to a low frame rate.

Table 1. Main features of the datasets used for evaluation

Dataset	Total length [m]	Revisited length [m]	Sampling	Subsampling	N° of images evaluated
Malaga 6L [20]	1192	162	7.5fps	3.25 fps	1737
City Centre [1]	2260	1570	1.5m	-	2474

Ground truth GPS coordinates are also provided in the two datasets. These measurements were used to derive ground truth loop closures, by using the following criteria: two images are assumed to be captured at the same location if the corresponding positions are within a distance of 6.5m and the robot headings do not differ more than $\pi/2$. For the samples where position measurements are not available due to GPS failure, loop closure was verified manually. For the following tests, LBP-gist parameters were experimentally set to R=9, P=8, τ_0=5, τ_1=0.03.

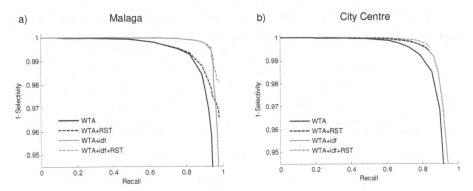

Fig. 2. 1-Selectivity versus recall for the a) Malaga and b) City Centre datasets

6.1   WTA Hashing Performance and Execution Times

To evaluate the performance of WTA hashing and the proposed modifications, we conducted image retrieval experiments on the two datasets. For these and subsequent experiments, hashing parameters K=10, T=4 and L=40 were empirically selected as good compromise between efficiency and accuracy of the algorithm. To measure hashing performance, LBP-gist descriptors are processed by each method and the corresponding similarity measure is analyzed in terms of recall and selectivity. Recall is defined as the ratio of the number of loop closures, detected for a given similarity threshold, and the actual number of loop closures existing. Selectivity was introduced in [21] and is defined as the fraction of items in the database that is retrieved for a given threshold. For ease of interpretation, in Fig. 2 we plot 1-selectivity versus recall, a curve that bears resemblances to the common precision-recall curves. The data in the y axis of the figure can therefore be interpreted as the percentage of items in the database that are ignored, while achieving a given recall rate. Results in Fig. 2 indicate that the idf method is very effective in reducing the number of irrelevant images retrieved. On the other hand, the impact of RST is not as consistent in our experiments. While its contribution in the City Centre dataset is beneficial, both when applied in isolation and in combination with idf, its effect in the Malaga dataset is modest.

Table 2 presents computation times for the all the stages of the loop closure detector, executed on a core 2 duo CPU, running at 2.2 GHz. These values correspond to experiments where images were retrieved by WTA hashing with the two proposed modifications. A value of 10 was chosen as the fixed threshold on the hashing similarity used to select the images for further processing. Images bearing a similarity to the test image above that value are then evaluated by the LBP-gist similarity measure. It can be seen that the hashing algorithm yields mean computation times lower than 2ms and is very effective in selecting only the relevant images, as it enabled fast LBP-gist similarity evaluation. For comparison purposes, we also present execution time of a fast BoF system, as reported in [3] (table 3). We note that these results were obtained on a larger dataset than the ones used in our work. Computation time of the BoF component, which is dependent on the dataset size, is therefore not comparable to our values. Most notable in the comparison of computation times is the difference

Table 2. Execution times [ms] of the components of the LBP-gist loop closure detector

Dataset	LBP-gist extraction		Insertion in Hashtable		Hashtable retrieval		LBP-gist similarity		Whole system	
	mean	max	mean	max	mean	max	mean	max	mean	max
Malaga	4.1	6.8	1.1	1.5	0.8	1.4	0.4	2.0	6.4	11.7
City Centre	4.1	6.8	1.1	1.8	0.8	1.6	0.1	0.9	6.1	11.1

Table 3. Mean execution times [ms] of a BoF based loop closure detector [3]

Feature extraction	Bag of Features	Geometric Verification	Whole system
14.35	6.9	1.6	21.60

Table 4. Recall rate at 100% precision obtained with LBP-gist, BRIEF-gist [12] and a BoF system [3]

Dataset	LBP-gist	BoF [3]	BRIEF-gist [12]
Malaga	82.1	74.8	-
City Centre	68.7	30.6	32

in the feature extraction component which in our case is about 10ms lower. BRIEF-gist, an alternative holistic feature is reported to have extracting cost of 1ms, however it shows lower accuracy (see next section).

6.2   Loop Closure Detection Accuracy

We now turn to the analysis of our system in terms of its accuracy in detecting loop closures. To this end, LBP-gist descriptors of the images retrieved by the hashing scheme are evaluated for their similarity with respect to the test image (eq. 4). A threshold on this similarity measure of 0.7 was found, so that a precision of 100% is guaranteed on the two datasets. Figure 2 shows the loop closures identified by the system, plotted over the robot trajectory. Note that, in the case of Malaga (Fig. 3.a), a number of locations are revisited with in opposite direction, thus are not considered valid loop closures by the criteria we applied. Fig 2.a shows that the system detected 4 out of 5 loop closing events, on the Malaga dataset, and Fig 2.b also confirms the good performance of the system, with loop closures detected being well distributed throughout the overlapping path. Table 4 compares LBP-gist performance with BRIEF-gist and a BoF system, using the values for these systems respectively reported in [12] and [3]. In the case of BRIEF-gist, only results for the City Centre dataset are available. These results show that LBP-gist outperforms both systems, offering performance gains that reach, in the case of City Centre, about a 37% recall difference.

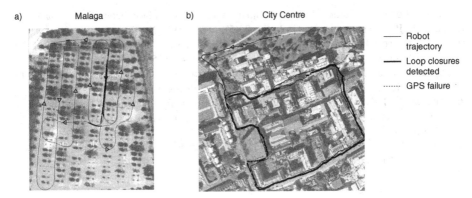

a)   Malaga   b)   City Centre

—— Robot
trajectory

—— Loop closures
detected

······ GPS failure

Fig. 3. Loop closures detected by our system, plotted over the map of the environment. The thin lines describe the robot trajectory and the revisited locations detected are indicated by the thicker lines.

7   Conclusions

This paper introduced LBP-gist, a holistic image feature developed for visual loop closure detection. The proposed model was tested on two outdoor datasets, where it showed accuracy gains over BRIEF-gist and a state of the art BoF system. In one of the datasets the performance improvement was highly significant.

When compared to BoF systems, computation cost of extracting LBP-gist is lower, which is achieved by removing the need to extract local features and relying on the fast computing LBP features instead. The computational requirements of the whole loop closure detector system were kept low by employing a suitable hashing scheme for fast image search. For this purpose we relied on WTA hashing, a simple hashing algorithm that was further modified to achieve better selectivity. Tests carried out on the two datasets have demonstrated the efficiency of the modified version of the algorithm. Notably, a modification based on idf weighting proved highly beneficial, by offering better similarity estimation.

Future work will evaluate the proposed system on a wider range of environments, including larger datasets and indoor environments. Additionally, it is our aim to investigate the benefits of combining LBP-gist with 3D information, which can currently be extracted with affordable sensors.

References

1. Cummins, M., Newman, P.: FAB-MAP: Probabilistic Localization and Mapping in the Space of Appearance. The Int. J. of Rob. Research 27(6), 647–665 (2008)
2. Angeli, A., Filliat, D., Doncieux, S., Meyer, J.: A Fast and Incremental Method for Loop-Closure Detection Using Bags of Visual Words. IEEE Transactions on Robotics, Special Issue on Visual Slam 24(5), 1027–1037 (2008)

3. Gálvez-López, D., Tardos, J.: Bags of binary words for fast place recognition in image sequences. IEEE Transactions on Robotics 28(5), 1188–1197 (2012)
4. Oliva, A., Torralba, A.: Modeling the shape of the scene: a holistic representation of the spatial envelope. Int. Journal of Comp. Vision 42(3), 145–175 (2001)
5. Ojala, T., Pietikäinen, M., Harwood, D.: A comparative study of texture measures with classification based on featured distribution. Pattern Recognition 29(1), 51–59 (1996)
6. Andoni, A., Indyk, P.: Near-optimal hashing algorithms for approximate nearest neighbor in high dimensions. Communications of the ACM - 50th Anniversary Issue 51(1), 117–122 (2008)
7. Yagnik, J., Strelow, D., Ross, D.A., Lin, R.: The power of comparative reasoning. In: International Conference on Computer Vision, pp. 2431–2438 (2011)
8. Siagian, C., Itti, L.: Biologically Inspired Mobile Robot Vision Localization. IEEE Transactions on Robotics 25(4), 861–873 (2009)
9. Murillo, A.C., Kosecka, J.: Experiments in place recognition using gist panoramas. In: IEEE Workshop on Omnidirectional Vision, Camera Netwoks and Non-Classical Cameras, ICCV, pp. 2196–2203 (2009)
10. Ni, K., Kannan, A., Criminisi, A., Winn, J.: Epitomic location recognition. IEEE Trans. on Pattern Analysis and Machine Intell. 31(12), 2158–2167 (2009)
11. Wu, J., Rehg, J.M.: CENTRIST: A Visual Descriptor for Scene Categorization. IEEE Trans. on Patt. Analysis and Machine Intell. 33(8), 1489–1501 (2011)
12. Sunderhauf, N., Protzel, P.: BRIEF-Gist - closing the loop by simple means. In: IEEE/RSJ Int. Conference on Intelligent Robots and Systems, pp. 1234–1241 (2011)
13. Sivic, J., Zisserman, A.: Video Google: a text retrieval approach to object matching in videos. In: IEEE Int. Conference on Computer Vision, pp. 1470–1477 (2003)
14. Mäenpää, T., Pietikäinen, M.: Multi-scale binary patterns for texture analysis. In: Bigun, J., Gustavsson, T. (eds.) SCIA 2003. LNCS, vol. 2749, pp. 885–892. Springer, Heidelberg (2003)
15. Topi, M., Timo, O., Matti, P., Maricor, S.: Robust texture classification by subsets of local binary patterns. In: Int. Conf. on Pattern Recognition, vol. 3, pp. 935–938 (2000)
16. Heikklä, M., Pietikäinen, M.: A texture-based method for modeling the background and detecting moving objects. IEEE Transactions on Pattern Analysis and Machine Intelligence 28(4), 657–662 (2006)
17. Zhao, G., Kellokumpu, V.-P., Pietikäinen, M., Li, S.Z.: Modeling pixel process with scale invariant local patterns for background subtraction in complex scenes. In: IEEE Conference on Computer Vision and Pattern Recognition, pp. 1301–1306 (2010)
18. Ahonen, T., Hadid, A., Pietikäinen, M.: Face recognition with local binary patterns. IEEE Trans. on Patt. Anal. and Machine Intell. 28(18), 2037–2041 (2006)
19. Datar, M., Immorlica, N., Indyk, P., Mirrokni, V.S.: Locality-sensitive hashing scheme based on p-stable distributions. In: Twentieth Annual Symposium on Computational Geometry - SCG 2004, pp. 253–262 (2004)
20. Blanco, J.-L., Moreno, F.-A., Gonzalez, J.: A collection of outdoor robotic datasets with centimeter-accuracy ground truth. Auton. Robots 27(4), 327–351 (2009)
21. Paulevé, L., Jégou, H., Amsaleg, L.: Locality sensitive hashing: A comparison of hash function types and querying mechanisms. Pattern Recognition Letters 31(11), 1348–1358 (2010)

Increasing Illumination Invariance of SURF Feature Detector through Color Constancy

Marcelo R. Petry[1,2], António Paulo Moreira[1], and Luís Paulo Reis[2,3]

[1] INESC TEC and Faculty of Engineering, University of Porto,
Rua Dr. Roberto Frias, s/n, 4200-465, Porto, Portugal
{marcelo.petry,amoreira}@fe.up.pt
[2] LIACC, University of Porto
[3] DSI/School of Engineering, University of Minho,
Campus de Azurem, 4800-058, Guimares, Portugal
lpreis@dsi.uminho.pt

Abstract. Most of the original image feature detectors are not able to cope with large photometric variations, and their extensions that should improve detection eventually increase the computational cost and introduce more noise to the system. Here we extend the original SURF algorithm increasing its invariance to illumination changes. Our approach uses the local space average color descriptor as working space to detect invariant features. A theoretical analysis demonstrates the impact of distinct photometric variations on the response of blob-like features detected with the SURF algorithm. Experimental results demonstrate the effectiveness of the approach in several illumination conditions including the presence of two or more distinct light sources, variations in color, in offset and scale.

Keywords: Feature detection, SURF, Color Invariance.

1 Introduction

Many computer vision tasks depend heavily on extraction and matching of corresponding points (features) over consecutive images. Applications of feature based algorithms include, but are not limited to, image classification, image segmentation, object recognition and camera calibration. In robotics, motion estimation methodologies like visual ociteetry and visual SLAM have been able to complement traditional navigation sensors (like global navigation satellite systems, encoders and inertial measurement units), offering smooth (30Hz) and locally highly accurate localization information.

In general, this problem is tackled by searching for image regions whose low-level characteristics (i.e size, shape, luminance, color, texture, binocular disparity) significantly differs from the background. As important as the feature distinctiveness, is its ability to be repeatedly identified in consecutive images. However, image RGB values are significantly influenced by variations in scene illuminant. Such variations introduce undesirable effects and negatively affect the

L. Correia, L.P. Reis, and J. Cascalho (Eds.): EPIA 2013, LNAI 8154, pp. 259–270, 2013.

performance of computer vision methods. For this reason, one of the most fundamental tasks of visual systems is to distinguish the changes due to underlying imaged surfaces from those changes due to the effects of the scene illumination. In order to increase the probability of image features to be re-detected in subsequent images, it is extremely important for them to be robust to noise and invariant with regard to geometric (changes in scale, translation, rotation, affine/projective transformation) and photometric variations (illumination direction, intensity, color and highlights).

The Speeded Up Robust Feature (SURF) [1] is a widely used feature detector in robotics motion estimation due to its low computation time. The algorithm provides features that are invariant to image scale and rotation, but only partially invariant to changes in viewpoint and illumination. In this paper the problem of SURF illumination invariance is addressed.

The remainder of the paper is organized as follows. The second section reports the related works under this topic. The third section briefly presents the theory regarding image formation and color invariance. In the fourth section we perform a mathematical analysis over the SURF feature detection function to identify its weakness regarding photometric variations. In the fifth section we present an approach to improve SURF illumination invariance by exploiting the local space average color descriptor as working space for feature detection. Next it is presented the experimental work and results comparing the repeatability rate of the proposed approach with the original SURF implementation. Finally some conclusions and directions for future work conclude the paper.

2   Related Work

Originally, most of the feature detectors and descriptors were designed to cope only with the image luminance. Later, in order to take advantage of illumination invariance properties of other color spaces, some researchers proposed extensions for the original algorithms. In [2], Ancuti and Bekaert proposed an extension to the SIFT descriptor (SIFT-CCH) that combines the SIFT approach with the color co-occurrence histograms (CCH) computed from the Nrgb color space. Their algorithm performs the same as SIFT in the detection step, but introduces one dimension to the descriptor. Thus, features are described by a two element vector that combines the SIFT and the CCH descriptor vectors. The main problem of such an approach is the increase in the computational effort during the feature matching due to the extra 128 elements added to the descriptor vector. The color-SURF proposed by Fan et al.[3] was maybe the first to approach suggesting the use of colors in SURF descriptors. Through a methodology similar to the SIFT-CCH, the authors propose the addition of a new dimension to the descriptor vector. This extra information corresponds to the color histogram computed from the YUV color space, and adds a 64-element vector for each feature descriptor. For this reason, just like in the SIFT-CCH, the extra elements in the descriptor vector increase the computational effort necessary during the matching step.

In [4], Abdel-Hakim and Farag uses the invariant property H (related to hue) of the Gaussian color model as working space. Thus, instead of using gray gradients to track SIFT features, they use the gradients of the color invariant to detect and describe features. Although the authors used the H invariant instead of the C invariant, the approach is called CSIFT in a reference to the introduction of color in the SIFT operator. [5], also use invariants derived from the Gaussian color model to reduce the photometric effects in SIFT descriptions. They compare the individual performance of four invariants with the original SIFT approach, with the CSIFT approach [4] and with the HSV-SIFT approach [6]. Their evaluation suggests that the C-invariant, which can be intuitively seen as the normalized opponent color space, outperforms the original SIFT description and all the other approaches. In reference to the results of the C-invariant, the combination of this invariant with the SIFT operator is called C-SIFT. Sande et al. [7] presents an evaluation of the different approaches that attempt to provide photometric invariance to SIFT like descriptors.

3 Image Theory

The geometric distribution of the body reflection is sometimes assumed to reflect light evenly in all directions. Therefore, the luminance in such isotropic surfaces, also known as Lambertian surfaces, is the same regardless of the viewing angle. Assuming that a scene contain surfaces which exhibits Lambertian reflectance properties, its resulting image I can be modelled in terms of the surface reflectance $S(\lambda, x_{obj})$ and the light spectral power distribution $E(\lambda, x_{obj})$ falling onto an infinitesimal small patch on the sensor array.

$$I(x_i) = \int E(\lambda, x_{obj})S(\lambda, x_{obj})p(\lambda)d\lambda \ . \tag{1}$$

Where $p(\lambda)$ is the camera spectral sensitivity of wavelength λ, x_{obj} is the object location in the world coordinate frame and x_i is its location in the image coordinate frame. Although each sensor responds to a range of wavelengths, the sensor is often assumed to respond to the light of a single wavelength. Thus, one can approximate the sensor response characteristics by Dirac's delta functions. Through the former assumption, it is possible to simplify the Equation (1) and express the intensity $I_k(x_i)$ measured by the sensor $k \in \{R, G, B\}$ in the position x_i as:

$$I_k(x_i) = E_k(x_{obj})S_k(x_{obj}) \ . \tag{2}$$

3.1 Diagonal Model

One of the most difficult problems when working with colors is that the object's apparent color varies unpredictably with variations in the intensity and temperature of the light source. A well-known example occur in outdoor environments with daylight variations, the color shift between sunny and cloudy days is simply not well modeled as Gaussian noise in RGB [8].

One of the most used models to describe those kind of variations is the von-Kries model, or Diagonal Model (DM), which corresponds to a diagonal transformation of the color space. According to Diagonal model, it is possible to map an observed image I_o taken under an unknown illuminant to a corresponding image I_c under a canonical illuminant through a proper transformation in order to render images color constant. Finlayson et al. [9] note that the DM model present shortcomings when mapping near saturated colors, and propose an extension that includes the "diffuse" light term by adding an offset. Such model is known as the Diagonal-Offset Model, and is given by:

$$
\begin{bmatrix} R_c \\ G_c \\ B_c \end{bmatrix} = \begin{bmatrix} a & 0 & 0 \\ 0 & b & 0 \\ 0 & 0 & c \end{bmatrix} \begin{bmatrix} R_o \\ G_o \\ B_o \end{bmatrix} + \begin{bmatrix} o_1 \\ o_2 \\ o_3 \end{bmatrix} . \tag{3}
$$

Using the Diagonal-offset model, illumination variations can be classified according to the values of the scalar and offset into five distinct categories [7]. In the light intensity change (LIC), the three RGB components of a given image varies equally by a constant factor, such that $a = b = c$ and $o_1 = o_2 = o_3 = 0$. Hence, when a function is invariant to light intensity changes, it is scale-invariant with respect to light intensity. In the light intensity shift (LIS), a constant shift affects equally all the RGB channels of a given image, such that $a = b = c = 1$ and $o_1 = o_2 = o_3 \neq 0$. The light intensity change and shift (LICS) is a combination of the two above mentioned categories, and also affect all three RGB channels equally, in such a way that $a = b = c$ and $o_1 = o_2 = o_3 \neq 0$. Thus, when a function is invariant to light intensity changes and to light intensity shift, it is known as scale-invariant and shift-invariant with respect to light intensity. The two remaining categories do not assume that RGB channels are equally affected by variations in the light source. The light color change (LCC) corresponds to the Diagonal model, and assumes that $a \neq b \neq c$ and $o_1 = o_2 = o_3 = 0$. Since images are able to vary differently in each channel, this category can model changes in the illuminant color temperature and light scattering. The last, light color change and shift (LCCS), corresponds to the full Diagonal-offset model and takes into consideration independent scales $a \neq b \neq c$ and offsets $o_1 \neq o_2 \neq o_3$ for each image channel.

3.2   Color Constancy

The ability to perceive color as constant under changing conditions of illumination is known as color constancy, and is a natural ability of human observers. The problem of computing a color constant descriptor based only on data measured by the retinal receptors is actually underdetermined, as both $E(\lambda, X_{obj})$ and $p_k(\lambda)$ are unknown. Therefore, one need to impose some assumptions regarding the imaging conditions. The most simple and general approaches to color constancy (i.e. White Patch [10] and the Gray World [11]) make use of a single statistic of the scene to estimate the illuminant, which is assumed to be uniform in the region of interest. Approaches like Gammut Mapping, on the other hand, make use of assumptions of the surface reflectance properties of the objects.

A more recent method is based on the Local Space Average color (LSAC), which can be defined as a computational model of how the human visual system performs averaging of image pixels [12]. The model proposed by Ebner makes two important assumptions. The first is that the essential processing required to compute a color constant descriptor in human observers is located in the visual area V4 of the extrastriate visual cortex [13]. The second is that gap junctions behave like resistors. Thus, Ebner models the gap junctions between neurons in V4 as a resistive grid, which can be used to compute Local Space Average color, and then color constant descriptors. Each neuron of this resistive grid computes the local space average color by iterating update equations indefinitely for all three bands. According to Ebner, the iterative computation of Local Space Average Color produces results which are similar to the convolution of the input image with a Gaussian kernel.

The Local Space Average Color alone is just a biologically inspired theory that tries to explain how the brain averages image pixels. However, when combined with the Gray World hypothesis, LSAC can provide means to derive color invariant descriptors. The advantage of Ebner's work is that if we consider the Gray World assumption in a local perspective, it is possible to estimate the color of the illuminant at each image pixel. For a more detailed theoretical formulation we may refer to [12]. Given the local space average color a_k, one can derive a local color invariant descriptor O_k through:

$$O_k(x,y) = \frac{I_k(x,y)}{2a_k(x,y)} \approx \frac{S_k(x,y)E_k(x,y)}{E_k(x,y)} \approx S_k(x,y) \ . \tag{4}$$

4   Analysis of Photometric Variations in SURF

In order to understand the effects of the light source variation in SURF responses consider an observed single channel image I_o with pixel intensity $I_o(x,y)$ at a given point $X = (x,y)$. Through the central difference method it is possible to express the second derivatives of $I_o(x,y)$ as:

$$\frac{\partial^2 I_o(x,y)}{\partial x^2} = I_o(x+1,y) - 2I_o(x,y) + I_o(x-1,y) \ . \tag{5}$$

Now, consider that I_o has a corresponding image I_u, taken under unknown illuminant. Assuming the Diagonal-offset model these two images are related by a linear transformation determined by a scalar constant α and an offset β. Therefore, the pixel intensity $I_u(x,y)$ of the image I_u at the same point $X = (x,y)$ can be modeled as:

$$I_u(x,y) = \alpha I_o(x,y) + \beta \ . \tag{6}$$

Thus, it is possible to conclude that the second derivatives of $I_u(x, y)$ with respect to x is:

$$\frac{\partial^2 I_u(x, y)}{\partial x^2} = \alpha \frac{\partial^2 I_o(x, y)}{\partial x^2} \quad . \tag{7}$$

The same applies to the second derivatives with respect to y and XY. When computing the derivatives, the diffuse term β is canceled out, causing no impact on the final result. However, by varying the illumination with a scalar α the second derivatives vary proportionally with the scalar. Feature localization is a three-step process that starts disregarding points in which blob-response is lower than a fixed threshold value. Thus, if the detector responses vary with the illumination, a given feature that is detected in a bright image may not be detected in a corresponding image with lower illumination levels. SURF detector response $R_u(x, y)$ of a given pixel $I_u(x, y)$ is given by the determinant of the Hessian matrix:

$$R_u(x, y) = \frac{\partial^2 I_u(x, y)}{\partial x^2} \frac{\partial^2 I_u(x, y)}{\partial y^2} - \left(\frac{\partial^2 I_u(x, y)}{\partial xy}\right)^2 \quad . \tag{8}$$

Replacing (7) into (8), the filter response R_u can be expressed in terms of the SURF response $R_o(x, y)$ of the I_o:

$$R_u(x, y) = \alpha \frac{\partial^2 I_o(x, y)}{\partial x^2} \alpha \frac{\partial^2 I_o(x, y)}{\partial y^2} - \left(\alpha \frac{\partial^2 I_o(x, y)}{\partial xy}\right)^2 = \alpha^2 R_o \quad . \tag{9}$$

The degree of the scalar (α^2) in (9) provides the theoretical explanation to why even small variations in the scene illuminant cause significant variations in the magnitude of the detector response.

5   The Proposed LSAC SURF Approach

Among color constancy methods, gamut mapping is referred in literature as one of the most successful algorithms [14]. It has demonstrated good results in different datasets of several works. The method is though computationally quite complex. Its implementation requires the computation two convex hulls, which is a difficult problem when using finite precision arithmetic. Another drawback is that the algorithm requires an image data set with known light sources to estimate the canonical gamut (learning phase) that will be used to compute the transformation matrix, and thus estimate the illuminant (testing phase). In practice, such methodology is not viable for robotic vision systems since robots are not constrained to one specific scenario, but subjected to multiple and dynamic environments.

Low level color constant algorithms, on the other hand, are less complex, faster and only slightly outperformed by the gamut mapping [15]. These characteristics make them perfect candidates for improving robotic vision systems.

One limitation of the Gray World assumption is that it is only valid in images with sufficient amount of color variations. Only when the variations in color are random and independent, the average value of the R, G, and B components of the image would converge to a common gray value. This assumption is, however, held very well in several real world scenarios, where it is usually true that there are a lot of different color variations.

Another limitation of most color constancy algorithms is that they are modeled with the assumption that the scene is uniformly illuminated. Since in practice multiple illuminants are present in the scene, the illumination is not uniform, and thus the premise is not fully verified. For instance, some daylight may be falling through a window while an artificial illuminant may be switched on inside the room. In fact, that may be the main advantage of the descriptors derived from the Local Space Average color methodology. Since LSAC estimates the illuminant locally for each point of the scene, its descriptors are better prepared to deliver color constancy in real world images.

Most color invariant feature detectors proposed combines the original detector with some sort of color space mapping. Our approach to achieve photometric invariant feature responses (LSAC SURF) consists on taking advantage of the invariant properties of the LSAC descriptor, using it as working space to perform SURF feature extraction. The inclusion of this pre-processing step adds a small computational load, but may provide a significant increase in feature detection robustness.

The size of the window that LSAC is computed plays a important role in the robustness of the feature detection. Empirical observation demonstrated that feature repeatability tends to perform better when LSAC is computed over small neighborhoods. In fact, due to the multiple illumination sources the values of α and β tends to vary significantly in distant image pixels, which makes the assumption that $E_k(x,y) \approx 2a_k(x,y)$ to be valid only for a small regions.

When a pixel reaches saturation, it does not present the same variation as its neighbors, causing non linear variations in the response of the feature detector and decreasing the probability to be correctly matched in subsequent images. Therefore, features which pixel intensities are close to saturation are not good candidates for matching. However, such features can not simply be ignored since under certain illumination variations their pixel intensity can move away from saturation, and make them good candidates for matching in subsequent images. For this reason, each detected feature is classified into hard and soft features according to their pixel intensities. If the pixel intensity of a distinct image region is lower than an upper threshold and higher than a lower threshold the feature is classified as hard feature, on the contrary, the feature is classified as soft feature. The choice of the proper upper and lower threshold values might be determined according to the expected variation in the scene illumination.

Since hard features are more likely to be found in subsequent images, we can reduce the search space and match only the current hard features with the subsequent set of features. In this context, soft features are used only to support matching of previous hard features, while hard features are used in the computation of sensitive visual tasks.

6   Results

This section presents the experimental validation of the proposed method. First, the influence of several illumination conditions on the performance of the proposed method is studied using controlled indoor images, Fig. 1. Next, the proposed method is applied on a data set of real images, Fig. 3. In the following experiments the optimum parameters of SURF [1] were assigned to both algorithms. The LSAC was approximated with a Gaussian kernel of size 15x15, with standard deviation $\sigma = 4$. The lower and upper thresholds used for feature classification into hard and soft were set to 20 and 220 respectively.

6.1   Performance Measurement

To evaluate the performance of our approach we adopted the repeatability criterion similar to the proposed by Schmid et al. [16]. The repeatability rate evaluates the ratio between the number of point-to-point correspondences that can be established for detected points in all images of the same scene $C(I_1, ..., I_n)$ and the total number of features detected in the current image m_i. Therefore, the higher the repeatability, the more likely features are to be matched and the better the matching results tend to be.

$$R_i = \frac{C(I_1, ..., I_n)}{m_i} \ . \tag{10}$$

Where i denotes the image under analysis and n the number of images of the same scene. The repeatability rate of our approach was compared with the repeatability rate of the SURF algorithm implemented in the OpenCV library.

6.2   Controlled Image Set

First experiments are performed on images available in the Amsterdam Library of Object Images (ALOI) [17]. ALOI provides several image collections, like the light color change (ALCC) collection. ALCC is a collection of images in which the color of the illumination source was varied from yellow to white, according to the voltage v_0 of the lamps (where $v_0 = 12i/255$ volts and $i \in \{110, 120, 130, 140, 150, 160, 170, 180, 190, 210, 230\}$). Since among the ALOI collections only the ALCC collection have a direct correspondence to the illumination variations modeled through the Diagonal offset model, three new controlled collections were artificially created: LIC, LIS, and LCC collections.

To create the controlled collections we selected all images from the ALCC collection with color temperature of 2750k, and performed the proper transformations. Thus, all collections contain a set of 9.000 images of 1.000 objects designed to evaluate the effects of specific variations in the scene illuminant. The LIS collection was created by shifting all the color channels equally by an offset $\beta \in \{-20, -15, -10, -5, 0, 5, 10, 15, 20\}$. The LIC collection was created scaling all the RGB channels equally by a factor $\alpha \in \{1/2.0, 1/1.5, 1/1.2, 1/1.1, 1, 1.1, 1.2, 1.5, 2.0\}$. Finally, the LCC collection, which mimics the effect of a light variation from bluish to white, was created by scaling both the Red and Green channels by a factor $\alpha \in \{0.2, 0.3, 0.4, 0.5, 0.6, 0.7, 0.8, 0.9, 1.0\}$, while keeping $\alpha \in \{1.0\}$ for the Blue channel.

The repeatability results in the LIS collection, Fig. 2a, confirm the theoretical analysis and demonstrate that offset variations indeed do not affect the performance of the detection algorithms. The results demonstrates a good performance for both algorithms, in which mean repeatability remained above 90%.

Variations of scalar order, on the other hand, greatly impact SURF repeatability performance. Fig. 2b demonstrates the low repeatability rate of SURF algorithm in the LIC collection. LSAC SURF demonstrated a much higher and constant mean repeatability rate, presenting a significant improvement in the mean repeatability for all values of α. Note that a higher mean repeatability rate occurs for the smaller values of α, in both SURF and LSAC SURF, due to the tendency to find a smaller number of features in darker images.

Results of the mean repeatability rate of the ALCC collection, Fig. 2c, demonstrate that the mean repeatability rate was not significantly affected by the illumination color variation, presenting a mean repeatability rate above 90% for both algorithms. This result can be justified by the weight of the color components in the grayscale conversion. When varying the light source from yellow to white, only the blue component of the RGB model varies. Since the weight of the blue component (0.114) is considerably lower than the weight of the red (0.299) and green components (0.587), the variation in this color channel does not cause a sufficient large photometric variation to impact the grayscale image used in the feature detection. However, when varying the color of the light source from bluish to white (LCC collection) the mean repeatability rate of LSAC SURF significantly outperformed SURF. Fig. 2d demonstrates once again the the low repeatability rate of

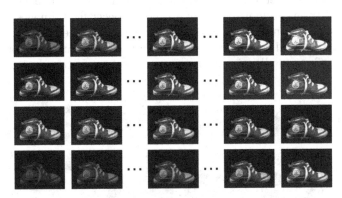

Fig. 1. Samples of the controlled indoor image set. From top row to bottom: samples of one object in the LIS, LIC, ALCC and LCC collections.

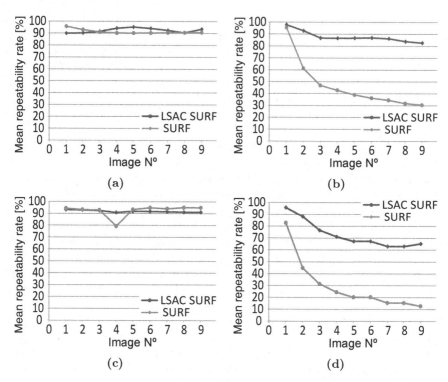

(a)   (b)

(c)   (d)

Fig. 2. Mean repeatability of (a)LIS, (b)LIC, (c)ALCC and (d)LCC collections

SURF algorithm and its tendency to decrease with higher values of α, while LSAC SURF presented a much higher and constant mean repeatability rate.

6.3 Real World Image Set

Here, the proposed method is tested on a data set of real images. This data set consists of twelve indoor images of a robotic soccer field. Images were taken with the camera mounted in a fixed position, while the scene illumination was varied through several combinations of individually regulated ceiling lightings. This dataset offers a challenging environment for robust feature detection since it

Fig. 3. Dataset of real images: robotic soccer field

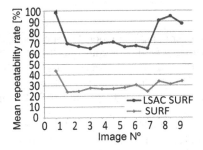

Fig. 4. Repeatability of the Robotic Soccer Field collection

contains non-uniform illumination due to multiple sources (different bulb lamps in the ceiling and natural illumination from the windows), as well as variations in the color of the illuminant, shading, shadows, specularities, and interreflections.

The theoretical improvement offered by the LSAC SURF in the robustness of feature detection can be experimentally verified through the Fig. 4. This result demonstrates the low repeatability rate of SURF algorithm in the Robotic Soccer Field collection, which remained around 29% for all images. LSAC SURF, on the other hand, demonstrated a much higher repeatability rate, not lower than 65%, presenting a significant improvement in the repeatability score in all illumination conditions.

7   Conclusion

In this paper, the LSAC SURF approach was introduced as an extension of SURF algorithm. The methodology proposed has shown to be able to improve feature detection repeatability rates in scenarios where the uniform light-source assumption is too restrictive. The theoretical analysis demonstrated which variations in the illuminant affects images derivatives and SURF responses. We demonstrated that SURF response is proportional to the square of the scalar variation of the illuminant.

The experimental results validate the theoretical invariance property of the proposed approach. We have shown that LSAC SURF detection can be as accurate as the original SURF detection when the light source is (approximately) uniform. Furthermore, when the illumination conditions vary significantly (presence of two or more distinct light sources, variations in color, in scale, etc.) in an image, the proposed methodology is able to overcome the performance of the existing algorithm considerably. Future works will concern with conducing experimental tests in real robot localization datasets, comparing the drift of the proposed methodology with the original SURF algorithm.

Acknowledgments. This work was funded by the ERDF European Regional Development Fund through the COMPETE Programme and by National Funds through FCT - Portuguese Foundation for Science and Technology within project IntellWheels, RIPD/ADA/109636/2009. The first author thanks FCT for his PhD Grant SFRH/BD/60727/2009.

References

1. Bay, H., Ess, A., Tuytelaars, T., Van Gool, L.: Speeded-up robust features (surf). Computer Vision and Image Understanding 110(3), 346–359 (2008)
2. Ancuti, C., Bekaert, P.: Sift-cch: Increasing the sift distinctness by color co-occurrence histograms. In: International Symposium on Image and Signal Processing and Analysis, Istanbul, Turkey, September 27-29, pp. 130–135 (2007)
3. Fan, P., Men, A.D., Chen, M.Y., Yang, B.: Color-surf: A surf descriptor with local kernel color histograms. In: IEEE International Conference on Network Infrastructure and Digital Content, Beijing, China, November 6-8, pp. 726–730 (2009)
4. Abdel-Hakim, A.E., Farag, A.A.: Csift: A sift descriptor with color invariant characteristics. In: IEEE Computer Society Conference on Computer Vision and Pattern Recognition - CVPR, June 17-22, vol. 2, pp. 1978–1983. IEEE Computer Society, New York (2006)
5. Burghouts, G.J., Geusebroek, J.M.: Performance evaluation of local colour invariants. Computer Vision and Image Understanding 113, 48–62 (2009)
6. Bosch, A., Zisserman, A., Muñoz, X.: Scene classification via pLSA. In: Leonardis, A., Bischof, H., Pinz, A. (eds.) ECCV 2006. LNCS, vol. 3954, pp. 517–530. Springer, Heidelberg (2006)
7. van de Sande, K.E.A., Gevers, T., Snoek, C.G.M.: Evaluating color descriptors for object and scene recognition. IEEE Transactions on Pattern Analysis and Machine Intelligence 32(9), 1582–1596 (2010)
8. Buluswar, S.D., Draper, B.A.: Color recognition in outdoor images. In: International Conference on Computer Vision, Bombay, India, January 04-07, pp. 171–177 (1998)
9. Finlayson, G.D., Hordley, S.D., Xu, R.: Convex programming colour constancy with a diagonal-offset model. In: IEEE International Conference on Image Processing - ICIP, Genova, Italy, September 11-14, vol. 3, pp. III–948–III–951 (2005)
10. van de Weijer, J., Gevers, T., Gijsenij, A.: Edge-based color constancy. IEEE Transactions on Image Processing 16(9), 2207–2214 (2007)
11. Buchsbaum, G.: A spatial processor model for object color-perception. Journal of the Franklin Institute-Engineering and Applied Mathematics 310(1), 1–26 (1980)
12. Ebner, M.: How does the brain arrive at a color constant descriptor? In: Mele, F., Ramella, G., Santillo, S., Ventriglia, F. (eds.) BVAI 2007. LNCS, vol. 4729, pp. 84–93. Springer, Heidelberg (2007)
13. Zeki, S., Marini, L.: Three cortical stages of colour processing in the human brain. Brain 121, 1669–1685 (1998)
14. Gijsenij, A., Gevers, T., van de Weijer, J.: Generalized gamut mapping using image derivative structures for color constancy. International Journal of Computer Vision 86(2-3), 127–139 (2010)
15. Barnard, K., Cardei, V., Funt, B.: A comparison of computational color constancy algorithms - part i: Methodology and experiments with synthesized data. IEEE Transactions on Image Processing 11(9), 972–984 (2002)
16. Schmid, C., Mohr, R., Bauckhage, C.: Evaluation of interest point detectors. International Journal of Computer Vision 37(2), 151–172 (2000)
17. Geusebroek, J.M., Burghouts, G.J., Smeulders, A.W.M.: The amsterdam library of object images. International Journal of Computer Vision 61(1), 103–112 (2005)

Intelligent Wheelchair Manual Control Methods

A Usability Study by Cerebral Palsy Patients

Brígida Mónica Faria[1,2,3,4], Luís Miguel Ferreira[2,3], Luís Paulo Reis[4,5], Nuno Lau[2,3], and Marcelo Petry[6,7]

[1] ESTSP/IPP - Escola Superior Tecnologia de Saúde do Porto, Instituto Politécnico do Porto, Vila Nova de Gaia, Portugal
[2] DETI/UA - Dep. Electrónica, Telecomunicações e Informática da Universidade de Aveiro, Aveiro, Portugal
[3] IEETA - Inst. Engenharia Electrónica e Telemática de Aveiro, Aveiro, Portugal
[4] LIACC/UP – Lab. Inteligência Artificial e Ciência de Computadores da Universidade do Porto, Porto, Portugal
[5] EEUM - Escola de Engenharia da Universidade do Minho, Departamento de Sistemas de Informação, Guimarães, Portugal
[6] FEUP - Faculdade de Engenharia, Universidade do Porto, Porto, Portugal
[7] INESC TEC - INESC Tecnologia e Ciência, Porto, Portugal
btf@estsp.ipp.pt, {luismferreira,nunolau}@ua.pt,
lpreis@dsi.uminho.pt, marcelo.petry@fe.up.pt

Abstract. Assistive Technologies may greatly contribute to give autonomy and independence for individuals with physical limitations. Electric wheelchairs are examples of those assistive technologies and nowadays each time becoming more intelligent due to the use of technology that provides assisted safer driving. Usually, the user controls the electric wheelchair with a conventional analog joystick. However, this implies the need for an appropriate methodology to map the position of the joystick handle, in a Cartesian coordinate system, to the wheelchair wheels intended velocities. This mapping is very important since it will determine the response behavior of the wheelchair to the user manual control. This paper describes the implementation of several joystick mappings in an intelligent wheelchair (IW) prototype. Experiments were performed in a realistic simulator using cerebral palsy users with distinct driving abilities. The users had 6 different joystick control mapping methods and for each user the usability and the users' preference order was measured. The results achieved show that a linear mapping, with appropriate parameters, between the joystick's coordinates and the wheelchair wheel speeds is preferred by the majority of the users.

Keywords: Intelligent Wheelchair, Manual Controls, Cerebral Palsy, Usability.

1 Introduction

The scientific community gives high importance to the real application of new discoveries. In the area of assistive technologies, robotics performs an important role.

L. Correia, L.P. Reis, and J. Cascalho (Eds.): EPIA 2013, LNAI 8154, pp. 271–282, 2013.

In particular, electric wheelchairs are now more intelligent due to the implementation of algorithms that assists the driving user.

Most electric wheelchairs are manually steered with a joystick, although there are several other possibilities for the interface between a user and the wheelchair [1-4]. A conventional joystick maps the position of the handle into a Cartesian coordinate system where normalized axis (x, y) range from minus one to plus one, the x axis is oriented towards the right and coordinates at the origin correspond to the central (resting) position. An electric wheelchair is typically driven by two individually powered wheels which rotate around a horizontal axis, and another two non-powered caster wheels, which besides rotating around a horizontal axis, also have the ability to rotate around a vertical axis. This vertical rotation axis allows the non-powered wheels to steer freely, minimizing friction during direction change. Assuming the terrain is flat and there are no obstacles, when the speed is the same on both powered wheels, the wheelchair moves in a straight line. Steering is determined by the velocity difference of the powered wheels, specifically the wheelchair will rotate towards the wheel with the lower speed, and rotate around itself when the wheels rotate in opposite directions. The radius of curvature of the wheelchair is dependent on the wheel spacing and also on the traveled distances of each wheel.

The mapping of joystick positions to individual wheel speed can be done in an infinite number of combinations, and it is this mapping that will determine the response behavior to manual control. Several of these mappings were implemented, out of which a few were selected for inclusion in this paper. The level of satisfaction of the volunteers that tested different joystick mappings was measured and some interesting conclusions about mapping were achieved based on the users' feedback.

This paper is organized in six sections. The first section is composed by this introduction. The second section reports the related work and related issues that are under study. The third section briefly presents the IntellWheels project. The implementation of the proposed algorithms is described in section four. Next, the experimental work and results are presented. Finally some conclusions and directions for future work conclude the paper.

2   Related Work

There are a significant number of scientific works related to robotic wheelchairs or intelligent/smart wheelchairs [5] [6]. The study about the joystick mapping is also an issue of investigation. Choi et al. [7] describes a more intuitive human interface by changing the prior mapping method of the joystick, which considers the consecutive operations of a motor of a wheelchair, to a new mapping method that corresponds to the internal model of a human being. They divided the existing joystick mapping method into two degrees of freedom, one in the vertical axis that can control the velocity and the other, in the horizontal axis for direction control, and controlled the wheelchair with an electromyography (EMG) signal. Dicianno et al. [8] concluded that individually customized isometric devices may be superior to commercially available proportional control for individuals with tremor, even though the used filtering did not improve the wheelchair driving as expected. Fattouh et al. presents [9] an evaluation of the use of force feedback joysticks with a powered wheelchair is performed and it was concluded that on people without disabilities there were some advantages (less collisions). Niitsuma et al. [10] introduced a vibrotactile interface that could be controlled by users even if they had not used one before, and although through the experiments for the vibration stimuli, users could not detect exact orientation of obstacles, it was possible to detect a direction of the obstacle movement. In [11] a reactive shared control system was presented which allows a semi-autonomous navigation in unknown and dynamic environments using joystick or voice commands.

3   IntellWheels Project

The IntellWheels project aims at developing an intelligent wheelchair platform that may be easily adapted to any commercial wheelchair and aid any person with special mobility needs [12]. The project main focus is the research and design of a multi-agent platform, enabling easy integration of different sensors, actuators, devices for extended interaction with the user, navigation methods and planning techniques and methodologies for intelligent cooperation to solve problems associated with intelligent wheelchairs [12-13].

A real prototype (Figure 1) was created by adapting a typical electric wheelchair. Two side bars with a total of 16 sonars, a laser range finder and two encoders were incorporated.

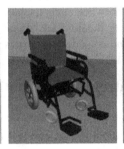

Fig. 1. The real and virtual prototype of the IW

In order to test the algorithms and methodologies a simulator was also developed. With this simulator a virtual world can be created where a user can drive an IW with behavior similar to the real prototype. The virtual wheelchair (Figure 1) was modeled with 3D Studio Max [14], the virtual environment was modeled with 3D UnrealEditor and USARSim [15] was the platform chosen for the simulation of robots and environments. USARSim is based on the Unreal Tournament game engine [15] and is intended as a general purpose research tool with various applications from human computer interfaces to behavior generation for groups of heterogeneous robots [15].

The purpose of this simulator is essentially to support the test of algorithms, analyze and test the modules of the platform and safely train users of the IW in a simulated environment [16].

A multimodal interface was also developed that allows driving the wheelchair with several inputs such as joystick, head movements or more high level commands such as voice commands, facial expressions, and gamepad or even with a combination among them. For example it is possible to blink an eye and say "go" for the wheelchair to follow a right wall [17-20].

Figure 2 presents the first person view and the multimodal interface available to the user.

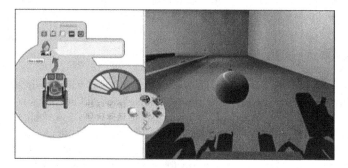

Fig. 2. Multimodal Interface and first person view

It is also possible to observe in the multimodal interface (Figure 2) the indications about the velocity and direction of the intelligent wheelchair.

The joystick mapping is important, because this mapping will determine the response behavior of the wheelchair to the user manual control. For that reason, different ways of mapping the joystick signal were tested. The next section presents different alternatives and considerations regarding the used mapping algorithms.

4   Manual Control Implementations

The mapping of joystick positions to individual IW wheel speeds can be performed in an infinite number of combinations and it is this mapping that will determine the manual control behavior. For that reason, several of these mappings were implemented tested with real users in a simulated environment and based on the users' feedback some interesting conclusions about mappings were achieved [16]. The achieved results show that a linear mapping, with appropriate parameters, between the joystick's coordinates and the wheelchair wheel speeds is preferred by the majority of the users.

Considering that the joystick handle position is represented in a Cartesian coordinate system, with two axis, x and y, which vary between -1 and 1. These (x, y) coordinates can be used to determine the distance of the handle to the central (resting) position of the joystick $(0, 0)$ and an angle relating to a reference vector (which is usually $(0, 1)$). The desired speed of the left wheel (L) or the right wheel (R) is represented by normalized values (between -1 and 1). With positive values the wheels rotate forward and with negative values the wheels rotate backward.

All mappings should meet the following conditions, that were defined based on an intuitive judgment of what should be the manual control behavior: θ is measured in relation to vector $(0, 1)$, that is, the yy axis; when θ is undefined, namely when $(x, y) = (0, 0)$, both wheels will be stopped; when $\theta = 0$, both wheels move forward at the same speed; when $\theta = \pm\pi$, both wheels move backward at the same speed and when $\theta = \pm\pi/2$, the wheels rotate in opposite directions at the same speed.

4.1   Original Mapping

The first mapping algorithm is very simple. The steering and power mapping is achieved through the use of both x and y in all instances and the wheelchair can rotate around itself when the joystick is pushed to the sides. Assuming that x and y are normalized and vary between -1 and 1 the system of equations that describes the mapping can be observed in Equation 1:

$$\begin{cases} R = y - x \\ L = y + x \end{cases} \tag{1}$$

Note that at the central position $(0, 0)$ no power is sent to the wheels. When x is near 0, and the joystick is pushed forward/backward, the speed increases proportionally to y, and the wheelchair moves forward/backward. There is no rotation because x equals zero, thus left speed and right speed are equal.

For all other values of x, when the joystick is pushed to either left or right, the right speed is proportional to $(y-x)$ and left speed is proportional to $(y+x)$. It is important to notice that the right speed and left speed are clipped when $(y-x)$ or $(y+x)$ are above the maximum (or below the minimum) normalized values. This implies a loss of the useable joystick area. Additionally, some filtering was added, so that minimal (x, y) variation near the axis is ignored. When the wheels rotate in opposite directions, speed is halved on both wheels in order to turn the narrow curves more controllable.

With this algorithm, steering the wheelchair requires accurate precision. The wheel speed variations that result from lateral joystick movements are quite steep, and when moving forward, little variations on the joystick horizontal axis result in a big adjustment of the direction.

A derived mapping can be obtained when the input variables are squared in order to attenuate the steep changes in wheelchair direction. A significant improvement in terms of range extension was verified with this modification.

4.2   Proportional Mapping

In the proportional mapping the distance of the handle to the center of the joystick (ρ) is proportional the maximum wheel speed and the angle (θ) of the joystick relating to the vector $(0, 1)$, determines how the speed is distributed to the wheels. In order to keep ρ in the 0 to 1 range, the value is clipped when ρ is above one. Assuming that x and y are normalized and vary between -1 and 1, the control follows these conditions (Equation 2 and Equation 3):

$$R = \begin{cases} \rho, if\ 0 \le \theta \le \dfrac{\pi}{2} \\[2mm] \rho.\dfrac{-\theta + 3.\pi/4}{\pi/4}, if\ \dfrac{\pi}{2} \le \theta \le \pi \\[2mm] -\rho, if\ -\pi < \theta < -\dfrac{\pi}{2} \\[2mm] \rho.\dfrac{\theta + \pi/4}{\pi/4}, if\ -\dfrac{\pi}{2} \le \theta \le 0 \end{cases} \quad (2)$$

$$L = \begin{cases} \rho.\dfrac{-\theta + \pi/4}{\pi/4}, if\ 0 \le \theta \le \dfrac{\pi}{2} \\[2mm] -\rho, if\ \dfrac{\pi}{2} \le \theta \le \pi \\[2mm] \rho.\dfrac{\theta + 3.\pi/4}{\pi/4}, if\ -\pi < \theta < -\dfrac{\pi}{2} \\[2mm] \rho, if\ -\dfrac{\pi}{2} \le \theta \le 0 \end{cases} \quad (3)$$

where $\rho = \sqrt{x^2 + y^2}$ and $\theta = \mathrm{atan2}\,(-x, y)$. If $\rho > 1$, ρ is clipped to 1 and if $\rho < -1$, ρ is clipped to -1.

When compared to the previous algorithm this provides a more pleasant driving experience, however steering is still not as smooth as desired. The joystick axis values were also attenuated with a quadratic function creating a new mapping that was tested in the real user experiments.

4.3   Intuitive Mapping

This mapping alternative is an updated version of the first algorithm purposed. Assuming that x and y are normalized and vary between -1 and 1, the Equation 4 is:

$$\begin{cases} R = y - nx \\ L = y + nx \end{cases} \quad (4)$$

the value nx follows the Equation 5:

$$nx = \begin{cases} u_1 c_{point} + (x - c_{point}) \times u_2\ if\ x > c_{point} \\[1mm] -u_1 c_{point} + (x + c_{point}) \times u_2\ if\ x < -c_{point} \\[1mm] u_1 x\ if\ -c_{point} \le x \le c_{point} \end{cases} \quad (5)$$

where $c_{point} \in [0,1]$; $u_1 \in [0,1]$ and $u_2 \in [0,1]$. The tested values were $c_{point} = 0.2$; $u_1 = 0.5$ and $u_2 = 0.25$. The first slope u_1 allows a fast curve and the next slope u_2 after the cut point (c_{point}) should allow a slower curve.

5   Experiments and Results

The mapping of joystick positions to individual wheel speed determines the response behavior to manual control. Six of these mappings were implemented and tested. The usability level was achieved based on users' feedback after testing the different mappings.

5.1   Simulated Environment

The implemented control algorithms were tested with real users, in a simulated environment, in order to verify the usability of the different mappings. A circuit was developed using the simulator. Figure 3 shows the overall circuit.

Fig. 3. Circuit and tasks used for testing the IW

A simple serious game was created. The game objective was to follow a specific track collecting eight blue balls and at the end a yellow star. The green arrows just indicate the route and the blue balls were obligatory checkpoints, since they could be collected by passing near them. The people added to the environment were dynamic obstacles that should be avoided.

5.2   System Usability Questionnaire

A quasi-experimental study was performed. The volunteers had an explanation of the purpose of the study and signed the inform consent.

Users were asked to fill a questionnaire composed of four parts: user identification; experience with videogames and joysticks; questions adapted from the Computer System Usability Questionnaire (CSUQ) [20] for each tested option and a final question about the preference order of the tested options. The questions from the CSUQ were measured in a *Likert* scale in order to obtain a final score from 0 to 100. The questions were:

- Overall, I am satisfied with how easy it is to use this control.
- It was simple to use this control.
- I can effectively complete this task using this control.
- I am able to complete this task quickly using this control.
- I am able to efficiently complete this task using this control.
- I feel comfortable using this control.
- It was easy to learn to use this control.
- Whenever I make a mistake using the control, I recover easily and quickly.
- Overall, I am satisfied with this control.
  Two more specific questions were asked:
- I felt I had control of the wheelchair.
- It is easy to drive the wheelchair in narrow spaces.

The users drove the simulated wheelchair with the six alternatives of joystick mappings: the algorithms A, B, C that are respectively the Original Mapping, Proportional and Intuitive Mapping and all of them with the quadratic function (D, E and F respectively). The order of the experiments was randomly set to avoid the bias relative to the experience of the user. After each round the volunteers answered to several questions related to each kind of mapping.

5.3   Cerebral Palsy Users' Results

Patients suffering from cerebral palsy that use joysticks to drive their wheelchairs were invited to test all the control algorithms. This sample is characterized for having five males and three females, with a mean of age of 29 years old. All had experience with the joystick of their electric wheelchair although the experience with video games was low, except in one case that answered always play videogames. Table 1 shows the summary of statistics measures about the final score for all the mapping options.

Table 1. Summary of statistical measures of the adapted CSUQ score

| *Statistics* | Adapted CSUQ – Final Score | | | | | |
	A	B	C	D	E	F
Mean	72.8	77.9	88.7	86.7	84.9	83.9
Median	81.7	88.1	**99.2**	96.0	**98.4**	**99.2**
Std. Deviation	27.7	28.2	19.8	20.6	23.5	24.5
Minimum	30.2	25.4	46.0	41.3	36.5	39.7
Maximum	100	100	100	100	100	100

The results about the score also confirm the tendency to the order of preference as can be observed in Table 2. The median of the preferences shows that options F and C are preferred.

Table 2. Summary statistics about the order of preference

	Order of preference (1- Best to 6- Worst)					
Statistics	*A*	*B*	*C*	*D*	*E*	*F*
Median	5	5	2	4	3	2
Minimum	1	1	1	1	1	1
Maximum	6	6	5	6	6	6

The questions regarding the specific behavior during gameplay have their answers distribution in Figure 4 and Figure 5 for each tested option.

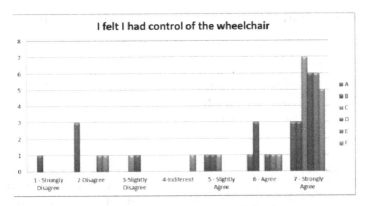

Fig. 4. Responses about the experiment when driving in narrow spaces

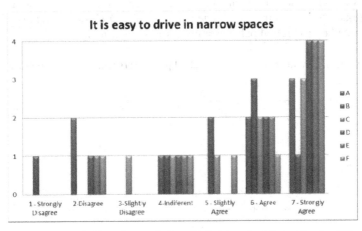

Fig. 5. Responses about the feeling of control when driving the wheelchair using the different options

It is interesting to observe that the number of answers Strongly Agree about the feeling of control of the IW was higher for option C with 7 cases. The controls which suffer attenuation are also positively viewed as a good way for controlling the wheelchair. Even when the objective is to drive the wheelchair in narrow spaces the distribution of answers are more positive when using these attenuated controls.

To confirm that the controls are significantly different the statistical Friedman test was applied to the variables "Feeling about control of driving the wheelchair" and "easy to drive in narrow spaces". In the case of the "feeling about control of driving the wheelchair" the p value obtained was 0.01, so there are statistical differences between the distributions of answers among each joystick mappings. The p value for the second question was 0.631 higher than the significance level therefore there are not statistical evidences that the distributions of answers among each joystick mappings are different in terms of "driving the wheelchair in narrow spaces". The same multiple comparisons of means of orders analysis were performed for the first question. Table 3 shows the corresponding p values.

Table 3. Multiple comparisons of the feeling about controlling the wheelchair

	Multiple Comparisons LSD – Feeling of control (p values)				
	A	*B*	*C*	*D*	*E*
B	0.389	--	--	--	--
C	**0.008**	0.058	--	--	--
D	**0.001**	**0.013**	0.517	--	--
E	**0.006**	**0.046**	0.914	0.589	--
F	**0.028**	0.165	0.589	0.238	0.517

The feeling about having control of the wheelchair with the option A is statistically different with all other options except with the option B. The feeling about having control of the wheelchair with the option B it is also statistically different from all the other options except F and C, although in this last case the p value is near the significance level which also denotes a tendency to be different. It is also interesting to notice a better attitude to the controls with attenuation. Table 4 shows examples of the circuits performed by a cerebral palsy user using the six alternatives.

Table 4. Circuits performed by cerebral palsy users with different controls

Circuits Images		
Original	Proportional	Intuitive
Original with attenuation	Proportional with attenuation	Intuitive with attenuation

These images reveal the difference between the Original, Proportional and Intuitive Manual Controls and with or without the attenuation function. In fact, it is noticeable that the trace of the circuit using the Original and Proportional mappings without attenuation are more irregular than with attenuation and even with the Intuitive solution. The results about the usability and feeling of control of the IW can be confirmed with the analysis of these circuit images.

6   Conclusions and Future Work

Usually, users control electric wheelchairs using conventional analog joysticks. Thus, they need appropriate methodologies to map the position of the joystick to the wheelchair motor velocities. This paper described the implementation of several joystick mappings and appropriate experiments performed in a realistic simulator in order to analyze the usability of the joystick mapping methods.

The study conducted enabled us to verify that the intuitive mapping method achieved was preferred by the majority of the users. Using this method, users could keep the control of the Intelligent Wheelchair, even in narrow spaces.

Future work will be concerned with conducting a deeper study of the control methods by testing different configuration parameters for each control type and test the controls with an even broader sample of wheelchair users. Another experiment will be concerned with the use of data mining algorithms to create user driving models and using them to create automatic control methods based on real user behavior.

Acknowledgments. This work was funded by the ERDF – European Regional Development Fund through the COMPETE Programme (operational programme for competitiveness) and by National Funds through FCT - Portuguese Foundation for Science and Technology within project «INTELLWHEELS - Intelligent Wheelchair with Flexible Multimodal Interface, RIPD/ADA/109636/2009», project Cloud Thinking (funded by the QREN Mais Centro program, ref. CENTRO-07-ST24-FEDER-002031) and project FCOMP-01-0124-FEDER-022682 (FCT ref. PEst-C/EEI/UI0127/2011). The first author would also like to acknowledge FCT for the PhD Scholarship FCT/SFRH/BD/44541/2008.

References

1. Nakanishi, S., Kuno, Y., Shimada, N., Shirai, Y.: Robotic wheelchair based on observations of both user and environment. In: Filipe, J., Fred, A. (eds.) Int. Conf. on Intelligent Robots and Systems (1999)
2. Matsumoto, Y., Ino, T., Ogasawara, T.: Development of Intelligent Wheelchair System with Face and Gaze Based Interface. In: 10th IEEE Int. Workshop on Robot and Human Communication, Bordeaux (2001)
3. Ju, S., Shin, Y., Kim, Y.: Intelligent wheelchair (iw) interface using face and mouth recognition. In: 13th Int. Conference on Intelligent User Interfaces, Canary, Islands (2009)
4. Sasou, A., Kojima, H.: Noise robust speech recognition applied to voice-driven wheelchair. Journal on Advances in Signal Processing 1(41) (2009)

5. Simpson, R.: Smart wheelchairs: A literature review. J of Rehabilitation Research & Development, 423–435 (2005)
6. Braga, R.A.M., Petry, M., Moreira, A.P., Reis, L.P.: Concept and Design of the Intellwheels Platform for Developing Intelligent Wheelchairs. In: Cetto, J.A., Ferrier, J.-L., Filipe, J. (eds.) Informatics in Control, Automation and Robotics. LNEE, vol. 37, pp. 191–203. Springer, Heidelberg (2009)
7. Choi, K., Sato, M., Koike, Y.: A new human-centered wheelchair system controlled by the EMG signal. In: I. Joint Conf. Neural Networks, Vancouver, BC, Canada, pp. 16–21 (2006)
8. Dicianno, B.E., Sibenaller, S., Kimmich, C., Cooper, R.A., Pyo, J.: Joystick use for virtual power wheelchair driving in individuals with tremor: Pilot study. J. Rehabilitation Research & Development 46(2), 269–276 (2009)
9. Fattouh, A., Sahnoun, M., Bourhis, G.: Force Feedback Joystick Control of a Powered Wheelchair: Preliminary Study. In: IEEE Int. Conf. on Systems, Man and Cybernetics, vol. 3, pp. 2640–2645 (2004)
10. Niitsuma, M., Ochi, T., Yamaguchi, M., Hashimoto, H.: Interaction between a User and a Smart Electric Wheelchair in Intelligent Space. In: Int. Symposium on Micro-Nano Mechatronics and Human Science (MHS), pp. 465–470 (2010)
11. Pires, G., Nunes, U.: A wheelchair steered through voice commands and assisted by a reactive fuzzy-logic controller. J. Intelligent and Robotics System 34, 301–314 (2002)
12. Braga, R., Petry, M., Reis, L.P., Moreira, A.P.: IntellWheels: Modular development platform for intelligent wheelchairs. Journal of Rehabilitation Research & Development 48(9), 1061–1076 (2011)
13. Petry, M., Moreira, A.P., Braga, R., Reis, L.P.: Shared Control for Obstacle Avoidance in Intelligent Wheelchairs. In: 2010 IEEE Conference on Robotics, Automation and Mechatronics (RAM 2010), Singapore, pp. 182–187 (2010)
14. Murdock, K.L.: 3ds Max 2011 Bible. John Wiley & Sons, Indianapolis (2011)
15. Carpin, S., Lewis, M., Wang, J., Balakirsky, S., Scrapper, C.: USARSim: a robot simulator for research and education. In: Proceedings of the IEEE International Conference on Robotics and Automation, Roma, Italy (2007)
16. Braga, R., Malheiro, P., Reis, L.P.: Development of a Realistic Simulator for Robotic Intelligent Wheelchairs in a Hospital Environment. In: Baltes, J., Lagoudakis, M.G., Naruse, T., Ghidary, S.S. (eds.) RoboCup 2009. LNCS (LNAI), vol. 5949, pp. 23–34. Springer, Heidelberg (2010)
17. Faria, B.M., Vasconcelos, S., Reis, L.P., Lau, N.: A Methodology for Creating Intelligent Wheelchair Users' Profiles. In: ICAART 2012 - 4th International Conference and Artificial Intelligence, Algarve, pp. 171–179 (2012)
18. Reis, L.P., Braga, R.A.M., Sousa, M., Moreira, A.P.: IntellWheels MMI: A Flexible Interface for an Intelligent Wheelchair. In: Baltes, J., Lagoudakis, M.G., Naruse, T., Ghidary, S.S. (eds.) RoboCup 2009. LNCS, vol. 5949, pp. 296–307. Springer, Heidelberg (2010)
19. Faria, B.M., Vasconcelos, S., Reis, L.P., Lau, N.: Evaluation of Distinct Input Methods of an Intelligent Wheelchair in Simulated and Real Environments: A Performance and Usability Study. Assistive Technology: The Official Journal of RESNA 25(2), 88–98 (2013)
20. Faria, B.M., Reis, L.P., Lau, N., Soares, J.C., Vasconcelos, S.: Patient Classification and Automatic Configuration of an Intelligent Wheelchair. In: Filipe, J., Fred, A. (eds.) ICAART 2012. CCIS, vol. 358, pp. 268–282. Springer, Heidelberg (2013)
21. Lewis, J.R.: IBM Computer Usability Satisfaction Questionnaires: Psychometric Evaluation and Instructions for Use. International Journal of Human-Computer Interaction 7(1), 57–78 (1995)

Omnidirectional Walking and Active Balance for Soccer Humanoid Robot

Nima Shafii[1,2,3], Abbas Abdolmaleki[1,2,4], Rui Ferreira[1], Nuno Lau[1,4], and Luís Paulo Reis[2,5]

[1] IEETA - Instituto de Engenharia Eletrónica e Telemática de Aveiro, Universidade de Aveiro
[2] LIACC - Laboratório de Inteligência Artificial e Ciência de Computadores, Universidade do Porto
[3] DEI/FEUP - Departamento de Engenharia Informática, Faculdade de Engenharia, Universidade do Porto
[4] DETI/UA - Departamento de Eletrónica, Telecomunicações e Informática, Universidade de Aveiro
[5] DSI/EEUM - Departamento de Sistemas de Informação, Escola de Engenharia da Universidade do Minho
{nima.shafii,rui.ferreira}@fe.up.pt, {abbas.a,nunolau}@ua.pt, lpreis@dsi.uminho.pt

Abstract. Soccer Humanoid robots must be able to fulfill their tasks in a highly dynamic soccer field, which requires highly responsive and dynamic locomotion. It is very difficult to keep humanoids balance during walking. The position of the Zero Moment Point (ZMP) is widely used for dynamic stability measurement in biped locomotion. In this paper, we present an omnidirectional walk engine, which mainly consist of a Foot planner, a ZMP and Center of Mass (CoM) generator and an Active balance loop. The Foot planner, based on desire walk speed vector, generates future feet step positions that are then inputs to the ZMP generator. The cart-table model and preview controller are used to generate the CoM reference trajectory from the predefined ZMP trajectory. An active balance method is presented which keeps the robot's trunk upright when faced with environmental disturbances. We have tested the biped locomotion control approach on a simulated NAO robot. Our results are encouraging given that the robot has been able to walk fast and stably in any direction with performances that compare well to the best RoboCup 2012 3D Simulation teams.

Keywords: Bipedal Locomotion, Gait Generation, Foot Planner, Active Balance.

1 Introduction

In robotics, bipedal locomotion is a form of locomotion where robot moves by means of its two legs. This movement includes any type of omnidirectional walking or running. A humanoid robot is a robot with its overall appearance is based on the human body, allowing the robot to move on its two legs.

Humanoid robots must have capability to adjust speed and direction of its walk to perform its tasks. Due to having a huge amount of controller design space, as well as being an inherently nonlinear system, it is very difficult to control the balance of the

L. Correia, L.P. Reis, and J. Cascalho (Eds.): EPIA 2013, LNAI 8154, pp. 283–294, 2013.

humanoid robot during walking. The ZMP [1] criterion is widely used as a stability measurement in the literature. For a given set of walking trajectories, if the ZMP trajectory keeps firmly inside the area covered by the foot of the support leg or the convex hull containing the support legs, this biped locomotion will be physically feasible and the robot will not fall while walking.

A Biped walking trajectory can be derived from a predefined ZMP reference by computing feasible body swing or CoM trajectory. The CoM trajectory can be calculated by a simple model, approximating the bipedal robot dynamics, such as Cart-table model or linear inverted pendulum model [2].

There is not a straightforward way to compute CoM from ZMP by solving the differential equations of the cart-table model. The approaches presented previously, on how to tackle this issue, are organized into two major groups, optimal control approaches and analytical approaches. Kajita has presented an approach to find the proper CoM trajectory, based on the preview control of the ZMP reference, which makes the robot able to walk in any direction [3]. This is a dominant example of the optimal control approach. Some analytical approaches were also developed based on the Fourier series approximation technique, which can create straight walking reference [4].

Although making a humanoid robot walk in a straight or curved line is very important and motivating for researchers, generating other types of walking, such as side and diagonal walking, and being able to change the direction of the walk, can improve the ability of a humanoid robot to maneuver with more agility in a dynamic environment as, for example, a soccer field. Recently, several researches focused on the locomotion of humanoid soccer robots have been published, which are mainly based on the ZMP control approaches [5][6][7]. Although many of them have a brief explanation of their walk engine, the foot planner and ZMP controller parts have not been explained in detail. In addition, most of them do not include an active balance method.

In this paper, we present an omnidirectional walk engine to develop an autonomous soccer humanoid robot, which mainly consist of a Foot planner, a ZMP and Center of Mass (CoM) generator and an Active balance loop. The Foot Planner generates footprint based on desired walk speed and direction. The ZMP generator generates the ZMP trajectory, which is the position of the center of the support foot in each walk step. The CoM reference trajectory is obtained by using a ZMP preview controller on the predefined ZMP trajectory. Position trajectories of the swing foot are generated through Bézier curves based on predefined footsteps. The swing foot orientation is also kept parallel to the ground to reduce the effect of the contact force. An active balance method is used to modify the foot position in order to keep the robot stable during the walk in face of external perturbations. Finally, leg joint angles are calculated based on swing foot positions, CoM references and support foot position by using inverse kinematics, and then joints are controlled by simple independent PID position controllers.

The reminder of the paper is organized as follows. The next section outlines our proposed omnidirectional walk engine. Then, the foot planner algorithm, which is used to plan the next feet positions and generate the ZMP trajectory, will be explained. Section 2.2 explains the preview controller applied to the biped walking to generate the feasible CoM reference trajectory from predefined ZMP trajectory. Our approach for the active balance, in order to keep the robot stable during walking, in the face of the external perturbation, is described in section 2.3. Experimental results on simulated NAO robot are presented and discussed in Section 3. General conclusions and future works are discussed in the last section.

2   Omnidirectional Walking Engine

This section, presented the design of the walking controllers to enable the robot with an omnidirectional walk. Developing an omnidirectional walk gait is a complex task made of several components. In order to get a functional omnidirectional walk, it is necessary to decompose it in several modules and address each module independently. Figure 1, shows the general architecture of walk engine modules and how they interact with each other to achieve stable walking.

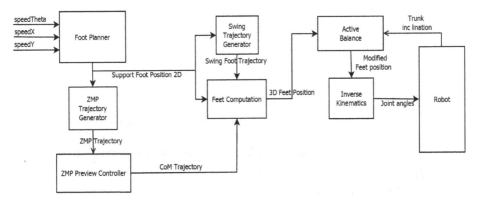

Fig. 1. Architecture of the Walk engine modules

In the following sections, the main modules such as foot planner, active balance and ZMP preview controller will be explained in detail.

2.1   Foot Planner

The footstep planner generates the support foot future positions that the robot will use. During the walking process, we need to plan future steps based on current state of feet, desired walk vector and the preview controller look ahead duration. At the initial state, the robot is in double support and CoM is located at the center of the line that connects the feet centers. After changing the CoM to the area of the support foot, the robot lifts its swing foot and moves it to next footprint. The CoM moves from its initial location to a new position. Figure 2 shows how we generate the next position and orientation of right (swing) foot based on velocity vector (V_x, V_y, w) in XY plane. We first calculate the next position of CoM by multiplying the time duration of one step by the input velocity. This gives the linear position and orientation change that can be used to determine the next position and orientation of CoM *(x,y,θ)*. Then, we calculate a reference point, which is used to compute the next swing foot position. The reference point has a 55mm distance (half distance of two legs in NAO robot), at 90 degree, from the support foot. From the previous reference point and the CoM position, the next reference point can be determined, from which the next support position can be derived. In order to rotate CoM θ degrees we rotate the target footstep θ degrees relative to the CoM frame.

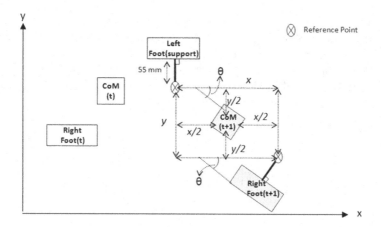

Fig. 2. Next step Calculation of foot planner for walk vector (x,y,θ)

We also take into consideration two constraints when calculating target foot:

1. Foot reachability
2. Self-Collision of feet

Foot reachability means whether the robot is capable to move its foot to the calculated target footprint or not, and self-collision means whether the feet is colliding with each other or not. We consider these constraints by defining maximum and minimum distances of feet in x and y axis, which should not be violated by new planned step and can be shown as follows.

$$A < Rightfoot.x - Leftfoot.x < B$$
$$C < Rightfoot.y - Leftfoot.y < D$$

A, B, C and D are constant parameters that represent the maximum and minimum relative feet positions in the global frame.

2.2   Cart-Table Model and CoM Reference Generator

Many popular approaches used for bipedal locomotion are based on the ZMP stability indicator and cart-table model. ZMP cannot generate reference walking trajectories directly but it can indicate whether generated walking trajectories will keep the balance of a robot or not. Nishiwaki proposed to generate walking patterns by solving the ZMP equation numerically [7]. Kajita assumed that biped walking is a problem of balancing a cart-table model [2], since in the single supported phase, human walking can be represented as the Cart-table model or linear inverted pendulum model [3].

Biped walking can be modeled through the movement of ZMP and CoM. The robot is in balance when the position of the ZMP is inside the support polygon. When the ZMP reaches the edge of this polygon, the robot loses its balance. Biped walking trajectories can be derived from desired ZMP by computing the feasible CoM. The possible body swing can be approximated using the dynamic model of a Cart-on-a-table.

Cart-table model has some assumptions and simplifications in its model [3]. First, it assumes that all masses are concentrated on the cart. Second, it assumes that the support leg does not have any mass and represents it as a massless table. Although these assumptions seem to be far from reality, modern walking robots usually have heavy trunks with electronics circuits and batteries inside. Therefore, the effect of leg mass is relatively small. Figure 3 shows how robot dynamics is modeled by a cart-on-a-table and its schematic view.

 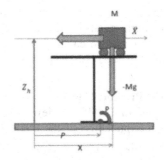

Fig. 3. Schematic view of Cart-table model and a humanoid robot

Two sets of cart-table are used to model 3D walking. One is for movements in frontal plane; another is for movements in coronal plane. The position of Center of Mass (CoM) M is x and Z_h defined in the coordinate system O. Gravity g and cart acceleration create a moment T_p around the center of pressure (CoP) point P_x. The Equation (1) provides the moment or torque around P.

$$T_p = Mg(x - P_x) - M\ddot{x}z_h \tag{1}$$

We know from [8] that when the robot is dynamically balanced, ZMP and CoP are identical, therefore the amount of moment in the CoP point must be zero, $T_p=0$. By assuming the left hand side of equation (1) to be zero, equation (2) provides the position of the ZMP. In order to generate proper walking, the CoM must also move in coronal plane, hence another cart-table must be used in y direction. Using the same assumption and reasoning equation (3) can be obtained. Here, y denotes the movement in y.

$$P_x = x - \frac{Z_h}{g}\ddot{x} \tag{2}$$

$$P_y = y - \frac{Z_h}{g}\ddot{y} \tag{3}$$

In order to apply cart-table model in a biped walking problem, first the position of the foot during walking must be planned and defined, then based on the constraint of ZMP position and support polygon, the ZMP trajectory can be designed. In the next step, the position of the CoM must be calculated using differential equations (2) (3). Finally, inverse kinematics is used to find the angular trajectories of each joint based on the planned position of the foot and calculated CoM. Two different inverse kinematic approaches, which were applied on the NAO humanoid soccer robot can be found in [8] [9].

The main issue of applying Cart-table model is how to solve its differential equations. Even though theoretically CoM trajectory can be calculated by using the exact solution of the Cart-table differential equations, applying calculated trajectory is not straightforward in a real biped robot walking because the solution consists of unbounded *cosh* functions, and the obtained CoM trajectory is very sensitive to the time step variation of the walking.

An alternative robust CoM trajectory generation method can be found in [4], in which the solution of the Cart-pole model differential equation is approximated based on Fourier representation of the ZMP equation. Kajita et. al also presents a very applicable approach to calculate the position of the CoM from the cart-table model. This approach is based on the preview control of the ZMP reference.

2.2.1 ZMP Preview Controller

In this section, the ZMP Preview Control approach proposed by Kajita et al [3] and an extended explanation of the method by Park [10] is presented. The jerk x of the cart areas of the system is assumed as input u of the cart table dynamics ($\frac{d\ddot{x}}{dt}=u$). Considering this assumption, the ZMP equations (2) (3) can be converted to a strongly appropriate dynamical system which is presented in equation (4), where P is the position of the ZMP and $X = (x, \dot{x}, \ddot{x})$ is the state of CoM.

$$\frac{d}{dt}\begin{pmatrix}x\\\dot{x}\\\ddot{x}\end{pmatrix} = \begin{pmatrix}0 & 1 & 0\\0 & 0 & 1\\0 & 0 & 0\end{pmatrix}\begin{pmatrix}x\\\dot{x}\\\ddot{x}\end{pmatrix} + \begin{pmatrix}0\\0\\1\end{pmatrix}u \tag{4}$$

$$P = \begin{pmatrix}1 & 0 & -\frac{Z_h}{g}\end{pmatrix}\begin{pmatrix}x\\\dot{x}\\\ddot{x}\end{pmatrix}$$

Using this Cart table model, a digital controller is designed which allows the system output to follow the input reference. The discretized system of the equation (5) is given below

$$\begin{aligned}x(k + 1) &= AX(k) + Bu(k)\\P(k) &= CX(k)\end{aligned} \tag{5}$$

Where

$$A = \begin{bmatrix} 1 & \Delta t & \Delta t^2/2 \\ 0 & 1 & \Delta t \\ 0 & 0 & 1 \end{bmatrix}, B = \begin{bmatrix} \Delta t^3/6 \\ \Delta t^2/2 \\ \Delta t \end{bmatrix}, C = \begin{bmatrix} 1 & 0 & -\frac{z_h}{g} \end{bmatrix}$$

By assuming the incremental state $\Delta X(k) = X(k) - X(k-1)$, the state is augmented as $\tilde{X} = \begin{bmatrix} p(k) \\ \Delta X(k) \end{bmatrix}$, and consequently the equation (6) is rewritten as

$$\tilde{X}(k+1) = \tilde{A}\tilde{X}(k) + \tilde{B}u(k)$$
$$P(k) = \tilde{C}\tilde{X}(k)$$
(6)

Where

$$\tilde{A} = \begin{bmatrix} 1 & CA \\ 0 & A \end{bmatrix}, \tilde{B} = \begin{bmatrix} CB \\ B \end{bmatrix}, \tilde{C} = \begin{bmatrix} 1 & 0 & 0 & 0 \end{bmatrix}$$

Calculation of digital controller input and error signal are presented in equation (7). Figure (4) shows the block diagram of the system.

$$u(k) = -G_i \sum_{i=0}^{k} e(i) - G_x x(k)$$
(7)

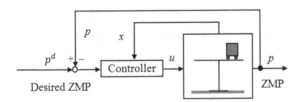

Fig. 4. ZMP Preview control Diagram

Here, G_i and G_x are assumed as the gain for the ZMP tracking error and the gain for state feedback respectively. The ZMP tracking error is $e(i) = P - P^d$ and k denoted the kth sample time. It was reported that the controller of (7) was not able to follow the reference ZMP sufficiently. The main cause of this issue is the inherent phase delay. For addressing this problem, the original digital controller is redesigned in equation (8).

$$u(k) = -G_i \sum_{i=1}^{k} e(i) - G_x x(k) - \sum_{j=1}^{NL} G_p P^d(k+j)$$
(8)

The third term consists of the planed ZMP reference up to NL samples in future. Since this controller uses future information, it is called a preview controller and the gain $G_p(i)$ is called the preview gain. Experience shows that one second of future desired walking is sufficient for the Preview controller to generate a smooth trajectory, therefore, parameter NL can be calculated based on the incremental time step, $NL = \frac{1}{\Delta t}$.

A detailed explanation of the preview control approach as an optimal controller technique can be found in [11]. The optimal gain, G_i, G_x, are calculated by solving discrete algebraic Riccati equation.

$$\tilde{P} = \tilde{A}^T \tilde{P} \tilde{A} - \tilde{A}^T \tilde{P} \tilde{B} (R + \tilde{B}^T \tilde{P} \tilde{B})^{-1} \tilde{B}^T \tilde{P} \tilde{A} + \tilde{Q} \tag{9}$$

Where $\tilde{Q} = diag\{Q_e, Q_x\}$. Then, the optimal gain is defined by

$$\tilde{G} = (R + \tilde{B}^T \tilde{P} \tilde{B})^{-1} \tilde{B}^T \tilde{P} \tilde{A} = [G_i \quad G_x] \tag{10}$$

The optimal preview gain is recursively computed as follows.

Considering $\widetilde{A_c} = \tilde{A} - \tilde{B} \tilde{G}$

$$G_p(i) = (R + \tilde{B}^T \tilde{P} \tilde{B})^{-1} \tilde{B}^T \tilde{X}(i - 1)$$
$$\tilde{X}(i) = \widetilde{A_c^T} \tilde{X}(i - 1) \tag{11}$$

Where $G(1) = -G_i \begin{bmatrix} 1 \\ 0 \end{bmatrix}$, $\tilde{X}(1) = -\widetilde{A_c^T} \tilde{P} \begin{bmatrix} 1 \\ 0 \end{bmatrix}$, $Q_e = 1, Q_x = 0$ and $R = 1 \times 10^{-6}$, are assumed. Figure (5) shows G_p profile towards the future. We can observe that the magnitude of the preview gain quickly decreases, thus the ZMP reference in the far future can be neglected.

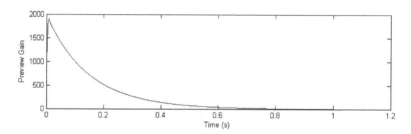

Fig. 5. Gain calculated based on the preview controller

2.3   Active Balance

The cart-table model has some simplifications in biped walking dynamics modeling; in addition, there is inherent noise in leg's actuators. Therefore, keeping walk balance generated by cart-table model cannot be guaranteed. In order to reduce the risk of falling during walking, an active balance technique is applied.

The Active balance module tries to maintain an upright trunk position, by reducing variation of trunk angles. One PID controller is designed to control the trunk angle to be Zero. Trunk angles, pitch and roll, can be derived from inertial measurement unit of the robot. When the trunk is not upright, instead of considering a coordinate frame attached to the trunk of biped robot, position and orientation of feet are calculated with respect to a coordinate frame center, which is attached to the CoM position and makes the Z axes always perpendicular to the ground plane.

The transformation is performed by rotating this coordinate frame with the rotation angle calculated by the PID controller. This makes the Z axes always perpendicular to the ground plane. The feet orientation is kept constantly parallel to ground. The Transformation formulation is presented in equation (12).

$$Foot = T_{Foot}^{CoM}(pitchAng, rollAng) \times Foot \tag{12}$$

The *pitchAng* and *rollAng* are assumed to be the angles calculated by the PID controller around y and x axis respectively. Figure 6 shows the architecture of the active balance unit.

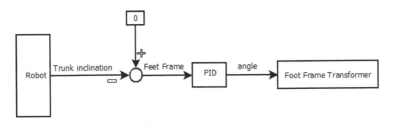

Fig. 6. Active Balance controller

3 Results and Discussion

In this study, a simulated NAO robot is used in order to test and verify the approach. The NAO model is a kid size humanoid robot that is 58 cm high and 21 degrees of freedom (DoF). The link dimensions of the NAO robot can be found in [12]. The simulation is carried out using the RoboCup soccer simulator, rcsssever3d, which is the official simulator released by the RoboCup community, in order to simulate humanoids soccer match. The simulator is based on Open Dynamic Engine and Simspark [13]. The Robocup simulator was studied and compared whit another humanoid robot simulator in [14].

In order to test the behavior and evaluate its performance, several walking scenarios were designed. The parameters of walking engine used in the walking scenarios are presented in Table 1.

Table 1. Walking scenarios parameters

Parameters	Value
Step Period	0.2 s
Step Height of the swing foot	0.02 m
Speed in X direction	0.5,0,-0.5 m/s
Speed in Y direction	0.3, 0,-0.3 m/s
Percentage of the Double Support Phase (DSP) to the whole step time	15 %
Height of the inverted pendulum (Z_h)	0.21 cm
Incremental sample time used in ZMP preview controller (Δt)	0.002 s

In the first scenario, the robot must walk forward with speed of 0.5 m/s for 10 seconds, then change to a sidewalk that lasts for 10 seconds with the speed of 0.3 cm/s, then he must do a diagonal back walk for another10 seconds with the speed of - 0.5 m/s and -0.3 m/s in x and y directions, respectively. At the end, the robot must stop at the starting point. In order to show the ability of the developed behavior to execute the scenario, generated CoM trajectory and footprint projected on the ground plane are shown in Figure 7. Figure 8 also shows a generated curved walking, in which the robot walks with the speed of 0.2 m/s and rotates 6 degree in each walk step (*w*=30 deg/s).

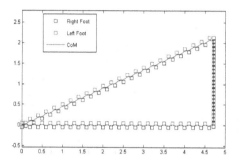

Fig. 7. CoM and footprint of the proposed walking scenario

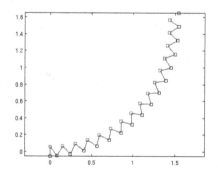

Fig. 8. CoM and footprint of a curved walk

The planned ZMP trajectory based on the first 5 walk steps footprints of the proposed walking scenario (forward walking) and the followed ZMP trajectory, achieved using the preview controller, are shown in figure 9.

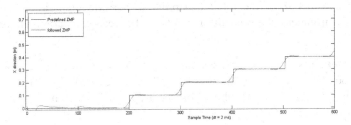

Fig. 9. ZMP trajectories for proposed the forward walk

Table 2 lists the values of some of the relevant results for the developed gait implemented on our simulated NAO compared with corresponding behavior from other RoboCup 3D simulation teams. We collect times and distances/angles for forward, backward, side and diagonal (45 degrees) walk and rotate in place. For walk without rotation the agent walked for 20 seconds and recorded its position and time at each cycle. For rotate in place, the agent completed 10 full turns and recorded its time. By analyzing the results of the last RoboCup competition, the reported results are approximated based on the average speeds of the locomotion skills for teams participating in the final round of the RoboCup 2012 competition. It is also impressive that the robot can change walking direction without reducing its speed, which for other teams is a very difficult task.

Table 2. Comparison between the performance of the locomotion generated by the proposed walk engine, with the skills performance of other teams that participating in final round of the RoboCup 2012 competition

Motion	Our Agent	UT Austin Villa 3D	RoboCanes	Bold Hearts	magmaOffenburg
Forward walk (m/s)	0.66	0.7	0.69	0.59	0.54
Backward Walk(m/s)	0.59	-	0.61	0.47	0.48
Side Walk (m/s)	0.51	-	0.47	0.42	-
Turn in Place (deg/s)	125	100	110	-	110

4   Conclusions and Future Work

Creating a good walk engine in RoboCup soccer competitions is a critical and important task for all participant teams. The walk engine should enable robots to walk stably and fast in any direction. This paper presents an implementation of an omnidirectional walk engine augmented with an active balance method for the NAO humanoid robot. The cart-table model and preview control are used to generate the CoM reference trajectory from the predefined ZMP trajectory. And the active balance method, keeps the robot's trunk upright in case of environmental disturbances. The biped locomotion control approach has been tested on a simulated NAO robot. Our results show that the robot has been able to walk fast and stably in any direction with performances that compare well with the best RoboCup 2012 teams. It is a proper approach in order to be applied on the humanoid soccer robot.

In future work, the proposed method will be tested and implemented on a real humanoid robot. Our aim is to develop a fully autonomous soccer humanoid team, in which the proposed walk engine will be used for their locomotion. Although the experimental results show the robot capability to perform omnidirectional walk in simulation environment, there is still a small gap between simulation and real soccer environment.

Acknowledgements. The first and second authors are supported by FCT under grant SFRH/BD/66597/2009 and SFRH/BD/81155/2011, respectively. This work was supported by project Cloud Thinking (funded by the QREN Mais Centro program, ref. CENTRO-07-ST24-FEDER-002031).

References

1. Vukobratovic, M., Stokic, D., Borovac, B., Surla, D.: Biped Locomotion: Dynamics, Stability, Control and Application, p. 349. Springer (1990)
2. Kajita, S., Kanehiro, F., Kaneko, K., Yokoi, K., Hirukawa, H.: The 3D linear inverted pendulum mode: a simple modeling for a biped walking pattern generation. In: IEEE/RSJ International Conference on Intelligent Robots and Systems, pp. 239–246 (2001)
3. Kajita, S., Kanehiro, F., Kaneko, K., Fujiwara, K.: Biped walking pattern generation by using preview control of zero-moment point. In: IEEE International Conference on Robotics and Automation, ICRA 2003, pp. 1620–1626 (2003)
4. Erbatur, K., Kurt, O.: Natural ZMP Trajectories for Biped Robot Reference Generation. IEEE Transactions on Industrial Electronics 56(3), 835–845 (2009)
5. Graf, C., Rofer, T.: A closed-loop 3D-LIPM gait for the RoboCup Standard Platform League humanoid. In: Fourth Workshop on Humanoid Soccer Robots in IEEE Conference of Humanoid Robots, pp. 15–22 (2010)
6. Gouaillier, D., Collette, C., Kilner, C.: Omni-directional Closedloop Walk for NAO. In: IEEE-RAS International Conference on Humanoid Robots, pp. 448–454 (2010)
7. Strom, J., Slavov, G., Chown, E.: Omnidirectional walking using ZMP and preview control for the NAO humanoid robot. In: Baltes, J., Lagoudakis, M.G., Naruse, T., Ghidary, S.S. (eds.) RoboCup 2009. LNCS, vol. 5949, pp. 378–389. Springer, Heidelberg (2010)
8. Ferreira, R., Reis, L.P., Moreira, A.P., Lau, N.: Development of an Omnidirectional Kick for a NAO Humanoid Robot. In: Pavón, J., Duque-Méndez, N.D., Fuentes-Fernández, R. (eds.) IBERAMIA 2012. LNCS (LNAI), vol. 7637, pp. 571–580. Springer, Heidelberg (2012)
9. Domingues, E., Lau, N., Pimentel, B., Shafii, N., Reis, L.P., Neves, A.J.R.: Humanoid Behaviors: From Simulation to a Real Robot. In: Antunes, L., Pinto, H.S. (eds.) EPIA 2011. LNCS (LNAI), vol. 7026, pp. 352–364. Springer, Heidelberg (2011)
10. Park, J., Youm, Y.: General ZMP preview control for bipedal walking. In: 2007 IEEE International Conference on Robotics and Automation, pp. 2682–2687 (April 2007)
11. Katayama, T., Ohki, T., Inoue, T., Kato, T.: Design of an optimal controller for a discrete-time system subject to previewable demand. International Journal of Control 41(3), 677–699 (1985)
12. Gouaillier, D., Hugel, V., Blazevic, P., Kilner, C., Monceaux, J., Lafourcade, P., Marnier, B., Serre, J., Maisonnier, B.: Mechatronic design of NAO humanoid. In: Proceedings of the IEEE International Conference on Robotics and Automation, pp. 769–774 (2009)
13. Boedecker, J., Asada, M.: SimSpark – Concepts and Application in the RoboCup 3D Soccer Simulation League. In: Autonomous Robots, pp. 174–181 (2008)
14. Shafii, N., Reis, L.P., Rossetti, R.J.: Two humanoid simulators: Comparison and synthesis. In: 2011 6th Iberian Conference on Information Systems and Technologies (CISTI), pp. 1–6 (2011)

Online SLAM Based on a Fast Scan-Matching Algorithm

Eurico Pedrosa, Nuno Lau, and Artur Pereira

University Of Aveiro,
Institute of Electronics and Telematics Engineering of Aveiro, Portugal
{efp,nunolau,artur}@ua.pt
http://www.ieeta.pt

Abstract. This paper presents a scan-matching approach for online simultaneous localization and mapping. This approach combines a fast and efficient scan-matching algorithm for localization with dynamic and approximate likelihood fields to incrementally build a map. The achievable results of the approach are evaluated using an objective benchmark designed to compare SLAM solutions that use different methods. The result is a fast online SLAM approach suitable for real-time operations.

Keywords: scan-matching, localization, mapping, simultaneous localization and mapping (SLAM), real-time.

1 Introduction

A basic requirement of an autonomous service robot is the capability to selflocalize in the real-world indoor and domestic environment where it operates. These type of environments are typically dynamic, cluttered and populated by humans (e.g. Figure 1). Not only should the robot act adequately and in due time to the perceived dynamic changes in the environment, it should also be robust to inconsistencies between the collected sensorial information and its internal world representation. The latter is of utmost importance for localization in dynamic environments like the ones already mentioned.

Providing the robot with the required representation of the environment (i.e. obstacles, free space and unknown space) can be a cumbersome task, even unfeasible, when done manually. Simultaneous Localization and Mapping (SLAM) provides the means towards a truly autonomous robot, by minimizing manual input. A robot carrying out SLAM has, in principle, the capability to autonomously construct an internal representation (or map) of the environment. The resulting map captures the structure of the environment that have been sensed at least once.

A common approach to SLAM in real-world applications is to manually direct the robot through the environment while collecting the acquired sensory information. The collected data is then processed offline to construct a representation of the world that can be used by the robot for localization in subsequent tasks. However, there are several tasks that require the robot to construct its internal

L. Correia, L.P. Reis, and J. Cascalho (Eds.): EPIA 2013, LNAI 8154, pp. 295–306, 2013.
© Springer-Verlag Berlin Heidelberg 2013

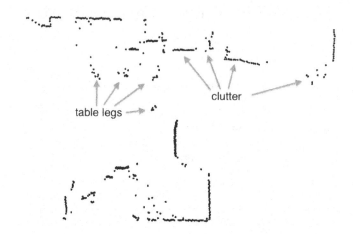

Fig. 1. Example of scan acquired in an office space

representation of the environment during operation, like, for instance, following a person through unknown space. The applied algorithm needs to be efficient so that it can be used during operation and to allow for further sensor data processing in real-time.

In this paper, we present a fast and practical online SLAM algorithm suitable for real-time usage. The overall algorithm is based on a fast, scan-matching based, localization algorithm for mobile robots. The remainder of this paper is organized as follow: in the next section related work is presented; in section 3 the localization algorithm is described; in section 4 the necessary changes and additions to the base localization algorithm for performing SLAM are detailed; section 5 presents practical experiments and adequate evaluation of the proposed SLAM algorithm; finally, section 6 presents the conclusion of this paper.

2   Related Work

There has been a fair amount of research into the SLAM problem in the last two decades. The probabilistic SLAM trend provide us with solutions that explicitly handle uncertainties about pose estimations and sensory information by estimating a probability distribution over the potential solutions. Example of such approaches are the Extended Kalman Filters (EKFs) [17], Unscented Kalman Filters (UKFs) [3], Sparse Extended Information Filters (SEIFs) [25], and Rao-Blackwellized Particle Filters (RBPFs) [9]. Notwithstanding their robust and accurate results, the necessary computational effort makes it difficult to use the previous techniques for real-time applicability.

SLAM algorithms based on graphs [10,21] compose the problem in terms of a graph where nodes are poses of the robot during mapping and edges are spacial constraint between the nodes. They are typically data association problem solvers with highly efficient algorithms used to find the most probable configuration of nodes given their constraints.

Iterative Closest Point (ICP) [26] and Iterative Closest Line (ICL) (e.g. [2]) are widely used in scan-matching. They both try to find the rigid-body transformation that best align the reference and the query points. Lu and Milios [18] describe methods using ICP and LIDAR scans for localization and map building. Peter Biber proposed the Normal Distribution Transform (NDT) [1] for SLAM. Using a grid map, a normal distribution is assigned to each cell which locally models the probability of measuring a point. This representation is then used for localization, making use of a gradient descent algorithm, and incrementally building a map. Holz et al. [11] proposed a fast ICP-based method for SLAM. It is based on range image registration using ICP and sparse point maps. Kohlbrecher et al. [12] proposed the combination of scan-matching using a LIDAR system with a 3D attitude estimation. A multi-resolution occupancy grid maps and approximation of map gradients are used for 2D localization and mapping.

3   The Perfect Match Algorithm

The original Perfect Match (PM) algorithm was developed by Lauer et al. [16] to provide a robust real time self-localization, in a highly dynamic but structured environment, to a mobile robot using a camera system to sense the world.

To the best of our knowledge, an initial adaptation of PM to indoor localization was proposed by Gouveia et al. [8] using range data from a LIDAR. Cunha et al. [4] proposes the use of PM using depth information. Both follow the original implementation of the algorithm with hand built maps that are not practical for a general use because of the geometric diversity of indoor environments. Here, we assume that the provided map was built in an automatic way, e.g. with SLAM. This can be seen as a prelude for map building using PM.

Without loosing the original properties of the algorithm, PM will be presented within a probabilistic formulation, a trending and well grounded point of view [24]. Furthermore, global localization and pose tracking will be omitted, for they are not relevant to this paper.

3.1   Probabilistic Formulation

Let s_t be the robot state at time t, furthermore, let z_t be the measurement at time t. The control u_t at time t determines the change of state in the time interval $(t-1, t]$, and the map m models the environment. The posterior probability $p(s_t | z_{1:t}, u_{1:t}, m)$ is given by the following recursive update equation [23]:

$$p(s_t | z_{1:t}, u_{1:t}, m) = \eta \, p(z_t | s_t, m) \int p(s_t | u_t, s_{t-1}, m) \\ p(s_{t-1} | z_{1:t-1}, u_{1:t-1}, m) ds_{t-1} \qquad , \qquad (1)$$

where η is a normalizer. Furthermore, assuming that the robot motion does not depend of the map m, and assuming that s_{t-1} is known at time $t-1$, the posterior for localization is:

$$p(s_t|z_{1:t}, u_{1:t}, m) \propto (p(z_t|s_t, m)p(s_t|\hat{s}_{t-1}, u_t)) . \tag{2}$$

The first term, $p(z_t|s_t, m)$, is the observation model, and the second term, $p(s_t|u_t, s_{t-1})$, is the motion model.

The motion model $p(s_t|u_t, s_{t-1})$ describes the probability of the robot pose s_t after executing the control u_t, that is, a probabilistic generalization of the robot kinematics. This model is typically known in terms of a multivariate Gaussian.

The measurement model $p(z_t|s_t, m)$, physically models the sensors of the robot. Assuming that the robot pose s_t and the map m are known, the measurements model specifies the probability that z_t is measured. It is common, for sensors equipped in robots, to generates more than one measurement at time t. Thus, the result calculation is the collection of probabilities $p(z_t^i|s_t, m)$, where z_t^i is the ith measurement. Assuming conditional independence between the measurements, the resulting probability is given by:

$$p(z_t|s_t, m) = \prod_i p(z_t^i|s_t, m) . \tag{3}$$

3.2   Likelihood Field for Range Finders

The inception of PM has assumed scans extracted from processed images [19], in a way it mimics a range finder (e.g. sonar or LIDAR). Therefore, using range data from a LIDAR, a more precise sensor, is a change without cost.

The adopted model for measurement is the *likelihood field* [23]. It is a model that lacks a plausible physical explanation, like the beam model [5]. However, it works well in practice and its computation is more efficient.

While the *beam model* applies a *ray casting* function to find the "true" z_t^{i*} range of the object measured by z_t^i, the *likelihood field* projects the endpoint of z_t^i into global coordinates of the map m to find the distance to the nearest object. Let $s_t = (x\ y\ \theta)^T$ denote the robot pose at time t, $(x_{i,\text{sens}}\ y_{i,\text{sens}})^T$ the relative position of the sensor on frame of the robot, and $\theta_{i,\text{sens}}$ the angular orientation of the sensor relative to the heading of the robot. The endpoint of the measurement z_t^i, in global coordinates, is given by the transformation $\mathcal{T}$:

$$\begin{pmatrix} x_{z_t^i} \\ y_{z_t^i} \end{pmatrix} = \begin{pmatrix} x \\ y \end{pmatrix} + \begin{pmatrix} \cos\theta & -\sin\theta \\ \sin\theta & \cos\theta \end{pmatrix} \left[\begin{pmatrix} x_{i,\text{sens}} \\ y_{i,\text{sens}} \end{pmatrix} + z_t^i \begin{pmatrix} \cos\theta_{i,\text{sens}} \\ \sin\theta_{i,\text{sens}} \end{pmatrix} \right] \tag{4}$$

The noise of the process is modeled by a probability density function (pdf) that requires finding the nearest object in the map. Let Δ_i denote the Euclidean distance between $(x_{z_t^i}\ y_{z_t^i})^T$ and the nearest object in the map m. Then, measurement noise can be modeled by

$$p(z_t^i|s_t, m) = \eta_i \frac{\sigma^2}{\sigma^2 + \Delta_i^2} , \tag{5}$$

a derivation of the error function presented in the original PM [16]. Furthermore, the value measured by a range sensor is, in practice, bounded by the interval

$[0, z_{max}]$, where z_{max} is the maximum value that a measurement yields. Hence, the normalizer η is given by

$$\eta_i = \left(\int_0^{z_{max}} \frac{\sigma^2}{\sigma^2 + \Delta_i^2} \, dz_t^i \right)^{-1} . \tag{6}$$

The value of Δ_i is computed by a search function defined by

$$\mathcal{D}(x_{z_t^i}, y_{z_t^i}) = \min_{x', y'} \left(\sqrt{(x_{z_t^i} - x')^2 + (y_{z_t^i} - y')^2} \mid \langle x', y' \rangle \text{occupied in } m \right) . \tag{7}$$

To speed up this search, $\mathcal{D}$ is a look-up table created by computing the Euclidean distance transform of the map. First, the map is converted to a binary occupancy map, then a simple and fast method is used to compute the distance transformation [7]. This method operates over discrete grids, therefore, the obtained result is only an approximation. To improve this approximation an interpolation scheme, using bilinear filtering, is employed to estimate a more accurate Δ_i [12].

3.3 Maximum Likelihood Pose Estimation

The *maximum likelihood* approach for pose estimation, although non-probabilistic, is simpler and easier to calculate than the posterior (1). The idea is simple: given a measurement and odometry reading, calculate the most likely pose. Mathematically speaking, the s_t pose is obtained as the maximum likelihood estimate of (2):

$$\hat{s}_t = \arg\max_{s_t} \ p(z_t|s_t, m) \, p(s_t|\hat{s}_{t-1}, u_t) \tag{8}$$

To summarize, in time $t - 1$ the (non-probabilistic) estimate of $\hat{s}_{t-1}$ is given to the robot. As u_t is executed and a new z_t is obtained, the most likely pose $\hat{s}_t$ is calculated by the robot.

In this approach $\hat{s}_t$ is found using a gradient ascent algorithm in log likelihood space. It is common to maximize the log likelihood instead of the likelihood because it is mathematically easier to handle, and the maximization is justified by the fact that it is a strict monotonic function. Thus, this approach tries to calculate

$$\hat{s}_t = \arg\max_{s_t} \ \ln[p(z_t|s_t, m) \, p(s_t|\hat{s}_{t-1}, u_t)] . \tag{9}$$

Taking advantage of the properties of the logarithm, this expression can be decomposed into additive terms:

$$\hat{s}_t = \arg\max_{s_t} \ \ln \, p(z_t|s_t, m) + \ln \, p(s_t|\hat{s}_{t-1}, u_t) . \tag{10}$$

The required differentiation with respect to the pose s_t are:

$$\nabla_{s_t} L = \nabla_{s_t} \ln \, p(z_t|s_t, m) + \nabla \ln \, p(s_t|\hat{s}_{t-1}, u_t) , \tag{11}$$

where L is the log likelihood. To simplify the calculation of $\nabla_{s_t} L$ the motion model, in this approach, is assumed to have constant probability, therefore the calculation of $\nabla \ln p(s_t|\hat{s}_{t-1}, u_t)$ is not required. Consequently, the required gradient is given by:

$$\nabla_{s_t} L = \nabla_{s_t} |z_t| \ln \eta + 2|z_t| \ln \sigma - \sum_i \ln(\sigma^2 + \Delta^2)$$

$$= -\sum_i \frac{2\Delta}{\sigma^2 + \Delta^2} \nabla_{\mathcal{T}} \mathcal{D}(x_{z_t^i}, x_{z_t^i})^T \begin{bmatrix} 1 & 0 & (-\hat{x}_{z_t^i} \sin\theta - \hat{y}_{z_t^i} \cos\theta) \\ 0 & 1 & (\hat{x}_{z_t^i} \cos\theta - \hat{y}_{z_t^i} \sin\theta) \end{bmatrix}, \quad (12)$$

where $(\hat{x}_{z_t^i} \, \hat{y}_{z_t^i})^T$ is the z_t^i endpoint relative to the robot, and $\nabla_{\mathcal{T}} \mathcal{D}(x_{z_t^i}, x_{z_t^i})$ is the gradient of the Euclidean *distance transform* with respect to $\mathcal{T}$ (see (4)). This gradient can not be presented in closed form, but is calculated with the *sobel* operator.

The gradient ascent maximizes the log likelihood interactively changing the pose s_t in the direction of the gradient. The aforementioned maximization is computed with the RPROP algorithm [22], which is capable of providing practical result in 10 iterations.

4   Mapping with Perfect Match

Mapping is the problem of generating a map from measurement values, which can be easy to execute if done with known locations [23]. The localization problem with a map is also relatively simple. However when both are combined the problem has a much higher dimensionality, one could argue that the problem has infinite dimensions [24].

The PM is a fast and robust algorithm for localization based on scan-matching. This type of algorithm has already been augmented for SLAM (e.g. [18,12]). Thus, knowing that PM is fast (i.e. execution times of about $1ms$ in our experiments), the goal is to create an also fast and practical online SLAM solution using PM.

4.1   Concurrent Mapping and Localization

To construct the map, a function for incrementally building maps with knowledge of the poses of the robot is required. The function $\hat{m}(s_{1:t}, z_{1:t})$ is used to incrementally build the map as a probabilistic occupancy grid [6,20].

Let us follow the same probabilistic framework presented for localization. By augmenting the state s that is being estimated by the map m, and maintaining the same assumptions already made, the posterior can be defined by:

$$p(s_t, \hat{m}(s_{1:t}, z_{1:t})|z_{1:t}, u_{1:t}) \propto p(z_t|s_t, \hat{m}(s_{1:t}, z_{1:t}))p(s_t|\hat{s}_{t-1}, u_t) . \quad (13)$$

The change of m by $\hat{m}(s_{1:t}, z_{1:t})$ also propagates to the estimation of the robot pose $\hat{s}_t$ (see subsection 3.3). The implication of this formulation, is that PM is used for localization using the map constructed so far, i.e. $\hat{m}(s_{1:t-1}, z_{1:t-1})$.

The initial distribution $p(s, m)$ is not known, but it is straightforward. The initial pose is set to $s = \langle 0, 0, 0 \rangle$, and the occupancy grid that models the map is initialized with an uniform prior. However, by having an uniform map the localization would fail at the very first execution due to the lack of a reference. Therefore, the first estimation of the robot pose $\hat{s}$ is equal to the initial s.

4.2 Dynamic Likelihood Field

PM requires a known map to work, and its speed relies in the use of the *likelihood field* model and the RPROP algorithm. In essence, the computation of the robot pose s_t with PM is the result of several transformations $\mathcal{T}$ for table look-ups (*likelihood field*) during the RPROP iterations to find a local maximum. The created tables are static over time, therefore, the overhead time introduced by their creation is negligible because it happens only once.

In the presented approach the map is built incrementally, thus as a result, the map changes over time and the static assumption falls through. For the PM to work, each time the map changes the Euclidean distance map $\mathcal{D}$ and gradient must be recalculated. For small maps their computation time can be ignored, however, as the map grows in dimension PM looses its *fast* and *real-time* "badges".

To solve the execution time degradation a dynamic update of the Euclidean distance map is employed. Let $\hat{\mathcal{D}}$ denote the Euclidean distance map with relation to $\hat{m}(s_{1:t}, z_{1:t})$. The implemented algorithm was presented by Lau *et al.* [15], and it seek to update only the cells affected by the change of state (i.e. from occupied to free or vice versa) of the cells involved in the update of the map. Additionally, the same interpolation scheme used in section 3.2 is employed.

The gradient $\nabla_{\mathcal{T}}\hat{\mathcal{D}}(x_{z_t^i}, y_{z_t^i})$ also needs to be calculated. Applying a *sobel* operator each time $\hat{\mathcal{D}}$ changes creates the same kind of temporal overhead already discussed for updating $\mathcal{D}$. The adopted solution is the one presented in [12], where the gradient is an interpolation of the difference of the four closest samples in the map along x- and y-axis.

5 Experiments and Evaluation

The initial experiments were carried out on an office type environment dataset with a fair amount of clutter, chairs and tables, and people walking. The legs of the chairs and tables create a considerable amount of misreadings from the LIDAR. It has a long corridor, but it lacks a loop. The obtained map was visually compared with a map generated with GMapping (see Figure 2), a state-of-the-art RBPF implementation [9], and both are very similar, meaning that our proposal can achieve practical results. However since it is a subjective evaluation, further tests were done.

Kümmerle *et al.* [14] proposes an objective benchmark for evaluating SLAM solutions. Instead of using the map itself in the evaluation procedure, the poses of the robot during data acquisition are used. The benchmark allows to compare different algorithm independently of the type of map representation they use,

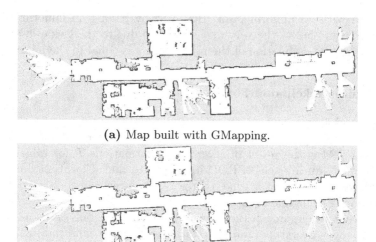

(a) Map built with GMapping.

(b) Map built with the proposed approach.

Fig. 2. Visual comparison of the same map built from the same dataset but with different SLAM approaches

such as occupancy grids or feature maps. The metric is based on relative displacement between robot poses $\delta_{i,j} = s_j \ominus s_i$ instead of comparing poses in the global coordinates. The comparison is made with the mean error in translation $\bar{\epsilon}_{\text{trans}}$ and mean error in rotation $\bar{\epsilon}_{\text{rot}}$ calculated from the obtained trajectory relation $\delta_{i,j}$ and the ground truth relation $\delta_{i,j}^*$:

$$
\epsilon(\delta) = \underbrace{\frac{1}{N} \sum_{i,j} \text{trans}(\delta_{i,j} \ominus \delta_{i,j}^*)^2}_{\bar{\epsilon}_{\text{trans}}} + \underbrace{\frac{1}{N} \sum_{i,j} \text{rot}(\delta_{i,j} \ominus \delta_{i,j}^*)^2}_{\bar{\epsilon}_{\text{rot}}} , \tag{14}
$$

where the functions *trans* and *rot* draw the translational and rotational part of $(\delta_{i,j} \ominus \delta_{i,j}^*)$, respectively.

To compare the PM for map building described in section 4 with other SLAM approaches, selected experiences from [14] were repeated using the datasets and evaluation software available in [13]. The obtained results are summarized in Table 1 and Table 2. The presented results also include the ones reported by Kümmerle *et al.* [14] and Holz *et al.* [11].

For each one of the selected datasets, the ACES building at the University of Texas in Austin (Figure 3a), the Intel Research Lab (Figure 3b), and the Freiburg Building 079 (Figure 3c), a consistent map was built using the proposed approach. The translational and rotational errors are, in the majority of the cases, lower than those from the other approaches. Additionally, an approximate distribution of the errors is visually presented in Figure 4 and Figure 5.

As expected, the PM algorithm for map building is also fast. The measured execution time per scan ($\bar{t}_{\text{scan}}$) provides an approximate trend (see Figure 6) showing that the proposed method can be applied online for most, if not all,

Table 1. Quantitative results of different datasets/approaches on the absolute translational error

	PM (proposed) $\bar{\epsilon}_{trans}/m$	ICP [11] $\bar{\epsilon}_{trans}/m$	Scanmatching [14] $\bar{\epsilon}_{trans}/m$	RBPF [14] $\bar{\epsilon}_{trans}/m$	GraphMapping [14] $\bar{\epsilon}_{trans}/m$
ACES	0.038 ± 0.036	0.060 ± 0.055	0.173 ± 0.614	0.060 ± 0.049	0.044 ± 0.044
INTEL	0.025 ± 0.037	0.043 ± 0.058	0.220 ± 0.296	0.070 ± 0.083	0.031 ± 0.026
FR79	0.037 ± 0.028	0.057 ± 0.043	0.258 ± 0.427	0.061 ± 0.044	0.056 ± 0.042

Table 2. Quantitative results of different datasets/approaches on the absolute rotational error

	PM (proposed) $\bar{\epsilon}_{rot}/def$	ICP [11] $\bar{\epsilon}_{rot}/deg$	Scanmatching [14] $\bar{\epsilon}_{rot}/deg$	RBPF [14] $\bar{\epsilon}_{rot}/deg$	GraphMapping [14] $\bar{\epsilon}_{rot}/deg$
ACES	0.55 ± 1.15	1.21 ± 1.61	1.2 ± 1.5	1.2 ± 1.3	0.4 ± 0.4
INTEL	1.03 ± 2.92	1.50 ± 3.07	1.7 ± 4.8	3.0 ± 5.3	1.3 ± 4.7
FR79	0.51 ± 0.74	1.49 ± 1.71	1.7 ± 2.1	0.6 ± 0.6	0.6 ± 0.6

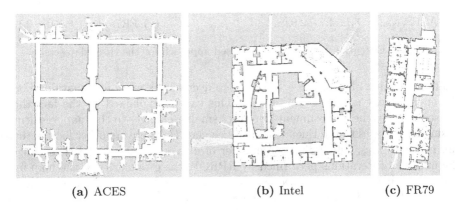

(a) ACES   (b) Intel   (c) FR79

Fig. 3. Constructed occupancy grid maps for the three datasets using the proposed approach. From visual inspection all maps all globally consistent.

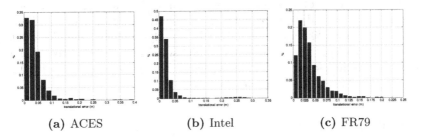

(a) ACES   (b) Intel   (c) FR79

Fig. 4. Distribution of the translational errors of the proposed approach for the three datasets

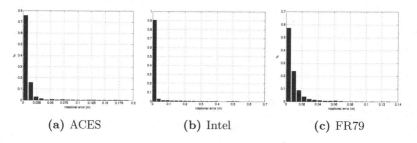

(a) ACES   **(b)** Intel   **(c)** FR79

Fig. 5. Distribution of the rotational errors of the proposed approach for the three datasets

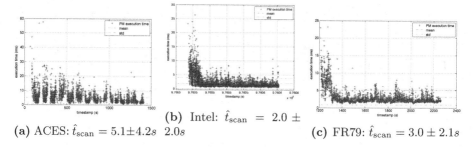

(a) ACES: $\hat{t}_{\text{scan}} = 5.1 \pm 4.2s$ **(b)** Intel: $\hat{t}_{\text{scan}} = 2.0 \pm 2.0s$ **(c)** FR79: $\hat{t}_{\text{scan}} = 3.0 \pm 2.1s$

Fig. 6. Execution times of the proposed approach for the three datasets.

scans. The value of $\bar{t}_{\text{scan}}$ has a tendency to decrease as more information about the environment is added to the map. This can be explained by the fact that revised areas of the environment introduce smaller changes to the map, resulting in smaller updates to the Euclidean distance map $\hat{\mathcal{D}}$. There are other factors that can influence $\bar{t}_{\text{scan}}$, including the number of measurements z_t^i and their range, the number of iterations performed by RPROP, the granularity of the map $\hat{m}$, and obviously the hardware used for computation. In the conducted experiments, a MacBook Pro with an Intel Core i7 2.8GHz and 4GB of RAM was used for the construction of the maps with $0.05m$ of resolution updated with measurements z_t with a maximum range of $40m$ after 30 optimizations of RPROP.

It should also be noticed that no pre-processing was applied to the used datasets, which was not true in the experiments of Kümmerle *et al.* [14] and Holz *et al.* [11]. Instead, to cope with the unreliable odometry of the datasets, shorter accumulated displacement, $0.01m$ for translation and $0.5rad$ for rotation, was used as threshold for update. As result, smaller errors had to be handle. For comparison, in the office dataset, an accumulated displacement of $1.0m$ for translation and $0.5rad$ for rotation was used.

6   Conclusion

In this paper an incremental map building algorithm, based on the scan-matching algorithm PM, was presented. Several modification were made to the original algorithm, including the use of a dynamic *likelihood field* capable of a fast adaptation to the changes of the map being builr, while maintaining the necessary properties for the localization part of SLAM.

In the experimental evaluation, it is shown that the obtained results are comparable with those of the extensively used GMapping, and others, to construct a suitable model of the perceived environment. Furthermore, the resulting execution times show that the method in question has run-time characteristics that makes it practical for concurrent operations with real-time restrictions.

Acknowledgments. This research is supported by: FEDER through the Operational Program Competitiveness Factors - COMPETE and by National Funds through FCT in the context of the project FCOMP-01-0124-FEDER-022682 (FCT reference PEst-C/EEI/UI0127/2011); the project Cloud Thinking (funded by the QREN Mais Centro program, ref. CENTRO-07-ST24-FEDER-002031); and PRODUTECH-PTI, QREN 13851, by National Funds through FCT/MCTES (PIDDAC) by FEDER through COMPETE – POFC.

References

1. Biber, P., Strasser, W.: The normal distributions transform: a new approach to laser scan matching. In: Proceedings of the 2003 IEEE/RSJ International Conference on Intelligent Robots and Systems (IROS 2003), vol. 3, pp. 2743–2748 (2003)
2. Censi, A.: An ICP variant using a point-to-line metric. In: 2011 IEEE International Conference on Robotics and Automation (ICRA), pp. 19–25. IEEE (2008)
3. Chekhlov, D., Pupilli, M., Mayol-Cuevas, W., Calway, A.: Real-time and robust monocular slam using predictive multi-resolution descriptors. In: Bebis, G., et al. (eds.) ISVC 2006, Part II. LNCS, vol. 4292, pp. 276–285. Springer, Heidelberg (2006), http://dx.doi.org/10.1007/11919629_29
4. Cunha, J., Pedrosa, E., Cruz, C., Neves, A.J., Lau, N.: Using a depth camera for indoor robot localization and navigation. In: RGB-D: Advanced Reasoning with Depth Cameras - RSS Workshop (2011), http://www.cs.washington.edu/ai/Mobile_Robotics/rgbd-workshop-2011
5. Dellaert, F., Fox, D., Burgard, W., Thrun, S.: Monte carlo localization for mobile robots. In: Proceedings of the 1999 IEEE International Conference on Robotics and Automation, vol. 2, pp. 1322–1328 (1999)
6. Elfes, A.: Occupancy grids: a probabilistic framework for robot perception and navigation. Ph.D. thesis, Carnegie Mellon University (1989)
7. Felzenszwalb, P., Huttenlocher, D.: Distance Transforms of Sampled Functions. Tech. rep., Cornell University (2004)
8. Gouveia, M., Moreira, A.P., Costa, P., Reis, L.P., Ferreira, M.: Robustness and precision analysis in map-matching based mobile robot self-localization. In: New Trends in Artificial Intelligence: 14th Portuguese Conference on Artificial Intelligence, pp. 243–253 (2009)
9. Grisetti, G., Stachniss, C., Burgard, W.: Improved Techniques for Grid Mapping With Rao-Blackwellized Particle Filters. IEEE Transactions on Robotics 23(1), 34–46 (2007)
10. Grisetti, G., Stachniss, C., Grzonka, S., Burgard, W.: A tree parameterization for efficiently computing maximum likelihood maps using gradient descent. In: Proc. of Robotics: Science and Systems, RSS (2007)

11. Holz, D., Behnke, S.: Sancta simplicitas - on the efficiency and achievable results of SLAM using ICP-based incremental registration. In: 2011 IEEE International Conference on Robotics and Automation (ICRA), pp. 1380–1387 (2010)
12. Kohlbrecher, S., von Stryk, O., Meyer, J., Klingauf, U.: A flexible and scalable SLAM system with full 3D motion estimation. In: 2011 IEEE International Symposium on Safety, Security, and Rescue Robotics (SSRR), pp. 155–160. IEEE (2011)
13. Kümmerle, R., Steder, B., Dornhege, C., Ruhnke, M., Grisetti, G., Stachniss, C., Kleine, A.: SLAM benchmarking (2009),
    http://kaspar.informatik.uni-freiburg.de/~slamEvaluation/datasets.php
14. Kümmerle, R., Steder, B., Dornhege, C., Ruhnke, M., Grisetti, G., Stachniss, C., Kleiner, A.: On measuring the accuracy of SLAM algorithms. Autonomous Robots 27(4), 387–407 (2009)
15. Lau, B., Sprunk, C., Burgard, W.: Improved updating of Euclidean distance maps and Voronoi diagrams. In: 2010 IEEE/RSJ International Conference on Intelligent Robots and Systems (IROS), pp. 281–286 (2010)
16. Lauer, M., Lange, S., Riedmiller, M.: Calculating the perfect match: An efficient and accurate approach for robot self-localization. In: Bredenfeld, A., Jacoff, A., Noda, I., Takahashi, Y. (eds.) RoboCup 2005. LNCS (LNAI), vol. 4020, pp. 142–153. Springer, Heidelberg (2006)
17. Leonard, J.J., Feder, H.J.S.: A Computationally Efficient Method for Large-Scale Concurrent Mapping and Localization. In: International Symposium of Robotics Research (2000)
18. Lu, F., Milios, E.: Robot Pose Estimation in Unknown Environments by Matching 2D Range Scans. Journal of Intelligent and Robotic Systems 18(3), 249–275 (1997)
19. Merke, A., Welker, S., Riedmiller, M.: Line Based Robot Localization under Natural Light Conditions. In: ECAI 2004 Workshop on Agents in Dynamic and Real Time Environments (2004)
20. Moravec, H.P.: Sensor Fusion in Certainty Grids for Mobile Robots. AI Magazine 9(2), 61 (1988)
21. Olson, E., Leonard, J., Teller, S.: Fast iterative alignment of pose graphs with poor initial estimates. In: 2011 IEEE International Conference on Robotics and Automation (ICRA), pp. 2262–2269. IEEE (2006)
22. Riedmiller, M., Braun, H.: A direct adaptive method for faster backpropagation learning: the RPROP algorithm. In: IEEE International Conference on Neural Networks, vol. 1, pp. 586–591 (1993)
23. Thrun, S.: A Probabilistic On-Line Mapping Algorithm for Teams of Mobile Robots. The International Journal of Robotics Research 20(5), 335–363 (2001)
24. Thrun, S.: Probabilistic Algorithms in Robotics. AI Magazine 21(4), 93 (2000)
25. Thrun, S., Liu, Y., Koller, D., Ng, A.Y., Durrant-Whyte, H.: Simultaneous Localization and Mapping with Sparse Extended Information Filters. The International Journal of Robotics Research 23(7-8), 693–716 (2004)
26. Zhang, Z.: Iterative point matching for registration of free-form curves and surfaces. Int. J. Comput. Vision 13(2), 119–152 (1994)

Towards Extraction of Topological Maps from 2D and 3D Occupancy Grids

Filipe Neves Santos, António Paulo Moreira, and Paulo Cerqueira Costa

INESC TEC (formerly INESC Porto) and Faculty of Engineering, University of Porto,
Campus da FEUP, Rua Dr. Roberto Frias, 378, Porto, Portugal
{fbnsantos,amoreira,paco}@fe.up.pt

Abstract. Cooperation with humans is a requirement for the next generation of robots so it is necessary to model how robots can sense, know, share and acquire knowledge from human interaction. Instead of traditional SLAM (Simultaneous Localization and Mapping) methods, which do not interpret sensor information other than at the geometric level, these capabilities require an environment map representation similar to the human representation. Topological maps are one option to translate these geometric maps into a more abstract representation of the the world and to make the robot knowledge closer to the human perception. In this paper is presented a novel approach to translate 3D grid map into a topological map. This approach was optimized to obtain similar results to those obtained when the task is performed by a human. Also, a novel feature of this approach is the augmentation of topological map with features such as walls and doors.

Keywords: SLAM, Semantic, topological maps.

1 Introduction

The well known Simultaneous Localization And Mapping (SLAM) problem for mobile robots was widely explored by the robotics community in the last two decades. Several approaches were proposed and explored. Nowadays there are valuable solutions based on these approaches that makes possible to the robots operate in crowed places and in buildings without special requirements. Rhino by [1], Robox by [2], Minerva by [3] and RoboVigil by [4] are some of those robots that can create a map and use it to localize and navigate through the environment. These robots rely on 2D or 3D accurate metric representations of the environment, derived from SLAM techniques.

The concept behind majority of SLAM approaches is collect natural or artificial features in the environment through the observations of the robot sensors. These features are tagged with a Cartesian position and its uncertainty. Recurring to special robot trajectories, Bayes theorem and Bayesian derivative techniques is possible to relate the position of each feature to the position of others features and reduce the uncertainty in that relative relative position. These features are useful to correct the estimated robot localization and also to build

L. Correia, L.P. Reis, and J. Cascalho (Eds.): EPIA 2013, LNAI 8154, pp. 307–318, 2013.

the grid map map. In [5], [6], [7] and [8] is possible to find different approaches to SLAM using the same concept. Correct feature extraction and association is the most hard issue.

Using these SLAM approaches is possible for the robot gets autonomously an accurate 2D or 3D grid map of the environment. However considering true that robots are moving from high tech factories to our homes, offices, public spaces and small factories, this will require that some of these robots should work and cooperate alongside us. In this scenario the information about cell occupation in the grid map is not enough to associate a human word to a place name or even to an object. Also this knowledge, in the form of a grid map, is not easily communicable. So taking this in mind, we have proposed an extension to SLAM called HySeLAM - HYbrid SEmantic Localization And Mapping system, described in [9]. This extension creates two new layers over this grid map, the topological layer and the semantic layer. The topological layer describes the places and places connectivity, detailed in section 3. The semantic layer relates Object/place in the 3D space. In the semantic layer, each object is an instance of a generic object described in Object dictionary.

The focus of this paper is to show an approach that translates a grid map obtained from SLAM into a topological map. Section 2 presents a global overview to the problem and different approaches to translate a grid map into a topological map. Section 3 presents the formal definition for our augmented topological map. Section 4 presents our approach to translate a gridmap into a topological map. Section 5 presents the obtained results. Section 6 presents the paper conclusions.

2   Build a Topological Map from a Grid Map

A grid map is a metric map ([10],[11]) that makes discretization of the environment into 2D or 3D cells. The grid map consists of empty cells, $m(x,y) = 0$, which represent free space and occupied cells, $m(x,y) = 1$, where obstacle exists. 2D grid mao are the most common but 3D grid map is growing in popularity due low cost of 3D acquisition system, as low cost RGB-D cameras and 3D laser range finder scan solutions. These 3D grid map are extremely expensive in terms of memory size so often these grid map are store in a form of octomaps.

A topological map ([12], [13]) describes the world using vertices and edges. The vertices represent the significant spaces in the environment, and the edges represent the connection between the different spaces. In this work, the edge should represent real or virtual door and the vertices should represent a room or a distinctive place.

Several researchers have tried to extract topological models from grid maps to solve global path planning and to help perform navigation and localization in local areas, [14] [15] [16] [17].

One way to get a topological map is extracting the Voronoi diagram from the grid map. A Voronoi diagram is a way of dividing space into a number of regions which are delimited by the Voronoi segments. The segments of the Voronoi diagram are all the points in the plane that are equidistant to the two nearest sites. The Voronoi vertices (vertices) are the points equidistant to three (or more) sites. In [18] we found an approach to extract the Voronoi diagram from a dynamic grid map, this approach was been optimized for online Voronoi diagram extraction.

Thrun in [14] extracts from the grid map the Voronoi diagram and critical points. Critical points are used to delimit regions and the Voronoi edges are used to relate the regions connectivity. These regions and edges are used to build the topological map.

In [19] and [20], graph partitioning methods are used to divide a grid map into several vertices. Buschka and Saffiotti in [21], Room-like spaces are extracted in grid maps by using fuzzy morphological opening and watershed segmentation. Even though those methods show successful topology extraction from grid maps, they are not easy to apply directly in home environments because they are suitable for corridor environments or considers only narrow passages to extract topological model.

In [15], Virtual door is defined as the candidates of real door, and the virtual doors are detected as edges of the topological map by extracting corner features from the occupancy grid-map; using this method, an initial topological map is generated, which consists of vertices and edges. The final topological map is generated using a genetic algorithm to merge the vertices and reduce the edges. As a result, the generated topological map consists of vertices divided by virtual doors and edges located in real doors. The proposed methods provide a topological map for the user interaction and the cleaning robot, and the topological map can be used to plan more efficient motion including room-to-room path planning and covering each room.

In [22] build the topological modeling using only low-cost sonar sensors. The proposed method constructs a topological model using sonar grid map by extracting subregions incrementally. A confidence for each occupied grid is evaluated to obtain reliable regions in a local grid map, and a convexity measure is used to extract subregions automatically.

3   Topological Map in HySeLAM

The augmented topological map, a layer of HySeLAM [9], defines a place by its delimitations (real or virtual walls), location, visual signature and by human words. The edges store the connectivity between places and are labeled with virtual or real doors. These doors are defined by their size, location and human words. This map will be part of the HySeLAM, managed by the topological engine. This is a map layer between grid map and semantic map. The semantic engine will relate each place from this map to a semantic label (human words). The topological engine will be able to buildup the map using the grid map and the descriptions received from other entities (other robots/people).

Therefore, the topological map $\mathcal{M}_t$ is defined by an attributed graph :

$$\mathcal{M}_t = (\mathcal{P}, \mathcal{C}) \tag{1}$$

where: $\mathcal{P}$ is the set of vertices (places) and $\mathcal{C}$ the edges ($\mathcal{C} \subseteq \mathcal{P} \times \mathcal{P}$). The places are augmented with five attributes: semantic words, geometric description, visual appearances, area and central position. A place is defined as:

$$p_i = \{\mathcal{SP}, \mathcal{W}, \mathcal{V}, A_r, X_c\} \tag{2}$$

Where: $\mathcal{SP}$ is a semantic set of words labeling the place, $\mathcal{W}$ defines the real and/or virtual delimitation of the place with a set of wall's ($\mathcal{W} = \{w_0, w_1, ..., w_{nw}\}$); $\mathcal{V}$ is the visual appearance described by a set of local panoramic views ($\mathcal{V} = \{v_0, v_1, ..., v_{nv}\}$); A_r is a real number which defines the place area; and X_c is the centric position ($X = [x, y, z]$) of the place.

The parameter wall w_i is defined by:

$$w_i = (X, \overrightarrow{V}, L, S, e) \tag{3}$$

Where: $X = ([x, y, z], \Sigma_X)$ defines the position of the center of mass of the wall and the uncertainty associated to this observed position, this position is related to the origin of SLAM referential frame; $\overrightarrow{V} = ([vx, vy, vz], \Sigma_V)$ contains the normal vector which defines the wall direction and the uncertainty associated to that vector; $L = ([vl, vh], \Sigma_V)$ defines the length (height and width) of the wall and the uncertainty associated to the length;$S = [sl, sh]$ defines the shape curvature of the wall; e defines the existence of the plan, (0 when the wall is virtual, and 1 when the wall is a brick wall).

The visual appearance parameter v_i is defined by:

$$v_i = (\mathcal{I}, X, La, t) \tag{4}$$

Where:$\mathcal{I}$ is a matrix containing the image, $X = ([x, y, z], \Sigma_X)$ is the center of image acquisition and the associated uncertainty described by a covariance matrix (Σ_X); La is the angular aperture ($L = [Lx, Ly]$); and t is time of acquisition.

The edges ($\mathcal{C} \subseteq \mathcal{P} \times \mathcal{P}$) are labeled with two attributes: semantic words and doorway set, as follow:

$$\{\mathcal{SD}, \mathcal{D}\} \tag{5}$$

Where: $\mathcal{SD}$ is a semantic set of words labeling the edge, and $\mathcal{D}$ is the doorway definition.

The parameter doorway $\mathcal{D}$ is defined by:

$$\mathcal{D} = (X, Ld, e, o, I) \tag{6}$$

Where: $X = ([x, y, z], \Sigma_X)$ defines the position of the center of mass of the door and the uncertainty associated to this observed position, this position is related to the origin of SLAM referential frame (Σ_X); Ld defines the length of the plan ($L = [vl, vh]$); e defines the existence of the door (0 when the door is virtual, and 1 when the door is real); o is the last door state observed (open or closed); I stores an visual appearance of the door.

4   Gr2To - 3D Grid Map to Topological Map Conversion

Gr2To Algorithm defines our approach to convert the grid map into the augmented topological map and it has been optimized to work in indoor robots. Gr2To is divided into five stages.

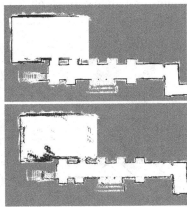

Fig. 1. At left, the 3D grid map obtained by the RobVigil [4]. The red points are occupied cells that are higher than 1.80 meters, the white are occupied cells bellow. This map was obtained in the first floor of the building I of Engineering Faculty From Porto University. At right top, the 2D grid map obtained using the 3D grid map and Gr2To compression stage. At right bottom, the 2D grid map obtained directly from a 2D SLAM approach with the laser range finder at 1.20 meter from the floor. Using a 3D grid map and Gr2To compression stage was possible to remove the furniture present in the environment and in the gridmap of 2D SLAM approach.

First stage) Compression from 3D grid map to 2Dgrid map and Filtering. This stage is considered in order to take the advantages when 3D grid map is available. One of those advantages is to make possible to filter the furniture and other objects/things from the grid-map. Furnitures or other kind of objects present in the environment are considered noise to our algorithm because they do not define the boundaries of a room/place. When the input is a 3D map, this map is compressed into a 2D map, figure 1. With this compression is possible to remove from the final map the furniture/noise present in the environment.

The compression algorithm compress all cells in Z-axis into a single cell, as follow:

$$map_{2D}(x,y) = \begin{cases} 0, & \text{if } NO > NO_{min} \text{ and } NF < NF_{max} \\ 1, & \text{if } NF > NF_{max} \\ 0.5, & \text{otherwise} \end{cases} \quad (7)$$

Where, NO is function that returns the number of occupied cells from $map_{3D}(x,y,z_{min})$ to $map_{3D}(x,y,z_{max})$, NF is function that returns the number of empty cells from $map_{3D}(x, y, z_{min})$ to $map_{3D}(x, y, z_{max})$. The compression algorithm requires six parameters, $(z_{min}, z_{max}, NF_{max}, NO_{min}, P_{Max}, P_{Min})$. Where, z_{min} and z_{max} defines the interval of 3D map to be compressed in a 2D map. NF_{max} and NO_{min} defines the maximum number of free cell and minimum number of occupied cells to consider a vertical occupied cell, P_{Max} defines the maximum probability for a cell to be consider empty cell and P_{Min} defines the minimum probability for a cell to be consider occupied cell.

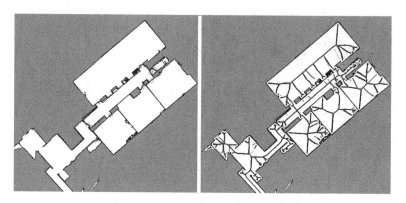

Fig. 2. In the left gridmap, the Gr2To has marked with red circles locations with higher probability for door existence. Blue lines represent the door in a closed state. In the right gridmap, the Gr2To draws the obtained Voronoi graph diagram over the gridmap. The gray squares represents the nodes.

Second stage) Door detection In this stage is created a map of distances, map_{dist} , from the filtered map. This map of distances contains in each cell the euclidean distance to the nearest occupied cell. After, door detection algorithm search for locations with probability for door existence, figure 2. This search have two parameters $Door_{min}$ and $Door_{max}$, which defines the minimum and maximum door size, in meters. These values are converted to pixel units, $Door_{minPixel} = \frac{Door_{min}}{2 \times Gridmap_{resolution}}$ and $Door_{maxPixel} = \frac{Door_{max}}{2 \times Gridmap_{resolution}}$. The search finds in the distance map for cells that satisfy the condition $Door_{minPixel} < map_{dist}(x, y) < Door_{maxPixel}$. For the cells that satisfy this condition, the algorithm takes eight samples from distance map, using this formula:

$$v(i) = map_{dis}\left(x + d \times cos\left(\frac{2\pi i}{8}\right), y + d \times sin\left(\frac{2\pi i}{8}\right)\right) \qquad (8)$$

Where, d is the distance parameter, in our tests this parameter takes the value of $Door_{minPixel}$, i is an integer number between 0 and 7. If exist two samples that satisfy the condition $v(i) > map_{dist}(x, y) \land v(j) > map_{dist}(x, y)$ with $2 <| i - j |< 6$, this place is consider with higher probability for door existence.

The closer cells with higher probability for door existence are used to estimate a central location. This central location is stored in vector of door locations. Closer cells are those with a distance under $Door_{minPixel}$ pixels.

Third stage) Voronoi Graph Diagram extraction In this stage is constructed the Voronoi Graph diagram from the distance map, left map of figure 2. The development of this algorithm was based on previous works, as [18].

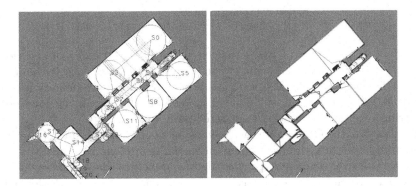

Fig. 3. In the left map, the Gr2To has marked with: red circles the main places (vertices), cyan circles the locations of the doors, and with black lines to show the connections between the main places. The yellow circles are the critical points. The green lines are the boundaries defined by the critical points. In the right map, the Gr2To has marked with random colors the delimitations of segmented places.

Fourth stage) Topological map construction and door validation To construct the topological map we have considered the definition of critical point from previous works of Thrun [14]. In this stage, the critical point is used to validate the doors found in the second stage. After this validation, the algorithm travels by the cells that belongs to the Voronoi diagram. In each one of this cell, the algorithm gets three parameters to define a circle, the circle location $\overrightarrow{r}_c = (x_c, y_c)$ which is the same as cell location and the circle radius $r_c = map_{dist}(x, y)$. With this circle definition the algorithm search in all stored circles if the condition $r_c(i) + r_c < \sqrt{(x_c - x_c(i))^2 + (y_c - y_c(i))^2}$ is satisfied. If this condition is not satisfied this circle is stored, if this condition is satisfied, the algorithm select the circle with bigger radius for the stored circles and drops the circle with smaller radius. After, the algorithms verify the next conditions for all circles:

$$\begin{cases} \sqrt{2}r_c(i) + \sqrt{2}r_c(j) < \sqrt{(x_c(j) - x_c(i))^2 + (y_c(j) - y_c(i))^2} \\ min(r_c(i), r_c(j)) > map_{dist}\left(\frac{x_c(j) - x_c(i)}{2}, \frac{y_c(j) - y_c(i)}{2}\right) \end{cases} \qquad (9)$$

if these two conditions are satisfied, the algorithm drops the circle with smaller radius. After, the algorithm find the circles connections using the the Voronoi Diagram Graph, the final result is in the figure 3.

Fifth stage) Optimization of space delimitation and topological map augmentation with real and virtual walls. In this stage the algorithm draws a closed door in all places that have a confirmed door. Then, it is associated one polygon of eight vertices to which stored circle. The eight vertices are placed over the circle edge and equally spaced. Each vertex of the polygon is updated to the farthest occupied cell. This update occurs inside of one beam, the beam origin is the circle center and it have an aperture of $\frac{2\pi}{8}$ radians and is aligned to the vertex. This update is limited to a maximum distance from the center circle, defined by:

$$dist = \frac{r_c(i)}{r_c(i) + r_c(j)} \sqrt{(x_c(j) - x_c(i))^2 + (y_c(j) - y_c(i))^2} \qquad (10)$$

Where, i is the index of the actual circle and j is the index of the closest circle that is inside of actual beam. Then, the algorithm test each polygon edge and if it lies over 80% of occupied cells is updated the place definition in the topological map with a real wall defined by the polygon edge. If the condition is not verified the place definition is updated with a virtual wall.

At this stage, the Gr2To approach over segments the most common corridors. This happens because the most common corridors are long rectangular spaces. To reduce this over segmentation, this algorithm merges three or more collinear places, that are directly connected by a Voronoy edge and which are defined by the circles with same radius.

At the final step, the Gr2To approach stores this augmented topological map into the robot memory and into a XML (eXtensible Markup Language) file. The vertices and door properties are related to the map referential defined by the SLAM approach.

5   Experimental Results

This approach Gr2To was tested using three grid-maps obtained in two real scenarios and in one virtual scenario. The output results were compared to a human segmentation. The three grid-maps, input of Gr2To, were printed and given to eleven persons. It was asked, to each person individually, to place a mark in each door, room and corridor of the map. The number of each item identified and the time to complete the task is shown in the table 1. Also, these three maps were processed by Gr2To in a computer with a processor Intel Pentium Dual Core T4300 at 2,16 GHz and with 2GB of memory.

The first grid-map was obtained from the robot RoboVigil using a 3D SLAM approach [4], in the ground floor of build I of Engineering Faculty of Porto University (FEUP). In figure 2 and 3 it is possible to see intermediate steps and final place segmentation done by Gr2To.

The second grid-map was obtained in TRACLabs facility, which is available in [23]. The third grid-map was obtained using the gazebo simulator with a turtlebot robot, Hector SLAM, and a virtual building. In figures 4 and 5 it is possible to see intermediate steps and final place segmentation done by Gr2To.

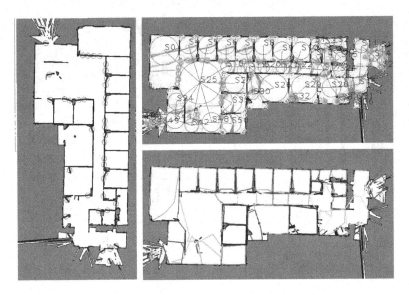

Fig. 4. Using the grid map of TRACLabs facility, available in [23]. In the left map, the Gr2To has marked with red circles locations with higher probability for door existence. In the right top map, the Gr2To draws red circles in the identified places, cyan circles in critical points with high probability for door existence and black lines the connectivity between vertex. In the right bottom map, the Gr2To draws places delimitation with random colors.

Table 1. Human segmentation Versus Gr2To approach segmentation

Grid Map	Segmentation by Human				Segmentation by Gr2To			
	Rooms	Corridors	Doors	t(s)	Rooms	Corridors	Doors	t(s)
FEUP	7 [5,8]	3 [2,4]	9 [5,10]	34 [26,47]	11	2 (7)	9 (2,2)	8.7
TRACLabs	17 [16,21]	2 [1,3]	18 [15,30]	43 [37,54]	29	1 (5)	39 (3,7)	8.4
Virtual	18 [17,21]	1 [1,2]	18 [18,20]	41 [33,53]	21	1 (9)	18 (0,0)	8.9

Table 1 summarizes the number of rooms, corridors and doors counted/identified by eleven persons and by Gr2To. Also, it shows the time taken to complete this task by humans and Gr2TO. In the side of human results, the three values in each cell are: the value of mean of the samples rounded to the nearest integer, and inside of brackets the minimum and maximum value of the samples. In the corridors row, of Gr2To results, there are two values, the first number is the number of corridors detected and inside of the brackets the number of vertices merged . In the doors column, of Gr2To results, there are three values: the first number is the number of doors detected, the second number is the number of missed doors, and the third number is the number of wrong doors detection.

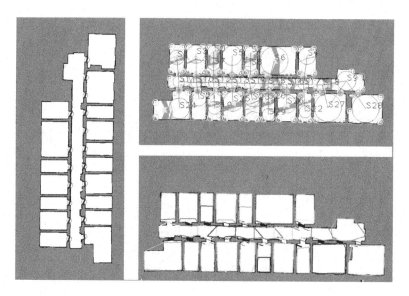

Fig. 5. The intermediate steps and final place segmentation done by Gr2To with the grid map obtained from the virtual scenario

From table 1 is possible to see that in the virtual scenario the algorithm have detected the same number of doors, rooms and corridor as humans, and it took less time to process. In the real scenarios, the Gr2To have failed to detect the doors without standard width. To make these doors detectable by Gr2To it was required to increase the value of $Door_{max}$ parameter, however it have also increased the number of outliers in the door detection. Although, these outliers can be removed using visual doors detectors. The number of doors detected by Gr2To in the TRACLabs is more higher when compared to the human average, this happens because humans have counted several pairs of doors as one door.

In the two real scenarios, the number of rooms detected by Gr2To is more higher then the number of rooms detected by the average human. One approach to reduce the number of segmented places and make it closer to the human is merge all vertices between a door and a terminal vertex or door. However, from RoboVigil experience we found that the over segmentation appears in long rectangular places. These long rectangular places are sometimes divided by humans in two distinctive places. So, if the Gr2To can fail to detect a door and these rectangular places are sometimes divided by humans, this vertices merge should happen only when the robot gets more information about the place. The augmented topological map in the HySeLAM framework is a dynamic map and it is updated during the human robot interaction (voice), so the robot should only infer that a set of vertices are the same place when the robot gets the same human word for all vertices.

6   Conclusion

The contribute made in this work was the development of a novel approach to translate a grid map into a topological map. This approach is able to compress a 3D grid map into a 2D grid map and filter objects present in the environment from the map. Another contribute of this work, was the optimization of Gr2TO in order to obtain similar results to those obtained when the task is performed by a human.

This work shows that is possible to translate the robot knowledge, stored in the form of occupation grid map, to closer the human perception. Indeed, the RoboVigil with Gr2To was able to segment the obtained gridmap and ask in each place for the human word that tags the place. Gr2To simplifies the human-robot interaction and it makes possible to the robot understand simple missions, as " robot go to Sam office and then go to robot charger place".

Acknowledgments. This work is financed by the ERDF European Regional Development Fund through the COMPETE Programme (operational programme for competitiveness) and by National Funds through the FCT Fundação para a Ciência e a Tecnologia (Portuguese Foundation for Science and Technology) within project FCOMP - 01-0124-FEDER-022701.

References

1. Burgard, W., Cremers, A.B., Fox, D., Hähnel, D., Lakemeyer, G., Schulz, D., Steiner, W., Thrun, S.: Experiences with an interactive museum tour-guide robot. Artificial Intelligence 114(1-2), 3–55 (1999)
2. Siegwart, R., Arras, K.O., Bouabdallah, S., Burnier, D., Froidevaux, G., Greppin, X., Jensen, B., Lorotte, A., Mayor, L., Meisser, M., Philippsen, R., Piguet, R., Ramel, G., Terrien, G., Tomatis, N.: Robox at Expo.02: A large-scale installation of personal robots. Robotics and Autonomous Systems 42(3-4), 203–222 (2003)
3. Thrun, S., Beetz, M., Bennewitz, M., Burgard, W., Cremers, A.B., Dellaert, F., Fox, D., Rosenberg, C., Roy, N., Schulte, J., Schulz, D.: Probabilistic Algorithms and the Interactive Museum Tour-Guide Robot Minerva. Journal of Robotics Research, 972–999 (2000)
4. Pinto, M., Moreira, A.P., Matos, A., Santos, F.: Fast 3D Matching Localisation Algorithm. Journal of Automation and Control Engineering 1(2), 110–115 (2013) ISSN:2301-3702
5. Bailey, T., Durrant-Whyte, H.: Simultaneous localization and mapping (SLAM): part II. IEEE Robotics & Automation Magazine 13(3), 108–117 (2006)
6. Thorpe, C.: Simultaneous localization and mapping with detection and tracking of moving objects. In: Proceedings of the 2002 IEEE International Conference on Robotics and Automation (Cat. No.02CH37292), pp. 2918–2924. IEEE (2002)
7. Kohlbrecher, S., Meyer, J., von Stryk, O., Klingauf, U.: A Flexible and Scalable SLAM System with Full 3D Motion Estimation. In: Proc. IEEE International Symposium on Safety, Security and Rescue Robotics (SSRR). IEEE (2011)
8. Montemerlo, M., Thrun, S.: The FastSLAM Algortihm for Simultaneous Localization and Mapping. Springer Tracts in Advanced Robotics (2007)

9. Santos, F.: HySeLAM - Hybrid Semantic Localization and Mapping (2012), http://hyselam.fbnsantos.com (accessed: May 20, 2013)
10. Elfes, A.: Occupancy grids: a probabilistic framework for robot perception and navigation. Phd, Carnegie Mellon University (1989)
11. Chatila, R., Laumond, J.: Position referencing and consistent world modeling for mobile robots. In: Proceedings of the 1985 IEEE International Conference on Robotics and Automation, vol. 2, pp. 138–145. Institute of Electrical and Electronics Engineers (1985)
12. Mataric, M.J.: A Distributed Model for Mobile Robot Environment-Learning and Navigation, Msc. MIT (1990)
13. Kuipers, B., Byuń, Y.T.: A Robot Exploration and Mapping Strategy Based on a Semantic Hierarchy of Spatial Representations. Journal of Robotics and Autonomous Systems 8, 47–63 (1991)
14. Thrun, S.: Learning metric-topological maps for indoor mobile robot navigation. Artificial Intelligence 99(1), 21–71 (1998)
15. Joo, K., Lee, T.K., Baek, S., Oh, S.Y.: Generating topological map from occupancy grid-map using virtual door detection. In: IEEE Congress on Evolutionary Computation, pp. 1–6. IEEE (July 2010)
16. Fabrizi, E., Saffiotti, A.: Extracting topology-based maps from gridmaps. In: Proceedings of the 2000 IEEE International Conference on Robotics and Automation, ICRA. Millennium Conference Symposia Proceedings (Cat. No.00CH37065), vol. 3, pp. 2972–2978. IEEE (2000)
17. Myung, H., Jeon, H.-M., Jeong, W.-Y., Bang, S.-W.: Virtual door-based coverage path planning for mobile robot. In: Kim, J.-H., et al. (eds.) FIRA 2009. LNCS, vol. 5744, pp. 197–207. Springer, Heidelberg (2009)
18. Lau, B., Sprunk, C., Burgard, W.: Improved Updating of Euclidean Distance Maps and Voronoi Diagrams. In: IEEE International Conference on Intelligent RObots and Systems (IROS), Taipei, Taiwan (2010)
19. Brunskill, E., Kollar, T., Roy, N.: Topological Mapping Using Spec-tral Clustering and Classification. In: Proc. of IEEE/RSJ International Conference on Intelligent Robots and Systems, pp. 3491–3496 (2007)
20. Zivkovic, Z., Bakker, B., Krose, B.: Hierarchical Map Building and Planning based on Graph Partitioning. In: Proc. of IEEE International Conference on Robotics and Automation, pp. 803–809 (2006)
21. Buschka, P., Saffiotti, A.: A Virtual Sensor for Room Detection. In: Proc. of IEEE/RSJ International Conference on Intelligent Robots and Systems, pp. 637–642 (2002)
22. Choi, J., Choi, M., Chung, W.K.: Incremental topological modeling using sonar gridmap in home environment. In: 2009 IEEE/RSJ International Conference on Intelligent Robots and Systems, pp. 3582–3587. IEEE (October 2009)
23. Kortenkamp, D.: TRACLabs in new facility (2012), http://traclabs.com/2011/03/traclabs-in-new-facility/ (accessed: May 20, 2013)

Trajectory Planning and Stabilization
for Formations Acting in Dynamic Environments

Martin Saska, Vojtěch Spurný, and Libor Přeučil

Department of Cybernetics, Faculty of Electrical Engineering, Czech Technical
University in Prague, Technická 2, 166 27 Prague 6, Czech Republic
{saska,preucil}@labe.felk.cvut.cz, spurnvoj@fel.cvut.cz
http://imr.felk.cvut.cz/

Abstract. A formation driving mechanism suited for utilization of multi-
robot teams in highly dynamic environments is proposed in this paper.
The presented approach enables to integrate a prediction of behaviour of
moving objects in robots' workspace into a formation stabilization and
navigation framework. It will be shown that such an inclusion of a model
of the surrounding environment directly into the formation control mechanisms facilitates avoidance manoeuvres in a case of fast dynamic objects
approaching in a collision course. Besides, the proposed model predictive
control based approach enables to stabilize robots in a compact formation
and it provides a failure tolerance mechanism with an inter collision avoidance. The abilities of the algorithm are verified via numerous simulations
and hardware experiments with the main focus on evaluation of performance of the algorithm with different sensing capabilities of the robotic
system.

Keywords: Obstacle avoidance, formation control, mobile robots, trajectory planning, model predictive control.

1 Introduction

Increasing field of applications of multi robot systems with their advanced abilities in faster acting and motion requires to design new principles of formation
driving in such high dynamic missions. This paper is focussed on the coordination of groups of autonomous vehicles closely cooperating together in a dynamic
environment, in which moving extraneous obstacles, but also team members
themselves, have to be efficiently avoided with minimal effect on the running
mission. The novelty of the formation driving approach lies in the possibility of
inclusion of movement prediction of objects in the robots' proximity into the
control and trajectory planning methods. It will be shown that even a rough
estimation of obstacles movement significantly improve performance of the formation stabilization and its navigation into a desired target region.

To be able to contextualize the contribution of the proposed method let us
briefly summarize recent achievements in the formation driving community. The
formation driving methods may be divided into the three approaches: Virtual

L. Correia, L.P. Reis, and J. Cascalho (Eds.): EPIA 2013, LNAI 8154, pp. 319–330, 2013.

structures [1–3], behavioral [4–6], and leader-follower methods [7–9]. Our approach is based on the leader-follower method which is often used in applications of nonholonomic mobile robots. In leader follower approach, a robot or even several robots are designated as leaders, while the others are following them. We rely on an adaptation of the classical leader-follower method, where the leading entity is a virtual object, which increases robustness of the method.

The formation driving methods are aimed mainly at tasks of a formation stabilization in desired shapes [10, 4, 11] and a formation following predefined paths [12, 1, 13]. In these methods, it is supposed that the desired trajectory is designed by a human operator or by a standard path planning method modified for the formation requirements. However, there is a lack of methods providing not only a local control but including a global trajectory planning with the dynamic avoidance ability, which is our target problem.

Our method enables to include the prediction of obstacles movement into the formation planning using a Model Predictive Control (MPC) (also known as a receding horizon control). MPC is an optimization based control approach suitable for stabilizing nonlinear dynamic systems (see a survey of nonlinear MPC methods in [14] and references therein). This technique is frequently used for the control of nonholonomic formations due to its possibility to achieve the desired system performance and to handle the system constraints. The MPC approaches are again employed mainly for the tasks of trajectory tracking and formation stabilization in the desired shape [15], [16], [17], [8].

We apply the MPC approach for the followers stabilization in the desired positions behind the leaders as e.g. in [8] or [17], but also for the virtual leaders trajectory planning to a desired goal area. Our concept of MPC combines the trajectory planning to the desired goal region and the immediate control of the formation into one optimization process. We propose an extension of the MPC control horizon with an additional planning horizon, which respects information on global structure of the workspace and effectively navigates the formation into the target region. It enables to integrate the obstacle movement prediction into the trajectory planning and to respond to changing environment. Besides, we provide a study of tolerance of the system performance to the precision of moving object perception and consequently prediction of their movement.

The presented paper extends the theory presented in our previous publication [18]. Here, we provide a methodology for integration of a prediction of obstacles movement into the formation stabilization as the main contribution. Besides, we present a hardware experiment showing practical applicability of the proposed method and additional simulations and statistical comparisons.

2   Problem Statement and Notation

In this paper, the problem of a autonomous formation of n_r nonholonomic mobile robots reaching a target region in a workspace with ν dynamic obstacles is tackled. The formation is led by a virtual leader (denoted by symbol L), which is positioned in front of the formation. The robots in the formation are equipped with a system of a precise relative localization (e.g. [19]), while the rough estimation of the formation position relatively to the target region is achieved by any available global positioning system. The robotic team is equipped by sensors for obstacle detection and their velocity estimation. In the presented experiments, a system based on analysing of differences in consequent range-finder scans using SyRoTek robots [20] is employed. Finally, we assume that the robots are able to communicate to share the information on obstacles movement and to distribute plan of the virtual leader within the team.

In the paper, configurations of robots (followers and the virtual leader) at time t is denoted as $\psi_j(t) = \{x_j(t), y_j(t), \theta_j(t)\}$, with $j \in \{1, \ldots, n_r, VL\}$. The Cartesian coordinates $(x_j(t), y_j(t))$ define the position $\bar{p}_j(t)$ and $\theta_j(t)$ denotes heading. For each robot, a circular detection boundary with radius $r_{s,j}$ and a circular avoidance boundary with radius $r_{a,j}$ are defined ($r_{s,j} > r_{a,j}$). Obstacles detected outside the region with radius $r_{s,j}$ are not considered in the avoidance function and distance to obstacles less than $r_{a,j}$ is considered as inadmissable. For the virtual leader, the radius $r_{a,VL}$ has to be bigger than radius of the formation to ensure that the obtained leader's plan is collision free also for all followers.

2.1 Kinematic Constraints of the Formation

The proposed formation driving approach is based on a model predictive technique, where complete trajectories for the virtual leader and all followers are computed to respect a given model of the robots. We use the following simple nonholonomic kinematic model for all robots: $\dot{x}_j(t) = v_j(t)\cos\theta_j(t)$, $\dot{y}_j(t) = v_j(t)\sin\theta_j(t)$, $\dot{\theta}_j(t) = K_j(t)v_j(t)$ In the model, the velocity $v_j(t)$ and curvature $K_j(t)$ represent control inputs $\bar{u}_j(t) = (v_j(t), K_j(t)) \in \mathbb{R}^2$. For the prediction of obstacles movement, it is assumed that the system is able to detect i) actual position of obstacles, ii) position and its actual speed, or iii) position, speed and actual curvature of its movement trajectory with a given precision.

For the virtual leader's trajectory planning, let us define a time interval $[t_0, t_{N+M}]$ containing a finite sequence with $N+M+1$ elements of nondecreasing times. The constant N corresponds to length of the short MPC control horizon and M corresponds to length of the additional planning horizon (details are given in Section 3). Let us assume that in each time sub-interval $[t_k, t_{k+1})$ with length $t_{k+1} - t_k$ (not necessarily uniform) the control inputs are held constant. For simplification of the notation, let us refer to t_k with its index k. By integrating the kinematic model over the interval $[t_0, t_{N+M}]$, the model presented in [18] for the *transition points* at which control inputs change is derived.

For the virtual leader's planning the control limits are restricted by capabilities of all followers depending on their position within the formation. The proposed method relies on a leader-follower method [21, 22], where followers track the leader's trajectory which is distributed within the group. The followers are maintained in a relative distance to the leader in curvilinear coordinates with two axes p and q, where p traces the leader's trajectory and q is perpendicular to p.

Fig. 1. Scheme of the integration of the obstacle prediction into the planning

2.2   Model Predictive Control

As mentioned in the introduction, the ability of the dynamic obstacle avoidance is enabled due to the fast replanning under the RHC scheme. The main idea of the receding horizon control is to solve a finite horizon optimal control problem for a system starting from a current configuration over a time interval $[t_0, t_f]$ under a set of constraints on the system states and control inputs. The length $t_f - t_0$ is known as the control horizon. After a solution from the optimization problem is obtained on a control horizon, In each planning step, a portion of the computed control actions obtained as a result of the optimization problem is applied on the interval $[t_0, \Delta tn + t_0]$, known as the receding step. The number of applied constant control inputs n is chosen according to computational demands. This process is then repeated on the interval $[t_0 + \Delta tn, t_f + \Delta tn]$ as the control horizon moves by the receding steps.

3   Method Description

As mentioned in [18], we have extended the standard RHC method with one control horizon into an approach with two finite time intervals T_N and T_M. This enables us to integrate the information on prediction of obstacle movement into the formation driving here. The first shorter time interval provides immediate control inputs for the formation regarding the local environment. The difference $\Delta t(k + 1) = t_{k+1} - t_k$ is kept constant (later denoted only Δt) in this time interval. The second time interval enable to integrate information on dynamics of the workspace between the end of the first control interval and the desired target region. In this interval, values $\Delta t(k + 1) = t_{k+1} - t_k$ are varying. The complete trajectory from the actual position of the virtual leader until the target region can be described by the optimization vector Υ_{VL}. The vector consists of transition points (states) of both intervals and control inputs applied in-between them. Besides, the values $\Delta t(N + 1), \ldots, \Delta t(N + M)$ are part of the vector Υ_{VL}.

As mentioned, the dynamic obstacle avoidance is enabled due to the additional planning horizon, which acts as a trajectory planning for the virtual leader of the formation. The trajectory planning with the prediction of objects movement is transformed to the minimization of a single cost function $\lambda_{VL}(\Upsilon_{VL})$ subject to set of equality constraints and inequality constraints. The cost function $\lambda_{VL}(\Upsilon_{VL})$ consists of two parts, whose influence is adjusted by a constant α as

$$\lambda_{VL}(\Upsilon_{VL}) = \lambda_{time}(\Upsilon_{VL}) + \alpha \sum_{j=1}^{\nu} \lambda_{obst}(\Upsilon_{VL}, \Upsilon_{obst,j}). \tag{1}$$

The first part of the multi-objective optimization is important for the trajectory planning and it forces the formation to reach the goal as soon as possible:

$$\lambda_{time}(\Upsilon_i) = \sum_{k=N+1}^{N+M} \Delta t(k). \tag{2}$$

The second objective of the planning process is to integrate the prediction of the obstacles movement. Therefore, the second term penalises solutions, which could cause a collision with a detected obstacle in future. The term is a sum of modified avoidance functions applied for each known object in the environment. The avoidance function contributes to the final cost if an object is closer to the trajectory of the i-th robot than $r_{s,i}$ and it will approach infinity if distance $r_{a,i}$ to the obstacle is reached:

$$\lambda_{obst}(\Upsilon_i, \Upsilon_{obst,j}) = (\min\{0, (dist(\Upsilon_i, \Upsilon_{obst,j}) - r_{s,i})/(dist(\Upsilon_i, \Upsilon_{obst,j}) - r_{a,i})\})^2. \tag{3}$$

Function $dist(\Upsilon_i, \Upsilon_{obst,j})$ provides the shortest Euclidean distance between pairs of positions on the predicted trajectory of obstacle j and the trajectory of i-th robot that should be reached at the same time (see Fig. 1 for demonstration).

The equality constraints of the optimization guarantee that the obtained trajectory stays feasible with respect of kinematics of nonholonomic robots. The equality is satisfied if $\psi_j(k+1)$ is obtained by substituting the vectors $\psi_j(k)$ and $\bar{u}_j(k+1)$ into the discretized kinematic model for all $k \in \{0, \dots, N-1\}$. The inequality constraints characterize bounds on the control constraints velocity and curvature of the virtual leader. Finally, there is a stability constraint, which is broken if the last state $\psi_{VL}(N+M)$ of the plan is outside the target region. The constraint is growing with distance from the border of this region.

The trajectory obtained solving the optimization problem $\min \lambda_{VL}(\Upsilon_{VL})$ for the virtual leader is used in the leader-follower scheme as an input of the trajectory tracking for the followers. For each of the followers, the solution is transformed using the leader-follower approach (discussed in details in [21, 22]) to obtain a sequence $\psi_{d,i}(k) = (\bar{p}_{d,i}(k), \theta_{d,i}(k))$, $k \in \{1, \dots, N\}$, $i \in \{1, \dots, n_r\}$, of desired states behind the virtual leader. To enable a response to dynamic events in the environment behind the actual position of the leader and to failures of neighbors in the formation, the mechanism designed to track these states is also enhanced by obstacle avoidance functions with integrated prediction of obstacles movement.

The discrete-time trajectory tracking is again transformed to minimization of cost function $\lambda_i(\Upsilon_i)$ subject to equality constraints and inequality constraints with the same meaning as the constraints applied for the virtual leaders's planning. The optimization vector Υ_i consists of transition points and control inputs applied on the first shorter time interval T_N for each follower i. The cost function consists of two segments adjusted by a constant β as

$$\lambda_i(\Upsilon_i) = \lambda_{diff}(\Upsilon_i) + \beta \left(\sum_{j=1}^{\nu} \lambda_{obst}(\Upsilon_i, \Upsilon_{obst,j}) + \sum_{j \in \bar{n}_n} \lambda_{obst}(\Upsilon_i, \Upsilon_j^\star) \right). \quad (4)$$

The first part of the cost function penalizes solutions with states deviated from the desired positions $\bar{p}_{d,i}(k)$, $\forall k \in \{1, \ldots, N\}$, as follows:

$$\lambda_{diff}(\Upsilon_i) = \sum_{k=1}^{N} \|(\bar{p}_{d,i}(k) - \bar{p}_i(k))\|^2. \quad (5)$$

The obstacle avoidance abilities are integrated in the second part of the cost function. The first sum penalizes solutions that could lead to a collision with dynamic or lately detected obstacles that could not be sufficiently avoided by the virtual leaders planning. The second sum of avoidance functions is crucial for the formation stabilization and failure tolerance. In this term, the other members of the team are considered also as dynamic obstacles with possibility of prediction of their movement. This part enables to stabilize the formation mainly in case of an unexpected behaviour of defective neighbours or in a formation shrinking enforced by obstacles (see Fig. 4 for examples). The vector $\Upsilon_j^\star$ represents actually exercised plan of j-th neighbour for $j \in \bar{n}_n$, where $\bar{n}_n = \{1, \ldots, i-1, i+1, \ldots, n_r\}$. In case of a j-th robot failure, when its plan does not respect its real motion, a prediction of its motion have to rely on on-board sensors of other followers, similarly as it is realized with "external" obstacles.

Table 1. Statistics showing importance of the obstacle movement prediction included in the formation driving scheme (300 runs of the algorithm for each variant were used). Variant 1 does not consider movement of the obstacle. Variant 2 considers only the actual velocity of the obstacle. Variant 3 enables to predict position of obstacles in future based on its velocity and estimated curvature of its movement.

Different sensing capabilities:	variant 1	variant 2	variant 3
Number of collisions:	200	14	0
Mean time to the target:	12.65	12.53	11.87
Mean minimum distance to the obstacle:	0.24	0.66	0.98

4 Discussion on the Algorithm Performance and Experimental Results

The aim of this section is to verify the ability of the proposed method to incorporate prediction of the obstacle movement into the formation driving and to demonstrate a failure tolerance, a response of the followers to changing environment and a possibility to autonomously shrink the formation in a narrow corridor. In the first experiment, a triangular formation with three robots is approaching a target region (Fig. 2(a)). In the scenario, two obstacles (one static and one dynamic) are positioned between the initial position of the formation and the target. A set of 300 situations with position of the formation randomly initialized in the square depicted in Fig. 2(a) is used. The dynamic obstacle is always moving along a circle with a course, which is colliding with the straight direction from the initial position of the formation to the target.

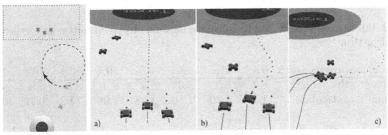

(a) Experiment (b) Variant 1 of the algorithm setting - no movement
setup.   prediction included.

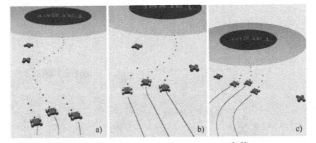

(c) Variant 3 of the algorithm setting - full movement
prediction included.

Fig. 2. Experiment with different knowledge on the obstacle movement

Three different sensing capabilities in utilization of the proposed algorithm are compared in table 1. In the first basic variant, the dynamic obstacle is in each planning step considered as a static obstacle. Any prediction of the obstacle position in future is not possible and the avoidance ability relies only on the replanning as the new position of the obstacle is detected in each planning step (the variant 1 is equivalent to the algorithm presented in [18]). In the variant 2, the system is able to estimate actual velocity (speed and movement direction) of the obstacle, but the curvature of the movement trajectory is unknown. In the obstacle movement prediction, it is assumed that the obstacle follows a straight line. In the short term receding horizon, such a simplification is relatively precise, which is beneficial for the collision avoidance (see the drop of collisions and increase of the mean minimum distance to the obstacle between the variants

1 and 2 in table 1). In the long term planning horizon, the difference between the predicted position of the obstacle and the real position in future may be already significant. This may prolong the time needed for reaching the target region in comparison with system being able to estimate the trajectory of the obstacle more precisely (variant 3 in table 1). In the variant 3, the motion model of the obstacle is known completely. Not only the actual velocity but also the curvature of the obstacle's trajectory can be estimated by formation's sensors. Therefore, the plan of the formation towards the target region is changed only slightly during the motion and the team is approaching the target in the most efficient way already from the beginning of the planning.

Table 2. Influence of precision of the obstacle movement perception on the performance of the algorithm

Error of curvature measure:	0%	2%	5%	10%	20%	50%	70%	100%
Mean time to the target:	8.3	8.3	8.3	8.3	8.3	8.5	8.9	9.7
Mean min. dist. to obst.:	1.11	1.10	1.10	1.08	1.08	1.02	0.79	0.38

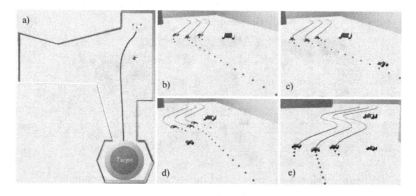

Fig. 3. Avoidance of a dynamic obstacle and two detected static obstacles

An example from the set of 300 situations used in table 1 is shown in Fig. 2(b) (the variant 1) and Fig. 2(c) (the variant 3). In Fig. 2(b), the obstacle is moving in front of the formation and in each planning step the proposed algorithm responds to the new detected position of the obstacle. Since, the system cannot predict the obstacle movement, its influence is "pushing" the plan of the formation to the side, which leads even into a collision between followers as shown in the last snapshot in Fig. 2(b). The same situation is solved by the algorithm with the obstacle movement prediction included in Fig. 2(c). The collision free trajectory that does not have to be changed during the whole movement of the formation is found already at the beginning of the mission. In the first snapshot, the plan is heading towards the obstacle, which seemingly looks as an illogical manoeuvre. In the second snapshot, it is already clear that the plan goes through the space, which is freed due to the obstacle's movement.

In multi-robot applications, it is difficult to precisely determine the movement model of obstacles. Mainly an estimation of parameters of the predicted trajectory requires a long term measurement and it is often biased. Therefore, we have studied influence of imprecisely determined curvature of obstacle movement in table 2. Similarly as in the previous experiment, a set of 100 situations with randomly initialized position of the formation is evaluated by the proposed method. Again, the dynamic obstacle is following a circle with radius equal to 1m. In this experiment, an uncertainty in measurement of curvature of the obstacle movement is introduced. The error in determination of the curvature is in range of 0%, which corresponds with the variant 3 in table 1, and 100%, which is equivalent to the variant 1 (no information about the curvature is provided). The statistic in table 2 shows, that the algorithm works well even with imprecisely determined parameters of the obstacle movement. Even, if the error in measurement of the curvature is 50%, the algorithm is able to solve all the situations without any collision and with only a slight increase of the mean total time to the target region and slight decrease of mean minimal distance to the obstacle.

A complex situation with a static known obstacle, detected dynamic obstacle and an unknown later detected obstacle is presented in Fig. 3. In the initial plan (snapshot a)), only the first static obstacle is included into the plan. The dynamic obstacle is detected in snapshot c), which results in the deviation of the plan in comparison with the snapshot b). The plan is surprisingly deviated

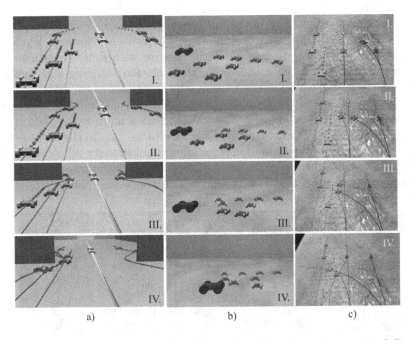

Fig. 4. Simulations of a formation shrinking enforced by the environment. a) Passage through a narrow corridor. b) Dynamic obstacle avoidance. c) Simulation of a follower's failure.

towards the obstacle, since it is predicted that the obstacle is moving across the leader's trajectory and it is more efficient to avoid it from the left as one can see in snapshot d). The obstacle avoidance is successfully proceeded even if the error in measurement of the curvature is 50% and the obstacle is randomly changing its movement direction, which verifies utilization of the algorithm in real applications, where the movement of obstacles can be chaotic. In snapshot e), a replanning of the virtual leader's trajectory due to a detected static obstacle is shown.

Examples of temporarily shrinking of the formation as an emergent result of the formation driving through a narrow corridor, in case of an obstacle avoidance realized by followers and due to a failure of a follower are shown in Fig. 4. In the snapshots in Fig. 4 a), the followers are temporarily deviating from their desired positions within the formation to be able to go through the doorway. The obtained solutions of the followers' trajectory planning is a compromise between the effort to follow the desired trajectory (the deviating from the trajectory is penalised in the second term of equation (4)) and to avoid the obstacles (the proximity of obstacles is penalized in the third term of equation (4)). The sequence of snapshots presented in Fig. 4 b) shows an avoidance manoeuvre around an obstacle that was static until the first snapshot in Fig. 4 b). Once the virtual leader is passing by the obstacle, the primarily static obstacle starts its movement towards the formation. This unforeseen behaviour could not be included into the plan of the leader and therefore the followers' trajectory planning has to react at the price of temporarily changing the desired shape of the formation. A similar process is described in the last sequence of snapshots in Fig. 4, where a failure of one of the followers is simulated. The broken follower is leaving its position within the formation in a collision course with other team members. They are able to avoid the collisions by their motion planning with included prediction of the broken follower's motion (the third part of equation (4)).

Finally, the ability of the formation to smoothly avoid dynamic obstacles is verified in Fig. 5. This experiment was realized with a multi-robot platform Sy-RoTek, which consists of a fleet of autonomous vehicles [20]. In the first snapshot (Fig. 5 a)), the plan of the formation is avoiding the stationary yellow robot from the right side. Once the formation starts its movement, also the yellow robot accelerates in a direction that is orthogonal with the actual plan of the group.

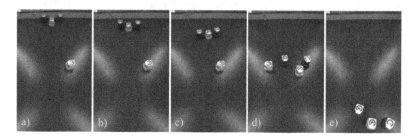

Fig. 5. Hardware experiment showing performance of the algorithm with prediction of the movement of the dynamic obstacle

The obstacle motion is detected by the robots and the plan of the formation is changed. With the new information about the dynamic obstacle, it is optimal to pass by the robot from the left as is visualised in the following snapshots.

5 Conclusions

In this paper a novel formation driving approach designed for dynamic environments is presented. The proposed method enables to incorporate a prediction of positions of moving objects into the formation control and trajectory planning. In numerous simulations and hardware experiment, it was verified that such a motion prediction based on data from on-board sensors improves performance of the formation stabilization and decreases probability of collisions. Besides, it was shown that the motion prediction included in to the model predictive control is useful even in case of not precisely determined model of obstacles movement.

Acknowledgments. The work of M. Saska was supported by GAČR under his postdoc grant no. P10312/P756, L. Přeučil was supported by MŠMT project Kontakt II no. LH11053.

References

1. Ghommam, J., Mehrjerdi, H., Saad, M., Mnif, F.: Formation path following control of unicycle-type mobile robots. Robotics and Autonomous Systems 58(5), 727–736 (2010)
2. Ren, W.: Decentralization of virtual structures in formation control of multiple vehicle systems via consensus strategies. European Journal of Control 14, 93–103 (2008)
3. Beard, R., Lawton, J., Hadaegh, F.: A coordination architecture for spacecraft formation control. IEEE Transactions on Control Systems Technology 9(6), 777–790 (2001)
4. Hengster-Movrić, K., Bogdan, S., Draganjac, I.: Multi-agent formation control based on bell-shaped potential functions. Journal of Intelligent and Robotic Systems 58(2) (2010)
5. Langer, D., Rosenblatt, J., Hebert, M.: A behavior-based system for off-road navigation. IEEE Transactions on Robotics and Automation 10(6), 776–783 (1994)
6. Lawton, J., Beard, R., Young, B.: A decentralized approach to formation maneuvers. IEEE Transactions on Robotics and Automation 19(6), 933–941 (2003)
7. Min, H.J., Papanikolopoulos, N.: Robot formations using a single camera and entropy-based segmentation. Journal of Intelligent and Robotic Systems (1), 1–21 (2012)
8. Chen, J., Sun, D., Yang, J., Chen, H.: Leader-follower formation control of multiple non-holonomic mobile robots incorporating a receding-horizon scheme. Int. Journal Robotic Research 29, 727–747 (2010)
9. Das, A., Fierro, R., Kumar, V., Ostrowski, J., Spletzer, J., Taylor, C.: A vision-based formation control framework. IEEE Transactions on Robotics and Automation 18(5), 813–825 (2003)
10. Dong, W.: Robust formation control of multiple wheeled mobile robots. Journal of Intelligent and Robotic Systems 62(3-4), 547–565 (2011)

11. Abdessameud, A., Tayebi, A.: Formation control of vtol unmanned aerial vehicles with communication delays. Automatica 47(11), 2383–2394 (2011)
12. Do, K.D., Lau, M.W.: Practical formation control of multiple unicycle-type mobile robots with limited sensing ranges. Journal of Intelligent and Robotic Systems 64(2), 245–275 (2011)
13. Xiao, F., Wang, L., Chen, J., Gao, J.: Finite-time formation control for multi-agent systems. Automatica 45(11), 2605–2611 (2009)
14. Boscariol, P., Gasparetto, A., Zanotto, V.: Model predictive control of a flexible links mechanism. Journal of Intelligent and Robotic Systems 58(2), 125–147 (2010)
15. Chao, Z., Zhou, S.L., Ming, L., Zhang, W.G.: Uav formation flight based on nonlinear model predictive control. Mathematical Problems in Engineering 2012(1), 1–16 (2012)
16. Zhang, X., Duan, H., Yu, Y.: Receding horizon control for multi-uavs close formation control based on differential evolution. Science China Information Sciences 53, 223–235 (2010)
17. Shin, J., Kim, H.: Nonlinear model predictive formation flight. IEEE Transactions on Systems, Man and Cybernetics, Part A: Systems and Humans 39(5), 1116–1125 (2009)
18. Saska, M., Mejia, J.S., Stipanovic, D.M., Schilling, K.: Control and navigation of formations of car-like robots on a receding horizon. In: Proc. of 3rd IEEE Multi-conference on Systems and Control (2009)
19. Faigl, J., Krajník, T., Chudoba, J., Přeučil, L., Saska, M.: Low-cost embedded system for relative localization in robotic swarms. In: IEEE International Conference on Robotics and Automation, ICRA (2013)
20. Kulich, M., Chudoba, J., Kosnar, K., Krajnik, T., Faigl, J., Preucil, L.: Syrotek - distance teaching of mobile robotics. IEEE Transactions on Education 56(1), 18–23 (2013)
21. Barfoot, T.D., Clark, C.M.: Motion planning for formations of mobile robots. Robotics and Autonomous Systems 46, 65–78 (2004)
22. Saska, M., Hess, M., Schilling, K.: Efficient airport snow shoveling by applying autonomous multi-vehicle formations. In: Proc. of IEEE International Conference on Robotics and Automation (May 2008)

Clustering and Selecting Categorical Features

Cláudia Silvestre, Margarida Cardoso, and Mário Figueiredo

Escola Superior de Comunicação Social, Lisboa, Portugal
ISCTE, Business School, Lisbon University Institute, Lisboa, Portugal
Instituto de Telecomunicações, Instituto Superior Técnico, Lisboa, Portugal
csilvestre@escs.ipl.pt,
margarida.cardoso@iscte.pt,
mario.figueiredo@lx.it.pt

Abstract. In data clustering, the problem of selecting the subset of most relevant features from the data has been an active research topic. Feature selection for clustering is a challenging task due to the absence of class labels for guiding the search for relevant features. Most methods proposed for this goal are focused on numerical data. In this work, we propose an approach for clustering and selecting categorical features simultaneously. We assume that the data originate from a finite mixture of multinomial distributions and implement an integrated expectation-maximization (EM) algorithm that estimates all the parameters of the model and selects the subset of relevant features simultaneously. The results obtained on synthetic data illustrate the performance of the proposed approach. An application to real data, referred to official statistics, shows its usefulness.

Keywords: Cluster analysis, finite mixtures models, EM algorithm, feature selection, categorical variables.

1 Introduction

Feature selection is considered a fundamental task in several areas of application that deal with large data sets containing many features, such as data mining, machine learning, image retrieval, text classification, customer relationship management, and analysis of DNA micro-array data. In these settings, it is often the case that not all the features are useful: some may be redundant, irrelevant, or too noisy. Feature selection extracts valuable information from the data sets, by choosing a meaningful subset of all the features. Some benefits of feature selection include reducing the dimensionality of the feature space, removing noisy features, and providing better understanding of the underlying process that generated the data.

In supervised learning, namely in classification, feature selection is a clearly defined problem, where the search is guided by the available class labels. In contrast, for unsupervised learning, namely in clustering, the lack of class information makes feature selection a less clear problem and a much harder task.

L. Correia, L.P. Reis, and J. Cascalho (Eds.): EPIA 2013, LNAI 8154, pp. 331–342, 2013.

An overview of the methodologies for feature selection as well as guidance on different aspects of this problem can be found in [1], [2] and [3].

In this work, we focus on feature selection for clustering categorical data, using an embedded approach to select the relevant features. We adapt the approach developed by Law et al. [4] for continuous data that simultaneous clusters and selects the relevant subset of features. The method is based on a minimum message length (MML) criterion [5] to guide the selection of the relevant features and an *expectation-maximization* (EM) algorithm [6] to estimate the model parameters. This variant of the EM algorithm seamlessly integrates model estimation and feature selection into a single algorithm. We work within the commonly used framework for clustering categorical data that assumes that the data originate from a multinomial mixture model. We assume that the number of components of the mixture model is known and implement a new EM variant following previous work in [7].

2   Related Work

Feature selection methods aim to select a subset of relevant features from the complete set of available features in order to enhance the clustering analysis performance. Most methods can be categorized into four classes: filters, wrappers, hybrid, and embedded.

The filter approach assesses the relevance of features by considering the intrinsic characteristics of the data and selects a feature subset without resorting to clustering algorithm. Some popular criteria used to evaluate the goodness of a feature or of a feature subset are distance, information, dependency, or consistency measures. Some filter methods produce a feature ranking and use a threshold to select the feature subset. Filters are computationally fast and can be used in unsupervised learning.

Wrapper approaches include the interaction between the feature subset and the clustering algorithm. They select the feature subset, among various candidate subsets of features that are sequentially generated (usually in a forward or backward way), in an attempt to improve the clustering algorithm results. Usually, wrapper methods are more accurate than filters, but even for algorithms with a moderate complexity, the number of iterations that the search process requires results in a high computational cost.

Hybrid methods aim at taking advantage of the best of both worlds (filters and wrappers). The main goal of hybrid approaches is to obtain the efficiency of filters and the accuracy of wrappers. Usually, hybrid algorithms use a filter method to reduce the search space that will subsequently be considered by the a wrapper. Hybrid methods are faster than wrappers, but slower than filters.

In embedded methods, the feature selection is included into the clustering algorithm, thus fully exploiting the interplay between the selected features and the clustering task. Embedded methods are reported to be much faster than wrappers, although their performance also depends on the clustering algorithm [8].

In clustering problems, feature selection is both challenging and important. Filters used for supervised learning can be used for clustering since they do not resort to class labels. The vast majority of work on feature selection for clustering has focused on numerical data, namely on Gaussian-mixture-based methods (e. g. [9], [4], and [10]). In contrast, work on feature selection for clustering categorical data is relatively rare [11].

Finite mixture models are widely used for cluster analysis. These models allow a probabilistic approach to clustering in which model selection issues (e.g., number of clusters or subset of relevant features) can be formally addressed. Some advantages of this approach are: it identifies the clusters, it is able to deal with different types of features measurements, and it outperforms more traditional approaches (e.g., k-means). Finite mixture models assume specific intra-cluster probability functions, which may belong to the same family but differ in the parameter values. The purpose of model estimation is to identify the clusters and estimate the parameters of the distributions underlying the observed data within each cluster. The maximum likelihood estimators cannot be found analytically, and the EM algorithm [6] has been often used as an effective method for approximating the estimates. To our knowledge there is only one proposal [11] within this setting for clustering and selecting categorical features. In his work, Talavera presents a wrapper and a filter to select categorical features. The proposed wrapper method, EM-WFS (EM wrapper with forward search), combines EM with forward feature selection. Assuming that the feature dependencies play a crucial role in determining the feature importance for clustering, a filter ranker based on a mutual information measure, EM-PWDR (EM pairwise dependency ranker), is proposed. In supervised learning, filter approaches usually measure the correlation of each feature with the class label by using distance, information, or dependency measures [12]. Assuming that, in the absence of class labels, we can consider as irrelevant those features that exhibit low dependency with the other features [13]. Under this assumption, the proposed filter considers as good candidates to be selected the highly correlated features with other features. Feature subset evaluation criteria like scatter separability or maximum likelihood seem to be more efficient for the purpose of clustering than the dependence between features. In our work, we propose an embedded method for feature selection, using a minimum message length model selection criterion to select the relevant features and a new EM algorithm for performing model-based clustering.

3   The Model

Let $Y = \left[\underline{y}_1, \ldots, \underline{y}_n\right]'$ be a sample of n independent and identically distributed random variables/features, where $\underline{y} = (Y_1, \ldots, Y_L)$ is a L-dimensional random vector. It is said that $\underline{y}$ follows a K component finite mixture distribution if its log-likelihood can be written as

$$\log \prod_{i=1}^{n} f(\underline{y}_i|\underline{\theta}) = \sum_{i=1}^{n} \log \sum_{k=1}^{K} \alpha_k f(\underline{y}_i|\theta_k)$$

where $\alpha_1, \ldots, \alpha_K$ are the mixing probabilities ($\alpha_k \geq 0, k = 1, .., K$ and $\sum_{k=1}^{K} \alpha_k = 1$), $\underline{\theta} = (\underline{\theta}_1, .., \underline{\theta}_K, \alpha_1, .., \alpha_K)$ the set of all the parameters of the model and $\underline{\theta}_k$ is the set of parameters defining the k-th component. In our case, for categorical data, $f(.)$ is the probability function of a multinomial distribution.

Assuming that the features are conditionally independent given the component-label, the log-likelihood is

$$\log \prod_{i=1}^{n} f\left(\underline{y}_i|\underline{\theta}\right) = \sum_{i=1}^{n} \log \sum_{k=1}^{K} \alpha_k \prod_{l=1}^{L} f(\underline{y}_{l_i}|\underline{\theta}_{lk})$$

The maximum likelihood estimators cannot be found analytically, and the EM algorithm has been often used as an effective method for approximating the corresponding estimates. The basic idea behind the EM algorithm is regarding the data Y as incomplete data, clusters allocation being unknown. In finite mixture models, variables $Y_1, \ldots, Y_L$ (the incomplete data) are augmented by a component-label latent variables $\underline{z} = (Z_1, \ldots, Z_K)$ which is a set of K binary indicator latent variables, that is, $\underline{z}_i = (Z_{1i}, \ldots, Z_{Ki})$, with $Z_{ki} \in \{0, 1\}$ and $Z_{ki} = 1$ if and only if the density of $y_i \in C_k$ (component k) implying that the corresponding probability function is $\overline{f}(\underline{y}_i|\theta_k)$. Assuming that the $Z_1, \ldots, Z_k$ are i.i.d., following a multinomial distribution of K categories, with probabilities $\alpha_1, \ldots, \alpha_K$, the log-likelihood of a complete data sample $(\underline{y}, \underline{z})$, is given by

$$\log f(\underline{y}_i, \underline{z}_i|\underline{\theta}) = \sum_{i=1}^{n} \sum_{k=1}^{K} z_{ki} \log \left[\alpha_k f(\underline{y}_i|\underline{\theta}_k)\right]$$

The EM algorithm produces a sequence of estimates $\hat{\underline{\theta}}(t)$, $t = 1, 2, \ldots$ until some convergence criterion is met.

3.1   Feature Saliency

The concept of feature saliency is essencial in the context of the feature selection methodology. There are different definitions of feature saliency/(ir)relevancy. Law et al. [4] adopt the following definition: a feature is irrelevant if its distribution is independent of the cluster labels i.e. an irrelevant feature has a common to all clusters probability function.

Lets denote the probability function of relevant and irrelevant features by p(.) and q(.), respectively. For categorical features, p(.) and q(.) refer to multinomial distributions. Let $B_1, \ldots, B_L$ be the binary indicators of the features relevancy, where $B_l = 1$ if the feature l is relevant and zero otherwise.

Using this definition of feature irrelevancy the log-likelihood becomes

$$\log \prod_{i=1}^{n} f\left(\underline{y}_i | \underline{\theta}\right) = \sum_{i=1}^{n} \log \sum_{k=1}^{K} \alpha_k \prod_{l=1}^{L} \left[p\left(\underline{y}_{li} | \theta_{lk}\right) \right]^{B_l} \left[q\left(\underline{y}_{li} | \theta_l\right) \right]^{1 - B_l}$$

Defining *feature saliency* as the probability of the feature being relevant, $\rho_l = P(B_l = 1)$ the log-likelihood is (the proof is in [4]):

$$\log \prod_{i=t}^{n} f\left(\underline{y}_i | \underline{\theta}\right) = \sum_{i=1}^{n} \log \sum_{k=1}^{K} \alpha_k \prod_{l=1}^{L} \left[\rho_l p\left(\underline{y}_{li} | \theta_{lk}\right) + (1 - \rho_l) q\left(\underline{y}_{li} | \theta_l\right) \right]$$

The features' saliencies are unknown and they are estimated using an EM variant based on the MML criterion. This criterion encourages the saliencies of the relevant features to go to 1 and of the irrelevant features to go to zero, pruning the features' set.

4   The Proposed Method

We propose an embedded approach for clustering categorical data, assuming that the data are originate from a multinomial mixture and the number of mixture components is known. The new EM algorithm is implemented using an MML criterion to estimate the mixture parameters, including the features' saliencies. This work extends that of Law et al. [4] dealing with categorical features.

4.1   The Minimum Message Length (MML) Criterion

The MML-type criterion chooses the model providing the shortest description (in an information theory sense) of the observations [5]. According to Shannon's information theory, if Y is some random variable with probability distribution $p(y | \underline{\theta})$, the optimal code-length for an outcome y is $l(y | \underline{\theta}) = \log_2 p(y | \underline{\theta})$, measured in bits and ignoring that $l(y)$ should be integer [14]. When the parameters, $\underline{\theta}$, are unknown they need to be encoded, so the total message length is given by $l(y, \underline{\theta}) = l(y | \underline{\theta}) + l(\underline{\theta})$, where the first part encodes the observation y, and the second the parameters of the model.

Under the MML criterion, for categorical features, the estimate of $\underline{\theta}$ is the one that minimizes the following description length function:

$$l(\underline{y}, \underline{\theta}) = -\log f(\underline{y} | \underline{\theta}) + \frac{K + L}{2} \log n + \sum_{l=1, \rho_l \neq 0}^{L} \frac{c_l - 1}{2} \sum_{k=1}^{K} \log(n \alpha_k \rho_l)$$

$$+ \sum_{l=1, \rho_l \neq 1}^{L} \frac{c_l - 1}{2} \log(n(1 - \rho_l))$$

where c_l is the number of categories of feature Y_l.

A Dirichlet-type *prior* (a natural conjugate *prior* of the multinomial) is used for the saliencies,

$$p(\rho_1, \ldots, \rho_L) \propto \prod_{l=1}^{L} \rho_l^{\frac{-Kc_l}{2}} (1 - \rho_l)^{\frac{c_l}{2}}.$$

As a consequence, the MAP (*maximum a posterior*) parameters estimators are obtained when minimizing the proposed description length function, $l(\underline{y}, \theta)$.

4.2   The Integrated EM

To estimate all the parameters of the model, we implemented a new version of the EM algorithm integrating clustering and feature selection - the *integrated Expectation-Maximization* (iEM) algorithm. This algorithm complexity is the same as the standard EM for mixture of multinomials. The iEM algorithm to maximize $[-l(\underline{y}, \theta)]$ has two steps:

E-step: Compute

$$P[Z_{ki} = 1 | \underline{Y}_i, \theta] = \frac{\alpha_k \prod_{l=1}^{L} \left[\rho_l p \left(\underline{y}_{li} | \theta_{lk} \right) + (1 - \rho_l) q \left(\underline{y}_{li} | \theta_l \right) \right]}{\sum_{k=1}^{K} \alpha_k \prod_{l=1}^{L} \left[\rho_l p \left(\underline{y}_{li} | \theta_{lk} \right) + (1 - \rho_l) q \left(\underline{y}_{li} | \theta_l \right) \right]} \quad (1)$$

M-step: Update the parameter estimates according to

$$\hat{\alpha}_k = \frac{\sum_i P[Z_{ki} = 1 | \underline{Y}_i, \theta]}{n}, \quad (2)$$

$$\hat{\theta}_{lkc} = \frac{\sum_i u_{lki} y_{lci}}{\sum_c \sum_i u_{lki} y_{lci}}, \quad (3)$$

$$\hat{\rho}_l = \frac{\max \left(\sum_{ik} u_{lki} - \frac{K(c_l - 1)}{2}, 0 \right)}{\max \left(\sum_{ik} u_{lki} - \frac{K(c_l - 1)}{2}, 0 \right) + \max \left(\sum_{ik} v_{lki} - \frac{c_l - 1}{2}, 0 \right)} \quad (4)$$

where

$$u_{lki} = \frac{\rho_l p \left(\underline{y}_{li} | \theta_{lk} \right)}{\rho_l p \left(\underline{y}_{li} | \theta_{lk} \right) + (1 - \rho_l) q \left(\underline{y}_{li} | \theta_l \right)} P[Z_{ki} = 1 | \underline{Y}_i, \theta]$$

$$v_{lki} = P[Z_{ki} = 1 | \underline{Y}_i, \theta] - u_{lki}$$

After running the iEM, usually the saliencies are not zero or one. Our goal is to reduce the set of initial features, so we check if pruning the feature which has the smallest saliency produces a lower message length. This procedure is repeated until all the features have their saliencies equal to zero or one. At the end, we will choose the model having the minimum message length value. The proposed algorithm is summarized in Figure 1.

Input: data $Y = \left[\underline{y}_1, \ldots, \underline{y}_n \right]'$ where $\underline{y} = (Y_1, \ldots, Y_L)$
the number of components K
mimimum increasing threshold for the likelihood function δ
Ouput: feature saliencies $\{\rho_1, \ldots, \rho_L\}$
mixture parameters $\{\underline{\theta}_{1k}, \ldots, \underline{\theta}_{LK}\}$ and $\{\alpha_1, \ldots, \alpha_K\}$
parameters of common distribution $\{\underline{\theta}_1, \ldots, \underline{\theta}_L\}$

Initialization: initialization of the parameters resorts to the empirical distribution:
set the parameters $\underline{\theta}_{lk}$ of the mixture components
$\mathrm{p}(\underline{y}_l | \underline{\theta}_{lk})$, $(l = 1, \ldots, L \; ; \; k = 1, \ldots, K)$
set the common distribution parameters $\underline{\theta}_l$, to cover all the data
$\mathrm{q}(\underline{y}_l | \underline{\theta}_l)$, $(l = 1, \ldots, L)$
set all features saliencies
$\rho_l = 0.5 \; (l = 1, \ldots, L)$
store the initial log-likelihood
store the initial message length (iml)
$mindl \leftarrow iml$
continue $\leftarrow 1$
while continue **do**
    while increases on log-likelihood are above δ **do**
        M-step according to (2), (3) and (4)
        E-step according to (1)
        if (feature l is relevant) $\rho_l = 1, \mathrm{q}(\underline{y}_l | \underline{\theta}_l)$ is pruned
        if (feature l is irrelevant) $\rho_l = 0, \mathrm{p}(\underline{y}_l | \underline{\theta}_{lk})$ is pruned for all k
        Compute the log-likelihood and the current message length (ml)
    end while
    if $ml < mindl$
        $mindl \leftarrow ml$
        update all the parameters of the model
    end if
    if there are saliencies, $\rho_l \notin \{0, 1\}$
        prune the variable with the smallest saliency
    else
        continue $\leftarrow 0$
    end if
end while
The best solution including the saliencies corresponds to the final $mindl$ obtained.

Fig. 1. The iEM algorithm for clustering and selecting categorical features

5   Numerical Experiments

For a L-variate multinomial we have

$$f\left(\underline{y}_i|\underline{\theta}\right) = \prod_{l=1}^{L}\left[n!\prod_{c=1}^{c_l}\frac{\theta_{lc}^{y_{lci}}}{(y_{lci})!}\right]$$

where c_l is the number of categories of feature Y_l.

5.1   Synthetic Data

We use two types of synthetic data: in the first type the irrelevant features have exactly the same distribution for all components. Since with real data, the irrelevant features could have little (non relevant) differences between the components, we consider a second type of data where we simulate irrelevant features with similar distributions between the components. In both cases, the irrelevant features are also distributed according to a multinomial distribution. Our approach is tested with 8 simulated data sets. We ran the proposed EM variant (iEM) 10 times and chose the best solution. According to the obtained results using the iEM, the estimated probabilities corresponding to the categorical features almost exactly match the actual (simulated) probabilities. Two of our data sets are presented in tables 1 and 2.

In Table 1 results refer to one data set with 900 observations, 4 categorical features and 3 components with 200, 300 and 400 observations. The first two features are relevant with 2 and 3 categories respectively, the other two are irrelevant and have 3 and 2 categories each. These irrelevant features have the same distribution for the 3 components. In Table 2 the data set has 900 observations and 5 categorical features. The features 1, 4 and 5 have 3 categories each and the features 2 and 3 have 2 categories. The first three features are relevant and the last two are irrelevant, with similar distributions between components.

5.2   Real Data

An application to real data referred to european official statistics (EOS) illustrates the usefulness of the proposed approach. This EOS data set originates from a survey on perceived quality of life in 75 european cities, with 23 quality of life indicators (clustering base features). For modeling purposes the original answers - referring to each city respondents- are summarized into: Scale 1)- *agree* (including strongly agree and somewhat agree) and *disagree* (including somewhat disagree and strongly disagree) and Scale 2)- *satisfied* (including very satisfied and rather satisfied) and *unsatisfied* (including rather unsatisfied and not at all satisfied).

A two-step approach is implemented: firstly, the number of clusters is determined based on MML criterion - see [15]; secondly, the proposed iEM algorithm is applied 10 times and the solution that has the lower *message length* is chosen.

Table 1. iEM results for a synthetic data set where irrelevant features have the same distributions between components

	Synthetic data			The algorithm's results			Saliency
	Component			Component			mean
	1	2	3	1	2	3	std. dev.
	Dim. 200	Dim. 300	Dim. 400	$\alpha = 0.22$	$\alpha = 0.33$	$\alpha = 0.45$	(of 10 runs)
	$\alpha = 0.22$	$\alpha = 0.33$	$\alpha = 0.45$				
Feature 1	0.20	0.70	0.50	0.20	0.70	0.50	$\bar{x} = .99$
relevant	0.80	0.30	0.50	0.80	0.30	0.50	$s = .01$
Feature 2	0.20	0.10	0.60	0.20	0.10	0.60	$\bar{x} = .97$
relevant	0.70	0.30	0.20	0.70	0.30	0.20	$s = .11$
	0.10	0.60	0.20	0.10	0.60	0.20	
Feature 3	0.50			0.50			$\bar{x} = 0$
irrelevant	0.20			0.20			$s = 0$
	0.30			0.30			
Feature 4	0.40			0.40			$\bar{x} = 0.10$
irrelevant	0.60			0.60			$s = 0.11$

Table 2. iEM results for a synthetic data set where irrelevant features have similar distributions between components

	Synthetic data		The algorithm's results		Saliency
	Component		Component		mean
	1	2	1	2	std. dev.
	Dim. 400	Dim. 500	$\alpha = 0.44$	$\alpha = 0.56$	(of 10 runs)
	$\alpha = 0.44$	$\alpha = 0.56$			
Feature 1	0.70	0.10	0.70	0.10	$\bar{x} = 1$
relevant	0.20	0.30	0.20	0.30	$s = 0$
	0.10	0.60	0.10	0.60	
Feature 2	0.20	0.70	0.20	0.69	$\bar{x} = 1$
relevant	0.80	0.30	0.80	0.31	$s = 0$
Feature 3	0.40	0.60	0.40	0.60	$\bar{x} = .99$
relevant	0.60	0.40	0.60	0.40	$s = .01$
Feature 4	0.5	0.49	0.50		$\bar{x} = .04$
irrelevant	0.20	0.22	0.20		$s = .06$
	0.30	0.29	0.30		
Feature 5	0.30	0.31	0.31		$\bar{x} = .04$
irrelevant	0.30	0.30	0.30		$s = .07$
	0.40	0.39	0.39		

Features' saliencies mean and standard deviations over 10 runs are presented in Table 3.

Applying the iEM algorithm to group the 75 European cities into 4 clusters, 2 quality of life indicators are considered irrelevant: *Presence of foreigners is good for the city* and *Foreigner here are well integrated*, meaning that the opinions

Table 3. Features' saliencies: mean and standard deviation of 10 runs

	Saliency	
Features	mean	std. dev.
Satisfied with sport facilities	0.95	0.16
Satisfied with beauty of streets	0.99	0.03
City committed to fight against climate change	0.99	0.04
Satisfied with public spaces	0.93	0.18
Noise is a big problem here	0.74	0.25
Feel safe in this city	0.78	0.34
Feel safe in this neighborhood	0.71	0.35
Administrative services help efficiently	0.99	0.02
Satisfied with green space	0.62	0.17
Resources are spent in a responsible way	0.72	0.21
Most people can be trusted	0.74	0.21
Satisfied with health care	0.64	0.11
Poverty is a problem	0.54	0.34
Air pollution is a big problem here	0.65	0.16
It is easy to find a good job here	0.51	0.21
This is a clean city	0.73	0.20
Satisfied with outdoor recreation	0.77	0.23
Easy to find good housing at reasonable price	0.44	0.09
City is healthy to live in	0.54	0.21
Satisfied with cultural facilities	0.5	0.22
Satisfied with public transport	0.28	0.14
Foreigner here are well integrated	0.26	0.24
Presence of foreigners is good for the city	0.38	0.4

regarding these features are similar for all the clusters. In fact, most of the citizens (79%) agree that the presence of foreigners is good for the city but they do not agree that foreigners are well integrated (only 39 % agree). Clustering results along with features' saliencies are presented in Table 4 - reported probabilities regard the agree and satisfied categories.

According to the obtained results we conclude that most respondents across all surveyed cities feel safe in their neighborhood and in their city. In cluster 1 cities air pollution and noise are relevant problems and it is not easy to find good housing at reasonable price. It is not easy to find a job in cities of cluster 2. Citizens of cities in cluster 3 have higher quality of life than the others e.g. they feel more safe, are more committed to fight against climate change and are generally satisfied with sport facilities, beauty of the streets, public spaces and outdoor recreation. Air pollution and noise are major problems of cities in cluster 4; in this cluster, cities are not considered clean or healthy to leave in.

Table 4. Features' saliencies and clusters probabilities regarding the agree and satisfied categories

Features	Saliency	Cluster 1 0.44	Cluster 2 0.12	Cluster 3 0.13	Cluster 4 0.31
	α				
Satisfied with sport facilities	1	0.78	0.67	0.84	0.52
Satisfied with beauty of streets	1	0.68	0.63	0.80	0.44
City committed to fight against climate change	1	0.58	0.53	0.69	0.35
Satisfied with public spaces	1	0.81	0.73	0.87	0.53
Noise is a big problem here	1	0.72	0.64	0.44	0.85
Feel safe in this city	1	0.80	0.82	0.94	0.66
Feel safe in this neighborhood	1	0.86	0.89	0.97	0.79
Administrative services help efficiently	1	0.67	0.57	0.68	0.42
Satisfied with green space	0.91	0.81	0.71	0.88	0.43
Resources are spent in a responsible way	0.88	0.53	0.49	0.63	0.35
Most people can be trusted	0.86	0.55	0.57	0.81	0.34
Satisfied with health care	0.81	0.82	0.73	0.89	0.49
Poverty is a problem	0.77	0.55	0.63	0.56	0.69
Air pollution is a big problem here	0.76	0.80	0.66	0.49	0.87
It is easy to find a good job here	0.53	0.49	0.26	0.47	0.39
This is a clean city	0.52	0.49	0.67	0.79	0.22
Satisfied with outdoor recreation	0.49	0.79	0.70	0.87	0.47
Easy to find good housing at reasonable price	0.49	0.24	0.48	0.59	0.33
City is healthy to live in	0.41	0.57	0.74	0.91	0.28
Satisfied with cultural facilities	0.32	0.95	0.91	0.57	0.68
Satisfied with public transport	0.31	0.81	0.70	0.88	0.40
Foreigner here are well integrated	0	0.39			
Presence of foreigners is good for the city	0	0.79			

6   Conclusions and Future Research

In this work, we implement an integrated EM algorithm to simultaneously select relevant features and cluster categorical data. The algorithm estimates the importance of each feature using a saliency measure. In order to test the performance of the proposed algorithm, two kinds of data sets were used: synthetic and real data sets. Synthetic data sets were used to test the ability to select the (previously known) relevant features and discard the irrelevant ones. The results clearly illustrate the ability of the proposed algorithm to recover the ground truth on data concerning the features' saliency and clustering. On the other hand, the usefulness of the algorithm is illustrated on real data based on European official statistics.

Results obtained with the data sets considered are encouraging. In the near future, an attempt to integrate both the selection of the number of clusters and of the relevant categorical features based on a similar approach will be implemented. Recently, this integration was successfully accomplished on synthetic data [16], but still offers some challenges when real data is considered.

References

[1] Guyon, I., Elisseeff, A.: An introduction to variable and feature selection. The Journal of Machine Learning Research 3, 1157–1182 (2003)

[2] Dy, J., Brodley, C.: Feature selection for unsupervised learning. The Journal of Machine Learning Research 5, 845–889 (2004)

[3] Steinley, D., Brusco, M.: Selection of variables in cluster analysis an empirical comparison of eight procedures. Psychometrika 73(1), 125–144 (2008)

[4] Law, M., Figueiredo, M., Jain, A.: Simultaneous feature selection and clustering using mixture models. IEEE Transactions on Pattern Analysis and Machine Intelligence 26, 1154–1166 (2004)

[5] Wallace, C., Boulton, D.: An information measure for classification. The Computer Journal 11, 195–209 (1968)

[6] Dempster, A., Laird, N., Rubin, D.: Maximum likelihood estimation from incomplete data via the EM algorithm. Journal of Royal Statistical Society 39, 1–38 (1997)

[7] Silvestre, C., Cardoso, M., Figueiredo, M.: Selecting categorical features in model-based clustering using a minimum message length criterion. In: The Tenth International Symposium on Intelligent Data Analysis - IDA 2011 (2011)

[8] Vinh, L.T., Lee, S., Park, Y.T., d'Auriol, B.J.: A novel feature selection method based on normalized mutual information. Applied Intelligence 37(1), 100–120 (2012)

[9] Constantinopoulos, C., Titsias, M.K., Likas, A.: Bayesian Feature and Model Selection for Gaussian Mixture Models. IEEE Transactions on Pattern Analysis and Machine Intelligence 28, 1013–1018 (2006)

[10] Zeng, H., Cheung, Y.: A new feature selection method for gaussian mixture clustering. Pattern Recognition 42, 243–250 (2009)

[11] Talavera, L.: An evaluation of filter and wrapper methods for feature selection in categorical clustering. In: Famili, A.F., Kok, J.N., Peña, J.M., Siebes, A., Feelders, A. (eds.) IDA 2005. LNCS, vol. 3646, pp. 440–451. Springer, Heidelberg (2005)

[12] Dash, M., Liu, H., Yao, J.: Dimensionality reduction for unsupervised data. In: Ninth IEEE International Conference on Tools with AI - ICTAI 1997 (1997)

[13] Talavera, L.: Dependency-based feature selection for symbolic clustering. Intelligent Data Analysis 4, 19–28 (2000)

[14] Cover, T., Thomas, J.: Elements of Information Theory. In: Entropy, Relative Entropy and Mutual Information, ch. 2 (1991)

[15] Silvestre, C., Cardoso, M., Figueiredo, M.: Clustering with finite mixture models and categorical variables. In: International Conference on Computational Statistics - COMPSTAT 2008 (2008)

[16] Silvestre, C., Cardoso, M., Figueiredo, M.: Simultaneously selecting categorical features and the number of clusters in model-based clustering. In: International Classification Conference - ICC 2011 (2011)

Monitoring the Habits of Elderly People through Data Mining from Home Automation Devices Data

Julie Soulas, Philippe Lenca, and André Thépaut

Institut Mines-Telecom; Telecom Bretagne
UMR 6285 Lab-STICC
Technopôle Brest Iroise CS 83818 29238 Brest Cedex 3, France
Université Européenne de Bretagne
{julie.soulas,philippe.lenca,andre.thepaut}@telecom-bretagne.eu

Abstract. Monitoring the habits of elderly people is a great challenge in order to improve ageing at home. Studying the deviances from or the evolution of regular behaviors may help to detect emerging pathologies. Regular patterns are searched in the data coming from sensors disseminated in the elderly's home. An efficient algorithm, xED, is proposed to mine such patterns. It emphasizes the description of the variability in the times when habits usually occur, and is robust to parasite events. Experiment on real-life data shows the interest of xED.

Keywords: habits, regular patterns, episodes.

1 Introduction

The population in industrialized countries is getting older. It is expected that in 2025, a third of the European population will be over 60 years old [19]. Because elderly people are frailer and more prone to chronic diseases, it becomes more dangerous for them to live alone at home. Falls and malnutrition problems are frequently observed. These threats to their health and happiness cause anxiety for them and their families, which often leads them to moving to a nursing-home. Health is in fact the first reason why people over 75 move. However, most elderly people would rather continue to live in their own home [13]. Living at home is also cheaper for the society than living in a nursing home.

Enabling people to remain in their home longer and in better conditions is thus an active effort (see for example the surveys [4,17] and the assistive smart home projects [3,16]). Different approaches are used, ranging from the home visits of medical staff to the use of technical aids, like support bars and emergency pendants. Ageing at home requires to be able to live independently. The assessment of one's functional status is generally performed assessing the ability or inability to perform **A**ctivities of **D**aily **L**iving (ADLs) [12]. These ADLs are things we do as part of our daily life: feeding, taking care of one's personal hygiene, dressing, home-making, moving about... and being able to perform them

L. Correia, L.P. Reis, and J. Cascalho (Eds.): EPIA 2013, LNAI 8154, pp. 343–354, 2013.

is thus a necessary condition in order to live at home independently. Elderly people tend to follow precise and unchanging daily routines [2]: these habits help them to remain in control of their environment; and changes in the routines may indicate that performing everyday activities is becoming harder, or even reveal the onset of a disorder, like Alzheimer's disease. This is why we are interested in monitoring remotely the ADLs, and more particularly in describing the daily habits i.e. in the discovery of periodic and regular patterns.

Nowadays, home automation devices and sensors enable the non-intrusive collection of data about the user behavior in the home. Thus, we use them as a data source in order to find interesting patterns describing the routines. The assumption is that the observable information (the state of the sensor array) characterizes (at least partially) the activities being carried out in the home. For that purpose, we propose xED, an extended version of Episode Discovery [9], to discover regularly occurring episodes.

Next section describes some of the current research on the use of sensor dataset mining for improving ageing at home. Section 3 details how ED works, and highlights some of its limitations. Our proposal, xED, is detailed in section 4. Section 5 presents experiments on a real dataset. Some conclusions and discussions are proposed in section 6.

2   Litterature Review

We here briefly present representative works that focus on monitoring activities of daily living to discovery periodic patterns and/or anomalies in the behavior.

The Episode Discovery (ED) [9] algorithm works on temporal sequences to discover behavioral patterns. The interestingness of a pattern is described with its frequency, length and periodicity. The algorithm has been successfully used by the intelligent smart home environment architecture MavHome [3]. This interesting algorithm is the basis of our proposal and will be described with more details in Section 3.

[20] introduced the simultaneous tracking and activity recognition problem. The authors exploit the synergy between location and activity to provide the information necessary for automatic health monitoring. An interesting aspect of this work is that the authors consider multiple people in a home environment.

Suffix Trees as an activity representation to efficiently extract structure of activities by analysing their constituent event-subsequences is proposed in [7]. The authors then exploit the properties of the Suffix Trees to propose an algorithm to detect anomalous activities in linear-time. [10] proposed an approach to discover temporal relations and apply it to perform anomaly detection on the frequent events by incorporating temporal relation information shared by the activity.

Popular k-means algorithm [14] and association rules framework have also been applied to mine temporal activity relations [15]. The authors first extract temporal features (the start time and the duration of the activities) and then cluster theses features to construct a mixture model for each activity. The Apriori algorithm [1] is then applied on the canonical temporal features to extract temporal relations between activities.

3   Episode Discovery Algorithm

Episode Discovery (ED) [9] works on a dataset containing timestamped sensor data: the *events*; and aims at describing the *episodes* (i.e. sets of events), that occur regularly over the dataset. ED is based on the *Minimum Description Length* principle [18], which states that the best model to represent data is the one with the shortest description, because it captures the main trends of the data. Periodically reoccurring episodes are replaced by a single header describing how and when the episodes occur, thus making a shorter version of the dataset. The missing occurrences (expected, but not observed) are explicitly characterized: they are reintroduced in the dataset. ED rewrites iteratively the dataset in more compact representations, repeating two steps: first, a list of candidate episodes is generated; second, the candidates are analyzed, in order to find the periodic episodes. The patterns leading to the smallest description of the dataset are considered as most interesting. The *periodicity* of an episode is described as a repeating cycle of time intervals between the successive occurrences of this episode. ED constrains the episodes using two parameters: their maximal duration, and the maximal number of events they contain (*episode capacity*). The event order in the episode is not taken into account, and duplicate events are considered only once.

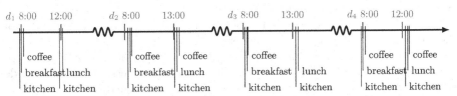

d1 7:50 kitchen; d1 8:00 breakfast; d1 8:15 coffee; d1 11:50 kitchen;
d1 12:10 lunch; d2 8:20 kitchen; d2 8:35 breakfast; ...

Fig. 1. Dataset example

Table 1. Episode analysis for the dataset from figure 1

Episode	Intervals (in hours)	Cycle	Errors	# fact. events
kitchen, breakfast, coffee	[24, 24, 24]	(24h)	∅	12
kitchen, lunch, coffee	[47]	(47h)	∅	4
kitchen	[4, 20, 5, 19, 5, 19, 4]	–	–	0
kitchen, lunch	[25, 24, 23]	–	–	0
kitchen, coffee	[24, 5, 19, 24, 4]	–	–	0

d_1 8:00 : kitchen, breakfast, coffee	
d_1 12:00 : kitchen, lunch	
d_2 8:00 : kitchen, breakfast, coffee	
d_2 13:00 : kitchen, lunch, coffee	
d_3 8:00 : kitchen, breakfast, coffee	
d_3 13:00 : kitchen, lunch	
d_4 8:00 : kitchen, breakfast, coffee	
d_4 12:00 : kitchen, lunch, coffee	

The episode that enables the greatest compression of the dataset is {kitchen, breakfast, coffee} (reduction by 12 events). The occurrences of this episode are thus removed.

d_1 8:00 → d_4, (24h), kitchen, breakfast, coffee
d_1 12:00 : kitchen, lunch
d_2 13:00 : kitchen, lunch, coffee
d_3 13:00 : kitchen, lunch
d_2 12:00 : kitchen, lunch, coffee

A header is added, telling that the episode occurs from 8:00 on d_1 to d_4, every 24 hours. There is no missing occurrence to describe.
The next iteration of the algorithm will lead to the removal of episode {kitchen, lunch, coffee}, and the insertion of d_2 13:00 → d_4, (47h), kitchen, lunch, coffee in the header.

Fig. 2. Rewriting of the dataset from table 1 using ED

The sample dataset described in figure 1 will be used as an example throughout this paper: it simulates four days (d_1 to d_4) of data, and contains 27 events, representing four activities: arrive in kitchen, have breakfast, have lunch, take a cup of coffee. We mine the example in figure 1, constraining the episode length to one hour. The episode capacity is not bounded. Table 1 describes the episodes of the dataset. For instance, the first line summarizes the characteristics of the occurrences for episode {kitchen, breakfast, coffee}, which is observed 4 times, separated by 24 hours. There is no missing occurrence, and describing this episode reduces the size of the dataset by 12 events (4 occurrences of the length-3-episode). Figure 2 shows the rewriting of the dataset using ED.

The approach developed in Episode Discovery is very interesting. The discovered periodic episodes are of special interest for habit monitoring, and the results were applied in concrete situations (e.g. in the MavHome project [3]). However, two main limitations appear: the strict definition for periodicity prevents the discovery of some habits, and the episode generation is slow when considering long episodes. We thus propose two solutions to address these problems.

Human life is not planned by the second. The main challenge when applying ED to real-life data comes from the variations inherent in human life. The solution used in ED is to preprocess the dataset to quantize the time values before mining: e.g. rounding each timestamp to the closest full hour. The choice of the quantization step size is no easy decision to make: it should prevent irrelevant time intervals from interfering with the results. But one should first know what can be considered as irrelevant, which is episode-dependant. For instance, the periodic daily episode *"Lunch at noon"* should be matched for the days when the user has lunch at noon of course, but also for the days when the user has lunch at 12:05, or even at 12:45, depending on the accepted tolerance.

As for the construction of the interesting episodes, ED uses set combinations (intersections, differences) on the episodes. There are two main issues with this way of generating the candidate episodes. First, the complexity is exponential. Second, when a constraint is set on the maximal number of events in an episode (capacity), there are risks to miss some of the interesting candidates. It would thus be interesting to leave the capacity unconstrained.

4   Towards an Extended Version of ED: xED

In this section, we present xED. All the aspects described subsequently were implemented in a prototype in Python, which can be used as a standalone program, or be imported as a module in any Python program.

Since we focus on habits, we are not interested here in very rare events. This is why a support threshold f_{min} is introduced: events need be more frequent than f_{min} to be part of an episode. Moreover, the early pruning of rare events and episodes speeds up the computation.

4.1   Flexibility on the Time Values

Adaptability of the Tolerances to Each Situation. The normal variability of an episode is activity-dependant. For instance, there may not be an underlying meaning to a relatively high variance when studying bedtime, which can be linked to the showing of an interesting film on television; but it is important to check that this variance stays low when studying the times when medication is taken. In order to allow flexibility on the time values, a new formalism is proposed (in comparison with ED). The descriptions of the events and episodes remain unchanged, but the periodic episodes are defined quite differently. Afterwards, we consider that:

- An episode *occurs around* time t (with standard deviation σ) if it occurs in the time interval $[t - a \cdot \sigma,\ t + a \cdot \sigma]$, where a is a user-chosen parameter.
- An episode E is *periodic* with period T on interval $\Delta t = [t_{start}, t_{end}]$ when there is a time distribution $\{(\mu_1, \sigma_1), ..., (\mu_m, \sigma_m)\}$, such that for all $i \in [1, m]$, there is an occurrence of E around $n \cdot T + \mu_i$ ($\pm \sigma_i$) for any integer n such that $n \cdot T + \mu_i \in \Delta t$. If occurrences are searched in two intersecting intervals, and an occurrence is found in the intersection, it is considered as matching only one of the two searched occurrences.
- The combination $t_{start} \rightarrow t_{end}, T, \{(\mu_1, \sigma_1), ..., (\mu_m, \sigma_m)\})$ is the *periodicity* of E.
  Let us consider the episode {kitchen} of the dataset presented in figure 1. As illustrated in figure 3, the episode {kitchen} is periodic, and its periodicity is $(d_1 \rightarrow d_4, 24h, \{(8:03, 13min), (12:40, 1h10)\})$.

The *periodic episodes* are those we are searching for, with constraints on the episode duration, the maximal standard deviation allowed, and the minimal

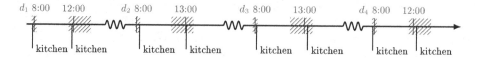

Fig. 3. Periodicity for episode {kitchen}

accuracy (see section 4.2). Thanks to the customized use of standard deviations, a periodic episode occurrence (expected to happen at date t_{exp}, with a standard deviation σ) is said to *happen as expected* if it is observed during the interval $[t_{exp} - a \cdot \sigma, \ t_{exp} + a \cdot \sigma]$. In our representation, each periodic episode is attached to its own list of (mean time, standard deviation) pairs: a highly episode-dependant characterization of the habits is thus allowed.

Periodicity Searching. To discover the periodic episodes with flexible occurrence time, the periodicity description is based on the density of the episode occurrences. The time distributions of the episodes are searched in the form of a Gaussian mixture model, using the expectation-maximisation algorithm [5]. This iterative method starts from random characteristics for the latent components of the mixture. Then for each occurrence time x of an episode and each component c of the mixture model describing the time distribution of the episode, the probability that x was generated by c is computed. The parameters of the components are tweaked to maximize the likelihood of the data point / component assignment. The number of components in the model is determined using DB-SCAN [6] (it is the number of clusters).

The relative occurrence times of the episodes in their period ($t \mod T$) are clustered with DBSCAN. This algorithm partitions the data points based on the estimated density of the clusters. A point x belongs to a cluster C when there are at least N_{min} points from C in the neighborhood of x. The neighborhood size controls the desired precision and tolerance for the periodic episodes. As for N_{min}, we use the support threshold defined previously to determine which episodes are frequent. A "general" activity (e.g. eating) can generate different clusters when considering its "specific" activities (e.g. lunch and dinner).

4.2   Interest Measures

Accuracy of a Rule. For a given periodic episode, we define its accuracy as the number of occurrences happening as expected divided by the number of expected occurrences. We use accuracy to choose which of several periodicity descriptions describes best the episode.

Compression Rate is the ratio between the sizes of the database before and after the compression of a periodic episode. The episode that enables the greatest compression is described first. The others are factorized on subsequent iterations, if they are still deemed interesting. When two periodic episodes allow the same compression rate, the one with the greatest accuracy is compressed first. But if the occurrences of the two episodes are disjoint, then the second compression can be done directly too: a new candidate study would show it to be the next most interesting periodic episode. That way, we save one iteration of the algorithm.

4.3 Alternative Candidate Search

When building the episodes with a constraint on the episode capacity C_{ep} (i.e. a constrained number of events in each episode), ED takes only the C_{ep} first events in an episode. But when in presence of parasite events, some interesting episodes might never be generated, and are thus never investigated. However, looking for candidate episodes is actually the same as looking for frequent itemsets (the sets of event labels describing the episodes: an itemset could be {"kitchen"} or {"lunch", "coffee"}) in a list of transactions (the maximal episodes, with $C_{ep} = \infty$). So the classic frequent itemset mining algorithms can be used to find the episodes. These algorithms are then guaranteed to find all frequent itemsets, and here, all the episodes that are more frequent than the frequency threshold. This threshold f_{min} is set quite low. Indeed, within all the frequent episodes, the periodic ones are not necessarily the most frequent ones. We used FP-growth, since it is very scalable and efficient [8]. A preprocessing step builds the transactions on which FP-growth works: each contains the events happening in an event-folding window of the same length than the maximal duration allowed for an episode.

4.4 Application to Example 1

Table 2 presents the results of the episode analysis step, using xED. The episodes {kitchen, breakfast, coffee} and {kitchen, coffee} both allow the removal of 12 events. But since the accuracy of episode {kitchen, breakfast, coffee} is higher, it is factorized first. At the next iteration, the episode {kitchen, lunch} is the most interesting one. The resulting database contains only two episode descriptions in its headers, and two coffee events (figure 4).

Table 2. Episode analysis for the dataset from figure 1 using xED

Episode	Description	Errors	# fact. events	A Accuracy
kitchen, breakfast, coffee	μ_1=8:03, σ_1=13min	$\emptyset$	12	100%
kitchen	μ_1=8:03, σ_1=13min μ_2=12:40, σ_2=1h10	$\emptyset$	8	100%
kitchen, lunch, coffee	μ_1=12:55, σ_1=44min	coffee event missing on d_3	4	67%
kitchen, lunch	μ_1=12:40, σ_1=1h10	$\emptyset$	8	100%
kitchen, coffee	μ_1=8:03, σ_1=13min μ_2=12:40, σ_2=1h10	coffee events missing on d_1 and d_3	12	75%

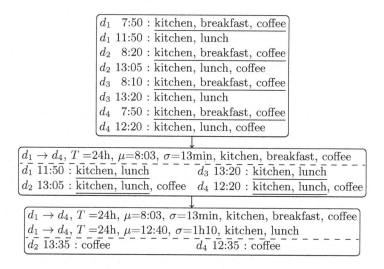

d_1 7:50 : kitchen, breakfast, coffee
d_1 11:50 : kitchen, lunch
d_2 8:20 : kitchen, breakfast, coffee
d_2 13:05 : kitchen, lunch, coffee
d_3 8:10 : kitchen, breakfast, coffee
d_3 13:20 : kitchen, lunch
d_4 7:50 : kitchen, breakfast, coffee
d_4 12:20 : kitchen, lunch, coffee

$d_1 \rightarrow d_4$, T =24h, μ=8:03, σ=13min, kitchen, breakfast, coffee
d_1 11:50 : kitchen, lunch d_3 13:20 : kitchen, lunch
d_2 13:05 : kitchen, lunch, coffee d_4 12:20 : kitchen, lunch, coffee

$d_1 \rightarrow d_4$, T =24h, μ=8:03, σ=13min, kitchen, breakfast, coffee
$d_1 \rightarrow d_4$, T =24h, μ=12:40, σ=1h10, kitchen, lunch
d_2 13:35 : coffee d_4 12:35 : coffee

Fig. 4. Compression of the dataset from table 2

When using only the original version of ED without time data preprocessing (i.e. without rounding off the timestamps), no episode can be discovered. Rounding to the closest hour allows the discovery of the morning routine. But the lunch activity, which is more prone to variability, is not detected. On the contrary, our version detects this activity, and tells us that the user eats at fairly irregular times. However, if our user eats lunch at 15:00 later in the experiment, it falls out of the known habits, and will be detected.

5   Experiment

xED is tested on both simulated and real-life data. The generated datasets were designed to validate the algorithm's behavior in various typical situations. We describe next the experiment on the real-life TK26M dataset [11] (https://sites.google.com/site/tim0306/datasets, consulted May, 29th 2013).

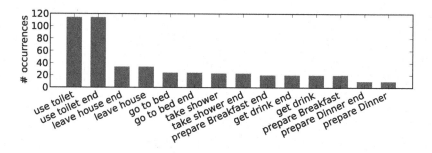

Fig. 5. Distribution of each event type frequency over TK26M dataset

Table 3. Regular episodes found in the dataset TK26M

#	Episode	Periodicity Start →End , Period, Description		A
1	use toilet, use toilet end, go to bed	Feb 21st →Mar 27th , 1 week,	(Mon 23:25, 4947s) (Tue 23:34, 3736s) (Wed 23:42, 4549s) (Thu 23:47, 1739s)	82 %
2	prepare breakfast, prepare breakfast end	Feb 26th →Mar 21st , 1 day ,	(09:22, 2306s)	74 %
3	take shower, take shower end	Feb 26th →Mar 22nd, 1 day ,	(09:45, 2694s)	71 %
4	leave house	Feb 26th →Mar 22nd, 1 day ,	(10:16, 3439s)	83 %
5	go to bed end	Feb 26th →Mar 22nd, 1 day ,	(08:55, 2470s)	79 %
6	leave house end	Feb 21st →Mar 27th , 1 week,	(Thu 20:29, 9537s) (Sun 20:06, 3932s) (Mon 19:17, 363s) (Tue 20:25, 8363s) (Wed 19:05, 4203s)	71 %
7	use toilet, use toilet end, go to bed	Mar 6th →Mar 20th , 1 week,	(Sat 01:32, 6120s) (Sat 10:00, 1s) (Sun 01:30, 1s)	75 %
8	get drink, get drink end, prepare Dinner end	Mar 6th →Mar 20th , 1 week,	(Sat 18:39, 1s) (Mon 20:13, 148s) (Tue 18:12, 1s)	75 %
9	get drink, get drink end, use toilet, use toilet end	Mar 7th →Mar 9th , 1 day ,	(08:27, 588s)	100 %
10	use toilet, leave house	Mar 11th→Mar 19th , 1 day ,	(19:16, 3224s)	60 %
11	leave house end, use toilet, use toilet end	Feb 28th →Mar 27th , 1 week,	(Fri 17:29, 5194s)	67 %
12	get drink, get drink end	Feb 25th →Feb 28th , 1 day ,	(20:27, 3422s)	100 %
13	missing episode 2	Feb 28th →Mar 13th , 1 week,	(Sun 09:22, 1s) (Tue 09:22, 1s)	100 %

TK26M. was recorded between February 25th 2008 and March 23rd 2008. Fourteen binary sensors were installed in a home, where a single 26-year-old man lived. The sensors include motion detectors between the rooms, and contact switches in the kitchen. A contact switch is also placed on the toilet flush. The 28-day experiment resulted in 2638 on / off sensor events.

The user annotated the higher-level activities being caried out using a Bluetooth headset. The activities that were recorded are: "Leave house", "Toileting", "Showering", "Sleeping", "Preparing breakfast", "Preparing dinner" and "Preparing a beverage", and each activity has a start and an end time. These records (490 in total, making now our *events*) are also available with the sensor data. We focus on finding habits in these activities. Their relative occurrence frequency over the dataset is given figure 5 (the least frequent appears 10 times).

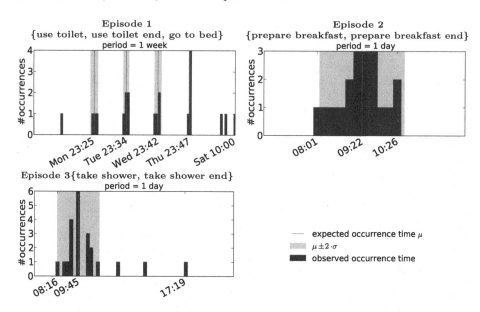

Fig. 6. Histograms for some of the periodic activities discovered in TK26M

We set $f_{min} = 3$, and $N_{min} = 3$ for DBSCAN in order to enable the discovery of weekly episodes (the dataset lasts four weeks). The episode length is 20 minutes (which is long enough to consider as part of one episode both the beginning and end of short activities, like preparing breakfast, or using the toilet). No constraint is set on the episode capacity. The a parameter is set to 2: if the time distribution for an episode follows a normal distribution (reasonable assumption when considering habits), this means that an average of 95% of the episode occurrences will be classified as expected occurrences.

Results. Table 3 and figures 6 and 7 show respectively the episodes that were found, the histograms of their occurrence time, and a timeline showing both the times at which an activity is expected to happen, and the events that were actually recorded during the experiment. In the histograms, the grayed area corresponds to the time interval $[\mu - 2 \cdot \sigma, \ \mu + 2 \cdot \sigma]$ (where μ and σ are the parameters of the components in the periodicity description). The habits (episodes 1 to 11) are found. Episodes 12 and 13 correspond to more exceptional situations. One can for instance notice (episode 1) that the user goes to the toilet before going to bed, which happens around 23:35 on week days (23:25 on Mondays, 23:34 on Tuesdays, ...). This same episode is also recognized outside of the expected intervals, as shows the first and last few bars in the top left histogram. Morning routines are also found, including waking up (episode 5), breakfast preparation (episode 2), showering (episode 3) and departure from home (episode 4) every morning. The second and third histograms in figure 6 show the occurrence times for the episodes 2 and 3: the breakfast episode occurs only in the expected interval, but that the shower activity is more likely to happen outside of the

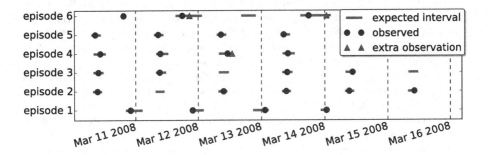

Fig. 7. The expected and observed occurrence times for the periodic episodes in TK26M

usual interval. Figure 7 allows day by day comparison, and highlights for for each episode how the expected episodes actually occurred. In particular, the expected but missing occurrences can be spotted where one can see an expected interval, but no (round dot) observation occurs in the interval.

The interest of xED appears especially when considering highly variable episodes, such as episode 6 "leave home end" (the inhabitant comes back at home). The intervals between the occurrences vary irregularly, as can be seen in figure 7. This otherwise regular and accurate episode cannot be detected as a regular episode using ED, even when quantizing the occurrence time.

6   Conclusions

The discovery of habits is of paramount importance for the monitoring of health of elderly people. Linked with anomaly detection tools, it enables the raise of alarms when dangerous or unusual situations are detected; or when the habits are shifting. We propose xED, which enables the efficient discovery of habits. This unsupervised algorithm characterizes periodic episodes, and quantifies the customary variability of these habits. Future work on this topic includes the extension of xED to the handling of data streams within the Human Ambient Assisted Living lab (http://labsticc.fr/le-projet-transverse-haal/). This would enable the real-time discovery and update of our knowledge about the habits, and the analysis of their evolution. In addition, an evaluation framework adapted to such problems should be investigated.

References

1. Agrawal, R., Imieliński, T., Swami, A.: Mining association rules between sets of items in large databases. SIGMOD Rec. 22(2), 207–216 (1993)
2. Bergua, V., Bouisson, J.: Vieillesse et routinisation: une revue de la question. Psychologie & NeuroPsychiatrie du vieillissement 6(4), 235–243 (2008)

3. Cook, D., Youngblood, M., Heierman III, E.O., Gopalratnam, K., Rao, S., Litvin, A., Khawaja, F.: MavHome: an agent-based smart home. In: IEEE International Conference on Pervasive Computing and Communications, pp. 521–524 (2003)
4. Demiris, G., Hensel, B.K.: Technologies for an aging society: A systematic review of "smart home" applications. Yearbook Med. Inf., 33–40 (2008)
5. Dempster, A., Laird, N., Rubin, D.: Maximum likelihood from incomplete data via the EM algorithm. Journal of the Royal Statistical Society. Series B, 1–38 (1977)
6. Ester, M., Kriegel, H.P., Sander, J., Xu, X.: A density-based algorithm for discovering clusters in large spatial databases with noise. In: KDD, pp. 226–231 (1996)
7. Hamid, R., Maddi, S., Bobick, A., Essa, M.: Structure from statistics - unsupervised activity analysis using suffix trees. In: IEEE International Conference on Computer Vision, pp. 1–8 (2007)
8. Han, J., Pei, J., Yin, Y.: Mining frequent patterns without candidate generation. SIGMOD Rec. 29(2), 1–12 (2000)
9. Heierman, E., Youngblood, M., Cook, D.J.: Mining temporal sequences to discover interesting patterns. In: KDD Workshop on Mining Temporal and Sequential Data (2004)
10. Jakkula, V.R., Crandall, A.S., Cook, D.J.: Enhancing anomaly detection using temporal pattern discovery. In: Kameas, A.D., Callagan, V., Hagras, H., Weber, M., Minker, W. (eds.) Advanced Intelligent Environments. Springer, US (2009)
11. van Kasteren, T., Englebienne, G., Kröse, B.: An activity monitoring system for elderly care using generative and discriminative models. Personal and Ubiquitous Computing 14, 489–498 (2010)
12. Katz, S.: Assessing self-maintenance: Activities of daily living, mobility, and instrumental activities of daily living. Journal of the American Geriatrics Society 31, 721–727 (2003)
13. Kotchera, A., Straight, A., Guterbock, T.: Beyond 50.05: A report to the nation on livable communities: Creating environments for successful aging. Tech. rep., AARP, Washington, D.C (2005)
14. MacQueen, J.: Some methods for classification and analysis of multivariate observations. In: Berkeley Symposium on Mathematical Statistics and Probability, California, USA, vol. 1, pp. 281–297 (1967)
15. Nazerfard, E., Rashidi, P., Cook, D.: Discovering temporal features and relations of activity patterns. In: IEEE ICDM Workshops, pp. 1069–1075 (2010)
16. Pigot, H.: When cognitive assistance brings autonomy in daily living: the domus experience. Gerontechnology 9(2), 71 (2010)
17. Rashidi, P., Mihailidis, A.: A survey on ambient assisted living tools for older adults. IEEE Journal of Biomedical and Health Informatics 17(3), 579–590 (2013)
18. Rissanen, J.: Stochastic complexity in statistical inquiry theory. World Scientific Publishing Co., Inc. (1989)
19. Sölvesdotter, M., ten Berg, H., Berleen, G.: Healthy aging – A challenge for Europe – ch. 2. Statistics. Tech. rep., European Commision (2007)
20. Wilson, D.H., Atkeson, C.: Simultaneous tracking and activity recognition (STAR) using many anonymous, binary sensors. In: Gellersen, H.-W., Want, R., Schmidt, A. (eds.) PERVASIVE 2005. LNCS, vol. 3468, pp. 62–79. Springer, Heidelberg (2005)

On the Predictability of Stock Market Behavior Using StockTwits Sentiment and Posting Volume

Nuno Oliveira[1], Paulo Cortez[1], and Nelson Areal[2]

[1] Centro Algoritmi/Dep. Information Systems, University of Minho,
4800-058 Guimarães, Portugal
nunomroliveira@gmail.com, pcortez@dsi.uminho.pt
[2] Department of Management, School of Economics and Management,
University of Minho, 4710-057 Braga, Portugal
nareal@eeg.uminho.pt

Abstract. In this study, we explored data from StockTwits, a microblogging platform exclusively dedicated to the stock market. We produced several indicators and analyzed their value when predicting three market variables: returns, volatility and trading volume. For six major stocks, we measured posting volume and sentiment indicators. We advance on the previous studies on this subject by considering a large time period, using a robust forecasting exercise and performing a statistical test of forecasting ability. In contrast with previous studies, we find no evidence of return predictability using sentiment indicators, and of information content of posting volume for forecasting volatility. However, there is evidence that posting volume can improve the forecasts of trading volume, which is useful for measuring stock liquidity (e.g. assets easily sold).

Keywords: Microblogging Data, Returns, Trading Volume, Volatility, Regression.

1 Introduction

Mining microblogging data to forecast stock market behavior is a very recent research topic that appears to present promising results [3,14,7,9]. In such literature, it is argued that a model that accounts for investor sentiment and attention can potentially be used to predict key stock market variables, such as returns, volatility and volume. Several arguments support this approach. For example, some studies have shown that individuals' financial decisions are significantly affected by their emotions and mood [10,8]. Also, the community of users that utilizes these microblogging services to share information about stock market issues has grown and is potentially more representative of all investors. Moreover, microblogging data is readily available at low cost permitting a faster and less expensive creation of indicators, compared to traditional sources (e.g. large-scale surveys), and can also contain new information that is not present in historical quantitative financial data. Furthermore, the small size of the message (maximum 140 characters) and the usage of cashtags (a hashtag identifier for financial

L. Correia, L.P. Reis, and J. Cascalho (Eds.): EPIA 2013, LNAI 8154, pp. 355–365, 2013.
© Springer-Verlag Berlin Heidelberg 2013

stocks) can make it a less noisy source of data. Finally, users post very frequently, reacting to events in real-time and allowing a real-time assessment that can be exploited during the trading day.

Regarding the state of the art, in 2004 the landmark paper of Antweiler and Frank [2] studied more than 1.5 million messages posted on Yahoo! Finance, suggesting the influence of post messages in the modeling of financial stock variables. More recently, Bollen et al. [3] measured collective mood states (e.g. "positive", "negative", "calm") through sentiment analysis applied to large scale Twitter data, although tweets were related with generic sentiment (e.g. "Im feeling") and not directly related to stock market. Still, they found an accuracy of 86.7% in the prediction of the Dow Jones Industrial Average daily directions. Sprenger and Welpe [14] have used sentiment analysis on stock related tweets collected during a 6-month period. To reduce noise, they selected Twitter messages containing cashtags of S&P 100 companies. Each message was classified by a Naïve Bayes method trained with a set of 2,500 tweets. Results showed that sentiment indicators are associated with abnormal returns and message volume is correlated with trading volume. Mao et al. [7] surveyed a variety of web data sources (Twitter, news headlines and Google search queries) and tested two sentiment analysis methods to predict stock market behavior. They used a random sample of all public tweets and defined a tweet as bullish or bearish only if it contained the terms "bullish" or "bearish". They showed that their Twitter sentiment indicator and the frequency of occurrence of financial terms on Twitter are statistically significant predictors of daily market returns. Oh and Sheng [9] resorted to a microblogging service exclusively dedicated to stock market. They collected 72,221 micro blog postings from stocktwits.com, over a period of three months. The sentiment of the messages was classified by a bag of words approach [12] that applies a machine learning algorithm J48 classifier to produce a learning model. They verified that the extracted sentiment appears to have strong predictive value for future market directions.

While this literature favors the use of microblogging data to forecast stock market behavior, the obtained results need to be interpreted with caution. According to Timmermann [16], there is in general scarce evidence of return predictability. And most of these studies do not perform a robust evaluation. For instance, only modeling (and not prediction) was addressed in [2] and [14], while very short test periods were performed in [3] (19 predictions), [7] (20 and 30 predictions) and [9] (8 predictions). Also, several works, such as [14][9], require a manual classification of tweets, which is prone to subjectivity and is difficult to replicate.

The main goal of this paper is to overcome the limitations of previous studies, by adopting a more robust evaluation of the usefulness of microblogging data for predicting stock market variables. Similarly to [9], and in contrast with other studies [2][3][14][7], we use StockTwits data. Such resource is more interesting, when compared with Twitter, since it is a social service specifically targeted for investors and traders. Also of note, we analyze a much larger data period (605 days) and adopt a robust fixed-sized rolling window (with different window sizes), leading to a test period that ranges from 305 to 505 predictions. Moreover, rather than predicting direction, such as performed in [9] (which is of lower informative value), we adopt a full regression approach and predict three market variable values for five large US companies and one relevant US index. Aiming to replicate the majority of previous works [14][7][3] we adopt automated methods, using the multiple regression as the base learner model and test five distinct predictive models (including a baseline that does not use microblog data). However, in contrast with previous studies (e.g. [3] and [7] only use MAPE), we use two error metrics (MAPE and RMSE) and adopt the equality of prediction statistical test to make inferences about the statistical significance of the results [4].

The rest of the paper is organized as follows. Section 2 describes the data and methods. Next, Section 3 presents and discusses the research results. Finally, Section 4 concludes with a summary and discussion of the main results.

2   Materials and Methods

2.1   StockTwits and Stock Market Data

Data was collected for five large US companies and one index: Apple (AAPL), Amazon (AMZN), Goldman Sachs (GS), Google (GOOG), IBM (IBM) and Standard and Poor's 500 Index (SPX). These stocks were chosen because they have a substantial posting volume on StockTwits and Chicago Board Options Exchange (CBOE) provides their implied volatility indexes. Therefore, we can process a significant amount of microblogging data that can be more indicative of investors' level of attention and sentiment on these stocks, and use a good volatility estimate. For each stock, we retrieved StockTwits and stock market data from June 1, 2010 to October 31, 2012, in a total of 605 trading days.

StockTwits (stocktwits.com) is a financial platform with more than 200,000 users that share information about the market and individual stocks. Similarly to Twitter, messages are limited to 140 characters and consist of ideas, links, charts and other data. We selected StockTwits content because it is exclusively about investing, resulting in a less noisy data set than collecting from a more generalist microblogging service. Messages were filtered by the company $TICKER tag, i.e., $AAPL, $AMZN, $GOOG, $GS, $IBM, $SPX. A $TICKER tag (cashtag) is composed by the company ticker preceded by the "$" symbol. We collected a larger number of messages (Figure 1), ranging from a total of 7283 (IBM) to 364457 (AAPL) tweets.

The stock market variables here considered are daily returns, volatility and trading volume. Price and volume data were collected from Thompson Reuters Datastream (http://online.thomsonreuters.com/datastream/) and volatility, as measured by their indexes, was collected from the Chicago Board Options Exchange CBOE (http://www.cboe.com/micro/EquityVIX/).

Market returns measure changes in the asset value. We used the adjusted close prices to calculate returns. Adjusted close price is the official closing price adjusted for capital actions and dividends. We computed market returns (R_t) using the following formula [7]:

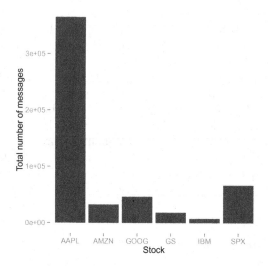

Fig. 1. Total number of StockTwits collected messages per stock

$$R_t = \ln(P_t) - \ln(P_{t-1}) \tag{1}$$

where P_t is the adjusted close price of day t and P_{t-1} is the adjusted close price of the preceding day. Returns provide useful information about the probability distribution of asset prices. This is essential for investors and portfolio managers as they use this information to value assets and manage their risk exposure.

Volatility (σ_t, for day t) is a latent measure of total risk associated with a given investment. Volatility can be estimated using different approaches. Previous studies have found that the model-free implied volatility index in an appropriate estimator of volatility [6]. Estimates of volatility are essential for portfolio selection, financial assets valuation and risk management.

Trading volume (v_t) is the number of shares traded in each day during a trading session. Volume can be used to measure stock liquidity, which in turn has been shown to be useful in asset pricing as several theoretical and empirical studies have identified a liquidity premium. Liquidity can help to explain the cross-section of expected returns [1].

2.2   Regression Models

Similarly to previous works [14][7][3], we adopt a multiple regression model, which is less prone to overfit the data:

$$\hat{y} = f(x_1, ..., x_I) = \beta_0 + \sum_{i=1}^{I} \beta_i x_i \tag{2}$$

where $\hat{y}$ is the predicted value for the dependent variable y (target output), x_i are the independent variables (total of I inputs) and $\beta_0, \ldots, \beta_i$ are the set of parameters to be adjusted, usually by applying a least squares algorithm. Due to its additive nature, this model is easy to interpret and has been widely used in Finance. Moreover, the learning process with such model is very fast, allowing an extensive experimentation of different input variables and a more robust evaluation, with different combinations of training set sizes (Section 2.3).

In this work, we test models that are quite similar to most previous works [14][7][3][9]. This means that for predicting returns we use sentiment data (i.e. the investors opinion about a given stock), while for predicting volatility and volume we explore posting volume indicators (which is a measure of attention). For each financial target, we test five different regression models, aiming to mimic the models proposed in previous works and also testing new variations.

As sentiment indicators, for each stock we count the daily number of messages that contain (case insensitive) the words "bullish" ($Bull_t$, for day t) or "bearish" ($Bear_t$) [7]. Using these two indicators, we compute other variables: bullishness index ($bind_t$) [2][14][9], Twitter Investor Sentiment (TIS_t) [7] and a ratio of TIS ($RTIS_t$, proposed here). These are given by:

$$
\begin{aligned}
bind_t &= \ln\left(\tfrac{1+Bull_t}{1+Bear_t}\right) \\
TIS_t &= \tfrac{Bull_t+1}{Bull_t+Bear_t+1} \\
RTIS_t &= \tfrac{TIS_t}{TIS_{t-1}}
\end{aligned}
\tag{3}
$$

The five tested regression models for predicting the returns are:

$$
\begin{aligned}
\hat{R}_t &= f(R_{t-1}) &&\text{(M1, baseline)} \\
\hat{R}_t &= f(R_{t-1}, \ln(TIS_{t-1})) &&\text{(M2)} \\
\hat{R}_t &= f(R_{t-1}, \ln(RTIS_{t-1})) &&\text{(M3)} \\
\hat{R}_t &= f(bind_{t-1}) &&\text{(M4)} \\
\hat{R}_t &= f(\ln(TIS_{t-1})) &&\text{(M5)}
\end{aligned}
\tag{4}
$$

Regarding the posting volume, we measure two indicators: n_t, the daily number of tweets (for day t); and $MA_t = \tfrac{1}{5}\sum_{k=t-5}^{t} n_k$, the moving average (when considering the last five days). Similarly to predicting the returns, and given that in some cases very high values are found for the target, we opt for modeling the natural logarithm values for volatility and trading volume. For predicting volatility, the tested models are:

$$
\begin{aligned}
\ln(\hat{\sigma}_t) &= f(\ln(\sigma_{t-1})) &&\text{(M1, baseline)} \\
\ln(\hat{\sigma}_t) &= f(\ln(\sigma_{t-1}), \ln(n_{t-1})) &&\text{(M2)} \\
\ln(\hat{\sigma}_t) &= f(\ln(\sigma_{t-1}), \ln(n_{t-1}), \ln(n_{t-2})) &&\text{(M3)} \\
\ln(\hat{\sigma}_t) &= f(\ln(\sigma_{t-1}), \ln(MA_{t-1})) &&\text{(M4)} \\
\ln(\hat{\sigma}_t) &= f(\ln(\sigma_{t-1}), \ln(MA_{t-1}), \ln(MA_{t-2})) &&\text{(M5)}
\end{aligned}
\tag{5}
$$

Finally, for predicting the trading volume, we test:

$$\ln(\hat{v}_t) = f(\ln(v_{t-1})) \qquad\qquad\qquad \text{(M1, baseline)}$$
$$\ln(\hat{v}_t) = f(\ln(v_{t-1}), \ln(\tfrac{n_{t-1}}{n_{t-2}})) \qquad\qquad \text{(M2)}$$
$$\ln(\hat{v}_t) = f(\ln(v_{t-1}), \ln(n_{t-1}), \ln(n_{t-2})) \qquad \text{(M3)} \qquad (6)$$
$$\ln(\hat{v}_t) = f(\ln(v_{t-1}), \ln(MA_{t-1}), \ln(MA_{t-2})) \ \text{(M4)}$$
$$\ln(\hat{v}_t) = f(\ln(v_{t-1}), \ln(\tfrac{MA_{t-1}}{MA_{t-2}})) \qquad\quad \text{(M5)}$$

2.3   Evaluation

To measure the quality of predictions of the regression models, we use two error metrics, Root-Mean-Squared Error (RMSE) and Mean Absolute Percentage Error (MAPE), given by [5]:

$$RMSE = \sqrt{\frac{\sum_{i=1}^{N}(y_i - \hat{y}_i)^2}{N}}$$
$$MAPE = \frac{1}{N}\sum_{i=1}^{N}\left|\frac{y_i - \hat{y}_i}{y_i}\right| \times 100\% \qquad (7)$$

where y_i and $\hat{y}_i$ are the target and fitted value for the i-th day and N is the number of days considered. The lower the RMSE and MAPE values, the better the model. For the sake of correct evaluation, we apply both metrics, which can be used according to diverse goals. RMSE and MAPE compute the mean error but RMSE is more sensitive to high individual errors than MAPE (e.g. useful to avoid large error estimates).

For achieving a robust evaluation, we adopt a fixed-size (of length W) rolling windows evaluation scheme [15]. Under this scheme, a training window of W consecutive samples (last example corresponds to day $t - 1$) is used to fit the model and then we perform one prediction (for day t). Next, we update the training window by removing the first observation in the sample and including the new observation for day t, in order fit the model and predict the value for $t + 1$, an so on. For a dataset of length L, a total of $N = L - W$ predictions (and model trainings) are performed. In this paper, three different window sizes (i.e. $W \in \{100, 200, 300\}$) are explored.

We measure the value of StockTwits based data if the respective regression model is better than the baseline method. As a baseline method, we adopt an $AR(1)$ model. This regression model has only one input: the previous day (t-1) observation. To test the forecasting ability of the models, we apply the equality of prediction statistical test [4], under a pairwise comparison between the tested model and the baseline for MAPE and RMSE metrics.

3   Results

All experiments here reported were conducted using the open source **R** tool [11] and the programming language Python running on a Linux server. The StockTwits posts were delivered in JSON format, processed in **R** using the rjson package and stored using the MongoDB database format by adopting the rmongodb package. The multiple regressions were run using the lm function of **R**.

The predictive errors for the returns are presented in Table 1. An analysis of this table reveals that the baseline is only outperformed in very few cases. Also, when better results are achieved, the differences tend to be small, specially for RMSE. In effect, there are only two cases where the differences are statistically significant (GS, $W = 300$; and SPX, $W = 300$), corresponding to the **M4** model and the MAPE metric.

Table 1. Returns predictive results (**bold** – better than baseline results, $\star$ – p-value < 0.05 and $\diamond$ – p-value < 0.10)

Stock	W	RMSE					MAPE (in %)				
		M1	M2	M3	M4	M5	M1	M2	M3	M4	M5
	100	0.0166	0.0167	**0.0165**	0.0166	0.0166	184	210	190	191	199
AAPL	200	0.0171	0.0172	0.0171	0.0171	0.0171	176	190	193	**174**	188
	300	0.0174	0.0174	0.0174	0.0174	0.0174	174	187	187	**171**	187
	100	0.0225	0.0228	0.0226	0.0225	0.0226	160	160	**156**	155	153
AMZN	200	0.0233	0.0236	0.0235	0.0233	0.0235	111	118	**109**	116	121
	300	0.0240	0.0243	0.0241	0.0241	0.0243	104	108	110	115	115
	100	0.0168	0.0169	0.0169	0.0171	0.0169	138	262	184	176	257
GOOG	200	0.0177	0.0179	0.0177	0.0177	0.0178	119	161	127	140	147
	300	0.0161	0.0161	**0.0160**	0.0161	0.0161	117	124	123	134	119
	100	0.0124	0.0127	0.0124	0.0124	0.0126	137	139	147	145	**135**
IBM	200	0.0127	0.0128	0.0127	0.0127	0.0128	145	**145**	**145**	154	**137**
	300	0.0130	0.0131	**0.0128**	**0.0128**	0.0130	129	**128**	137	145	**129**
	100	0.0213	0.0214	0.0213	**0.0212**	**0.0212**	129	146	140	131	141
GS	200	0.0225	0.0226	0.0225	0.0225	0.0225	119	135	124	**117**	124
	300	0.0234	0.0235	0.0235	0.0234	0.0234	116	124	**113**	**108***	110
	100	0.0119	0.0121	0.0119	0.0119	0.0120	141	182	**141**	169	173
SPX	200	0.0126	0.0126	**0.0125**	0.0126	0.0126	173	193	**159**	166	168
	300	0.0121	0.0121	**0.0119**	**0.0119**	**0.0119**	166	189	**138**	**142$^\diamond$**	151

Table 2 presents the forecasting results for volatility ($\ln(\sigma_t)$). For some stocks (e.g. AAPL and MAPE; AMZN and RMSE) there are several models that present lower errors when compared with the baseline. However, these differences are not statistically significant, since none of the tested models that include posting indicators outperformed the baseline.

The performances of the trading volume regressions is shown in Table 3. In contrast with the results obtained for the previous financial variables (returns and volatility), in this case there are some interesting results that are worth mentioning. There is a total of 16 models that statistically outperform the baseline under the RMSE metric. In particular, we highlight **M2**, which is statistically better than the baseline in 8 cases (corresponding to four stocks: AAPL, AMZN, GOOG and IBM), followed by **M3** (which outperforms the baseline in 5 cases, for AMZN and GOOG).

Table 2. Volatility predictive results (**bold** – better than baseline results)

Stock	W	RMSE					MAPE (in %)				
		M1	M2	M3	M4	M5	M1	M2	M3	M4	M5
	100	0.069	0.070	0.071	0.070	0.069	1.399	**1.397**	1.421	**1.392**	1.400
AAPL	200	0.070	0.070	0.070	0.070	**0.069**	1.389	**1.379**	1.389	**1.374**	1.391
	300	0.068	0.068	0.068	0.069	0.068	1.338	**1.334**	**1.337**	**1.335**	1.357
	100	0.062	**0.061**	**0.061**	**0.062**	**0.061**	1.090	1.121	1.121	1.119	1.135
AMZN	200	0.066	**0.064**	**0.063**	**0.065**	**0.063**	1.129	1.151	1.145	1.139	1.150
	300	0.067	**0.066**	**0.065**	**0.067**	**0.064**	1.132	1.142	1.140	**1.126**	1.144
	100	0.064	**0.064**	**0.064**	0.064	**0.064**	1.250	1.307	1.324	1.308	1.321
GOOG	200	0.065	**0.065**	**0.064**	0.065	**0.064**	1.230	1.318	1.326	1.279	1.301
	300	0.063	0.063	**0.063**	0.064	**0.063**	1.196	1.280	1.282	1.236	1.275
	100	0.070	**0.070**	**0.070**	**0.070**	**0.070**	1.323	1.334	1.336	1.335	1.338
GS	200	0.072	0.072	0.073	0.073	0.073	1.280	1.314	1.322	1.307	1.319
	300	0.065	0.065	0.065	0.066	0.065	1.171	1.212	1.208	1.212	1.207
	100	0.067	0.067	0.067	0.068	**0.067**	1.458	1.475	1.471	1.479	1.514
IBM	200	0.069	**0.069**	**0.069**	0.070	**0.068**	1.476	1.480	**1.471**	**1.471**	1.489
	300	0.055	**0.054**	**0.053**	**0.055**	**0.054**	1.248	**1.248**	**1.241**	**1.247**	1.267
	100	0.074	**0.074**	**0.074**	**0.074**	0.075	1.787	1.789	1.797	1.804	1.811
SPX	200	0.075	0.075	0.076	0.075	0.075	1.781	1.803	1.810	1.800	1.802
	300	0.069	0.070	0.070	0.069	0.069	1.695	1.734	1.736	1.723	1.725

Table 3. Trading volume predictive results (**bold** – better than baseline results, ⋆ – p-value < 0.05 and ⋄ – p-value < 0.10)

Stock	W	RMSE					MAPE (%)				
		M1	M2	M3	M4	M5	M1	M2	M3	M4	M5
	100	0.300	0.301	0.304	0.304	0.302	1.406	1.423	1.444	1.440	1.422
AAPL	200	0.279	**0.279⋆**	0.280	0.281	0.280	1.334	1.340	1.358	1.355	1.340
	300	0.281	**0.281⋆**	0.283	0.283	0.282	1.344	1.351	1.371	1.363	1.347
	100	0.322	**0.321**	0.324	0.322	**0.320**	1.599	1.616	1.631	1.645	1.635
AMZN	200	0.326	**0.324⋄**	**0.322⋆**	**0.323⋄**	**0.325⋄**	1.631	1.647	1.636	1.656	1.659
	300	0.327	**0.324⋆**	**0.324⋄**	0.328	**0.326**	1.633	1.644	1.647	1.676	1.659
	100	0.328	**0.325⋄**	**0.326⋆**	0.328	0.329	1.662	1.669	1.693	1.684	1.674
GOOG	200	0.330	**0.323⋆**	**0.323⋆**	**0.325⋄**	**0.325**	1.675	**1.657**	1.680	1.687	**1.671**
	300	0.315	**0.308⋆**	**0.311⋆**	**0.312**	**0.310**	1.628	**1.605**	1.650	1.648	**1.615**
	100	0.300	**0.299**	0.301	0.303	0.302	1.466	1.469	1.482	1.497	1.479
IBM	200	0.309	**0.308**	**0.307**	**0.309**	0.309	1.557	1.564	1.562	1.558	1.568
	300	0.319	**0.316⋆**	**0.317**	0.320	0.320	1.636	1.636	**1.636**	1.642	1.653
	100	0.323	0.324	0.325	0.329	0.327	1.568	1.573	1.577	1.596	1.590
GS	200	0.325	0.326	0.328	0.329	0.327	1.578	1.588	1.592	1.600	1.590
	300	0.321	0.322	0.322	0.323	0.321	1.583	1.585	1.588	1.601	1.588
	100	0.203	0.204	0.206	0.205	0.205	0.623	0.634	0.630	0.626	0.629
SPX	200	0.164	0.164	0.164	0.164	0.164	0.538	0.540	**0.535**	**0.532**	**0.536**
	300	0.160	0.160	0.161	0.160	0.160	0.526	0.528	0.531	**0.526**	**0.526**

For demonstration purposes, Figure 2 shows the quality of the fitted results for the best model for Google (**M2**, $W = 300$). The left of the plot shows the last fifty out-of-sample observations and predicted values (of a total of 305), revealing an interesting fit of the predictions. The right of the plot presents the scatter plot of observed versus predicted values (with all 305 predictions), showing the adequacy of the proposed linear model.

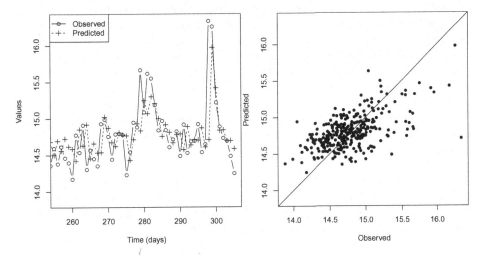

Fig. 2. Example of the last fifty trading volume ($\ln(v_t)$ out-of-sample values) and predictions for Google (GOOG), **M2** and $W = 300$ (left); and full scatter plot (with 305 predicted points) of observed versus predicted values (bottom, diagonal line denotes the perfect fit, natural logarithm values are used in both axis)

4 Conclusions

The main purpose of this study is to provide a more robust assessment about the relevance of microblogging data for forecasting three valuable stock market variables: returns, volatility and trading volume. We focused on a very recent and large dataset of messages collected from StockTwits, a social network service specifically targeted for communication about markets and their individual stocks. Following the recent related studies, we addressed two types of daily data from five large US companies and one major US index, sentiment indicators and posting volume, and explored four regression models that included input variables based on these microblogging indicators. However, and in contrast with previous works, we performed a more robust evaluation of the forecasting models, using fixed-sized rolling windows (with different window sizes), leading to much larger test periods (from 305 to 505 predictions). Also, we used two error metrics (MAPE and RMSE) and the equality of prediction statistical test, to compare the regression results with a baseline method (that uses only one input, the previous day stock market variable).

The results presented here suggest that predicting stock market variables using microblogging data, such as returns and volatility, is a much more complex and harder task than the previous and quite recent works presume. While this is an attractive research line, some caution is required when promising forecasting ability. This is not surprising considering the lack of support for return predictability in the Finance literature [16]. In this paper, we found scarce evidence for the utility of the tested sentiment variables when predicting returns, and of posting volume indicators when forecasting volatility. However, and aligned with the state of the art, we found interesting results when assessing the value of using posting volume for predicting trading volume, for two of the proposed regression models. We highlight that these results were obtained using the large test period so far, and much more stringent evaluation methods than previously.

In future work, we intend to explore more sophisticated parsers and lexicons, more adjusted to stock market terminology, to check if these can improve the investors' sentiment indicators. Also, we aim to address more complex regression models, using more time lags and complex base learners (e.g. Support Vector Machine [13]).

Acknowledgments. This work is funded by FEDER, through the program COMPETE and the Portuguese Foundation for Science and Technology (FCT), within the project FCOMP-01-0124-FEDER-022674. The also authors wish to thank StockTwits for kindly providing their data.

References

1. Amihud, Y., Mendelson, H., Pedersen, L.H.: Liquidity and Asset Prices. Foundations and Trends in Finance 1(4), 269–364 (2007)
2. Antweiler, W., Frank, M.Z.: Is all that talk just noise? the information content of internet stock message boards. The Journal of Finance 59(3), 1259–1294 (2004)
3. Bollen, J., Mao, H., Zeng, X.: Twitter mood predicts the stock market. Journal of Computational Science 2(1), 1–8 (2011)
4. Harvey, D., Leybourne, S., Newbold, P.: Testing the equality of prediction mean squared errors. International Journal of Forecasting 13(2), 281–291 (1997)
5. Hyndman, R.J., Koehler, A.B.: Another look at measures of forecast accuracy. International Journal of Forecasting 22(4), 679–688 (2006)
6. Jiang, G.J.: The Model-Free Implied Volatility and Its Information Content. Review of Financial Studies 18(4), 1305–1342 (2005)
7. Mao, H., Counts, S., Bollen, J.: Predicting financial markets: Comparing survey, news, twitter and search engine data. arXiv preprint arXiv:1112.1051 (2011)
8. Nofsinger, J.R.: Social mood and financial economics. The Journal of Behavioral Finance 6(3), 144–160 (2005)
9. Oh, C., Sheng, O.R.L.: Investigating predictive power of stock micro blog sentiment in forecasting future stock price directional movement. In: ICIS 2011 Proceedings (2011)
10. Peterson, R.L.: Affect and financial decision-making: How neuroscience can inform market participants. The Journal of Behavioral Finance 8(2), 70–78 (2007)
11. R Core Team: R: A Language and Environment for Statistical Computing. R Foundation for Statistical Computing, Vienna, Austria (2012) ISBN 3-900051-07-0

12. Schumaker, R.P., Chen, H.: Textual analysis of stock market prediction using breaking financial news: The azfin text system. ACM Transactions on Information Systems (TOIS) 27(2), 12 (2009)
13. Smola, A., Schölkopf, B.: A tutorial on support vector regression. Statistics and Computing 14, 199–222 (2004)
14. Sprenger, T., Welpe, I.: Tweets and trades: The information content of stock microblogs. Social Science Research Network Working Paper Series, pp. 1–89 (2010)
15. Tashman, L.J.: Out-of-sample tests of forecasting accuracy: an analysis and review. International Journal of Forecasting 16(4), 437–450 (2000)
16. Timmermann, A.: Elusive return predictability. International Journal of Forecasting 24(1), 1–18 (2008)

Predicting the Future Impact
of Academic Publications*

Carolina Bento, Bruno Martins, and Pável Calado

Instituto Superior Técnico, INESC-ID
Av. Professor Cavaco Silva
2744-016 Porto Salvo, Portugal
{carolina.bento,bruno.g.martins,pavel.calado}@ist.utl.pt

Abstract. Predicting the future impact of academic publications has many important applications. In this paper, we propose methods for predicting future article impact, leveraging digital libraries of academic publications containing citation information. Using a set of successive past impact scores, obtained through graph-ranking algorithms such as PageRank, we study the evolution of the publications in terms of their yearly impact scores, learning regression models to predict the future PageRank scores, or to predict the future number of downloads. Results obtained over a DBLP citation dataset, covering papers published up to the year of 2011, show that the impact predictions are highly accurate for all experimental setups. A model based on regression trees, using features relative to PageRank scores, PageRank change rates, author PageRank scores, and term occurrence frequencies in the abstracts and titles of the publications, computed over citation graphs from the three previous years, obtained the best results.

1 Introduction

Citations between articles published in academic digital libraries constitute a highly dynamic structure that is continuously changing, as new publications and new citations are added. Moreover, the graph of citations between publications can provide information for estimating the impact of particular publications, through algorithms such as PageRank, since highly cited papers are more likely to be influential and to have a high impact on their fields. Ranking papers according to their potential impact is thus a highly relevant problem, given that this can enable users to effectively retrieve relevant and important information from digital libraries. Having accurate prediction methods for estimating the impact of recently published papers is particularly important for researchers, since articles with high future impact ranks can be more attractive to read and should be presented first when searching for publications within digital libraries.

---

* This work was partially supported by Fundação para a Ciência e a Tecnologia (FCT), through project grants with references UTA-EST/MAI/0006/2009 (REACTION) and PTDC/EIA-EIA/109840/2009 (SInteliGIS), as well as through PEst-OE/EEI/LA0021/2013 (INESC-ID plurianual funding).

L. Correia, L.P. Reis, and J. Cascalho (Eds.): EPIA 2013, LNAI 8154, pp. 366–377, 2013.

In this paper, we propose a framework that enables the prediction of the impact ranking of academic publications, based on their previous impact rankings. Specifically, given a series of time-ordered rankings of the nodes (i.e., the individual publications) from a citation graph, where each node is associated with its ranking score (e.g., the PageRank score) for each time-stamp, we propose a learning mechanism that enables the prediction of the node scores in future times. Through the formalism of regression trees, we propose to capitalize on existing trends through the changes in impact rankings between different snapshots of the citation graph, in order to accurately predict future PageRank scores and the future number of downloads. Moreover, we also experimented with the use of features derived from the textual abstracts and titles of the publications, in an attempt to capture trending topics. We evaluate the prediction quality through the correlation between the predicted ranked lists and the actual ranking lists, and through the error rates computed between the predictions and the correct results. The obtained results show that there is a significant correlation between the predicted ranked lists and the actual impact ranking lists, therefore revealing that this methodology is suitable for impact score prediction.

The remaining of this paper is organized as follows: Section 2 presents related work. Section 3 describes the proposed approaches for predicting the future impact of academic publications, detailing both the computation of impact estimates at a given time, and the machine learning models for making predictions. Section 4 presents the experimental validation of the proposed approaches, describing the evaluation protocol and the obtained results. Finally, Section 5 presents our conclusions and points directions for future work.

2 Related Work

The PageRank algorithm is a well-known method for ranking nodes in a graph according to their importance or prestige. It was originally proposed in the context of the Google search engine, and it has been extensively studied [9]. Many authors have proposed the application of PageRank, or of adaptations of this particular algorithm, to measure impact in scientific publication networks encoding citation and/or co-authorship information [2, 18].

An interesting approach that aims at approximating PageRank values without the need of performing the computations over the entire graph is that of Chien et al. [3]. The authors propose an efficient algorithm to incrementally compute approximations to PageRank scores, based on the evolution of the link structure of the Web graph. Davis and Dhillon proposed an algorithm that offers estimates of cumulative PageRank scores for Web communities [4]. In our work we also propose algorithms for estimating PageRank scores, but instead focusing on the prediction of future scores, based on previous PageRank computations.

Specifically focusing on PageRank computation over time-dynamic networks encoding citations, Radicchi et al. divided the entire data period into homogeneous intervals, containing equal numbers of citations, and then applied a PageRank-like algorithm to rank papers and authors within each time slice, thereby enabling them to study how an author's influence changes over time [12]. Lerman et al. proposed a novel centrality metric for dynamic network analysis that is similar to PageRank, but exploiting the intuition that, in order for one node in a dynamic network to influence another over some period of time, there must exist a path that connects the source and destination nodes through intermediaries at different times [10]. The authors used their dynamic centrality metric to study citation information from articles uploaded to the theoretical high energy physics (hep-th) section of the arXiv preprints server, obtaining results contrasting to those reached by static network analysis methods.

Sayyadi and Getoor suggested a new measure for ranking scientific articles, based on future citations [13]. The authors presented the FutureRank model, based on publication time and author prestige, that predicts future citations. FutureRank implicitly takes time into account by partitioning data in the temporal dimension, using data in one period to predict a paper's ranking in the next period. The FutureRank scores are shown to correlate well with the paper's PageRank score computed on citation links that will appear in the future.

Kan and Thi have partially addressed the problem of predicting impact scores, by presenting a study that focused on Web page classification, based on URL features [6]. In their study, the authors also report on experiments concerning with predicting PageRank scores for graphs of hyperlinks between Web pages, using the extracted URL features and linear regression models.

The works that are perhaps more similar to ours are those of Vazirgiannis et al. [16] and of Voudigari et al. [17]. Vazirgiannis et al. presented an approach for predicting PageRank scores for Web pages, generating Markov Models from historical ranked lists and using them for making the predictions. Voudigari et al. extended this method, comparing models based on linear regression and high-order Markov models. Although both these works aimed at predicting PageRank for Web graphs, the authors evaluated their methods on co-authorship networks built from DBLP data. In this paper, we instead report on experiments made over citation networks built from DBLP data, aiming at the prediction of both PageRank scores and number of downloads for publications. We also relied on a highly-robust regression approach for learning to make the predictions, namely ensemble models based on Gradient Boosting Regression Tress (GBRT) [11].

3   Predicting the Future Impact of Publications

In this section, we present a learning method for predicting the future impact of academic publications, leveraging on patterns existing in the ranking evolution of the publications, and on the textual contents of the titles and abstracts. Given a set of successive snapshots for the graph of citations between publications, we generate, for each publication, a sequence of features that captures the trends of this publication through the previous snapshots. For each publication, we also generate features based on the publication date, and based on the words appearing on the title and abstract. We then use these features of previous snapshots as training data for a learning method, afterwards trying to predict the future impact of particular publications, based on the previous snapshots. Figure 1 presents a general overview on the proposed methodology.

Let G_{t_i} be a snapshot of the citation graph, capturing citations between papers published before the timestamp t_i that is associated to the snapshot. Let $N_{t_i} = |G_{t_i}|$ be the number of publications at time t_i. In the case of citation networks for academic publications, we have that $N_{t_i} \leq N_{t_{i+1}}$. We also assume the existence of a function $rank(p, t_i)$ that provides an influence estimate for a publication $p \in G_{t_i}$, according to some criterion. In this paper, we used the original PageRank algorithm, over the citation graphs, to compute $rank(p, t_i)$, although other impact estimates could also have been used instead.

The original PageRank formulation states that a probability distribution over the nodes in a graph, encoding importance scores for each of the nodes, can be computed by leveraging links through the following equation, which represents a random walk with restarts in the graph:

$$Pr(p_i) = \frac{1-d}{N} + d \sum_{p_j \in I(p_i)} \frac{Pr(p_j)}{L(p_j)} \qquad (1)$$

In the formula, $Pr(p_i)$ refers to the PageRank score of a node p_i, $I(p_i)$ is the set of nodes that link to p_i (i.e., the citations made to article p_i), $L(p_j)$ (i.e., the citations made in article p_j towards other articles) is the number of outbound links on node p_j, and N is the total number of nodes in the graph. The parameter d controls the random transitions to all nodes in the graph (i.e., the restarts), with a residual probability that is usually set to $d = 0.85$. In our experiments, the computation of PageRank relied on the implementation present in the WebGraph package[1], from the Laboratory of Web Algorithms of the University of Milan.

Let $x_p = (r_{p_1}, \ldots, r_{p_m})$ encode the *rank* values for a publication p at time points $t = (t_1, \ldots, t_m)$, and let $M = (N \times m)$ be a matrix storing all the observed *rank* values, so that each row corresponds to a publication and each column corresponds to a time point. Given these values, we wish to predict the *rank* value r_{p_*} for each publication at some time t_*, corresponding to a future time point. To do this, we propose to leverage on a regression model, using the k previous *rank* values as features, together with other features that capture (i) trends in the evolution of influence scores, and (ii) intrinsic properties of the publications, that can be used to group together similar sequences of impact scores (e.g., having the same authors or using the same textual terms).

Regarding the first group of features from the above enumeration, we propose to use the Rank Change Rate between consecutive snapshots of the citation graph, in order to capture trends in the evolution of the PageRank impact scores. The Rank Change Rate (Racer) between two instances t_{i-1} and t_i is given by the following equation:

$$\text{Racer}(p, t_i) = \frac{rank(p, t_i) - rank(p, t_i - 1)}{rank(p, t_i)} \qquad (2)$$

---

[1] Available at http://law.dsi.unimi.it/software.php#pagerank

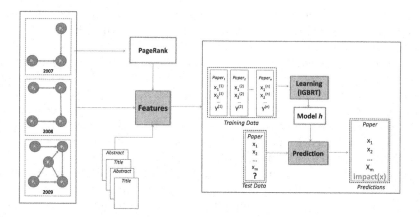

Fig. 1. The framework for predicting the future impact of academic publications

As for the second group of features, we used the assumption that publications from the same authors are likely to exhibit similar trends in the evolution of their impact scores. Thus, for each publication p, we computed the average and the maximum PageRank scores of all publications having an author in common with the set of authors of publication p.

To capture the intuition that impact metrics evolve differently for recent or older publications, we also used a feature that corresponds to the difference between the year for which we are making the predictions, and the year of publication. Finally, on what concerns the textual features, we used term frequency scores for the top 100 most frequent tokens in the abstracts and titles of the publications, not considering terms from a standard English stop-word list.

The above features were integrated into an ensemble regression approach based on the formalism of Gradient Boosting Regression Trees (GBRT).

We developed two different models, using the formulation of GBRT, for better understanding the impact of the combination of the aforementioned features, namely the (i) *Age Model*, and the (ii) *Text Model*. Both these regression models share a set of common features, which are:

- PageRank score of the publication in previous year(s) (Rank);
- Rank Change Rate (Racer) score towards the previous year(s);
- Average and maximum PageRank scores of all publications having an author in common with the set of authors of the publication (Auth);

In addition to the these common features, the *Age Model* also includes a feature that indicates the age of the publication, while the *Text Model* includes a feature indicating the age of the publication, as well as, features for the term frequency scores of the top 100 most frequent tokens in the abstracts and titles of the publications, up until the current date.

By experimenting with different groups of features, we can compare the impact of the amount of information that each model is given.

The actual learning approach that was used to build both the *Age* and *Text* models is named Initialized Gradient Boosting Regression Trees (IGBRT). This is an ensemble learning technique that, instead of being initialized with a constant function like in traditional Gradient Boosting approaches, it is initialized with the predictions obtained through the application of the Random Forests (RF) technique [1]. The IGBRT algorithm has, thus, a first step, in which the RF technique is applied, and then a final step, in which the traditional Gradient Boosting Regression Trees (GBRT) technique is applied [11]. The algorithm for GBRT evolved from the application of boosting methods to regression trees, through the idea that each step of the algorithm (i.e., the fitting of each of the regression trees for the final model) can involve the use of gradient descent to optimize any continuous, convex, and differentiable loss function (e.g., the squared loss). In the implementation that we have used, the individual regression trees that make up the boosting ensemble have a depth of 4, and they are are built through a version of the traditional CART algorithm, greedily building a regression tree that minimizes the squared-loss and that, at each split, uniformly samples k features and only evaluates those as candidates for splitting.

The general idea in GBRT is to compute a sequence of trees, where each successive tree is built over the prediction residuals of the preceding tree [5]. More specifically, the average y-value can be used as a first guess for predicting all observations (in the case of the IGBRT approach, the first guesses are instead given by the RF model). The residuals from the model are computed, and a regression tree is then fit to these residuals. The regression tree can then used to predict the residuals (i.e., in the first step, this means that a regression tree is fit to the difference between the observation and the average y-value, and the tree can then predict those differences), and these predictions are used to set the optimal step length (i.e., the weight of the current tree in the final model). The boosting regression model, consisting of the sum of all previous regression trees, is updated to reflect the current regression tree. The residuals are updated to reflect the changes in the boosting regression model, and a new tree is then fit to the new residuals, proceeding up to a given number of steps. Each term of the resulting regression model thus consists of a tree, and each tree fits the residuals of the prediction of all previous trees combined. The additive weighted expansions of trees can eventually produce an excellent fit of the predicted values to the observed values, even if the specific nature of the relationships between the predictor variables and the dependent variable of interest is very complex (i.e., nonlinear in nature).

4   Experimental Validation

To validate the proposed approach, we used a dataset[2] made available in the context of the ArnetMiner project, which encodes citation information between academic publications listed in the DBLP service [15]. The dataset contains information about 1,572,277 papers published until August 2011, having a total

---

[2] Available at http://www.arnetminer.org/citation.

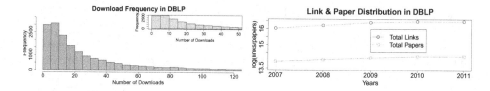

Fig. 2. Download frequency in the ACM DL for the papers in the dataset

Fig. 3. Distribution for the links and the total number of papers

of 2,084,019 citation relationships. We also enriched the original ArnetMiner dataset with information about the number of downloads associated to each paper, by collecting this information in January 2012 from the ACM Digital Library service. We collected download information for a total of 17,973 papers. Figure 2 presents the distribution for the number of papers associated with a given number of downloads, while Figure 3 presents the number of papers and the number of citations, collected for each different year.

The dataset was split into five different groups, thus generating five graphs, corresponding to citations between papers published until the years of 2007, 2008, 2009, 2010 and 2011, respectively. PageRank scores were computed for each of these five graphs. We then evaluated our results on the tasks of (i) predicting the PageRank scores for the years of 2010 and 2011, based on the PageRank scores from the previous k years (with $1 \leq k \leq 3$), and (ii) predicting the number of downloads for papers in 2012, based on the PageRank scores from the most recent and from the previous k years. In terms of the prediction method, we used an implementation of the IGBRT algorithm that is provided by the RT-Rank project[3] [11]. Table 1 presents a brief statistical characterization of the DBLP dataset and of the five considered yearly graphs, showing the number of publications, citations, and authors per year.

To measure the quality of the results, we used Kendall's τ rank correlation coefficient [8], which consists of a non-parametric rank statistic that captures the strength of the association between two variables, with sample size n. Kendall's τ rank correlation coefficient is given by the formula below:

$$\tau = \frac{|\text{concordant pairs}| - |\text{discordant pairs}|}{\frac{1}{2}n(n-1)} \tag{3}$$

Kendall's τ rank correlation coefficient varies from $+1$ through -1, being $+1$ if the two rankings are equal and -1 if they are the exact opposite.

We also used Spearman's Rank Correlation Coefficient to measure the quality of the obtained results [14]. In this case, if X_1 and X_2 are two variables with corresponding ranks $x_{1,i}$ and $x_{2,i}$, and if n is the sample size, then their Spearman Rank Correlation Coefficient is given by the following equation:

$$\rho = 1 - \frac{6 \times \sum_{i=1}^{n}(x_{1,i} - x_{2,i})^2}{n(n^2 - 1)} \tag{4}$$

---

[3] Available at https://sites.google.com/site/rtranking/

As in the case of Kendall's τ, the Spearman correlation corresponds to a value between $+1$ and -1. Notice that Kendall's τ penalizes dislocations in the ranked lists independently of the distance of the dislocation, whereas Spearman's ρ does this through the square of the distance. Thus, Kendall's τ penalizes two independent swaps as much as two sequential swaps, while Spearman's ρ gives a stronger penalty to the latter than to the former. Both Spearman's ρ and Kendall's τ measure the quality of the ranked lists independently of the actual impact scores that are produced as estimates (i.e., only the relative ordering is taken into account).

In order to measure the accuracy of the prediction models, we used the normalized root-mean-squared error (NRMSE) metric between our predictions and the actual ranks, which is given by the following formula:

$$\text{NRMSE} = \frac{\sqrt{\frac{\sum_{i=1}^{n}(x_{1,i}-x_{2,i})^2}{n}}}{x_{\max} - x_{\min}} \tag{5}$$

In the formula, $x_{\min}$ and $x_{\max}$ correspond, respectively, to the minimum and maximum values observed in the sample of objects being predicted, and n corresponds to the sample size. We also used the Mean Absolute Error (MAE) metric, which in turn corresponds to the formula bellow, where $x_{1,i}$ corresponds to the prediction and $x_{2,i}$ to the real value:

$$\text{MAE} = \frac{1}{n} \sum_{i=1}^{n} |x_{1,i} - x_{2,i}| \tag{6}$$

Tables 2 and 3 present the obtained results, respectively for the prediction of the PageRank scores for the years of 2010 and 2011, and for the prediction of the number of downloads for each paper in the dataset.

Considering the prediction of the PageRank scores for the year of 2010, both models (i.e., the *Age Model* and the *Text Model*) have provided very similar results. Both models are also improved if we consider more input information (i.e., comparing the three groups of features and also within the same groups,

Table 1. Statistical characterization of the DBLP dataset

	Publications	Citations	Authors	Papers with Downloads	Papers with Abstract	Average Terms Per Paper
Overall	1,572,277	2,084,019	601,339	17,973	529,498	104
2007	135,277	1,150,195	330,001	15,516	343,837	95
2008	146,714	1,611,761	385,783	17,188	419,747	98
2009	155,299	1,958,352	448,951	17,973	504,900	101
2010	129,173	2,082,864	469,719	17,973	529,201	103
2011	8,418	2,083,947	469,917	17,973	529,498	104

the quality of the results improves consistently). Only for the set of features that combines the PageRank score of one previous year (Rank $k = 1$) with its respective PageRank Change Rate (Racer) and the average and maximum PageRank score of the author (Auth), the *Age Model* is outperformed by the *Text Model*. Comparing the error rate for the same year, one can acknowledge that, for both models, as we add more information, the error rate increases, causing a deviation in the results. Nevertheless, for the first two groups of features, the *Text Model* has a lower error rate than the *Age Model*, while the opposite happens for the third group of features. Having computed the absolute error for all the groups of features in both models, the results show that, on average, the *Text Model* has always a lower absolute error than the *Age Model*.

For the year of 2011, as we add more information to the models, the *Text Model* outperforms the *Age Model*, as shown in the last two sets of features from the third group, i.e., PageRank score, Rank Change Rate score and Average and Maximum PageRank scores of all publications having an author in common with the set of authors of the publication. Also, in the scenario in which the models only have information about the immediately previous PageRank score, the *Age Model* is again outperformed by the *Text Model*. Nevertheless, when considering the error rate for both models for the year of 2011, the *Text Model* has an overall higher error rate than the *Age Model*, showing that, even though the quality of the predicted results is lower in the *Age Model*, the rankings are more accurate.

As occurred in the computation of the absolute error for the year 2010, in all the groups of features in both models, the results for the year of 2011 show that, on average, the *Text Model* has a lower NRMSE than the *Age Model*.

Table 2. Results for the prediction of future impact scores. Highlighted in bold are the best results for each metric, according to group of features, and for each year.

Model	Features	PageRank 2010			PageRank 2011		
		ρ	τ	NRMSE	ρ	τ	NRMSE
Age	Rank k = 1	0.97251	0.91640	**0.00032**	0.99299	0.98371	0.00011
	Rank k = 2	0.98365	0.93819	0.00062	**0.99991**	**0.99948**	**0.00001**
	Rank k = 3	**0.98907**	**0.95064**	0.00064	0.99990	0.99938	0.00048
	Racer + Rank k = 1	0.97245	0.91736	**0.00035**	0.99989	**0.99940**	0.00023
	Racer + Rank k = 2	0.98371	0.93876	0.00065	**0.99990**	0.99930	**0.00016**
	Racer + Rank k = 3	**0.98887**	**0.94937**	0.00066	0.99524	0.98662	0.00055
	Auth + Racer + Rank k = 1	0.96752	0.90985	**0.00054**	0.99985	0.99945	0.00025
	Auth + Racer + Rank k = 2	0.98405	0.93555	0.00083	0.99984	0.99934	0.00030
	Auth + Racer + Rank k = 3	**0.98925**	**0.94687**	0.00070	0.99380	0.98285	0.00053
Text	Rank k = 1	0.97087	0.91017	**0.00036**	0.99921	0.99797	**0.00025**
	Rank k = 2	0.98310	0.93104	0.00063	**0.99980**	**0.99924**	0.00045
	Rank k = 3	**0.98869**	**0.94515**	0.00063	0.99950	0.99834	0.00058
	Racer + Rank k = 1	0.97112	0.90989	**0.00055**	0.99943	0.99845	**0.00016**
	Racer + Rank k = 2	0.98320	0.93144	0.00067	**0.99973**	**0.99907**	0.00019
	Racer + Rank k = 3	**0.98880**	**0.94701**	0.00067	0.99941	0.99807	0.00064
	Auth + Racer + Rank k = 1	0.97052	0.99845	**0.00016**	0.99970	0.99906	**0.00025**
	Auth + Racer + Rank k = 2	0.98370	**0.99907**	0.00019	0.99986	0.99934	0.00028
	Auth + Racer + Rank k = 3	**0.98884**	0.99807	0.00064	**0.99988**	**0.99939**	0.00070

Regarding the prediction of download numbers, one can acknowledge that the *Text Model* shows evidence of better results. Moreover, in the *Age Model*, we can verify that adding information about PageRank Change Rates to the previous PageRank scores affects the results negatively, while combining previous PageRank scores with PageRank Change Rates, and average and maximum PageRank score of the author, provides better results, as well as, a lower error rate. From this fact, we can conclude that the *Age Model* provides a more accurate prediction as it includes more information, while the opposite happens in all groups of the *Text Model* (i.e., within the same group, as we add more information to the model, the quality of the results decreases, even though they are far better than the corresponding results in the *Age Model*).

We can also verify that the *Age Model*, for the groups of features that only include previous PageRank scores, and for the ones that combine previous PageRank scores with PageRank Change Rates and average and maximum PageRank score of the author, have a lower error rate than the corresponding groups in the *Text Model*. Even though with better overall results, the *Text Model* has a greater error rate than the *Age Model* for the prediction of download numbers. In what concerns the MAE, the results showed that, overall, the *Text Model* has a lower absolute error rate than the *Age Model*, in all groups except for the third.

From the results in Tables 2 and 3, we can see that predicting the number of downloads is a harder task than predicting the future PageRank scores. Also, predicting the future PageRank scores for 2011 turned out to be easier than making the same prediction for the year of 2010, which may be due to the combination of some aspects, namely the fact that we naturally took more papers into account while predicting the future PageRank scores for 2011 (i.e., we used

Table 3. Results for the prediction of future download numbers. Highlighted in bold are the best results for each metric, according to group of features, and for each year.

Model	Features	ρ	τ	NRMSE	MAE
Age	Rank k = 1	0.38648	0.27430	0.00806	31.78320
	Rank k = 2	0.42215	0.30015	0.00310	28.29388
	Rank k = 3	**0.43232**	**0.30810**	**0.00292**	**25.6090**
	Racer + Rank k = 1	**0.43966**	0.30766	**0.00767**	30.14214
	Racer + Rank k = 2	0.33702	0.47472	0.00784	29.38590
	Racer + Rank k = 3	0.33134	**0.46124**	0.00883	**27.01311**
	Auth + Racer + Rank k = 1	0.33776	0.25584	0.01547	37.21956
	Auth + Racer + Rank k = 2	0.53355	0.38949	0.00881	28.10313
	Auth + Racer + Rank k = 3	**0.54069**	**0.39625**	**0.00786**	**25.62784**
Text	Rank k = 1	0.52502	0.38370	**0.00912**	39.33091
	Rank k = 2	**0.52612**	**0.38496**	0.00913	39.30055
	Rank k = 3	0.50600	0.36748	0.00932	**36.39041**
	Racer + Rank k = 1	**0.53254**	**0.38880**	**0.00912**	39.33073
	Racer + Rank k = 2	0.52240	0.38230	0.00913	39.30457
	Racer + Rank k = 3	0.50874	0.37034	0.00932	**36.39390**
	Auth + Racer + Rank k = 1	**0.57098**	**0.42348**	**0.00912**	39.30848
	Auth + Racer + Rank k = 2	0.56513	0.41801	0.00913	39.29639
	Auth + Racer + Rank k = 3	0.56090	0.41486	0.00932	**36.36546**

more 20,768 papers than in 2010) providing, therefore, more information to the models, for training and for testing.

In sum, we have shown that the proposed framework based on ensemble regression models, offers accurate predictions, providing an effective mechanism to support the ranking of papers in academic digital libraries.

5   Conclusions and Future Work

In this paper, we proposed and evaluated methods for predicting the future impact of academic publications, based on ensemble regression models. Using a set of successive past top-k impact rankings, obtained through the PageRank graph-ranking algorithm, we studied the evolution of publications in terms of their impact trend sequences, effectively learning models to predict the future PageRank scores, or to predict the future number of downloads for the publications. Results obtained over a DBLP citation dataset, covering papers published in years up to 2011, show that the predictions are accurate for all experimental setups, with a model that uses features relative to PageRank scores, PageRank change rates, and author PageRank scores from the three previous impact rankings, alongside with the term frequency of the top 100 most frequent tokens in the abstracts and titles of the publications, obtaining the best results.

Despite the interesting results, there are also many ideas for future work. Our currently ongoing work is focusing on the application of the prediction mechanism to other impact metrics, perhaps better suited to academic citation networks. A particular example is the CiteRank method, which is essentially a modified version of PageRank that explicitly takes paper's age into account, in order to address the bias in PageRank towards older papers, which accumulate more citations [18]. Another interesting example of an impact metric for publications would be the Affinity Index Ranking mechanism proposed by Kaul et al., which models graphs as electrical circuits and tries to find the electrical potential of each node in order to estimate its importance [7]. We also plan on experimenting with the application of the proposed method in the context of networks encoding information from other domains, particularly on the case of online social networks (i.e., predicting the future impact of blog postings or twitter users) and location-based online social networks (i.e., predicting the future impact of spots and/or users in services such as FourSquare).

References

[1] Breiman, L.: Random Forests. Machine Learning 45(1) (2001)
[2] Chen, P., Xie, H., Maslov, S., Redner, S.: Finding scientific gems with Google's PageRank algorithm. Journal of Informetrics 1(1) (2007)
[3] Chien, S., Dwork, C., Kumar, R., Simon, D.R., Sivakumar, D.: Link evolution: Analysis and algorithms. Internet Mathematics 1(3) (2003)
[4] Davis, J.V., Dhillon, I.S.: Estimating the global PageRank of web communities. In: Proceedings of the ACM SIGKDD International Conference on Knowledge Discovery and Data Mining (2006)

[5] Friedman, J.H.: Greedy function approximation: A gradient boosting machine. Annals of Statistics 29(5) (2000)

[6] Kan, M.-Y., Thi, H.O.N.: Fast webpage classiffication using URL features. In: Proceedings of the ACM International Conference on Information and Knowledge Management (2005)

[7] Kaul, R., Yun, Y., Kim, S.-G.: Ranking billions of web pages using diodes. Communications of ACM 52(8) (2009)

[8] Kendall, M.G.: A new measure of rank correlation. Biometrika 30(1/2) (1938)

[9] Langville, A., Meyer, C.D.: Survey: Deeper inside PageRank. Internet Mathematics 1(3) (2003)

[10] Lerman, K., Ghosh, R., Kang, J.H.: Centrality metric for dynamic networks. In: Proceedings of the Workshop on Mining and Learning with Graphs (2010)

[11] Mohan, A., Chen, Z., Weinberger, K.Q.: Web-search ranking with initialized gradient boosted regression trees. Journal of Machine Learning Research 14 (2011)

[12] Radicchi, F., Fortunato, S., Markines, B., Vespignani, A.: Diffusion of scientific credits and the ranking of scientists. Physical Review (2009)

[13] Sayyadi, H., Getoor, L.: Future rank: Ranking scientific articles by predicting their future PageRank. In: Proceedings of the SIAM International Conference on Data Mining (2009)

[14] Spearman, C.: The proof and measurement of association between two things. American Journal of Psychology 15 (1904)

[15] Tang, J., Zhang, J., Yao, L., Li, J., Zhang, L., Su, Z.: ArnetMiner: Extraction and mining of academic social networks. In: Proceedings of the ACM SIGKDD International Conference on Knowledge Discovery and Data Mining (2008)

[16] Vazirgiannis, M., Drosos, D., Senellart, P., Vlachou, A.: Web page rank prediction with Markov models. In: Proceedings of the International Conference on World Wide Web (2008)

[17] Voudigari, E., Pavlopoulos, J., Vazirgiannis, M.: A framework for web Page Rank prediction. In: Iliadis, L., Maglogiannis, I., Papadopoulos, H. (eds.) EANN/AIAI 2011, Part II. IFIP AICT, vol. 364, pp. 240–249. Springer, Heidelberg (2011)

[18] Walker, D., Xie, H., Yan, K.-K., Maslov, S.: Ranking scientific publications using a simple model of network traffic. Technical Report CoRR, abs/physics/0612122 (2006)

SMOTE for Regression

Luís Torgo[1,2], Rita P. Ribeiro[1,2], Bernhard Pfahringer[3], and Paula Branco[1,2]

[1] LIAAD - INESC TEC
[2] DCC - Faculdade de Ciências - Universidade do Porto
[3] Department of Computer Science - University of Waikato
{ltorgo,rpribeiro}@dcc.fc.up.pt, bernhard@cs.waikato.ac.nz,
paobranco@gmail.com

Abstract. Several real world prediction problems involve forecasting rare values of a target variable. When this variable is nominal we have a problem of class imbalance that was already studied thoroughly within machine learning. For regression tasks, where the target variable is continuous, few works exist addressing this type of problem. Still, important application areas involve forecasting rare extreme values of a continuous target variable. This paper describes a contribution to this type of tasks. Namely, we propose to address such tasks by sampling approaches. These approaches change the distribution of the given training data set to decrease the problem of imbalance between the rare target cases and the most frequent ones. We present a modification of the well-known SMOTE algorithm that allows its use on these regression tasks. In an extensive set of experiments we provide empirical evidence for the superiority of our proposals for these particular regression tasks. The proposed SMOTER method can be used with any existing regression algorithm turning it into a general tool for addressing problems of forecasting rare extreme values of a continuous target variable.

1 Introduction

Forecasting rare extreme values of a continuous variable is very relevant for several real world domains (e.g. finance, ecology, meteorology, etc.). This problem can be seen as equivalent to classification problems with imbalanced class distributions which have been studied for a long time within machine learning (e.g. [1–4]). The main difference is the fact that we have a target numeric variable, i.e. a regression task. This type of problem is particularly difficult because: i) there are few examples with the rare target values; ii) the errors of the learned models are not equally relevant because the user's main goal is predictive accuracy on the rare values; and iii) standard prediction error metrics are not adequate to measure the quality of the models given the preference bias of the user.

The existing approaches for the classification scenario can be cast into 3 main groups [5, 6]: i) change the evaluation metrics to better capture the application bias; ii) change the learning systems to bias their optimization process to the goals of these domains; and iii) sampling approaches that manipulate the

L. Correia, L.P. Reis, and J. Cascalho (Eds.): EPIA 2013, LNAI 8154, pp. 378–389, 2013.
© Springer-Verlag Berlin Heidelberg 2013

training data distribution so as to allow the use of standard learning systems. All these three approaches were extensively explored within the classification scenario (e.g. [7, 8]). Research work within the regression setting is much more limited. Torgo and Ribeiro [9] and Ribeiro [10] proposed a set of specific metrics for regression tasks with non-uniform costs and benefits. Ribeiro [10] described system UBARULES that was specifically designed to address this type of problem. Still, to the best of our knowledge, no one has tried sampling approaches on this type of regression tasks. Nevertheless, sampling strategies have a clear advantage over the other alternatives - they allow the use of the many existing regression tools on this type of tasks without any need to change them. The main goal of this paper is to explore this alternative within a regression context. We describe two possible methods: i) using an under-sampling strategy; and ii) using a SMOTE-like approach.

The main contributions of this work are: i) presenting a first attempt at addressing rare extreme values prediction using standard regression tools through sampling approaches; and ii) adapting the well-known and successful SMOTE [8] algorithm for regression tasks. The results of the empirical evaluation of our contributions provide clear evidence on the validity of these approaches for the task of predicting rare extreme values of a numeric target variable. The significance of our contributions results from the fact that they allow the use of any existing regression tool on these important tasks by simply manipulating the available data set using our supplied code.

2   Problem Formulation

Predicting rare extreme values of a continuous variable is a particular class of regression problems. In this context, given a training sample of the problem, $\mathcal{D} = \{\langle \mathbf{x}, y \rangle\}_{i=1}^{N}$, our goal is to obtain a model that approximates the unknown regression function $y = f(\mathbf{x})$. The particularity of our target tasks is that the goal is the predictive accuracy on a particular subset of the domain of the target variable Y - the rare and extreme values. As mentioned before, this is similar to classification problems with extremely unbalanced classes. As in these problems, the user goal is the performance of the models on a sub-range of the target variable values that is very infrequent. In this context, standard regression metrics (e.g. mean squared error) suffer from the same problems as error rate (or accuracy) on imbalanced classification tasks - they do not focus on the rare cases performance. In classification the solution usually revolves around the use of the precision/recall evaluation framework [11]. Precision provides an indication on how accurate are the predictions of rare cases made by the model. Recall tells us how frequently the rare situations were signalled as such by the model. Both are important properties that frequently require some form of trade-off. How can we get similar evaluation for the numeric prediction of rare extreme values? On one hand we want that when our models predict an extreme value they are accurate (high precision), on the other hand we want our models to make extreme value predictions for the cases where the true value is an extreme (high recall).

Assuming the user gives us information on what is considered an extreme for the domain at hand (e.g. $Y < k_1$ is an extreme low, and $Y > k_2$ is an extreme high), we could transform this into a classification problem and calculate the precision and recall of our models for each type of extreme. However, this would ignore the notion of numeric precision. Two predicted values very distant from each other, as long as being both extremes (above or below the given thresholds) would count as equally valuable predictions. This is clearly counter-intuitive on regression problems such as our tasks. A solution to this problem was described by Torgo and Ribeiro [9] and Ribeiro [10] that have presented a formulation of precision and recall for regression tasks that also considers the issue of numeric accuracy. We will use this framework to compare and evaluate our proposals for this type of tasks. For completeness, we will now briefly describe the framework proposed by Ribeiro [10] that will be used in the experimental evaluation of our proposal[1].

2.1   Utility-Based Regression

The precision/recall evaluation framework we will use is based on the concept of utility-based regression [10, 12]. At the core of utility-based regression is the notion of relevance of the target variable values and the assumption that this relevance is not uniform across the domain of this variable. This notion is motivated by the fact that contrary to standard regression, in some domains not all the values are equally important/relevant. In utility-based regression the usefulness of a prediction is a function of both the numeric error of the prediction (given by some loss function $L(\hat{y}, y)$) and the relevance (importance) of both the predicted $\hat{y}$ and true y values. Relevance is the crucial property that expresses the domain-specific biases concerning the different importance of the values. It is defined as a continuous function $\phi(Y) : \mathcal{Y} \to [0, 1]$ that maps the target variable domain $\mathcal{Y}$ into a $[0, 1]$ scale of relevance, where 0 represents the minimum and 1 represents the maximum relevance.

Being a domain-specific function, it is the user responsibility to specify the relevance function. However, Ribeiro [10] describes some specific methods of obtaining automatically these functions when the goal is to be accurate at rare extreme values, which is the case for our applications. The methods are based on the simple observation that for these applications the notion of relevance is inversely proportional to the target variable probability density function. We have used these methods to obtain the relevance functions for the data sets used in the experiments section.

The utility of a model prediction is related to the question on whether it has led to the identification of the correct type of extreme and if the prediction was precise enough in numeric terms. Thus to calculate the utility of a prediction it is necessary consider two aspects: (i) does it identify the correct type of extreme? (ii) what is the numeric accuracy of the prediction (i.e. $L(\hat{y}, y)$)? This latter issue

---

[1] Full details can be obtained at Ribeiro [10]. The code used in our experiments is available at http://www.dcc.fc.up.pt/~rpribeiro/uba/.

is important because it allows for coping with different "degrees" of actions as a result of the model predictions. For instance, in the context of financial trading an agent may use a decision rule that implies buying an asset if the predicted return is above a certain threshold. However, this same agent may invest different amounts depending on the predicted return, and thus the need for precise numeric forecasts of the returns on top of the correct identification of the type of extreme. This numeric precision, together with the fact that we may have more than one type of extreme (i.e. more than one "positive" class) are the key distinguishing features of this framework when compared to pure classification approaches, and are also the main reasons why it does not make sense to map our problems to classification tasks.

The concrete utility score of a prediction, in accordance with the original framework of utility-based learning (e.g. [2, 3]), results from the net balance between its benefits and costs (i.e. negative benefits). A prediction should be considered beneficial only if it leads to the identification of the correct type of extreme. However, the reward should also increase with the numeric accuracy of the prediction and should be dependent on the relevance of the true value. In this context, Ribeiro [10] has defined the notions of benefits and costs of numeric predictions, and proposed the following definition of the utility of the predictions of a regression model,

$$U_\phi^p(\hat{y}, y) = B_\phi(\hat{y}, y) - C_\phi^p(\hat{y}, y)$$
$$= \phi(y) \cdot (1 - \Gamma_B(\hat{y}, y)) - \phi^p(\hat{y}, y) \cdot \Gamma_C(\hat{y}, y) \tag{1}$$

where $B_\phi(\hat{y}, y)$, $C_\phi^p(\hat{y}, y)$, $\Gamma_B(\hat{y}, y)$ and $\Gamma_C(\hat{y}, y)$ are functions related to the notions of costs and benefits of predictions that are defined in Ribeiro [10].

2.2   Precision and Recall for Regression

Precision and recall are two of the most commonly used metrics to estimate the performance of models in highly skewed domains [11] such as our target domains. The main advantage of these statistics is that they are focused on the performance on the target events, disregarding the remaining cases. In imbalanced classification problems, the target events are cases belonging to the minority (positive) class. Informally, precision measures the proportion of events signalled by the model that are real events, while recall measures the proportion of events occurring in the domain that are captured by the model.

The notions of precision and recall were adapted to regression problems with non-uniform relevance of the target values by Torgo and Ribeiro [9] and Ribeiro [10]. In this paper we will use the framework proposed by these authors to evaluate and compare our sampling approaches. We will now briefly present the main details of this formulation[2].

Precision and recall are usually defined as ratios between the correctly identified events (usually known as true positives within classification), and either the signalled events (for precision), or the true events (for recall). Ribeiro [10]

---

[2] Full details can be obtained in Chapter 4 of Ribeiro [10].

defines the notion of event using the concept of utility. In this context, the ratios of the two metrics are also defined as functions of utility, finally leading to the following definitions of precision and recall for regression,

$$recall = \frac{\sum\limits_{i:\hat{z}_i=1,z_i=1} (1 + u_i)}{\sum\limits_{i:z_i=1} (1 + \phi(y_i))} \tag{2}$$

and

$$precision = \frac{\sum\limits_{i:\hat{z}_i=1,z_i=1} (1 + u_i)}{\sum\limits_{i:\hat{z}_i=1,z_i=1} (1 + \phi(y_i)) + \sum\limits_{i:\hat{z}_i=1,z_i=0} (2 - p\,(1 - \phi(y_i)))} \tag{3}$$

where p is a weight differentiating the types of errors, while $\hat{z}$ and z are binary properties associated with being in the presence of a rare extreme case.

In the experimental evaluation of our sampling approaches we have used as main evaluation metric the F-measure that can be calculated with the values of precision and recall,

$$F = \frac{(\beta^2 + 1) \cdot precision \cdot recall}{\beta^2 \cdot precision + recall} \tag{4}$$

where β is a parameter weighing the importance given to precision and recall (we have used $\beta = 1$, which means equal importance to both factors).

3   Sampling Approaches

The basic motivation for sampling approaches is the assumption that the imbalanced distribution of the given training sample will bias the learning systems towards solutions that are not in accordance with the user's preference goal. This occurs because the goal is predictive accuracy on the data that is least represented in the sample. Most existing learning systems work by searching the space of possible models with the goal of optimizing some criteria. These criteria are usually related to some form of average performance. These metrics will tend to reflect the performance on the most common cases, which are not the goal of the user. In this context, the goal of sampling approaches is to change the data distribution on the training sample so as to make the learners focus on cases that are of interest to the user. The change that is carried out has the goal of balancing the distribution of the least represented (but more important) cases with the more frequent observations.

Many sampling approaches exist within the imbalanced classification literature. To the best of our knowledge no attempt has been made to apply these strategies to the equivalent regression tasks - forecasting rare extreme values. In this section we describe the adaptation of two existing sampling approaches to these regression tasks.

3.1   Under-Sampling Common Values

The basic idea of under-sampling (e.g. [7]) is to decrease the number of observations with the most common target variable values with the goal of better balancing the ratio between these observations and the ones with the interesting target values that are less frequent. Within classification this consists on obtaining a random sample from the training cases with the frequent (and less interesting) class values. This sample is then joined with the observations with the rare target class value to form the final training set that is used by the selected learning algorithm. This means that the training sample resulting from this approach will be smaller than the original (imbalanced) data set.

In regression we have a continuous target variable. As mentioned in Section 2.1 the notion of relevance can be used to specify the values of a continuous target variable that are more important for the user. We can also use the relevance function values to determine which are the observations with the common and uninteresting values that should be under-sampled. Namely, we propose the strategy of under-sampling observations whose target value has a relevance less than a user-defined parameter. This threshold will define the set of observations that are relevant according to the user preference bias, $\mathcal{D}_r = \{\langle \mathbf{x}, y \rangle \in \mathcal{D} : \phi(y) \geq t\}$, where t is the user-defined threshold on relevance. Under-sampling will be carried out on the remaining observations $\mathcal{D}_i = \mathcal{D} \setminus \mathcal{D}_r$.

Regards the amount of under-sampling that is to be carried out the strategy is the following. For each of the relevant observations in $\mathcal{D}_r$ we will randomly select n_u cases from the "normal" observations in $\mathcal{D}_i$. The value of n_u is another user-defined parameter that will establish the desired ratio between "normal" and relevant observations. Too large values of n_u will result in a new training data set that is still too unbalanced, but too small values may result in a training set that is too small, particularly if there are too few relevant observations.

3.2   SMOTE for Regression

Smote [8] is a sampling method to address classification problems with imbalanced class distribution. The key feature of this method is that it combines under-sampling of the frequent classes with over-sampling of the minority class. Chawla et. al. [8] show the advantages of this approach when compared to other alternative sampling techniques on several real world problems using several classification algorithms. The key contribution of our work is to propose a variant of Smote for addressing regression tasks where the key goal is to accurately predict rare extreme values, which we will name SMOTER .

The original Smote algorithm uses an over-sampling strategy that consists on generating "synthetic" cases with a rare target value. Chawla et. al. [8] propose an interpolation strategy to create these artificial examples. For each case from the set of observations with rare values ($\mathcal{D}_r$), the strategy is to randomly select one of its k-nearest neighbours from this same set. With these two observations a new example is created whose attribute values are an interpolation of the values of the two original cases. Regards the target variable, as Smote is applied to classification problems with a single class of interest, all cases in $\mathcal{D}_r$ belong to this class and the same will happen to the synthetic cases.

There are three key components of the SMOTE algorithm that we need to address in order to adapt it for our target regression tasks: i) how to define which are the relevant observations and the "normal" cases; ii) how to create new synthetic examples (i.e. over-sampling); and iii) how to decide the target variable value of these new synthetic examples. Regards the first issue, the original algorithm is based on the information provided by the user concerning which class value is the target/rare class (usually known as the minority or positive class). In our problems we face a potentially infinite number of values of the target variable. Our proposal is based on the existence of a relevance function (c.f. Section 2.1) and on a user-specified threshold on the relevance values, that leads to the definition of the set $\mathcal{D}_r$ (c.f. Section 3.1). Our algorithm will over-sample the observations in $\mathcal{D}_r$ and under-sample the remaining cases ($\mathcal{D}_i$), thus leading to a new training set with a more balanced distribution of the values. Regards the second key component, the generation of new cases, we use the same approach as in the original algorithm though we have introduced some small modifications for being able to handle both numeric and nominal attributes. Finally, the third key issue is to decide the target variable value of the generated observations. In the original algorithm this is a trivial question, because as all rare cases have the same class (the target minority class), the same will happen to the examples generated from this set. In our case the answer is not so trivial. The cases that are to be over-sampled do not have the same target variable value, although they do have a high relevance score ($\phi(y)$). This means that when a pair of examples is used to generate a new synthetic case, they will not have the same target variable value. Our proposal is to use a weighed average of the target variable values of the two seed examples. The weights are calculated as an inverse function of the distance of the generated case to each of the two seed examples.

Algorithm 1. The main SMOTER algorithm

---

function SMOTER($\mathcal{D}, t_E, o, u, k$)

    // $\mathcal{D}$ - A data set

    // t_E - The threshold for relevance of the target variable values

    // %o,%u - Percentages of over- and under-sampling

    // k - The number of neighbours used in case generation

    $rareL \leftarrow \{\langle \mathbf{x}, y \rangle \in \mathcal{D} : \phi(y) > t_E \wedge y < \tilde{y}\}$ // $\tilde{y}$ is the median of the target Y

    $newCasesL \leftarrow$ GENSYNTHCASES($rareL$, %o, k) // generate synthetic cases for rareL

    $rareH \leftarrow \{\langle \mathbf{x}, y \rangle \in \mathcal{D} : \phi(y) > t_E \wedge y > \tilde{y}\}$

    $newCasesH \leftarrow$ GENSYNTHCASES($rareH$, %o, k) // generate synthetic cases for rareH

    $newCases \leftarrow newCasesL \bigcup newCasesH$

    $nrNorm \leftarrow$ %u of $|newCases|$

    $normCases \leftarrow$ sample of $nrNorm$ cases $\in \mathcal{D} \backslash \{rareL \bigcup rareH\}$ // under-sampling

    return $newCases \bigcup normCases$

end function

Algorithm 2. Generating synthetic cases

---

function GENSYNTHCASES($\mathcal{D}, o, k$)

    $newCases \leftarrow \{\}$

    $ng \leftarrow \%o/100$ // nr. of new cases to generate for each existing case

    for all $case \in \mathcal{D}$ **do**

        $nns \leftarrow$ KNN($k, case, \mathcal{D}_r \setminus \{case\}$) // k-Nearest Neighbours of $case$

        for $i \leftarrow 1$ **to** ng **do**

            $x \leftarrow$ randomly choose one of the nns

            for all $a \in$ attributes **do** // Generate attribute values

                if ISNUMERIC(a) **then**

                    $diff \leftarrow case[a] - x[a]$

                    $new[a] \leftarrow case[a] +$ RANDOM($0, 1$) $\times diff$

                else

                    $new[a] \leftarrow$ randomly select among $case[a]$ and $x[a]$

                end if

            end for

            $d_1 \leftarrow$ DIST($new, case$) // Decide the target value

            $d_2 \leftarrow$ DIST(new, x)

            $new[Target] \leftarrow \frac{d_2 \times case[Target] + d_1 \times x[Target]}{d_1 + d_2}$

            $newCases \leftarrow newCases \bigcup \{new\}$

        end for

    end for

    return $newCases$

end function

Algorithm 1 describes our proposed SMOTER sampling method. The algorithm uses a user-defined threshold (t_E) of relevance to define the sets $\mathcal{D}_r$ and $\mathcal{D}_i$. Notice that in our target applications we may have two rather different sets of rare cases: the extreme high and low values. This is another difference to the original algorithm. The consequence of this is that the generation of the synthetic examples is also done separately for these two sets. The reason is that although both sets include rare and interesting cases, they are of different type and thus with very different target variable values (extremely high and low values). The other parameters of the algorithm are the percentages of over- and under-sampling, and the number of neighbours to use in the cases generation. The key aspect of this algorithm is the generation of the synthetic cases. This process is described in detail on Algorithm 2. The main differences to the original SMOTE algorithm are: the ability to handle both numeric and nominal variables; and the way the target value for the new cases is generated. Regards the former issue we simply perform a random selection between the values of the two seed cases. A possible alternative could be to use some biased sampling that considers the frequency of occurrence of each of the values within the rare cases. Regards the target value we have used a weighted average between the values of the two seed cases. The weights are decided based on the distance between the new case and these two seed cases. The larger the distance the smaller the weight.

R code implementing both the SMOTER method and the under-sampling strategy described in Section 3.1 is freely provided at `http://www.dcc.fc.up.pt/~ltorgo/EPIA2013`. This URL also includes all code and data sets necessary to replicate the experiments in the paper.

4   Experimental Evaluation

The goal of our experiments is to test the effectiveness of our proposed sampling approaches at predicting rare extreme values of a continuous target variable. For this purpose we have selected 17 regression data sets that can be obtained at the URL mentioned previously. Table 1 shows the main characteristics of these data sets. For each of these data sets we have obtained a relevance function using the automatic method proposed by Ribeiro [10]. The result of this method are relevance functions that assign higher relevance to high and low rare extreme values, which are the target of the work in this paper. As it can be seen from the data in Table 1 this results in an average of around 10% of the available cases having a rare extreme value for most data sets.

In order to avoid any algorithm-dependent bias distorting our results, we have carried out our comparisons using a diverse set of standard regression algorithms. Moreover, for each algorithm we have considered several parameter variants. Table 2 summarizes the learning algorithms that were used and also the respective parameter variants. To ensure easy replication of our work we have used the implementations available in the free open source R environment, which is also the infrastructure used to implement our proposed sampling methods.

Each of the 20 learning approaches (8 MARS variants + 6 SVM variants + 6 Random Forest variants), were applied to each of the 17 regression problems using 7 different sampling approaches. Sampling comprises the following approaches: i) carrying out no sampling at all (i.e. use the data set with the original imbalance); ii) 4 variants of our SMOTER method; and iii) 2 variants of under-sampling. The four SMOTER variants used 5 nearest neighbours for case generation, a relevance threshold of 0.75 and all combinations of $\{200, 300\}\%$ and $\{200, 500\}\%$ for percentages of under- and over-sampling, respectively (c.f. Algorithm 1). The two under-sampling variants used $\{200, 300\}\%$ for percentage of under-sampling and the same 0.75 relevance threshold. Our goal was to compare the 6 (4 SMOTER + 2 under-sampling) sampling approaches against the default of using the given data, using 20 learning approaches and 17 data sets.

All alternatives we have described were evaluated according to the F-measure with $\beta = 1$, which means that the same importance was given to both precision and recall scores that were calculated using the set-up described in Section 2.2. The values of the F-measure were estimated by means of 3 repetitions of a 10-fold cross validation process and the statistical significance of the observed paired differences was measured using the non-parametric Wilcoxon paired test.

Table 3 summarizes the results of the paired comparison of each of the 6 sampling variants against the baseline of using the given imbalanced data set. Each sampling strategy was compared against the baseline 340 times (20 learning

Table 1. Used data sets and characteristics (N: n. of cases; p: n. of predictors; $nRare$: n. cases with $\phi(Y) > 0.75$; $\%Rare$: $nRare/N$)

Data Set	N	p	$nRare$	$\%Rare$	Data Set	N	p	$nRare$	$\%Rare$
a1	198	12	31	0.157	dAiler	7129	6	450	0.063
a2	198	12	24	0.121	availPwr	1802	16	169	0.094
a3	198	12	34	0.172	bank8FM	4499	9	339	0.075
a4	198	12	34	0.172	cpuSm	8192	13	755	0.092
a5	198	12	22	0.111	dElev	9517	7	1109	0.116
a6	198	12	33	0.167	fuelCons	1764	38	200	0.113
a7	198	12	27	0.136	boston	506	14	69	0.136
Abalone	4177	9	679	0.163	maxTorque	1802	33	158	0.088
Accel	1732	15	102	0.059					

Table 2. Regression algorithms and parameter variants, and the respective R packages

Learner	Parameter Variants	R package
MARS	$nk = \{10, 17\}, degree = \{1, 2\}, thresh = \{0.01, 0.001\}$	**earth** [13]
SVM	$cost = \{10, 150, 300\}, gamma = \{0.01, 0.001\}$	**e1071** [14]
Random Forest	$mtry = \{5, 7\}, ntree = \{500, 750, 1500\}$	**randomForest** [15]

Table 3. Summary of the paired comparisons to the no sampling baseline (S - SMOTER ; U - under-sampling; ox - $x \times 100\%$ over-sampling; ux - $x \times 100\%$ under-sampling)

Sampling Strat.	Win (99%)	Win (95%)	Loss (99%)	Loss (95%)	Insignif. Diff.
S.o2.u2	164	32	5	6	99
S.o5.u2	152	38	5	1	110
S.o2.u3	155	41	1	8	101
S.o5.u3	146	41	5	4	110
U.2	136	39	6	4	121
U.3	123	44	5	4	130

variants times 17 data sets). For each paired comparison we check the statistical significance of the difference in the average F score obtained with the respective sampling approach and with the baseline. These averages were estimated using a 3×10-fold CV process. We counted the number of significant wins and losses of each of the 6 sampling variants on these 340 paired comparisons using two significance levels (99% and 95%).

The results of Table 3 show clear evidence for the advantage that sampling approaches provide, when the task is to predict rare extreme values of a continuous target variable. In effect, we can observe an overwhelming advantage in terms of number of statistically significant wins over the alternative of using the data set as given (i.e. no sampling). For instance, the particular configuration of using 200% over-sampling and 200% under-sampling was significantly better than the alternative of using the given data set on 57.6% of the 340 considered situations, while only on 3.2% of the cases sampling actually lead to a significantly worst model. The results also reveal that a slightly better outcome is obtained by the SMOTER approaches with respect to the alternative of simply under-sampling the most frequent values.

Figure 1 shows the best scores obtained with any of the sampling and no-sampling variants that were considered for each of the 17 data sets. As it can be seen, with few exceptions it is clear that the best score is obtained with some sampling variant. As expected the advantages decrease as the score of the baseline no-sampling approach increases, as it is more difficult to improve on results that are already good. Moreover, we should also mention that in our experiments we have considered only a few of the possible parameter variants of the two sampling approaches (4 in SMOTER and 2 with under-sampling).

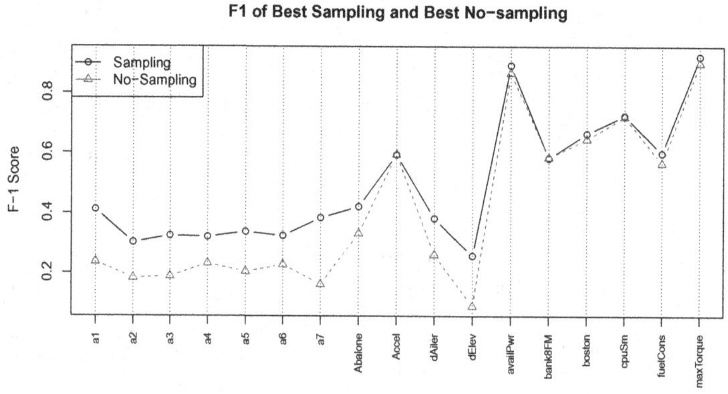

Fig. 1. Best Scores obtained with sampling and no-sampling

5   Conclusions

This paper has presented a general approach to tackle the problem of forecasting rare extreme values of a continuous target variable using standard regression tools. The key advantage of the described sampling approaches is their simplicity. They allow the use of standard out-of-the-box regression tools on these particular regression tasks by simply manipulating the available training data.

The key contributions of this paper are : i) showing that sampling approaches can be successfully applied to this type of regression tasks; and ii) adapting one of the most successful sampling methods (SMOTE) to regression tasks.

The large set of experiments we have carried out on a diverse set of problems and using rather different learning algorithms, highlights the advantages of our proposals when compared to the alternative of simply applying the algorithms to the available data sets.

Acknowledgements. This work is part-funded by the ERDF - European Regional Development Fund through the COMPETE Programme (operational programme for competitiveness), by the Portuguese Funds through the FCT (Portuguese Foundation for Science and Technology) within project FCOMP - 01-0124-FEDER-022701.

References

1. Domingos, P.: Metacost: A general method for making classifiers cost-sensitive. In: KDD 1999: Proceedings of the 5th International Conference on Knowledge Discovery and Data Mining, pp. 155–164. ACM Press (1999)
2. Elkan, C.: The foundations of cost-sensitive learning. In: IJCAI 2001: Proc. of 17th Int. Joint Conf. of Artificial Intelligence, vol. 1, pp. 973–978. Morgan Kaufmann Publishers (2001)
3. Zadrozny, B.: One-benefit learning: cost-sensitive learning with restricted cost information. In: UBDM 2005: Proc. of the 1st Int. Workshop on Utility-Based Data Mining, pp. 53–58. ACM Press (2005)
4. Chawla, N.V.: Data mining for imbalanced datasets: An overview. In: The Data Mining and Knowledge Discovery Handbook. Springer (2005)
5. Zadrozny, B.: Policy mining: Learning decision policies from fixed sets of data. PhD thesis, University of California, San Diego (2003)
6. Ling, C., Sheng, V.: Cost-sensitive learning and the class imbalance problem. In: Encyclopedia of Machine Learning. Springer (2010)
7. Kubat, M., Matwin, S.: Addressing the curse of imbalanced training sets: One-sided selection. In: Proc. of the 14th Int. Conf. on Machine Learning, pp. 179–186. Morgan Kaufmann (1997)
8. Chawla, N.V., Bowyer, K.W., Hall, L.O., Kegelmeyer, W.P.: Smote: Synthetic minority over-sampling technique. JAIR 16, 321–357 (2002)
9. Torgo, L., Ribeiro, R.: Precision and recall for regression. In: Gama, J., Costa, V.S., Jorge, A.M., Brazdil, P.B. (eds.) DS 2009. LNCS, vol. 5808, pp. 332–346. Springer, Heidelberg (2009)
10. Ribeiro, R.P.: Utility-based Regression. PhD thesis, Dep. Computer Science, Faculty of Sciences - University of Porto (2011)
11. Davis, J., Goadrich, M.: The relationship between precision-recall and roc curves. In: ICML 2006: Proc. of the 23rd Int. Conf. on Machine Learning, pp. 233–240. ACM ICPS, ACM (2006)
12. Torgo, L., Ribeiro, R.P.: Utility-based regression. In: Kok, J.N., Koronacki, J., Lopez de Mantaras, R., Matwin, S., Mladenič, D., Skowron, A. (eds.) PKDD 2007. LNCS (LNAI), vol. 4702, pp. 597–604. Springer, Heidelberg (2007)
13. Milborrow, S.: Earth: Multivariate Adaptive Regression Spline Models. Derived from mda:mars by Trevor Hastie and Rob Tibshirani (2012)
14. Dimitriadou, E., Hornik, K., Leisch, F., Meyer, D., Weingessel, A.: e1071: Misc Functions of the Department of Statistics (e1071), TU Wien (2011)
15. Liaw, A., Wiener, M.: Classification and regression by randomforest. R News 2(3), 18–22 (2002)

(TD)²PaM: A Constraint-Based Algorithm for Mining Temporal Patterns in Transactional Databases

Sílvia Moura Pina and Cláudia Antunes

Dep. Computer Science and Engineering
Instituto Superior Técnico, Technical University of Lisbon
Av Rovisco Pais, 1049-001 Lisboa
{silvia.pina,claudia.antunes}@ist.utl.pt

Abstract. The analysis of frequent behaviors regarding temporal issues begins to achieve some interest, in particular in the area of health care. However, existing approaches tend to ignore the temporal information and only make use of the order among events occurrence. In this paper, we introduce the notion of temporal constraint, and propose three instantiations of it: complete cyclic temporal constraints, partial cyclic temporal constraints and timespan constraints. Additionally, we propose a new algorithm – (TD)²PaM, that together with these constraints, makes possible to focus the pattern mining process on looking for cyclic and timespan patterns. Experimental results reveal the algorithm to be as efficient as its predecessors, and able to discover more informed patterns.

Keywords: Pattern Mining, Temporality, Constraints, Ontologies, Semantic Aspects of Data Mining.

1 Introduction

In modern society, and especially since the explosion of Internet use, we have been accumulating exponentially increasing amounts of data, usually tagged with some time reference. Methods to explore such massive amounts of data have been proposed on the last years, through the field of knowledge discovery and data mining, aiming for extracting new non-trivial information. However, the ability to explore temporal information remains a challenge.

In temporal datasets, a transaction time is attached to each transaction. In this context, hidden information or relationships among the items may be there, which does not necessarily exist or hold throughout the whole time period covered by the dataset, but only in some time intervals. Another feature of temporal data is that they may be of a periodic nature, thus repeating at regular intervals. The existing work in the field of temporal pattern mining presents several critical shortcomings. The most important of these are: (i) not being able to effectively deal with different time granularities; (ii) relying on non-robust and inflexible representations of time.

L. Correia, L.P. Reis, and J. Cascalho (Eds.): EPIA 2013, LNAI 8154, pp. 390–407, 2013.
© Springer-Verlag Berlin Heidelberg 2013

In this paper, we propose a method to aid the discovery of the regularities among temporal data recorded in transactional databases, namely *cycles* and *lifespans*. This is done within the D^{2}PM framework, through the definition of temporal constraints and a new method able to discover temporal patterns – the (TD)^{2}PaM algorithm (*Temporality in Transactional Domain-Driven Pattern Mining*).

The paper is organized as follows: next, in section 2, we overview the different approaches to temporal data, paying particular attention to the ones dedicated to pattern mining. In section 3, we propose a set of temporal constraints and then describe the new algorithm to deal with it (section 4). Following this, experimental results over real data assess the algorithm's performance and the quality of the discovered patterns. The paper concludes with a critical analysis of the results achieved and some guidelines for future work.

2   Background

Time is a compulsory dimension in any event in the real world, and is usually incorporated into databases by associating a timestamp to the event data [1]. Due to its importance and complexity, time has been a relevant research-topic in the area of Artificial Intelligence, playing a role in areas that span from logical foundations to knowledge-based systems applications. Particularly relevant is the area of *temporal reasoning*, consisting of a formalization of the notion of time and providing a form of representing and reasoning about the temporal aspect of knowledge. The work in this area is vast, and has made important contributions to the representation of time and its impact on the data stored. Of particular importance is the work by Allen [2], where the notions of time point and time interval, as long as a set of relations between temporal entities (points, intervals, or other) were defined.

Another important feature addressed in this field is *time granularity*, which can be defined as the resolution power of the temporal qualification of a statement, and refers to the existence of temporal values or objects at different levels (for example, hours, days, weeks, months) [1]. This involves not merely considering different time units, but moreover considering a layered temporal model, where associations can be formed between different layers through proper operators. Switching from one domain to a finer/ coarser one can alter interpretations of temporal statements.

In this work, we follow the definitions provided in the *Reusable Time Ontology* [3] – a time ontology based on the notion of the time line, and having as central classes *Time Point* and *Time Interval*. Another important concept is the *Time Unit* which encloses the granularity considered for each time point, and consequently for the duration of time intervals. Further down the hierarchy are the concepts of *Convex Time Interval*, which consist of a connected interval on the time line, and *Non Convex Time Interval*, corresponding to non-connected time intervals. A subclass of this is the *Regular Non Convex Time Interval*, which is composed of several convex time intervals corresponding to periodically occurring events. A simplified representation of this ontology is presented in Figure 1.

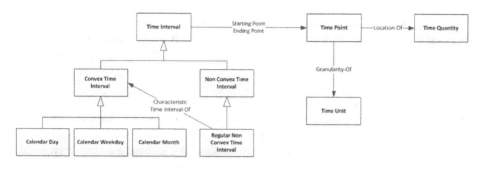

Fig. 1. Simplified representation of the *Reusable Time Ontology*

When talking about time, there are a few fundamental concepts. First *duration*, that refers to the persistence of a property over several time points or over an interval. Second, *order*: *which* expresses the occurrence of one event before or after another one, and can be applied to either time points or time intervals. *Concurrency* and *synchronicity* refer to the closeness or coincidence of two or more temporal events in time, regardless of their relative order. But also, *periodicity* which reveals the repetition of the same event with a given time period, and *evolution* to refer to the changes verified on an event or pattern occurrence over time.

2.1   Temporal Pattern Discovery

In the field of data mining, time has been addressed for decades, but mainly focused on the prediction of time series. In classification problems, time has rarely been considered as a special feature, and most always used as an ordering feature. In terms of pattern mining, time has deserved some attention, in particular in sequential pattern mining, where besides being used as the ordering key, it has also been used to impose constraints on the duration and time-distance among events [4]. Concerning temporal patterns, there are a few approaches, most of all extensions to the sequential pattern discovery ones. These methods usually combine the traditional pattern mining methods with temporal aspects, by using time information to describe the validity, periodicity or change of an association ([5]and [6]).

Besides the discovery of temporal patterns on sequential data, a few methods were proposed for dealing with transactional data, and most of all divide the temporal database into a set of partitions (according to the time granularity considered), and then find frequent temporal itemsets within these partitions. In this context, we can distinguish among the identification of *exhibition periods* or *lifespans* and *periodic patterns*. The first ones, also known as *lifespans*, correspond to the time duration from the partition when an item appears in the transaction database to the partition where the item no longer appears. The *PPM* [7] and the *Twain* [8] are examples of algorithms for this purpose.

On the other hand, *periodic patterns* are patterns that occur in regular periods or durations. In this context, Ozden et al. [9] introduced the idea of *cyclic association rules* – rules that may not hold during all time intervals, but repeat periodically from a fixed time interval until the last one considered. Relaxations to this formulation were proposed, for example through periodicity constraint relaxations specified as wildcards ([10], [11]).

An important extension to those approaches is the incorporation of calendars, as a mean to describe the information about time points and time intervals. In particular, the proposal by Ramaswamy et al. [12] allows users to define time intervals described in the form of a calendar algebraic expression. However, this approach requires that users have some prior knowledge about the temporal patterns he expects to find. In order to avoid reliance on user's prior knowledge, several approaches were proposed ([13], [14]). The approaches conceived for this purpose have, however, some limitations. First, the majority of those works are based on Allen's intervals, which present several drawbacks in terms of the expressivity of time intervals. Second, periodicity as it was defined is unable to capture some real-life concepts such as "the first business day of every month", since the distances between two consecutive business days are not constant. And third, the methods used are static in the sense that the periodicity is almost always defined ahead and then it remains the same throughout the mining process.

In this manner, it is clear that a better representation of time information, which allows for addressing those issues, should contribute significantly for simplifying and complementing the different approaches. Time ontologies are certainly such models.

3   Temporal Constraints and the D^{2}PM Framework

The D^{2}PM framework [15] has the goal of supporting the process of pattern mining with the use of domain knowledge, represented through a domain ontology, and encompasses the definition of pattern mining methods able to discover transactional and sequential patterns, guided by constraints. These constraints are the core of this framework, since they are the responsible for incorporating existing knowledge in the mining algorithm. Moreover, in this context, a constraint is much more than a predicate over the powerset of items, as usual.

> **Definition 1.** In the context of the D^{2}PM framework, a *constraint* is defined as a tuple $C = (\sigma, \mu, \psi, \varphi)$ where σ is the *minimum support threshold*, μ is a *mapping function* that links items in the dataset with the concepts in the ontology, ψ is a predicate called the *equivalence function* that defines the equivalence among items which contribute for the same support, and φ defines the *acceptance function*, specifying the satisfying conditions to consider some pattern valid.

Along with the definition of a generic constraint, a set of pre-defined constraints was suggested: *content constraints*, *structural constraints* and *temporal constraints*. Content constraints are used to limit the items present in discovered patterns, stating that items in the rule should possess some specific characteristic, among those constraints are considered the ones that impose the existence of known relations among the items in the pattern. Structural constraints are defined as content constraints that distinguish the different expressions of patterns and define the constraints that patterns need to satisfy to be considered to be transactional, sequential or structured. Last but not least, temporal constraints focus on the temporal characteristics of events, and aim for introducing dynamic temporal aspects, for example the existence of cycles, or the fact that some transactions are limited to certain time intervals.

3.1   Temporal constraints

We chose to use this framework because it combines pattern mining with domain knowledge, represented through powerful representations models – ontologies. Another strong feature of this framework is its extensibility potential, thus allowing the introduction of different kinds of constraints and algorithms. Next, we present some basic concepts to define temporal constraints in the framework.

Definition 2. Let $L = \{\ i_1, ..., i_n\ \}$ be a set of distinct literals, called *items*; a subset of these items is called an *itemset I*. A *transaction* is a tuple $T = (I, t)$, where I is an *itemset* and t a *timestamp*.

The itemset in a transaction corresponds to the set of elements that co-occur at a particular instant of time, denoted by the timestamp.

Definition 3. A transaction $T = (I, t_T)$ is said *to occur in a given time interval t* $= (t_i, t_f)$, if its timestamp is contained in the time interval t, which means that $t_i \leq t_T \leq t_f$.

A temporal constraint is a constraint that aims for introducing time information into the mining process and bases on the main concepts introduced in the reusable time ontology mentioned above.

Definition 4. A *temporal constraint* is a constraint in the D^2PM context, where the mapping function μ maps timestamps to time points at a user-defined time granularity, and the acceptance function φ filters off all the patterns that do not fit into a given time interval.

The first element in the constraint (see Definition 1) consists of the minimum support threshold σ, as defined for constraints in general in D^2PM framework, and so it is not specific to the proposed definition of temporal constraints. Similarly, the equivalence function ψ defines the equivalence among items that determines which items contribute for the same support.

However, some of the functions in the constraint definition are given different responsibilities. First, the mapping function μ has the purpose of mapping the timestamp of each transaction to a time point in the knowledge base, at the granularity chosen. In this manner, the mapping function makes the bridge between the data in the database and the domain knowledge represented in the ontology. Second, the acceptance function φ works as a filter that eliminates all the transactions that do not fit into a given time interval, and only accepting data at the granularity in consideration.

Based on the time ontology, we can further distinguish between two types of temporal constraints: *timespan constraints* and *cyclic temporal constraints*. Cyclic constraints can be further categorized into *complete cyclic temporal constraints* and *partial cyclic temporal constraints*, illustrated in Figure 1.

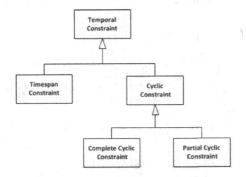

Fig. 2. Hierarchy of temporal constraints

Definition 5. A *timespan constraint* is a temporal constraint whose acceptance function only accepts patterns that occur in convex time intervals.

The timespan temporal constraint considers all patterns that occur in some convex time interval, which means that the pattern repeats over a contiguous partition of time, under the given granularity. Rules for describing items' lifespans [14] should be discovered under timespan constraints.

Definition 6. A *cyclic temporal constraint* is a temporal constraint whose acceptance function only accepts patterns that occur in non-convex time intervals.

Cyclic temporal constraints are useful to restrict the search to patterns that occur periodically, repeating from an initial time point to an ending one. These include both patterns in which every time point contributes to the cyclic behavior (Definition 7), i.e., exhibit full periodicity, and patterns which show a looser kind of periodicity where only some points contribute to it (Definition 8).

Definition 7. A *complete cyclic temporal constraint* is a cyclic temporal constraint where the acceptance function only considers open-ended non-convex time intervals.

Definition 8. A *partial cyclic temporal constraint* is a cyclic temporal constraint where the acceptance function considers non-convex time intervals with a limited duration.

The difference from complete cyclic constraint to partial cyclic constraint has respect to the duration of the time interval considered: while the first one only looks for patterns that occur periodically from a point in time until the last time point in the database, the second considers any non-convex time intervals.

4   (TD)²PaM Algorithm

Using these formulations, the research problem can now be stated as: to define a method for mining temporal patterns in the context of the D^2PM framework, that allows for the inclusion of temporal constraints to guide the mining process.

Table 1. Toy problem original dataset

Transactions	1	2	3	4	5	6	7	8	9	10	11	12	13	14	15	16	17	18	19	20
Timestamp	6/1/12 9:01	6/8/12 11:30	6/14/12 12:54	6/15/12 17:34	6/21/12 11:24	7/15/12 18:33	7/13/12 22:55	7/2/12 19:27	7/27/12 23:32	7/1/12 14:34	8/15/12 7:48	8/10/12 12:23	8/23/12 16:54	8/16/12 5:34	8/19/12 4:32	9/15/12 23:39	9/21/12 11:49	9/1/12 19:52	9/28/12 20:34	9/5/12 11:00
Timepoints for C_{Month}	2012-06	2012-06	2012-06	2012-06	2012-06	2012-07	2012-07	2012-07	2012-07	2012-07	2012-08	2012-08	2012-08	2012-08	2012-08	2012-09	2012-09	2012-09	2012-09	2012-09
Time Partition	D_0 (2012-06)					D_1 (2012-07)					D_2 (2012-08)					D_3 (2012-09)				
Items	30 31	30 31 32	30 31 32 34	30 32 33	31 33	30 31 32	32 33	31 32 33	31 33	33 34	30 31 32	30 32	31 32 33	31 33	30 31 33 34	30 32	30 32 33 34	30 31 32 33 34	30 31 35	31 33
(30 31)	Support = 3/5					Support = 1/5					Support = 2/5					Support = 1/5				

Table 2. Generated frequent itemsets with their candidate cycles in the toy example

Constraint	Level	Patterns	Candidate cycles
Timespan constraint	1-itemsets	pattern 0: 30	[Cycle(p=1, o=2, d=2)]
		pattern 1: 31	[Cycle(p=1, o=0, d=4)]
		pattern 2: 32	[Cycle(p=1, o=0, d=4)]
		pattern 3: 33	[Cycle(p=1, o=1, d=3)]
	2-itemsets	pattern 4: 31 32	[Cycle(p=1, o=0, d=2)]
Complete cyclic constraint	1-itemsets	pattern 0: 30	[Cycle(p=2, o=0)]
		pattern 1: 31	[Cycle(p=1, o=0), Cycle(p=2, o=0), Cycle(p=2, o=1)]
		pattern 2: 32	
		pattern 3: 33	[Cycle(p=1, o=0), Cycle(p=2, o=0), Cycle(p=2, o=1)]
			[Cycle(p=2, o=1)]
	2-itemsets	pattern 4: 30 31	[Cycle(p=2, o=0)]

The *Interleaved* algorithm proposed by Ozden et al. [9] is one of the few able to recognize periodic patterns, and so we propose to adapt it. The challenge is to incorporate time constraints deep into its mining process, and at the same time to extend it to deal with other patterns than those that exhibit complete periodicity. In particular, we extend the method to include timespan and partial cyclic constraints defined in the previous section. We refer the resulting method as (TD)^{2}PaM, standing for *Temporality* in *Transactional Domain-Driven Pattern Mining*.

The method devised has two main stages: in a first stage the temporal constraints are instantiated, and according to this, each transaction is assigned to a time partition (i.e., the dataset is partitioned into several time segments, according to the parameters defined in the temporal constraint), and in a second stage the algorithm runs in order to find the temporal patterns.

These stages are described next, using a toy dataset (Table 1) for illustrating its approach. Each transaction is a tuple (Timestamp, Itemset), where the transaction number is not taken into account. Additionally, consider the constraint $C_{Month} = (\sigma = 40\%, \mu(T=(I,t)) = Year(t):Month(t), \psi, \varphi)$, which specifies a calendric constraint, namely Calendar-Month, so the mapping function μ translates each timestamp into the $Year(t):Month(t)$ format.

4.1   Partitioning of the Data

The first stage, the partitioning of the data, receives the dataset file with each line representing one transaction, where one of the fields contains the timestamp and the others contain the items that compose the transaction. In this stage, data is mapped to the ontology and then partitioned into smaller datasets, according to the given time granularity.

In the first step, each transaction timestamp is mapped to the knowledge base (ontology) in use, through the constraint mapping function μ. In particular, from the timestamp in each transaction is created a time point at the granularity specified by the function. For example, the C_{Month} constraint would map all transactions to a specific month, since this is a calendric constraint it would differentiate between months of different years. The row "Time Points for C_{Month}" in Table 1 presents the complete transformation performed by the mapping function, and the dataset after the preprocessing.

Since we are interested in sets of transactions that occur within a given time period, we need to consider the Time Interval class, which by definition includes sets of two or more *Time Points*. This does not mean that instances of *Time Interval* have to correspond to connected intervals; as we have seen before, this class is a generalization and this distinction is made further down the hierarchy.

If the time intervals in consideration correspond to connected intervals, then they are instances of *Convex Time Interval*, such as *Calendar-Month*. This approach is similar to that presented in Ramaswamy et al. [12] for finding cyclic calendric rules. However, this is insufficient for our purposes, since the expressivity of these rules is somewhat limited.

For time intervals that do not correspond to connected intervals, then they are instanced as members of the *Non Convex Time Interval* class, used in the non-calendric mode of instantiation. An important subclass of this is *Regular Non Convex Time Interval*, which represents regularly occurring events and thus represents a cyclic temporal constraint as defined above (Definition 6). So, if we want to consider transactions that occur "every Friday in December" this would be an instance of *Regular Non Convex Time In*terval. This consists of four to five connected intervals, each one representing a Friday, which is in turn an instance of *Convex Time Interval*.

In the second step, the mapped data are partitioned into different sub-datasets, called *time partitions*.

Definition 9. A *time partition* D_i is the set of transactions that occur in the time interval $[i \times tg, (i+1) \times tg]$, where tg is the time granularity, defined by the μ mapping function in the temporal constraint in use.

Row "Time Partitions" in Table 1 shows the different partitions for the constraint considered. Four partitions are created, one for each month, since transactions span from June to September 2012. So, each transaction, which occurs in the interval defined by the time point, is added to the partition of its corresponding time segment. In order to segment the dataset into different partitions, it is enough to use different temporal constraints, in particular different mapping functions.

4.2   Finding Patterns

In the second stage of the algorithm, the goal is to find complete cyclic, partial cyclic and timespan patterns. After applying the data partitioning, the dataset consists of a list of sets of transactions $D_0, ..., D_{n-1}$, where D_i is a set of transactions that takes place in a particular time point i, with the time unit within a given granularity.

In the original Ozden et al.'s work [9], the problem of finding all cyclic association rules is solved by the Interleaved algorithm. In this context, a *cycle* is defined as a tuple (*p-period*, *o-offset*) that defines time intervals that start at the o time point and that repeat p after p time points. Given a minimum support threshold σ, a *cyclic pattern* is defined as a tuple (I, ρ), where I is an itemset and ρ is a cycle (p, o), and such I is frequent during the cycle ρ. For example, for a given granularity of hour, the cyclic pattern ({ highGlucoseLevels, insulin } (24, 7)) is discovered if the itemset displays a level of support above minimum threshold repeatedly for the 7-9 a.m. period every 24 hours.

The Interleaved algorithm is Apriori-based and works in two stages: first, the algorithm finds the frequent cyclic itemsets and in a second stage it generates the corresponding cyclic association rules. The algorithm uses three optimization techniques to improve cycle detection: *cycle skipping*, *cycle pruning* and *cycle elimination*.

The strict definition of cyclic patterns only allows for considering periodicity, ignoring duration and the other referred properties. In order to approach them consider the following definitions.

Definition 10. In the D^2PM context, a *cyclic pattern* is defined as a tuple I (p,o,d), where d defines the duration of the pattern, and the periodicity does not have to happen throughout the totality of temporal partitions in consideration, but only partially until the $o+d^{th}$ partition.

A particular case of a cyclic pattern is a *complete cyclic pattern*, which corresponds to the one considered by Interleaved, and which d corresponds to the total number of partitions. For partial cyclic patterns, the duration parameter d corresponds to the number of repetitions of the itemset.

Definition 11. A *timespan pattern* is defined as a temporal pattern, where the cycle is a tuple (p,o,d) defining non-convex time intervals, the period p assumes the value 1 and the duration parameter d defines the number of time partitions (counted from the offset o onwards) in which the pattern is present.

The temporal constraints defined previously are used to target the particular patterns that one aims to discover, in a straightforward way: using a timespan constraint leads to finding timespan patterns, and using a partial or complete cyclic constraint elicits patterns belonging to their corresponding pattern categories.

In order to find this new set of patterns, we propose an adaptation of the Interleaved method that is able to integrate the target temporal constraints in the mining process. We center our method in the discovery of patterns instead of generating association rules. This adaptation, besides allowing the discovery of complete cyclic patterns, contains also an extension for timespan and cyclic patterns with limited duration. The pseudo code for this is shown in Algorithm 1.

Algorithm 1. (TD)²PaM algorithm

```
Input:
- A temporal constraint C and a sequence D₀, ..., Dₙ₋₁, where Dᵢ is a set of itemsets
- Lmin > 0, Lmax > 0, Minimum level of support σ
Output:
- All cyclic, complete cyclic or timespan patterns (I (p, o, d))

Pattern Generation
    P¹ = Find-1-ItemsetPatterns()
    For (i = 2 ; (Pⁱ⁻¹ ≠ empty); i++)
        Tⁱ := Find-k-ItemsetCandidatePatterns(i);
        Pⁱ := CheckCandidatePatterns(C, Tⁱ)    //Employs optimization techniques

CheckCandidatePatterns:
    For each itemset in Ti:
        Cycle-skipping determines the set of k-itemsets for which support will be
        calculated in each D[i].
        If (Constraint C is a Timespan Constraint)
            If (itemset is frequent in consecutive partitions
                && Number of partitions left < minimum duration )
                Add cycle (1, i, duration)
        If (Constraint C is a Partial Cyclic Constraint)
            If (itemset has the minimum support in partition D[i])
                Check successively each (D[i] + period) until ((pattern is
                not frequent in D[i]) or (number of partitions left < period))
                If (number of repetitions of pattern < minimum duration)
                    Add cycle (period, i, duration)
```

The proposed algorithm receives as input the temporal constraint, as well as the partitioned dataset, and like Interleaved [9] works in two stages: first, it finds frequent 1-itemsets for each partition (function *Find-1-ItemsetPatterns*); and then, it generates frequent k-itemsets from the (k-1) itemsets for each partition (*Find-k-ItemsetCandidatePatterns*), performing pruning on each step and using the temporal constraint to filter out candidates (*CheckCandidatePatterns*). Indeed, like Interleaved, it follows the candidate generation and test strategy, exploring the anti-monotonicity property: if a j-cycle is frequent (i.e., has support above minimum value), all of the i-itemsets for i<j are also frequent.

In both stages, the candidate procedure is different for complete cyclic constraints on the one hand, and for partial cyclic and timespan constraints on the other. For the first, the complete set of candidate cycles is generated and then progressively eliminated using *cycle elimination*. For partial cyclic constraints, this initial cycle generation is not complete, since each cycle generated begins with duration value set to 0, and the cycle determination is only completed in the *CheckCandidatePatterns* procedure. Finally, for timespan constraints cycles are not generated ahead at all, but only determined also in the *CheckCandidatePatterns* stage.

The procedure for pruning is specified in the temporal constraint, and differs according to the type of constraint received as input. For complete cyclic temporal constraints, the *CheckCandidatePatterns* procedure corresponds to the application of *cycle skipping*, *pruning* and *elimination*. The distinction here is that the temporal constraint is responsible for this checking process which prunes some of the candidates to consider in the rule generation stage.

For partial cyclic and timespan constraints, for each possible period, the pruning procedure (*CheckCandidatePatterns*) checks if there is a repetition with at least duration *d* of a specified value. For example, if the minimum duration is 3, the itemset has to be frequent in at least 3 time partitions to be considered a pattern of interest. For partial cyclic constraints, this involves checking for all possible periods if there is a repetition from p to p time units starting in the partition specified in the offset.

The *CheckCandidatePatterns* procedure, both for timespan and cyclic patterns, only takes advantage of *cycle skipping* and *cycle pruning*. As for cycle elimination, it cannot be applied in these cases, since these are not of cyclic nature in a strict sense (period is assumed to be 1 for timespan patterns, and for cyclic patterns each cycle does not last throughout all the partitions), so for these patterns another optimization strategy is employed. This involves considering a threshold value for the duration parameter, which is used to stop considering a pattern as soon as the number of partitions to process is not enough to achieve a pattern with duration equal or higher than the duration threshold.

We will use the example above to illustrate the algorithm steps for two of the types of constraints considered. In this, the partitioned data serves as input to the algorithm. We use 0.4 as the minimum support threshold, 1 for Lmin and 2 for Lmax. The minimum and maximum length of the period correspond to Lmin and Lmax values.

Table 1 shows how to determine the support for the pattern (30 31), in each partition of the data. According to the support constraint, the pattern only stands in D_0 and D_2. The algorithm receives also as input a temporal constraint that specifies both the granularity and the type of temporal pattern of interest. The candidate patterns generated by the (TD)^{2}PaM procedure in the end of the pattern generation stage are shown in Table 2, with their respective candidate cycles for timespan and complete cyclic constraints.

The algorithm generates all the 1-itemsets and their corresponding possible cycles, which for complete cyclic patterns are: (p=1,o=0), (p=2,o=0) and (p=2,o=1). For itemset 30, observing that this itemset is frequent in all partitions except D_1 allows us to eliminate the first two candidate cycles, maintaining only cycle (p=2,o=0). For itemset 31, we cannot eliminate any cycle because this itemset is frequent in all partitions. So, for the candidate 2-itemset (30 31) we calculate the intersection of the candidate cycles for both 1-itemsets, which is (p=2,o=0) and only need to assess if this cycle is present, and thus calculate support in the corresponding partitions. For partial cyclic constraints, the same process would be applied, with some differences, namely taking into account cycle duration and performing cycle elimination as described above.

In the case of the timespan constraint, the cycles generated for each itemset correspond to contiguous partitions in which the itemset is frequent, and the temporal constraint defines a minimum duration (here we consider at least 2) that is considered to obtain these 1-period timespan cycles. For 1-itemset 30, the only possible timespan comprises the contiguous partitions D_2 and D_3, so the only candidate is (p=1,o=2,d=2), while for 1-itemsets 31 and 32 we have a timespan that includes all the partitions, with cycle (p=1,o=0,d=4). So, for the 2-itemset (30 31) by computing the intersection between both 1-itemset candidate cycles we only need to determine support in D_2 and D_3, and we find that this cycle is not present; for itemset (31 32) we need to verify that the candidate cycle is present, but since the itemset is only frequent in D_0 and D_1, we retain the cycle (p=1,o=0,d=2).

In the next sections the proposed (TD)^{2}PaM method is applied to a real-world dataset, in order to perform its evaluation and discuss some of its implications.

5   Experimental Results

This case study uses a dataset comprised of the listening habits of Lastfm users, and was created using the Last.fm API (http://www.lastfm.com.br/api/tos). It represents the listening habits from nearly 1000 users, spanning a time period of 5 years, and each record contains *user id, user gender, user country, timestamp, artist* and *track* listened. From these users, 100.000 records were randomly selected, and used for the analysis of algorithm performance and generation of temporal patterns.

In this section, we analyze the main results achieved by the (TD)^{2}PaM method, for the three types of temporal constraints considered in this work, namely: complete cyclic, partial cyclic and timespan constraints. We also perform tests, for comparison purposes, using the original version of the Interleaved algorithm [9] that allows finding complete cyclic patterns without resourcing to the time ontology used in this work, but performing each data partition in a sequential manner.

This analysis was conducted for several time granularities, and that are representative of the kinds of results that we can achieve. The granularities considered were *year*, *quarter*, *month*, *weekday* and *hour*, but we only present the results for *month* and *hour* granularities, mainly due to lack of space. The granularities chosen to present, raise two different kinds of time intervals. The *month* granularity corresponds to a convex time interval, the *calendar-month*, and maps timestamps to some particular month of a specific year. The *hour* granularity corresponds to a regular non-convex time interval, mapping timestamps to the hour of the day. For example *June 1st, 2013 9:01 am* and *July1st, 2013 9:51 am*, would map to two different time points at the *month* granularity (*June 2013* and *July 2013*), but to only one at the *hour* granularity (*9 am*).

Since the algorithm receives as input the maximum length of the period for the cases of complete and partial temporal constraints, this value also has to be parameterized but not presented. In this work, we use as minimum length the value 1 and as maximum the value given by (number of partitions / 2), rounded down to the nearest integer value, which seems to be a good rule of thumb, but other values could be considered.

The average number spent per pattern, obtained by dividing the total time spent by the number of patterns found is shown in Figure 3. We can see that the charts present the same general trend for all of the types of constraints used and for all time granularities, showing an increase in the average time per pattern when the support increases. This can be attributed to the fact that for higher minimum support values the number of patterns found is greatly reduced, so the average time spent in each one increases. Levels of support above 20% were not evaluated, since no patterns are found in the great majority of cases. Also, for both granularities, the average time spent per pattern in the case of partial patterns is tendentiously higher than for the other types of patterns, for minimum support levels above 1%.

In Figure 4, we present the results concerning the total running time of the algorithm, for each granularity considered. Time values were subjected to a logarithmic transformation such that $t = ln(t+1)$. The lines in the chart present the typical pattern found in pattern mining algorithms, showing an explosion in time (accompanied by an increase in the number of patterns found, as we will see next), for lower support levels (bellow 1%).

The total running time of the algorithm for partial cyclic and timespan patterns is higher across all granularities, when compared to the running time for the complete cyclic patterns. This, in turn, is almost the same as for the original version of the Interleaved algorithm that does not use a time ontology. This result shows that even though the proposed method uses an extra constraint instantiation step, this does not have a deleterious effect on the algorithm performance.

The disparity found between results for the timespan and partial cyclic patterns on the one hand and the complete cyclic patterns on the other can be partially explained by the higher number of patterns found when using timespan constraints, as we will see ahead. For partial cyclic constraints, even though in some cases the number of patterns discovered is not significantly higher than the number of patterns found when using complete cyclic constraints, the higher running time of the algorithm can be attributed to the fact that the number of partitions that we need to check in order to verify if a pattern is present represents a higher cost for the algorithm. Moreover, as seen in Figure 3, the average time spent per pattern seems to be higher for partial patterns than for the other types, especially if we consider higher support values (i.e. above 5%). In contrast, in the case of complete cyclic constraints, if we verify that an itemset is not above minimum support in a given time partition, we instantly eliminate all candidate cycles (j, i mod j), and do not have to count support for this itemset in the corresponding partitions (which is the step of the algorithm that is the most time-consuming).From a quantitative point of view, we will describe the number of patterns found according to different values of minimum support considered, and the maximum pattern length achieved.

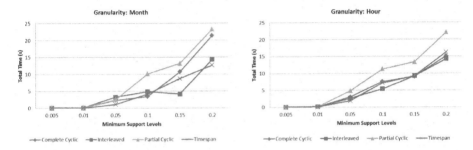

Fig. 3. Average time spent per pattern for (TD)²PaM and Interleaved algorithms for varying support levels

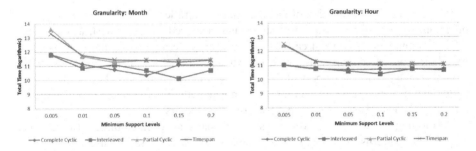

Fig. 4. Total running time for (TD)²PaM and Interleaved algorithms for varying support levels

Figure 5 shows the total number of patterns found for all types of patterns and granularities considered. For levels of minimum support above 10%, we can see that the number of patterns found is extremely reduced, and for the lowest levels (1% and especially for 0.5%), there is a great increase in the number of patterns. This is consistent also both with the results found for pattern mining algorithms in the literature, and with the running time results presented above (and showing an increase in time spent for lower support levels). If we compare both granularities shown in figure 5, we can see this explosion in the number of patterns found for lower support levels and the corresponding decrease in the number of patterns found with higher support.

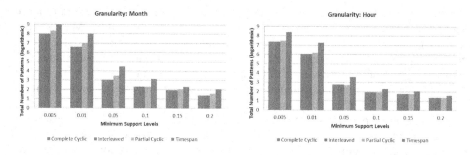

Fig. 5. Total number of patterns generated for $(TD)^2PaM$ (in a logarithmic scale) for varying support levels

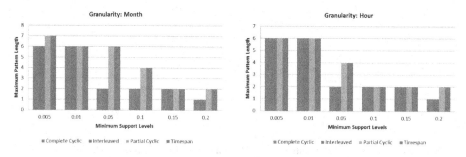

Fig. 6. Maximum pattern length for $(TD)^2PaM$ for varying support levels

Figure 6 shows the variation of the maximum pattern length. For the lower values of minimum support, the maximum pattern length achieved is the highest, across all the different granularities. In most cases, the maximum pattern length achieved is the same for all the types of patterns considered, and when it varies (for example in the case of the month granularity for a support of 5%), the maximum length is higher in the case of partial and timespan patterns than for the complete patterns. This result can be merely a reflection of the higher number of patterns found for the first two cases.

Performing a more qualitative analysis of the patterns revealed three different types of patterns: patterns containing only dimensions that refer to the users (containing attributes such as gender and country), patterns referring to artists and/or tracks, and patterns that contain items from both classes of dimensions. These last patterns relate items from the users with the tracks that they listened to. Examples of patterns containing only user dimensions and of patterns containing track dimensions can be seen in Table 3.

Looking at patterns with one item first, for example pattern P1 shows that women listened to tracks on the Lastfm website every month since the beginning of the dataset (the granularity in consideration is month granularity, which is calendric), since this is a complete cyclic pattern and its period is 1 and offset is 0. If we look at pattern P3, we can see that the artist "Blink-182" is listened to for 2 hours (duration) starting at 5 a.m., since the granularity considered is hour (this is a non-calendric granularity, since it aggregates every hour records in the dataset).

Table 3. Illustrative set of temporal patterns and their cycles (p= period, o= offset, d= duration)

	ID	Patterns	Cycles	Minimum Support	Constraint Type	Granularity
User dimensions	P1	Gender = female	(p=1,o=0)	15%	Complete cyclic	Month
	P2	Gender=male, Country=United Kingdom	(p=2,o=12, d=6) (p=3,o=15,d=3) (p=5,o=13,d=3)	5%	Partial Cyclic	Hour
Track dimensions	P3	Artist=Blink-182	(p=1,o=5,d=2)	1%	Timespan	Hour
	P4	Artist=Daft Punk, Track=Aerodynamic	(p=26,o=9) (p=25,o=10) (p=26,o=10)	1%	Complete cyclic	Month

Table 4. Illustrative set of temporal patterns and their cycles (p= period, o= offset, d= duration), with both items relating to the user and the tracks

ID	Patterns	Cycles	Minimum Support	Constraint Type	Granularity
P1	Gender=female, Country=United States, Artist=Elliott Smith	(p=4,o=0) (p=3,o=0) (p=4,o=3)	0.005	Complete cyclic	Semester
P2	Gender=male, Artist=Queen	(p=8,o=3) (p=9,o=2)	0.01	Complete cyclic	Hour
P3	Country=United States, Artist=Andrew Bird	(p=24,o=1,d=4) (p=26,o=2,d=3) (p=26,o=18,d=3)	0.005	Partial cyclic	Month
P4	Gender=female, Country=United States, Artist=Death Cab For Cutie	(p=2,o=2,d=4) (p=3,o=3,d=3) (p=7,o=2,d=3)	0.005	Partial cyclic	Hour
P5	Gender=female, Artist=The Gathering	(p=1, o=2, d=2)	0.01	Timespan	Weekday
P6	Gender=female, Country=United States, Artist=Daft Punk	(p=1, o=0, d=3)	0.005	Timespan	Weekday

Patterns P2 and P4 are examples of patterns with more than one item, revealing the co-occurrence of two or more items. In P2, we can see that male users from the United Kingdom listen to tracks with period 2, 3 and 5, respectively starting at 12, 15 and 13 hours (and with the duration pertaining to the number of times this happens during the 24 hours of the day). Even though this information can be hard to grasp at face value, it could be useful for a website that tracks user's listening habits to find such patterns in these periods, for example. In pattern P4, the track "Aerodynamic" is indeed from the artist Daft Punk, so in this case we could discard either artist name or track name from the pattern, or maintain this if intended.

Table 4 shows some illustrative examples of some of the patterns found using this method, which contain items from both user and tracks that they listen to. Looking at some of these patterns, we can see, for example (P2) that males listen to Queen starting at 3 and 2 a.m., with periodicity of 8 and 9 hours. Another example (P6), states that females from the United States listen to Daft Punk music, starting on Sunday (weekday 1, corresponding to the 0 offset), for three days of the week

(Sunday, Monday and Tuesday). So, the patterns presented allow the establishment of relations between both classes of items (user and track dimensions).

We have shown that the proposed method has indeed the potential to find patterns that incorporate the temporal dimension, and could be used as a component of a recommendation system or for advertising targeting purposes, among other applications.

6   Conclusions

The main purpose of this work consisted in developing an integrated and efficient method able to mine expressive temporal patterns in a time-stamped transactional dataset, through the incorporation of temporal constraints in the mining process. In this context, we extended the D^2PM framework, with the formalization of temporal constraints, and the proposal and definition of its subtypes, namely timespan and cyclic constraints.

In addition, we defined an efficient algorithm able to deal with these constraints: the $(TD)^2PaM$ algorithm. It adapts the Interleaved algorithm [9] for using time constraints defined based on the notions of time represented in a time ontology, and makes use of the same strategies to augment its performance. In this manner, our algorithm allows for the discovery of both lifespans and periodical patterns, according to any time granularity chosen, without any further pre-processing.

According to experimental results over real datasets, is clear that the most consuming step in the algorithm remains the support counting, which leads to the necessity of replace the candidate-based approach with a pattern-growth one. Another interesting line of research would be centered on automatically selecting the granularity levels and other parameters such as maximum period length for the cyclic patterns. To address the granularity choice, one strategy may be adopt the approach followed on mining generalized association rules [16], establishing an hierarchy of time granularities to be considered, and consider all the possibilities.

Acknowledgments. This work is partially supported by Fundação para a Ciência e Tecnologia under research project *D2PM* (PTDC/EIA-EIA/110074/2009).

References

1. Mitsa, T.: Temporal Data Mining. Chapman & Hall/CRC (2010)
2. Allen, J.: Maintaining knowledge about temporal intervals. Communications of the ACM 26(11), 832–843 (1983)
3. Zhou, Q., Fikes, R.: A Reusable Time Ontology. In: Press, A. (ed.) AAAI Workshop on Ontologies for the Semantic Web (2002)
4. Srikant, R., Agrawal, R.: Mining Sequential Patterns: Generalizations and Performance Improvements. In: Apers, P.M.G., Bouzeghoub, M., Gardarin, G. (eds.) EDBT 1996. LNCS, vol. 1057, pp. 3–17. Springer, Heidelberg (1996)
5. Chen, Y., Chiang, M., Ko, M.: Discovering time-interval sequential patterns in sequence databases. Expert Systems Applications 25, 343–354 (2003)

6. Mannila, H., Toivonen, H., Verkamo, A.: Discovery of Frequent Episodes in Event Sequences. Data Mining and Knowledge Discovery 1(3), 259–289 (1997)
7. Lee, C., Chen, M., Lin, C.: Progressive partition miner: an efficient algorithm for mining general temporal association rules. IEEE Transaction on Knowledge and Data Engineering 15(4), 1004–1017 (2003)
8. Huang, J., Dai, B., Chen, M.: Twain: Two-End Association Miner with Precise Frequent Exhibition Periods. ACM Transactions on Knowledge Discovery from Data 1(2) (2007)
9. Ozden, B., Ramaswamy, S., Silberschtaz, A.: Cyclic Association Rules. In: International Conference on Data Engineering, pp. 412–421 (1998)
10. Laxman, S., Sastry, P.S.: A survey of temporal data mining. Sadhana 31(2), 173–198 (2006)
11. Han, J., Dong, G., Yin, Y.: Efficient mining of partial periodic patterns in time series database. In: International Conference on Data Engineering, Sydney, Australia, pp. 106–115 (1999)
12. Ramaswamy, S., Mahajan, S., Silberschatz, A.: On the Discovery of Interesting Patterns in Association Rules. In: International Conference on Very Large Databases, New York, USA, pp. 368–379 (1998)
13. Li, Y., Ning, P., Wang, X., Jajodia, S.: Discovering calendar-based temporal association rules. Data Knowledge Engineering 44, 193–218 (2003)
14. Zimbrão, G., Souza, J., Almeida, V.T., Silva, W.: An Algorithm to Discover Calendar-based Temporal Association. In: Workshop on Temporal Data Mining at ACM SIGKDD Int'l Conf on Knowledge Discovery and Data Mining, Edmonton, Alberta, Canada (2002)
15. Antunes, C., Bebiano, T.: Mining Patterns with Domain Knowledge: a case study on multi-language data. In: International Conference on Information Systems, Shanghai, China, pp. 167–172 (2012)
16. Srikant, R., Agrawal, R.: Mining Generalized Association Rules. In: International Conference on Very Large Databases, Zurich, pp. 407–419 (1995)

Distributed Coalition Structure Generation with Positive and Negative Externalities

Daniel Epstein and Ana L.C. Bazzan

Instituto de Informática, Universidade Federal do Rio Grande do Sul
91501-970 - Porto Alegre, Brazil
{depstein,bazzan}@inf.ufrgs.br

Abstract. One research challenge in multi-agent systems is how to partition a set of agent into coalition structures. The number of different coalitions for a set of agents grows exponential with the number of agents and the number of different partitions grows even faster. The common approach for this problem is to search for the coalition structure that maximizes the system outcome (denoted CS^*). Until recently, most algorithms that solve this problem considered that the value of a coalition is given by a characteristic function, in where those values are not influenced by factors that are external to the coalition. More recently, several authors have focused on problems that consider the presence of externalities. In this case, centralized algorithms were developed to search for the CS^*, but no algorithm was developed to work in a distributed environment. This paper presents a distributed algorithm for searching coalition structures under presence of externalities.

Keywords: Game Theory, Multi-agent Systems.

1 Introduction

One of the main research issues in co-operative multi-agent system is to determine which division of agents into disjoint and exhaustive coalitions (i.e., a coalition structure or CS) maximizes the total payoff of the system [1]. The coalition structure generation problem (CSG) has a much larger complexity than simply dividing agents into coalitions. More precisely, the problem of distributing all agents in coalitions (for a set A of agents) has a complexity of $O(2^{|A|})$, while the problem of searching all possible coalition structures has a complexity of $O(|A|^{|A|})$. According to [2], the formal definition of a coalition structure is:

Definition 1. *A coalition structure $CS = \{C_1, C_2, ..., C_{|CS|}\}$ is an exhaustive partition of A into disjoint coalitions.*

The main objective of most algorithms designed to solve this problem is to find the CS that maximizes the gain of the system (denoted CS^*). For this, each coalition has a value associated with it. Until 2008, most of the research dealing with CSG in computer science used to deal with this problem as a characteristic function game (CFG). In a CFG each coalition has a value that depends exclusively on the agents that compose the coalition, i. e., other coalitions or agents

L. Correia, L.P. Reis, and J. Cascalho (Eds.): EPIA 2013, LNAI 8154, pp. 408–419, 2013.
© Springer-Verlag Berlin Heidelberg 2013

in the system do not affect the value of any coalition. However, there are many scenarios where this hypothesis is not true and one must consider the influence of one set of agents over others. In such scenarios, the CSG must be shaped as a Partition Function Game (*PFG*). In a PFG, the value of a coalition is defined not only by its members but also by the coalitions formed by other agents. The influence that one coalition plays over another is called an *externality*. Since in PFG the value of a coalition is affected by other coalitions in the system, it is not possible to associate a single value with a coalition. For instance, given two coalition structures $CS' = \{C_1, C_2, C_3\}$ e $CS'' = \{C_1, C_2 \bigcup C_3\}$, the value of C_1 may be different in each CS. One of the main reasons for neglecting externalities has been that the problem complexity grows from $O(2^{|A|})$ to $O(|A|^{|A|})$.

PFG has been widely studied in economics and other social sciences, where interdependency among coalition is an important factor. Examples for this can be found in collusion in oligopolies, where two companies merge in order to create a cheaper or better product, thus gaining market from other companies [2]. Also, [3] shows that if multiple countries join together to reduce pollution from carbon dioxide, it has a positive impact in all countries, even those that are not related to this particular coalition. [4] presented several examples with companies cooperation and [5] use the *Shapley's value* and the *core* in PFG. Generally, environmental policies of countries and overlapping or partially overlapping goals may cause positive externalities. On the other hand, collusion between companies and shared resources or conflicting goals may induce negative externalities [2].

Although PFGs are being widely studied in many researches areas, there is no distributed algorithm that can solve the CSG problem when modeled as a PFG. A distributed algorithm could split its processing and memory cost among the agents, even allowing agents with higher resources capability to have a bigger amount of workload. Furthermore, it addresses issues such as privacy and confidentiality. The agents in the system may not want (or may not be able) to share all the information gathered. [6] lists a series of advantages from distributed system, as well as desirable characteristics that a good distributed algorithm must have.

There is a need for a distributed algorithm that can solve the coalition structure generation problem in PFG scenarios. This paper proposes an algorithm capable of doing so. The *DCSGE* (Distributed Coalition Structure Generation with Externality) algorithm divides the burden of the calculations among the agents and addresses the CSG problem in PFG in a decentralized method. We show that DCSGE is capable of finding the CS^* without searching the entire search space, with a guarantee bound ratio from the optimal. Also, it is capable of solving problems in distributed and centralized environments, with positive or negative externalities and in CFG or PFG.

2   Related Work

Until recently, the search space representation employed by most algorithms for finding the CS^* was an undirected graph in which the vertices represent coalition structures [1]. A key turning point in CSG research was the search space partition defined by [7] and refined in [8], where a new representation has been proposed allowing a faster prune of subspaces that could not contain the CS^*.

The search space representation in most algorithms since [7] is a sub-space that contains coalition structures defined based on the integer partitions of the number of agents n. These integer partitions of an integer n are the sets of nonzero integers that add up to exactly n. For example, the five distinct partitions of the number 4 are $\{4\}, \{3,1\}, \{2,2\}, \{2,1,1\}$, and $\{1,1,1,1\}$. Each of those partitions is called a *configuration*. For instance, the coalition structures $\{\{a_1\}, \{a_2\}, \{a_3, a_4\}\}$ and $\{\{a_3\}, \{a_1\}, \{a_4, a_2\}\}$ are both associated with the configuration $\{1,1,2\}$. Considering a set A of agents, $n = |A|$.

Among the benefits of this representation, one must be emphasized: it is possible to calculate the upper bound of a configuration without evaluating all the CS associated with this particular configuration. To do so, one must divide the coalition into lists of the same size. For instance, one list containing all coalitions of size 1, one list with coalitions of size 2 and so on up to the list with only one coalition with size n (usually referred to as "*grand coalition*"). Knowing the value of those coalitions, the upper bound of a configuration can be evaluated by adding the values of the best coalitions that could be inside that configuration. For instance, in the configuration $\{2, 3\}$, the coalitions of size 2 and 3 that have the highest value compose the upper bound of this configuration. If an upper bound is lower than the value of the best CS found so far, then it can be pruned.

With this new representation of the state space, several algorithms were developed. In particular, the centralized algorithm currently considered the state-of-art is called *IDP-IP* and has been shown to exploit the strengths of both dynamic programming and anytime algorithms and, at the same time, avoid their main weaknesses. After that, the distributed algorithm *D-IP* was proposed [9]. Since our DCSGE algorithm is built upon D-IP, next we briefly explain the way D-IP works. We remark that D-IP does not deal with externalities.

2.1   D-IP

Michalak et. al. [9] presented the first distributed algorithm to solve the coalition structure generation problem optimally, called *D-IP*. Before them, the only algorithm existent to solve this problem distributed was presented by [10], but it had no guarantees to find the optimal solution. The algorithm D-IP is considered the state-of-art in finding the CS^* decentralized for CFG. The baseline for this algorithm comes from [6], where an optimal algorithm to split the coalition values calculation between agents is described, and [8], where an anytime optimal algorithm for searching for the CS^* is described.

In [6], Rahwan and Jennings have created an algorithm called DCVC (*Distributing Coalition Value Calculations*) that is used in the first step of D-IP. DCVC's objective is to split the calculation of coalitions equally among agents. To do so, the coalitions are separated in lists, each one containing coalition of a given size. For n agents there are n lists, named as L_1, L_2, ... , L_n such that $L_S : s \in \{1, 2, ..., n\}$ contains all lists of size s. Coalition in L_1 have size 1 (single agent coalitions), coalitions in L_2 have size 2 and so on. Each list is then divided into $\lfloor \frac{|L_s|}{n} \rfloor$ segments. DCVC assumes that each agent has a global identifier and that it is known by all agents. This identifier is used to indicate to the agent which segment it is responsible for. At last, the lists that are not into any segment are ordered and distributed among agents.

D-IP has an important modification when compared to DCVC. In DCVC, the coalitions are ordered lexicographically. In D-IP, the lists $L_i : i \in \{1, 2, ..., \lfloor \frac{n}{|2|} \rfloor - 1\}$ are ordered like in DCVC, but the list L_{n-i} is divided in a reverse order. This modification ensures that each agent has enough information to calculate the value of the coalition structures using only the coalitions from its share. An example of the division of coalition among agents can be seen in Figure 1.

Fig. 1. Example of the resulting assignment in D-IP for 6 agents. Figure from [9].

After splitting the coalitions and evaluating their values, each agent sends to all other agents the value of the singleton coalition (coalitions of a single agent) it calculated. By so doing, every agent would have the values of all singletons and they proceed to evaluate the values of all CS for which they know the coalition values. During this stage, the agents must keep track of the maximum and average values of their share, as well as the best CS found so far. At the end of this stage, agents exchange among themselves the maximum and average values of their segments, as well as the best CS found by each one.

After this first step, each agent has the information needed to calculate the maximum and average value of each possible configuration, pruning the subspaces that cannot contain the CS^*. Using a series of filters to avoid unwanted or unnecessary communication, the agents exchange the values of all coalitions that are present in the configuration with the highest upper bound. The final step is the search inside the configuration for the optimal CS. This search is also distributed and once it is done, agents proceed to the next configuration in case the CS^* has not been found.

2.2 $IP^{+/-}$

Recently, [2] proposed the first computational study about CSG problem with externalities (i.e., a PFG) in the multi-agent system context and an algorithm to solve CSG problem, called $IP^{+/-}$ (a reference to IP algorithm [8], now set on a context of externalities). Using the background established by [11], [2] shows that it is possible to set the upper and lower bounds for any coalition as well as a worst-case guarantee on solution quality.

A PFG generates a non-negative integer value $v(C; CS)$, where CS is a coalition structure of A and $C \in CS$. [11] defined four classes of PFG based on the value of the externality. These classes exist in environments where the externality is either positive or negative, but never both. That means that for any two coalition structures CS' and CS'', such that CS'' is created by one or more merges of coalitions in CS', the value of all coalition not involved on this merge is always bigger or is always smaller in CS'' than in CS'. The four classes are:

- super-additive games with positive externalities (PF^+_{sup});
- super-additive games with negative externalities (PF^-_{sup});
- sub-additive games with positive externalities (PF^+_{sub});
- sub-additive games with negative externalities (PF^-_{sub}).

They also proved that in PF^+_{sup} (PF^-_{sub}) the grand coalition (the coalition structure of singletons) always belong to CS^*. Therefore, the focus of their work and their successors were the remaining two classes: PF^+_{sub} and PF^-_{sup}.

To establish the upper and lower bound for each coalition in PF^+_{sub} and PF^-_{sup}, [2] proved the following:

Theorem 1. *In the class PF^-_{sup}, any coalition C_1 has its upper bound when inside the coalition structure $CS' = \{C_1, \{a_1\}, ..., \{a_{(n-|C_1|)}\}\}$ and has its lower bound when inside the coalition structure $CS'' = \{C_1, C_{A \setminus c_1}\}$.*

Theorem 2. *In the class PF^+_{sub}, any coalition C_1 has its lower bound when inside the coalition structure $CS' = \{C_1, \{a_1\}, ..., \{a_{(n-|C_1|)}\}\}$ and has its upper bound when inside the coalition structure $CS'' = \{C_1, C_{A \setminus c_1}\}$.*

As proved by [2], it is possible to obtain a worst case guarantee solution if some specific CS are searched. In some cases, only a sub-optimal solution is required, as long as it is within a finite ratio bound β from the optimal. This β is defined in Equation 1, with CS^*_N being the best solution found so far:

$$\beta \geq \frac{V(CS^*)}{V(CS^*_N)} \tag{1}$$

Theorems 3 and 4 explain how to obtain a solution with a finite ratio bound β from the optimal (the proofs can be found in [2]):

Theorem 3. *To establish a ratio bound β on PF_{sub}^{+} it is necessary to evaluate the value of all CS containing one or two coalitions. The total number of CS calculated is 2^{n-1} with $\beta = n$.*

Theorem 4. *To establish a ratio bound β on PF_{sup}^{-} it is necessary to evaluate the value of all CS belonging to the configurations: $\{[n]; [n-1,1]; [n-2,1,1]; ...; [1,1,1,...1]\}$. The total number of CS calculated is $2^{n} - n + 1$ with $\beta = \lceil \frac{n}{2} \rceil$.*

3 Distributed Coalition Structure Generation with Externality

Using the constructions created by D-IP and $IP^{+/-}$, we now present our distributed algorithm for solving PFG. The DCSGE algorithm is divided into two stages. The first one is responsible for evaluating the values of the coalitions inside certain CS and these values are then shared among the agents. The information retrieved in the first stage is then used in the second stage, when a configuration is selected by each agent and the CS^{*} is searched (if not found in the first stage). Next, both stages are explained.

3.1 First Stage: Coalition Values and Communication

Although this stage is similar to the one found in D-IP, it cannot proceed the same way, because coalitions do not possess a single value, like in the case with CFG. Instead, agents have to search for more than one value for each coalition. As explained in the previous section, it is possible to calculate the maximum and minimum values of a coalition in PFG as long as the problem belongs to one of the four classes mentioned in Section 2.2. These values are the ones when the coalition is at a coalition structure with $n - |C|$ other coalitions and with exactly one other coalition.

At first, the coalitions must be divided into lists, and these lists must be split into segments (just like in D-IP). It is important to use the same algorithm used in D-IP for this, once the value of a coalition are determined by the CS where it is inserted. Using the same segment division as D-IP guarantees that, when searching for the value of a coalition inside a CS with exactly two coalitions, we can also search for the value of the complementary coalition and the value of the CS itself. Therefore, at each evaluation, these three values are calculated simultaneously. For instance, given $n = 5$ and the coalition $C' = \{a_1, a_3\}$, when calculating its values (either maximum value or minimum value) we have to evaluate the $CS = \{C', C''\}$. This means that we also obtain the value of C'' (in this particular case, $C'' = \{a_2, a_4, a_5\}$) when inside this CS and, by summing both coalitions, the value of the CS itself.

Once the lists are distributed in the same way as in D-IP, each agent proceeds to evaluate its share. As in D-IP, we also assume the existence of a global identifier, which indicates for which segment each agent is responsible. When all agents are done with their calculation, they need to exchange the maximum and lower values (for PF^+sub) of each list. The maximum value of each list is used to estimate the upper bound of each configuration. The lower values serve to calculate the value of a coalition inside each CS (as explained in Section 4, the value of a coalition is given by a default value added for the externality in the CS). The default value of a coalition C' is the value of C' on a CS with only singletons coalitions and the coalition C'. While in PF^+_{sub} it is the lowest value for the coalition, in PF^-sup it is its maximum value, hence no need to exchange the lower value of the coalitions in PF^-sup.

Next, each agent may evaluate the CS composed only by singletons. At the end of this stage, all the CS containing exactly one coalition, two coalitions and n coalitions have their values calculated. In dealing with PF^+_{sub} we may then guarantee that the solution found is at a ratio $\beta = n$ from the optimum. If the problem belongs to PF^-_{sup}, the agents have to calculate additional coalition structures to guarantee a ratio $\beta = \lceil \frac{n}{2} \rceil$. To avoid redundant computations and unnecessary communication, each agent is responsible to calculate the value for the CS for which it has calculated the value of a coalition with the same cardinality as the coalition with the highest cardinality inside the CS. For example, in $CS = \{n-3, 1, 1, 1\}$, each agent with a known value for a coalition of size $n - 3$ calculates the value of the CS with the known coalition. The set of $CS = \{[A]; [A-1, 1]; [1, 1, ..., 1]\}$ does not need to be evaluated, as their values are already known.

3.2   Second Stage: Configurations

After all the information about the coalitions have been exchanged the agents must calculate the upper bound for each configuration. Once all agents have the same information about the maximum value for each list, it would be redundant to require that all agents evaluate all configurations upper bound. Therefore, the configurations are ordered (any order can be used) and separated into segments. Each agent is responsible for one segment and investigates its upper bound. If the upper bound for a configuration is lower than the CS_N^*, it is pruned.

The remaining configurations are ordered by their upper bound limits. Each agent broadcasts the value of its configuration with higher upper bound. In the case some agent has a configuration with an upper bound value higher than one of the configuration values transmitted, it broadcasts the value of this configuration. This ensures that at least the n higher upper bound configurations have been broadcasted and all agents know them. In the worst case scenario, the number of configuration broadcasted is $\frac{n*(1+n)}{2}$. After this communication, configurations are ordered by upper bound value and each agent is responsible for searching the CS^* inside a configuration.

To search inside a configuration, agents have to exchange the remaining known coalition values. Since the size of coalition lists is exponential, two filters (similar to the ones used in D-IP) are used to prune those coalitions that cannot be part of the CS^*. The first filter (Equation 2) allows removing coalitions where the sum of singletons coalitions of a set A' of agents is higher than the value of a coalition containing A' agents:

$$v_{max}(c) < \sum_{a_j \in c} v_{max}(\{a_j\}) \tag{2}$$

This filter can only be used in the PF_{sup}^- class. Once the externality for this class is ≤ 0, a coalition with higher cardinality only produce negative influence over the remaining coalition inside the CS. For PF_{sub}^+, the externality applied over the remaining coalition may overcome the benefits of splitting the coalition in singletons. Therefore, it is not valid for problems in PF_{sub}^+ class.

The second filter (Equation 3) allows to verify if a coalition could be in CS^* (in any specific configuration) and may be used in any situation or class.

$$v_{max}(c) + UB_i - Max_{|c|} < v(CS_N^*). \tag{3}$$

If by replacing the highest value coalition of the list $L_{|c|}$ with the value of c in the evaluation of a configuration upper bound and the result is lower than the best CS found so far, than c cannot be part of the CS^* in that configuration. It is important to remind that the values used for estimating the upper bound of a configuration are the maximum values a coalition may have. Hence, there is no scenario where this sum may reach a higher value.

After applying those two filters, the remaining coalitions that are part of some configuration are broadcasted. The value broadcasted is the value of the coalition without the influence of externalities, that is, the highest coalition value in PF_{sup}^- and the lowest coalition value in PF_{sub}^+. Given these data, agents proceed to search inside the configurations for the CS^*. This search follows the same procedures as in [8], each agent being responsible for a whole configuration. The reason for not splitting a configuration between many agents is because it would prevent them from knowing coalitions values that could help them prune even more configurations as well as yield a number of redundant calculations.

DCSGE only prunes the configurations where there is a guarantee that it cannot contain the CS^*, for its upper bound is lower than the best CS found until that moment. Once all remaining configurations are searched by the agents, it is possible to ensure that the optimal CS is found.

4   Experiments and Results

An important contribution from [2] is the equation used to calculate the influence of the externality over a coalition. Since this influence depends on a number of factors (such as the other coalitions within a CS, whether it is positive or negative and if the maximum influence is bounded or not), the need for an equation capable of describing the influence for any environment is essential. One important characteristic about this equation is that all coalitions must be affected by the same kind of externality (either positive or negative externality). For instance, given any two coalition structures CS' and CS'', such that CS'' is formed by a series of coalitions merges from CS', the value of all coalitions not involved on these merges and within CS'' cannot be greater (on PF^-) or smaller (on PF^+) than in CS'.

The equation proposed by [2] is used in this paper. It is important to highlight that v_C is the value of the coalition C and that e_c sets the maximum externality possible, but not all coalitions are influenced by the same externality value. More specifically, the value of e_c for a coalition C may be sampled from the uniform distribution ranging from 0 to $v_C \times (x/100)$, with x as the maximum externality possible.

Our experiments intend to measure how much search space can be pruned and how much it has to be searched before finding the CS^*. We propose two metrics. The first one is related to the number of configurations the first part of the algorithm is capable of pruning. In this first part not only it is possible to guarantee a ratio bound from the optimal, but it is also possible to reduce drastically the number of configurations that can contain the CS^*. The second measure refers to the percentage of coalition structure that were searched before the CS^* could be found.

The experiments are conducted with different number of agents (ranging from 10 to 16), positive externalities ranging from $e_c = 0.1$ to $e_c = 1.0$. Following standard distributions in the CSG literature, we assign the values of the coalition by means of the uniform distribution, where a coalition value may range from 0 to $|C|$, as well as by means of a normal distribution $N(\mu, \sigma^2)$, where coalition value may range from 0 to $|C| * N(\mu, \sigma^2)$, and $\mu = 1.0$ e $\sigma^2 = 0.1$. The β value is fixed at 0.98 and 1 (for lower values it is very common to find the solution on the first part of the algorithm and the results are not be interesting). Since every combination of the above represents a different setting (i.e., 12 agents, $\beta = 1.0$, $e_c = 0.5$, normal distribution), it is impossible to show all results on this paper. Hence, only the most representative combination are shown. The results are reported as an average of 50 simulation runs.

Figure 2 (A) shows the percentage of configurations pruned in the first stage of DCSGE, with uniform and normal distributions. As seen, almost half of the configurations can be pruned in the first stage. This represents a considerable reduction of the search space, as well as a reduction of calculation needed to establish upper bound values for the configurations. Furthermore, the number of exchanged coalition values also decreases, once some coalition lists may not be present on the remaining configurations. Figure 2 (B) shows the difference between the percentage of configuration pruned for different externality values. Once externalities affect all coalitions in the same way, a bigger externality does not necessary indicates a larger number of configurations pruned.

Figure 3 (A) indicates the percentage of the total coalition structures that had to be searched before finding the CS^*. As the number of agents increases, percentage gets smaller. That does not mean that agents are searching fewer CS, but that the relation between the searched space and total space is getting

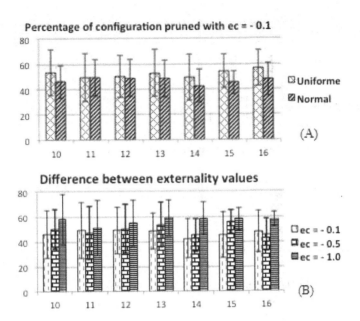

Fig. 2. Percentage of configurations pruned after the first part of DCSGE with uniform and normal distribution (A). Percent of configurations pruned with different externalities (B).

smaller. Since the number of CS grows in the order of $O(n^n)$, it grows much faster than the number of CS searched. For 16 agents, the total number of CS is 10.480.142.147 and the algorithm searched nearly 0.01% of this total to find the CS^*, with $\beta = 1.0$. That is a significant reduction of the problem size, as DCSGE avoided searching about 10.479.094.132 CS .

In many cases it is not necessary to find the optimal coalition structure. In some experiments we set the value of β to 0.98 in order to verify if the algorithm could find a result searching fewer CS. As a result, using normal distribution and $e_c = 1.0$, Figure 3 (B) shows that the experiments using a lower value for β searched fewest CS, as expected.

Finally, we compare the number of CS searched using different values for e_c. In Figure 3 (C) we show a comparison of the percentage of space searched by a normal distribution simulation with $e_c = 0.1$, $e_c = 0.5$ and $e_c = 1.0$. The higher the value of e_c, the greater is the number of CS searched. The reason may be that due to the large number of coalitions in the lists surrounding $L_{n/2}$, those lists have a higher probability of one of their coalitions to have its value increased due to externality. Therefore, the values of the configurations are bounded by those coalitions and are closer one to another. When searching inside the configurations for the CS^*, the agents prune the configurations that do not have an upper bound higher than the best CS found so far. If the values of the configurations are similar, they cannot prune many configurations and have to search them all.

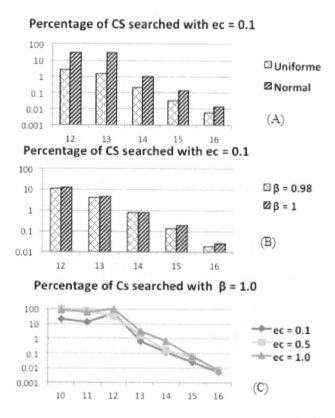

Fig. 3. Percentage of CS searched before finding the optimal solution based on the distribution (A). Percentage of CS searched based on the value of β (B). Percentage of CS searched based on the value of the externality (C).

5   Conclusion

The coalition structure generation problem has only recently gained attention from the AI community. In many real-world scenarios, it is imperative to deal with externalities. The size of the input (that grows from $O(2^n)$ to $O(n^n)$) makes it a much more difficult task than treating CSG as a CFG.

In this paper we presented a distributed algorithm for finding the optimal CS^* in the CSG problem with externalities. The basis for this algorithm is D-IP (that can be used in both distributed and centralized environment but can solve exclusively CFG) and IP_-^+ (which can solve PFG as well as CFG, but is a strictly centralized algorithm). Unlike the other algorithms, DCSGE can find the CS^* in both centralized and distributed environments and in CFG and PFG. Therefore it represents a more comprehensive algorithm.

Among the possible improvements, the one that seems to be most important is an efficient search for CS inside the configurations. The most time-consuming stage of the CSG problem is to search for the precise CS that contains the optimal value. One way to improve this process would be to allow agents to share their information about specific groups of coalitions that cannot occur together in the CS . Equation 3 allows removing single coalitions from the search space, but it does not check for multiple coalitions that cannot occur together. Further, a more comprehensive study about communication is required, as more information can be broadcasted among agents in order to prune additional parts of the search space.

References

1. Sandholm, T., Larson, K., Andersson, M., Shehory, O., Tohmé, F.: Coalition structure generation with worst case guarantees. Artificial Intelligence (1999)
2. Rahwan, T., Michalak, T., Wooldridge, M., Jennings, N.R.: Anytime coalition structure generation in multi-agent systems with positive or negative externalities. Artificial Intelligence 186, 95–122 (2012)
3. Finus, M., Rundshagen, B.: Endogenous coalition formation in global pollution control: a partition function approach. In: The Endogenous Formation of Economic Coalitions, pp. 199–243. Edward Elgar, Cheltenham (2003)
4. Catilina, E., Feinberg, R.: Market power and incentives to form research consortia. Review of Industrial Organization 28(2), 129–144 (2006)
5. Koczy, L.A.: A recursive core for partition function form games. Research Memoranda 031 (2006)
6. Rahwan, T., Jennings, N.R.: An algorithm for distributing coalitional value calculations among cooperating agents. Artificial Intelligence 8(171), 535–567 (2007)
7. Rahwan, T., Ramchurn, S.D., Dang, V.D., Jennings, N.R.: Near-optimal anytime coalition structure generation. In: Proc. of the Int. Joint Conf. on Art. Intelligence (IJCAI 2007), pp. 2365–2371 (January 2007)
8. Rahwan, T., Ramchurn, S.D., Jennings, N.R., Giovannucci, A.: An anytime algorithm for optimal coalition structure generation. J. Artif. Int. Res. 34, 521–567 (2009)
9. Michalak, T., Sroka, J., Rahwan, T., Wooldridge, M., McBurney, P., Jennings, N.R.: A distributed algorithm for anytime coalition structure generation. In: Proceedings of the 9th International Conference on Autonomous Agents and Multiagent Systems (2010)
10. Shehory, O., Kraus, S.: Methods for task allocation via agent coalition formation. Artificial Intelligence 101(1-2), 165–200 (1998)
11. Michalak, T., Dowell, A., McBurney, P., Wooldridge, M.: Optimal coalition structure generation in partition function games. In: Proceeding of the 2008 Conference on ECAI 2008: 18th European Conference on Artificial Intelligence (2008)

Enhancing Agent Metamodels
with Self-management for AmI Environments

Inmaculada Ayala, Mercedes Amor Pinilla, and Lidia Fuentes

Departamento de Lenguajes y Ciencias de la Computación
Universidad de Málaga, Spain
{ayala,pinilla,lff}@lcc.uma.es

Abstract. Ambient Intelligence (AmI) is the vision of a future in which environments support the people inhabiting them unobtrusively. The distributed nature the AmI, the autonomy, awareness and adaptation properties make software agents a good option to develop self-managed AmI systems. AmI systems demand the reconfiguration of their internal functioning in response to changes in the environment. So, AmI systems must behave as autonomic systems with the capacity for self-management. However, current agent metamodels lack adequate modeling mechanisms to cope with self-management. We propose to extend an existing agent metamodel with a number of specific modeling concepts and elements that facilitate the design of the self-management capabilities in agents.

Keywords: Self-management, AmI, AOSE, Metamodel.

1 Introduction

Ambient Intelligence (AmI) is the vision of a future in which environments support the people inhabiting them [1]. In this vision the traditional computer input and output media disappears and processors and sensors are integrated inside everyday objects. Different technologies contribute to the development of this vision such as distributed computing or human-computer interaction. In general, AmI systems are characterized by being composed of a large variety of heterogeneous devices, which are intended to observe, react and adapt to events that occur in the environment.

The distributed nature of AmI systems, the autonomy, awareness and adaptation properties make agents a good option for the development of AmI systems: agents are reactive, proactive and their communication is supported by distributed agent platforms. Different approaches have highlighted the suitability of the agent paradigm to support the different stages of the development of AmI systems [1]. Agents have been used as abstractions to model and implement the AmI system functionality, to encapsulate artificial intelligence techniques and to coordinate the different elements that compose the AmI system. Normally, the majority of AmI devices show symptoms of degradation, such as energy loss, which require explicit management action, for example saving energy to guarantee the system's survival. Additionally, the distribution of these applications

L. Correia, L.P. Reis, and J. Cascalho (Eds.): EPIA 2013, LNAI 8154, pp. 420–431, 2013.
© Springer-Verlag Berlin Heidelberg 2013

sometimes makes it difficult to directly control or manage the applications. Furthermore, AmI systems demand the reconfiguration of their internal functioning in response to changes in their environment, which means that AmI systems must behave as autonomic systems with the capacity for self-management [2].

So, agents integrated in an AmI environment must not only be aware of the physical world, but they must also demand the reconfiguration of their behaviors in response to changes. Each agent should be able to manage itself and the environment given a set of high level goals, without the intervention of human administrators. To perform self-management, the agent must be able to monitor the environment (self-situation), and itself (self-awareness); to detect changing circumstances (self-monitoring) and after that the agent must be able to adapt its behavior (self-adjusting), probably in coordination with other agents of the AmI environment (coordinated interaction). These activities are also known as autonomic functions (AFs) and are used to develop systems that include, as properties, self-configuring, self-optimizing, self-healing and self-protecting.

Current agent metamodels lack explicit modeling mechanisms to adequately cope with the design of these capabilities. Normally, the self-management capabilities of agents, if they exist, are intermingled with the agent architecture, making difficult both the specification and reasoning about self-management issues. This results in self-management specifications that are ambiguous, and aside many concerns such as detecting self-management policy conflicts. Therefore, our goal is to extend an existing agent metamodel with specific modeling elements that facilitate the design of self-management capabilities in a Multi-Agent System (MAS). Specifically, we are going to extend the PIM4Agents metamodel [3] with self-management concepts.

PIM4Agents is a generic metamodel that can be used to specify agents of the most representative architectural styles (BDI and reactive). This metamodel consists of eight different but related viewpoints, each one focusing on specific issues of an MAS. We have included a new viewpoint, named *Self-Management*, to represent and relate the self-management features and capabilities of MASs for AmI. So, the main contribution of this paper is a metamodel for agents that includes specific concepts for supporting self-management. In particular, we concern ourselves with the specification of policies to describe when and how to adjust the behavior of the agent and the MAS, by means of agent roles; and also by adding specific actions for self-awareness and self-adjusting.

The paper is structured as follows: Section 2 presents the case study that we are going to use to illustrate our approach. Section 3 presents the metamodel for self-managed agents. Section 4 presents some related work and the paper finishes with some conclusions.

2   Motivating Case Study

In order to illustrate our proposal, we are going to use a case study of AmI environments, which encompasses a Wireless Sensor Network (WSN) that interacts with smartphones and PCs to control the physical condition of a patient and the house where he/she lives. This case study has been partially implemented for Sun SPOT sensor motes [4] and Android devices, however it is not a contribution of the work presented here. A WSN consists of a set of sensor nodes monitoring physical or environmental conditions, such as temperature or sound. Sensed data is sent through the network to a device node or a sink node. Source sensor nodes generally operate on a resource of limited power-supply capacity such as a battery, and are compact and inexpensive. Sink nodes have more resources than source nodes, but they operate on a resource of a limited power-supply capacity.

Table 1. Self-management policies for the WSN

Name	Condition	Action
Decrease Sampling Frequency	$batteryLife <$ 10% $\land$ $Active(Task) \land$ $F < 0.003$ $\land$ $F > 3.33 * 10^{-5}$	1: Task task=getComponent(ID) 2: double s=getSampling() 3: task.setSampling(s - X)
Task Allocation	$batteryLife <$ 10%	1: send(REQUEST) and wait(ANSWER) 2: if answer=BAT_LIFE then candidates.store(sensor) 3: while $!candidates.isEmpty() \land !success$ do 4: select(sensorX) where $sensorX = \{sensor \in candidates \land$ $!\exists sensorY \in candidates\|$ $sensorY(BAT\_LIFE) > sensor(BAT\_LIFE)\}$ 5: send(PROPOSAL) to sensorX and wait(ANSWER) 6: if receive(ACCEPT) then removeRole(Task) and success=true
Sink Drop	$Communication$ $exception$	1: send(SINK_DROP) to AMS 2: if receive(SINK:YOU) then assignRole(Sink) elseif receive(SINK:ID) then setSink(ID) elseif receive(RETRY) then wait 10 seconds and retry

In our application, each sensor accomplishes different tasks related to the application-specific functionality, e.g. to monitor the light and send the value to a sink node. Sink nodes receive monitored data, process it, and send it to another device (e.g. a PC). These tasks are performed by agents running inside each sensor node or device. In addition to these tasks, these sensors accomplish other tasks related with self-management in order to extend the life of the system or recover from the failure of some of the sensors. For example, when any of these situations is detected, sensor agents can allocate additional tasks to agents in sensors with more battery life.

Therefore, apart from their application-specific tasks, agents have to be endowed with additional behavior which constantly checks the context and automatically adapts the agent behavior to changing conditions. The adaptation is ruled by policies that can be specified in terms of Event Condition Action (ECA), goals or utility functions [5]. An ECA policy dictates the action that should be taken whenever the system is in a given current state. A goal policy specifies either a single desired state, or one or more criteria that characterize an entire set of desired states. A utility function policy is an objective function that expresses the value of each possible state. Policies in the form of goals or utility functions require planning and are not adequate for lightweight devices like sensor motes, due to their limited resources. So, in order to model the self-management behavior of sensors, we choose ECA policies. A description of the rules that comprise the polices for the sensors of our case study is given in Table 1 in informal semantics. The first column (*Name*) is the name of the policy, the second column (*Condition*) describes the situation (expressed by a logical expression or the occurrence of an event) that causes the execution of actions depicted in the third column (*Action*). The first two rules correspond to self-optimizing policies, while, the third one is a self-healing policy. An interesting challenge at modeling level for the agent designer is how to write policies for self-managed systems and ensure they are appropriate for the application, however due to limitations of space this issue is beyond scope of this paper.

3   A metamodel for Self-managed Agents

This section is devoted to describing the main contribution of this work, a metamodel that integrates self-management concepts to be effectively used for the modeling of AmI applications based on agents. Our approach takes the PIM4Agents metamodel, and extends it to add a new viewpoint for the modeling of self-management, existing viewpoints are extended with new self-management related concepts.

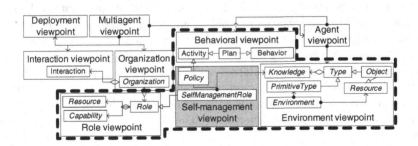

Fig. 1. Relationship between the Self-management and base metamodel viewpoints

PIM4Agents is a generic metamodel that can be used to specify agents of the most representative architectural styles (BDI and reactive), unifying the most common agent oriented concepts used in well-known agent frameworks. As stated, the basic PIM4Agents metamodel is structured as eight viewpoints (see Fig. 1) each focusing on a specific concern of an MAS: the *Multiagent* viewpoint contains the main building blocks of an MAS (roles, agents and so on) and their relationships; the *Agent* viewpoint describes agents, the capabilities they have to solve tasks and the roles they play within the MAS; the *Behavioral* viewpoint describes agent plans; the *Organization* viewpoint describes how single autonomous entities cooperate within the MAS and how complex organizational structures can be defined; the *Role* viewpoint covers the abstract representations of functional positions of autonomous entities within an organization or other social relationships; the *Interaction* viewpoint describes how the interaction in the form of interaction protocols takes place between autonomous entities or organizations; the *Environment* viewpoint contains any kind of resource that is dynamically created, shared, or used by the agents or organizations; and the *Deployment* viewpoint contains a description of the MAS application at runtime, including the types of agents, organizations and identifiers.

Fig. 2. Steps for designing Self-management

The new viewpoint, called *Self-management*, includes concepts of other viewpoints, and is designed to explicitly model self-management policies and the roles involved in self-management functions. However, the modeling of self-management also implies the incorporation and representation of new resources, knowledge and behaviors, which are described in the corresponding viewpoints. Fig. 1 provides a high-level description of the relationships and dependencies of existing viewpoints (big rectangles labeled *viewpoint*) and concepts (small rectangles) of the base metamodel and the *Self-management* viewpoint (grey rectangle) and related concepts (*Policy* and *SelfManagementRole* entities). In the following subsections, we explain the concepts that are contained in this new viewpoint and how they are used to model self-managed AmI systems. This explanation focuses on self-management capabilities, however a description of how to model MASs using the base metamodel is beyond the scope of this paper. The different modeling steps required by our proposal are depicted in Fig. 2. The proposed modeling process basically consists of six steps, it starts by including a new viewpoint and ends with the optional validation and correction of the policies that rule self-management. The following sections describe these steps.

3.1 Organizations and Self-management Roles

The first step in the design process of our application is to provide the *MAS* viewpoint (see Fig. 3), focusing on the concept of organization, which is presented in several metamodels [6] and which can be used to separate the different concerns of the MAS. These organizations have roles that represent the different functions required in the MAS inside an organization, e.g. the different monitoring tasks of the application. A role encapsulates a set of behaviors (i.e. capability), resources and participation in interaction protocols. An agent can accomplish some roles

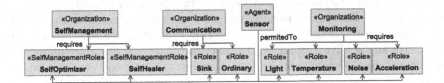

Fig. 3. UML class diagram corresponding to the *MAS* viewpoint

at runtime or stop fulfilling such a role. In our application, we have a single type of agent (*Sensor* in Fig. 3) that belongs to three organizations via the roles it performs. The self-management capability is represented in this viewpoint as an *Organization* (left hand side of Fig. 3), which requires two roles (*self-healing* and *self-optimizing*) representing the self-management policies and behavior required by sensors (depicted in Table 1).

At this stage of the modeling, organizations and roles are a good option for integrating self-management because organizations promote the separation of the application concerns and roles deal with dynamic functionality. Although by using these concepts we could model policies for distributed systems like AmI applications, we would still be lacking an important concept which should be integrated within roles, which is the knowledge required by the role. For example, if an agent is the sink of the WSN, i.e. it performs the *Sink* role in the *Communication* organization, it must store knowledge about the radio addresses of the agents it receives data from. However, in most agent metamodels (included PIM4Agents), knowledge can only be associated with agents, and not with roles [6]. Therefore, agents all have the same knowledge, regardless of the roles that they perform. Since an agent can adopt different roles at runtime, it would be better for it to include the knowledge associated with the role, and not only the behavior at the modeling stage. In our opinion this is not a serious limitation, but it is not conceptually correct. In order to overcome this, a new role is defined, the *SelfManagementRole*, which extends *Role* (in this way we extend the metamodel and overcome this limitation only for self-management behavior, while *Role* semantics remain the same). For example, the description of the *SelfHealer* role (see Fig. 3) would be included in the *SelfManagement* viewpoint. The policy associated with this role is given in Table 1.

The second viewpoint related with self-management is the *Environment* viewpoint. In this viewpoint the designer describes the internal information of the system which is relevant for self-management, e.g. the internal components of the agents. So, here we define data types associated with self-management and later, we use them to define the knowledge belonging to the *SelfManagement-Role*. For example, the knowledge required by the *SelfHealer* role concerns the agent that it must locate in case of sink failure.

In self-managed systems, it is important to distinguish between policies and activities that support the application of policies. In our metamodel, these activities are modeled using capabilities associated with the *Role* concept (see Fig. 1), and then it is possible to model a *SelfManagementRole* without policies. For example, we can have an organization with two roles, *Slave* and *Master*. The function of the *Slave* is to periodically send "is alive" signals to the *Master*. While the function of the *Master* is to restore a *Slave* if in a time span it does not receive an "is alive" signal. So, the actions of the role *Slave* are intended to support self-management, and its execution does not depend on an event or a condition. A typical action contained in capabilities is the monitoring behavior for self-management.

3.2   Policies Using APPEL Notation

The next step in our modeling approach is to define the policies for self-management. In order to model policies, one option is to use the existing viewpoints that are used to model the dynamic behavior of the system, which are the *Interaction* and the *Behavior* viewpoints. However, the principal limitation of using the *Behavior* viewpoint for the modeling of agent's plans [6] is that it does not have support for conflict detection. Conflicts between policies arise when two or more policies can be applied at the same time and those contain actions that can cause conflict, e.g. one policy requires an increase in the frequency a task, while the other states that the frequency of the same task must be decreased. As a solution, we propose to use a similar approach to that presented in [7]. This approach uses an ECA policy language named APPEL, which has been used for policy specification in environments related with AmI (e.g. management of sensor networks). The main advantage of APPEL for us is that this language has support for conflict detection using the UMC model checker.

Table 2. APPEL syntax

policy	::= polRuleGroup |polRuleGroup policy
polRuleGroup	::= polRule |polRuleGroup op polRuleGroup
op	::= **g**(condition) |**u** |**par** |**seq**
polRule	::= [**when** trigger] [**if** condition] **do** action
trigger	::= trigger |trigger **or** trigger
condition	::= condition |**not** condition |condition **and** condition
	|condition **or** condition
action	::= action |action actionop action
acntionop	::= **and** |**or** |**andthen** |**orelse**

We have integrated APPEL in our metamodel using its syntax to define *policies* (see Fig. 4) that any *SelfManagementRole* has to follow. Our metamodel uses the syntax depicted in Table 2, but it is adapted for the agent domain. This domain-specific language, which is based on APPEL, specifies the triggers, conditions and actions of the self-management policy. Triggers can be internal events of the agent (*EventTrigger*), or the sending and reception of a message (*MSGTrigger* and *Type*). The *Condition* concept can be simple or composed as in the APPEL notation, but it is related with agent knowledge (*Knowledge*) or an equality expression over the trigger of the rule (*TriggerPredicate*). The *TriggerPredicate* concept represents an equality expression between a value and an attribute (*Attribute* enumeration) that can be the type of the trigger (*instance*) or a field of the message associated with the trigger (*performative, protocol* or *content*). Actions associated with policies, i.e. *SMPlan*, are described in the next section. The policy *Sink Drop* (third row in Table 1) can be expressed in APPEL-based notation as follows:

```
seq(when CommunicationException do (send(SINK_DROP));
    par(when MSGTrigger.RCV if Content=SINK_YOU
        do (assignRole(Sink));
    when MSGTrigger.RCV if Content=ID
        do (setSink(ID));
    when MSGTrigger.RCV if Content=RETRY
        do (wait(10);send(SINK_DROP));););)
```

In order to validate this policy using the UMC model checker, it has been translated to a set of UML finite state machines (FSMs) using the mapping proposed in [7]. We do not show the whole process because it is not a contribution of the work presented here. The UMC framework takes the FSM and transform it in doubly labeled transition system (L^2TS), which is a formal model of the system evolution [8]. This L^2TS can be model checked using logic formulae expressed in UCTL, a UML-oriented-time temporal logic. When the translated model is introduced in the UMC tool, we can check whether or not a conflicting action is executed simultaneously. For example, the expression $AG\neg((increaseFrequency(true))\&(decreaseFrequency(true)))$ would be used to express that is not possible to increase and decrease frequency simultaneously.

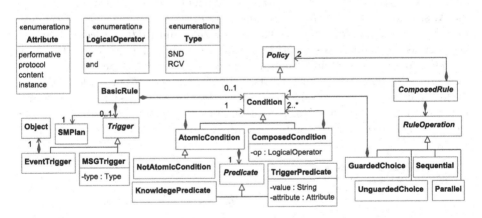

Fig. 4. Metamodel for *Policy* concept using APPEL notation

Distributed policies are a special type of policy, which are quite common in AmI systems. These policies require the communication and coordination of different self-management roles. In order to model these policies, we require the viewpoints to model the distributed behavior of the system (*Organization* and *Interaction*), in addition to the *SelfManagement* viewpoint. An example of a distributed policy is the *Task Allocation* policy (second row in Table 1), which requires the agents to interact in order to allocate a task. Firstly, we associate an interaction protocol (*TaskAllocationProtocol*) to the *SelfManagement* organization in the *Organization* viewpoint. Secondly, in the same viewpoint we model the actors required by the protocol (*Requester* and *Responder*) and the roles that can play these actors (*SelfOptimizer* in Fig. 3). Finally, in the *Interaction* viewpoint, the message exchange between actors is defined and described.

3.3 Actions for Self-management

The last step in our process is to define the actions required by the policies. The *Behavior* viewpoint describes plans associated with agents and capabilities of agents and roles. Due to the inadequacy of the *Behavior* viewpoint for modeling some concepts and activities specific to self-management (the Autonomic Functions (AFs) described in the Introduction Section), we introduce the concept of *SMPlan* (see Fig. 4). The *SMPlan* is similar to the *Plan* concept. A plan is composed of a set of actions, such as sending a message, which are related using complex control structures such as loops. At modeling, self-awareness and self-adjusting related actions, which are closely related with the agent architecture and are independent of the application domain, require a common vocabulary to avoid ambiguous specifications. So, it would be better to have specific purpose actions included in the *Behavior* viewpoint and to avoid ad-hoc solutions. With the definition of the *SMPlan* concept, we ensure that self-management actions cannot be used for plans related with other concerns of the application.

An *SMPlan* can include the same actions of any plan and actions for self-management to overcome the limitations of agent metamodels with regard to the AFs. Concretely, it includes the following actions: *GetBehaviors*, *GetRoles*, *RemoveResource*, *AddResource*, *RemoveKnowledge* and *AddKnowledge*. The first two actions are defined to support self-awareness, while the last four actions are intended to support self-adjusting. The base metamodel has actions that support self-management too, such as *AssignRole*, *RemoveRole* or *CalledBehavior*. Additionally, due to the restrictions imposed by the APPEL syntax, the control structures are limited to loops, parallel actions and decisions. In Fig. 5 we can see the *SMPlan* that corresponds to the *Task Allocation* policy (see Table 1).

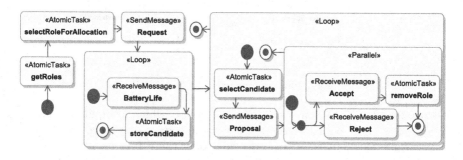

Fig. 5. UML state machine corresponding to the *SMPlan* of the *Task Allocation* policy

4   Related Work

Although the majority of agent technology contributions to the AmI focus on the implementation level, there are interesting approaches that focus on the modeling level as our approach does [1]. One remarkable example is [9], which uses Tropos to design an interactive tourist guide in an Intelligent Museum. As in our approach, in this proposal, role descriptions have been used to deal with the dynamic functionality of an AmI application. In [10], the SODA methodology is used to test how to design an agent-based application for controlling an intelligent home. This contribution focuses on the use of agents to design the coordination of a distributed and intelligent system with complex interaction patterns. Related with the AmI, [11] uses the MAS theory to model a home automation system. On the other hand, Agent-π [12] is a metamodel to design ubiquitous applications embedded in Android devices. The work [13] uses an agent metamodel enhanced with *aspects* to model important properties of AmI systems such as context-awareness. In these approaches, the use of agents to model the application is justified because of the distributed and dynamic nature of AmI systems. However, these approaches do not deal with any self-management capability of the agents in the AmI domain.

An interesting approach that joins together MAS and self-management is MaCMAS [14]. This approach uses agent-based modeling techniques and software product lines to model software systems that evolve over time. This evolution of a system is expressed by means of roles that an agent performs in a given time to achieve a set of goals. Roles are designed as a set of features that can be composed using different operators. However, this work focuses on the evolution of the system architecture, and does not offer mechanisms to validate the self-managed behavior. Another interesting approach is [15], which uses the Agent Modeling Language to design the entities that control and manage self-managed systems, which are also known as autonomic managers. However, these agents are not able to manage themselves, as our agents can.

On the other hand, there are approaches for the systematic modeling of self-managed systems in general. Although they do not consider particularities of the agent domain, they take into account typical agent concepts like reactive behavior or distributed interaction [16,2]. ASSL [16] provides a framework for the specification, validation, and code generation for self-management of complex systems. The self-managed behavior is provided by wrappers promoting the separation of concerns. This separation of concerns is also tackled by FORMS [2], a reference model for the formal specification of distributed autonomic systems. Specifically, it uses a reflective modeling perspective that, like our metamodel, allows reasoning about the architecture of the self-managed system. Although both approaches use a formal notation, they do not consider mechanisms to detect policy conflicts.

Finally, another area related with the work presented here is metareasoning [17]. Metareasoning is the process of reasoning about the agent reasoning cycle and is used to implement meta-level control. The goal of this kind of control is to improve the quality of agent decisions by spending some effort to decide what and how much reasoning to do as opposed to what actions to do. So, it is related with our approach in the sense that we reason upon the agent to affect agent actions. However, the role of the models and metamodels in this area is to be the input of the metareasoning engine. Our approach is focused on the modeling of the MAS and its self-management functionality.

5   Conclusions

Agents are a natural metaphor for the modeling of distributed applications with autonomous and intelligent behavior, like AmI systems. These environments are characterized by a high degree of unpredictability and dynamism in the execution context, that makes the application of techniques like self-management necessary. Agent metamodels offer an excellent basis for modeling self-managed AmI systems, however they have the following limitations: (i) poor specification of dynamic behavior; (ii) lack of support to validate the self-managed behavior; and (iii) ambiguous notation to express AFs. In order to overcome these limitations, we have extended the metamodel PIM4Agents in different ways in order to overcome the aforementioned limitations. We have defined a new modeling viewpoint called *Self-Management*, which allows modeling the roles for self-management (including the knowledge related with self-management) and the policies that drive the self-managed behavior of the AmI system. Policies are described using a domain specific language that follows the APPEL syntax. The use of APPEL allows conflict between policies to be detected using the UMC model checker. Finally, the new viewpoint also includes specific actions to model AFs, facilitating the modeling of self-awareness and self-adjusting functions. This paper has presented the extended metamodel and how the process to model self-management activities of a agent-based AmI application.

For future work, we are considering testing of our metamodel with different case studies in the context of AmI. Additionally, we plan to incorporate this metamodel into a code generation process for self-managed agents.

Acknowledgment. Work supported by the Andalusian project FamWare P09-TIC-5231, the European project INTER-TRUST FP7-317731 and the Spanish projects RAP TIN2008-01942 and MAVI TIN2012-34840.

References

1. Sadri, F.: Ambient intelligence: A survey. ACM Comput. Surv. 43(4), 36:1–36:66 (2011)
2. Weyns, D., Malek, S., Andersson, J.: Forms: Unifying reference model for formal specification of distributed self-adaptive systems. ACM Trans. Auton. Adapt. Syst. 7(1), 8:1–8:61 (2012)
3. Hahn, C., Madrigal-Mora, C., Fischer, K.: A platform-independent metamodel for multiagent systems. Autonomous Agents and Multi-Agent Systems 18, 239–266 (2009)

4. Labs, O.: Sun SPOT World (2013), http://www.sunspotworld.com/
5. Kephart, J., Walsh, W.: An artificial intelligence perspective on autonomic computing policies. In: POLICY 2004, pp. 3–12 (June 2004)
6. Bernon, C., Cossentino, M., Gleizes, M.P., Turci, P., Zambonelli, F.: A study of some multi-agent meta-models. In: Odell, J.J., Giorgini, P., Müller, J.P. (eds.) AOSE 2004. LNCS, vol. 3382, pp. 62–77. Springer, Heidelberg (2005)
7. Beek, M., Gnesi, S., Montangero, C., Semini, L.: Detecting policy conflicts by model checking UML state machines. In: ICFI X, pp. 59–74. IOS Press (2009)
8. Beek, M.H.T., Fantechi, A., Gnesi, S., Mazzanti, F.: A state/event-based model-checking approach for the analysis of abstract system properties. Science of Computer Programming 76(2), 119–135 (2011)
9. Penserini, L., Bresciani, P., Kuflik, T., Busetta, P.: Using tropos to model agent based architectures for adaptive systems: a case study in ambient intelligence. In: SWSTE 2005, pp. 37–46 (February 2005)
10. Molesini, A., Denti, E., Omicini, A.: HomeManager: Testing agent-oriented software engineering in home intelligence. In: Filipe, J., Fred, A., Sharp, B. (eds.) ICAART 2009. CCIS, vol. 67, pp. 205–218. Springer, Heidelberg (2010)
11. Morganti, G., Perdon, A., Conte, G., Scaradozzi, D.: Multi-agent system theory for modelling a home automation system. In: Cabestany, J., Sandoval, F., Prieto, A., Corchado, J.M. (eds.) IWANN 2009, Part I. LNCS, vol. 5517, pp. 585–593. Springer, Heidelberg (2009)
12. Agüero, J., Rebollo, M., Carrascosa, C., Julián, V.: Model-driven development for ubiquitous MAS. In: Augusto, J.C., Corchado, J.M., Novais, P., Analide, C. (eds.) ISAmI 2010. AISC, vol. 72, pp. 87–95. Springer, Heidelberg (2010)
13. Ayala, I., Pinilla, M.A., Fuentes, L.: Modeling context-awareness in agents for ambient intelligence: An aspect-oriented approach. In: Antunes, L., Pinto, H.S. (eds.) EPIA 2011. LNCS, vol. 7026, pp. 29–43. Springer, Heidelberg (2011)
14. Peña, J., Hinchey, M.G., Resinas, M., Sterritt, R., Rash, J.L.: Designing and managing evolving systems using a mas product line approach. Sci. Comput. Program. 66(1), 71–86 (2007)
15. Trencansky, I., Cervenka, R., Greenwood, D.: Applying a uml-based agent modeling language to the autonomic computing domain. In: OOPSLA 2006, pp. 521–529. ACM (2006)
16. Vassev, E., Hinchey, M.: Assl: A software engineering approach to autonomic computing. Computer 42(6), 90–93 (2009)
17. Cox, M.T., Raja, A.: Metareasoning: Thinking about thinking. MIT Press, Cambridge (2008)

Jason Intentional Learning:
An Operational Semantics

Carlos Alberto González-Alarcón[1], Francisco Grimaldo,
and Alejandro Guerra-Hernández[2]

[1] Departament d'Informàtica, Universitat de València
Avinguda de la Universitat s/n, (Burjassot) València, España
{carlos.a.gonzalez,francisco.grimaldo}@uv.es
[2] Departamento de Inteligencia Artificial, Universidad Veracruzana
Sebastián Camacho No. 5, Xalapa, Ver., México, 91000
aguerra@uv.mx

Abstract. This paper introduces an operational semantics for defining
Intentional Learning on *Jason*, the well known Java-based implementation of *AgentSpeak(L)*. This semantics enables *Jason* to define agents
capable of learning the reasons for adopting intentions based on their
own experience. In this work, the use of the term *Intentional Learning* is strictly circumscribed to the practical rationality theory where
plans are predefined and the target of the learning processes is to learn
the reasons to adopt them as intentions. Top-Down Induction of Logical Decision Trees (TILDE) has proved to be a suitable mechanism for
supporting learning on *Jason*: the first-order representation of TILDE is
adequate to form training examples as sets of beliefs, while the obtained
hypothesis is useful for updating the plans of the agents.

Keywords: Operational semantics, Intentional Learning, AgentSpeak(L),
Jason .

1 Introduction

In spite of the philosophical and formal sound foundations of the Belief-Desire-
Intention (BDI) model of rational agency [4,11,12], learning in this context
has received little attention. Coping with this, frameworks based on Decision
Trees [13,14] or First-Order Logical Decision Trees [7] have been developed to
enable BDI agents to learn about the executions of their plans.

JILDT[1] [8] is a library that provides the possibility to define Intentional
Learning agents in *Jason*, the well known Java-based implementation [3] of
AgentSpeak(L) [10]. Agents of this type are able to learn about their reasons to
adopt intentions, performing Top-Down Induction of Logical Decision Trees [1].
A plan library is defined for collecting training examples of executed intentions,
labelling them as succeeded or failed, computing logical decision trees, and using
the induced trees to modify accordingly the plans of learner agents. In this way,

---

[1] Available on http://jildt.sourceforge.net/

L. Correia, L.P. Reis, and J. Cascalho (Eds.): EPIA 2013, LNAI 8154, pp. 432–443, 2013.
© Springer-Verlag Berlin Heidelberg 2013

the Intentional Learning approach [9] can be applied to any *Jason* agent by declaring its membership to this type of agent.

The *AgentSpeak(L)* language interpreted by *Jason* does not enable learning by default. However it is possible to extend the language grammar and its semantics for supporting Intentional Learning. This paper focuses on describing this operational semantics, which enables *Jason* to define agents capable to learn the reasons for adopting intentions based on their own experience. Direct inclusion of the learning steps into the reasoning cycle makes this approach unique.

The organization of the paper is as follows: Section 2 briefly introduces the *AgentSpeak(L)* agent oriented programming language, as implemented by *Jason*. Section 3 introduces briefly the Top-Down Induction of Logical Decision Trees (TILDE) method. Section 4 describes the language grammar and operational semantics that define Intentional Learner agents on *Jason*. Finally, section 5 states the final remarks and discusses future work, particularly focusing on the issues related with social learning.

2   Jason and AgentSpeak(L)

Jason [3] is a well known Java-based implementation of the *AgentSpeak(L)* abstract language for rational agents. For space reasons, a simplified version of the language interpreted by *Jason* containing the fundamental concepts of the language that concerns this paper is shown in the Table 1 (the full version of the language is defined in [3]). An agent *ag* is defined by a set of beliefs *bs* and plans *ps*. Each belief $b \in bs$ can be either a ground first-order literal or its negation (a belief) or a Horn clause (a rule). Atoms *at* are predicates, where P is a predicate symbol and $t_1, \ldots, t_n$ are standard terms of first-order logic. Besides, atoms can be labelled with sources. Each plan $p \in ps$ has the form: @*lbl te* : *ct* ← *h*. @*lbl* is an unique atom that identifies the plan. A trigger event (*te*) can be any update (addition or deletion) of beliefs or goals. The context (*ct*) of a plan is an atom, the negation of an atom or a conjunction of them. A non empty plan body (*h*) is a sequence of actions, goals, or belief updates. Two kinds of goals are defined, achieve goals (!) and test goals (?).

Table 1. *Jason* language grammar. Adapted from [3].

ag	::=	$bs \ ps$		h_1	::=	$a \mid g \mid u \mid h_1; h_1$
bs	::=	$b_1 \ldots b_n$	$(n \geq 0)$	at	::=	$P(t_1, \ldots, t_n)$ $(n \geq 0,$
ps	::=	$p_1 \ldots p_n$	$(n \geq 1)$			$\mid P(t_1, \ldots, t_n)[s_1, \ldots, s_m]$ $m > 0)$
p	::=	@$lbl \ te : ct \leftarrow h$		s	::=	$\texttt{percept} \mid \texttt{self} \mid id$
te	::=	$+at \mid -at \mid +g \mid -g$		a	::=	$A(t_1, \ldots, t_n)$ $(n \geq 0)$
ct	::=	$ct_1 \mid \top$		g	::=	$!at \mid ?at$
ct_1	::=	$at \mid \neg at \mid ct_1 \wedge ct_1$		u	::=	$+b \mid -b$
h	::=	$h_1; \top \mid \top$				

The operational semantics of the language is given by a set of rules that define a transition system between configurations, as depicted in Figure 2(a). A configuration is a tuple $\langle ag, C, M, T, s \rangle$, where:

- ag is an agent program defined by a set of beliefs bs and plans ps.
- An agent circumstance C is a tuple $\langle I, E, A \rangle$, where: I is a set of intentions; E is a set of events; and A is a set of actions to be performed in the environment.
- M is a set of input/output mailboxes for communication.
- T is a tuple $\langle R, Ap, \iota, \varepsilon, \rho \rangle$ that keeps track of temporary information. R and Ap are the sets of relevant and applicable plans, respectively. ι, ε, ρ record the current intention, event and selected plan, respectively.
- s labels the current step in the reasoning cycle of the agent.

Transitions are defined in terms of semantic rules with form:

$$\frac{cond}{C \to C'} (\textbf{rule id})$$

where $C = \langle ag, C, M, T, s \rangle$ is a configuration that can become a new configuration C' if a *cond*ition is satisfied. Appendix A shows the operational semantic rules extracted from from [3,2] that are relevant for the purposes of this paper.

3   Top-down Induction of Logical Decision Trees

Top-down Induction of Logical DEcision Trees (TILDE) [1] is an Inductive Logic Programming technique adopted for learning in the context of rational agents [9]. The first-order representation of TILDE is adequate to form training examples as sets of beliefs, e.g., the beliefs of the agent supporting the adoption of a plan as an intention; and the obtained hypothesis is useful for updating the plans and beliefs of the agents.

A Logical Decision Tree is a binary first-order decision tree where: (*a*) Each node is a conjunction of first-order literals; and (*b*) The nodes can share variables, but a variable introduced in a node can only occur in the left branch below that node (where it is true). Unshared variables may occur in both branches.

Three inputs are required to compute a Logical Decision Tree: A set of training examples, the background knowledge of the agent and the language bias. Training examples are atomic formulae composed of an atom referring to the plan that was intended; the set of beliefs the agent had when the intention was adopted or when the intention failed; and the label indicating a successful or failed execution of the intention. Examples are collected every time the agent believes an intention has been achieved (success) or dropped (failure). The rules believed by the agent, constitute the background knowledge of the agent, i.e., general knowledge about the domain of experience of the agent. The language bias is formed by *rmode* directives that indicate which literals should be considered as candidates to form part of a Logical Decision Tree.

The TILDE algorithm is basically a first-order version of the well known C4.5 algorithm. The algorithm is not described in this paper, due to space limitations, but it is advisable to consult the original report of TILDE [1] or the version of the algorithm reported in [8] for further details.

4   Extending the Language: A TILDE-Learning Approach

The extension to the grammar that is required to incorporate the induction of Logical Decision Trees [8] into *Jason* is shown in Table 2. As any agent, a learner agent ag_{lrnr} is formed by a set of beliefs and a set of plans. Beliefs can be either normal beliefs *nbs* (as defined by *bs* in Table 1) or learning beliefs *lbs*. Learning beliefs are related to the learning process input and configuration. These beliefs can be of three types: *rmode* directives, *settings* and training *examples*. *rmode* literals are directives used to represent the language bias (as introduced in Section 3). *settings* literals customize the learning process configurations, e.g., the metrics used for building a new hypothesis. In turn, *examples* are literals used for defining training examples. A training example defines the relation between the label of the plan chosen to satisfy an intention, the perception of the environment, and the result of the execution of the plan (*successful* or *failed*) captured as the class of the examples. A plan can be either normal (non learnable) or *learnable*, i.e., a plan in which new contexts can be learned. To become a learnable plan (*lp*), a plan just need to annotate its label as such.

Table 2. Extension to the *Jason* grammar language enabling *Intentional Learning*

ag_{lrnr}	::=	*bs ps*		*class*	::=	$\texttt{succ} \mid \texttt{fail}$
bs	::=	*nbs lbs*		*ps*	::=	*nps lps*
nbs	::=	$b_1 \ldots b_n$	$(n \geq 0)$	*nps*	::=	$p_1 \ldots p_n$ $(n \geq 1)$
lbs	::=	$lb_1 \ldots lb_n$	$(n \geq 0)$	*lps*	::=	$lp_1 \ldots lp_n$ $(n \geq 1)$
lb	::=	$\texttt{rmode}(b)$		*lp*	::=	$@lbl[\texttt{learnable}]$
	|	$\texttt{settings}(t_1, \ldots, t_n)$	$(n \geq 2)$			$te : ct \leftarrow h$
	|	$\texttt{example}(lbl, bs, class)$				

Semantics is extended by adding a Learning component (L) into configurations $\langle ag, C, M, T, L, s \rangle$. L is a tuple $\langle \rho, Exs, Cnds, Bst, Build, Tree \rangle$ where:

- The plan that triggered the learning process is denoted by ρ,
- *Exs* is the set of training examples related to the executed intention,
- *Cnds* is the set of literals (or conjunctions of literals) that are candidates to form part of the induced logical decision tree,
- *Bst* is a pair $\langle at, gain \rangle$ (where $gain \in \mathbb{R}$) keeping information about the candidate that maximizes the gain ratio measure,
- *Build* is a stack of *building tree parameters (btp)* tuples. Every time a set of training examples is split, a new *btp* is added into *Build* for computing a new inner tree afterwards. Each *btp* is a tuple $\langle Q, Exs, Branch \rangle$, where : Q is the conjunction of literals in top nodes; *Exs* is the partition of examples that satisfies Q, which will be used for building the new tree; and $Branch \in \{left, right\}$ indicates where the tree being computed must be placed.

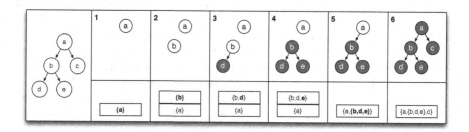

Fig. 1. Every time an inner tree is being built, a new list is added into the stack (1,2). When a leaf node is reached, this is added into the list on the top of the stack (3,4); in case of a right branch, the whole list is removed and added into the next list (5). If there is not a list under the top of the stack the main tree has been computed (6).

- *Tree* is a stack of lists. A tree can be represented as a list, where the first element denotes the node label and the remaining elements represent left and right branches, respectively. Figure 1 shows how this works.

For the sake of readability, we adopt the following notational conventions in semantic rules:

- We write L_ρ to make reference to the component ρ of L. Similarly for all the other components of a configuration.
- A stack is denoted by $[\alpha_1 \ddagger \ldots \ddagger \alpha_z]$, where α_1 is the bottom of the stack and α_z is the top of the stack. $\ddagger$ delimits the elements of the stack. $L_{Build}[\alpha]$ denotes the α-element on the top of stack L_{Build}. Similarly for L_{Tree}.
- If p is a plan on the form $@lbl\ te : ct \leftarrow h$, then $Label(p) = lbl$, $TrEv(p) = te$, $Ctxt(p) = ct$ and $Body(p) = h$.
- $Head(lst)$ and $Tail(lst)$ denote the head and the tail of a list, respectively.

The reasoning cycle is then extended for enabling Intentional Learning as can be seen in Figure 2(b). Two rules enable agents to collect training examples labelled as *succ* when the execution of a plan is successful ($ColEx_{succ}$) or *fail* otherwise ($ColEx_{fail}$).

Rule **ColEx$_{succ}$** adds a training example labelled as *succ* when the selected event T_ε is an achievement goal addition event, the selected plan T_ρ is a learnable plan, and the execution of an intention is done, (i.e., when the reasoning cycle is in the step $ClrInt$ and there is nothing else to be executed). This rule removes the whole intention T_ι like rule $ClrInt_1$ in the default operational semantics (Appendix A) but for learnable plans.

$$\frac{T_\varepsilon = \langle +!at, i\rangle \quad T_\rho \in ag_{lps} \quad T_\iota = [head \leftarrow \top]}{\langle ag, C, M, T, L, ClrInt\rangle \rightarrow \langle ag', C', M, T, L, ProcMsg\rangle} \text{ (\textbf{ColEx}$_{\textbf{succ}}$)}$$

s.t. $ag'_{lbs} = ag_{lbs} + example(Label(T_\rho), intend(at) \cup ag_{bs}, succ)$
   $C'_I = C_I \backslash \{T_\iota\}$

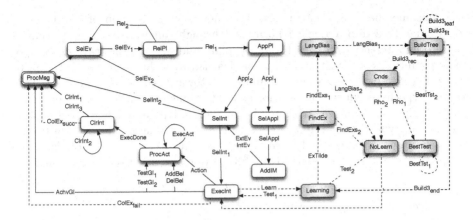

Fig. 2. Extended reasoning cycle. a) Unshaded states and solid lines define the basic reasoning cycle. b) Shaded states and dashed lines represent the extension of the reasoning cycle.

In a similar way, rule **ColEx$_{fail}$** adds a training example labelled as *fail* when the reasoning cycle is on the step *ExecInt*, the selected event T_ε is an achievement goal deletion event and the selected plan T_ρ is a learnable plan. Besides adding a new training example, this rule adds an achievement goal learning event. The current intention T_ι is suspended and associated to the new event. Since a new event is added, the reasoning cycle is moved towards *ProcMsg*, as it does the rule *AchvGl* in the default operational semantics (Appendix A).

$$\frac{T_\varepsilon = \langle -!at, i \rangle \quad T_\rho \in ag_{lps} \quad T_\iota = i[head \leftarrow h]}{\langle ag, C, M, T, L, ExecInt \rangle \to \langle ag', C', M, T, L', ProcMsg \rangle} \text{ (ColEx}_{fail})$$

s.t. $ag'_{lbs} = ag_{lbs} + example(Label(T_\rho), intend(at) \cup ag_{bs}, fail)$
$C'_E = C_E \cup \{\langle +!learning, T_\iota \rangle\}$
$L'_\rho = T_\rho$
$C'_I = C_I \backslash \{T_\iota\}$

The learning process starts in *ExecInt* by rule **Learn** when the selected event is $\langle +!Learning, i \rangle$. Rule **ExTilde** fires when L_{Tree} is an empty stack, and starts the construction of the Logical Decision Tree.

$$\frac{T_\varepsilon = \langle +!learning, i \rangle}{\langle ag, C, M, T, L, ExecInt \rangle \to \langle ag, C, M, T, L, Learning \rangle} \text{ (Learn)}$$

$$\frac{L_{Tree} = \top}{\langle ag, C, M, T, L, Learning \rangle \to \langle ag, C, M, T, L, FindEx \rangle} \text{ (ExTilde)}$$

Once the tree has been built, rules **Test$_1$** and **Test$_2$** check whether something new has been learned. Rule **Test$_1$** is fired when the new hypothesis does not subsume the prior context (i.e., it is not a generalization of it). Then, it parses the Tree into a Logical Formula (through function `parseLF`), that is used to update the context of the failed plan, and restarts the learning component for future calls. Instead, when the learned hypothesis subsumes the prior context, rule **Test$_2$** moves the learning cycle towards the *NoLearn* state. Note that a hypothesis subsuming a prior context means that nothing new has been learned, since learning was triggered because of a plan failure. The agent cycle is automatically lead from the *NoLearn* to the *ExecInt* state. Recovering from a situation in which individual learning has been unsuccessful is out of the scope of this paper and remains part of the future work, as discussed in Section 5.

$$\frac{L_\rho = @lbl\ te : ct \leftarrow h \quad \mathtt{parseLF}(L_{Tree}) = lct \quad lct \not\preceq ct}{\langle ag, C, M, T, L, Learning \rangle \rightarrow \langle ag', C', M, T, L', ExecInt \rangle}(\mathbf{Test_1})$$

s.t. $ag'_{ps} = \{ag_{ps} \backslash L_\rho\} \cup \{@lbl[learnable]\ te : lct :\leftarrow h\}$
$\quad L' = \langle \top, \{\}, \{\}, \top, [], [] \rangle$
$\quad C'_E = C_E \backslash T_\varepsilon$

$$\frac{L_\rho = te : ct \leftarrow h \quad \mathtt{parseLF}(\mathtt{L_{Tree}}) = lct \quad lct \preceq ct}{\langle ag, C, M, T, L, Learning \rangle \rightarrow \langle ag', C, M, T, L, NoLearn \rangle}(\mathbf{Test_2})$$

4.1   Semantics for Building Logical Decision Trees

This section presents the transition system for building a Logical Decision Tree. The first thing a learner agent needs for executing the TILDE algorithm is to get the set of training examples regarding the failed plan. Each example related to the failed plan is represented as $example(lbl, bs, class)$, where the first argument is the label of the failed plan; the rest of the arguments are the state of the world when the example was added and the label class of the example. Rule **FindExs$_1$** applies when at least one example regarding L_ρ is a logical consequence of the learning beliefs of the agent. Here, the set L_{Exs} is updated and the learning cycle goes forward the *LangBias* step. If there is no training example, rule **FindExs$_2$** moves the cycle forward the *NoLearn* step.

$$\frac{Exs = \{lbs | lbs = example(Label(L_\rho), bs, class) \wedge ag_{lbs} \models lbs\}}{\langle ag, C, M, T, L, FindEx \rangle \rightarrow \langle ag, C, M, T, L', LangBias \rangle}\ (\mathbf{FindExs_1})$$

s.t. $L'_{Exs} = Exs$

$$\frac{ag_{lbs} \not\models example(Label(L_\rho), bs, class)}{\langle ag, C, M, T, L, FindEx \rangle \rightarrow \langle ag, C, M, T, L, NoLearn \rangle}\ (\mathbf{FindExs_2})$$

The language bias is generated by rule **LangBias$_1$**, through the function `getLangBias()`, whose only parameter is the set of training examples in L_{Exs}. If the set of *rmode* directives is not empty, the cycle goes forward the step *BuildTree*. In this transition, the whole directives in LB are added as beliefs in the learning beliefs of the agent. Besides, a *building tree parameters* tuple is added in L_{Build}: the initial query is a literal indicating the intention the agent was trying to reach; the initial examples are those in L_{Exs}; and the symbol $\top$ denotes that this is the configuration for building the main node. The rule **LangBias₂** moves the cycle forward the *NoLearn* step when is not possible to generate the language bias (training examples had no information about the beliefs of the agent when these were added).

$$\frac{\texttt{getLangBias}(L_{Exs}) = LB}{\langle ag, C, M, T, L, LangBias \rangle \rightarrow \langle ag', C, M, T, L', BuildTree \rangle} \quad (\textbf{LangBias}_1)$$

s.t. $\forall (rmode(X) \in LB) . ag'_{lbs} = ag_{lbs} + rmode(X)$
$L_\rho = @lbl + !at : ct \leftarrow h$
$L'_{Build} = [\langle intend(at), L_{Exs}, \top \rangle]$

$$\frac{\texttt{getLangBias}(L_{Exs}) = \{\}}{\langle ag, C, M, T, L, LangBias \rangle \rightarrow \langle ag, C, M, T, L, NoLearn \rangle} \quad (\textbf{LangBias}_2)$$

At this point, the necessary data for executing the TILDE algorithm [1] has been processed. Next rules define the transitions for building a Logical Decision Tree. Rule **Build3_leaf** is applied when a stop condition is reached (e.g., the whole examples belong to the same class). A boolean function like `stopCriteria()` returns *true* if the examples in its argument satisfy a stop criteria, and *false* otherwise. The *leaf* node is obtained through the function `majority_class` and it is added into the list on the top of the L_{Tree} stack. If the leaf node is in a *Right* branch, the whole list on the top is removed from the top and added into the list under the top of the stack (see Figure 1). The stack L_{Build} is updated removing the tuple on the top of it. If no stop condition is found, the learning cycle moves towards the next step (**Build3_rec**).

$$\frac{L_{Build}[\langle Q, Exs, Branch \rangle] \quad \texttt{stopCriteria}(Exs) = true}{\langle ag, C, M, T, L, BuildTree \rangle \rightarrow \langle ag, C, M, T, L', BuildTree \rangle}(\textbf{Build3}_{leaf})$$

s.t. $leaf = \texttt{majority\_class}(Exs),$
$L_{Tree} = [T_z \ddagger \ldots \ddagger T_2 \ddagger T_1],$
$Tree = T_1 \cup \{leaf\},$
$$L'_{Tree} = \begin{cases} [T_z \ddagger \ldots \ddagger T_2 \ddagger Tree] & \text{if } Branch = Left, \text{ or} \\ & (Branch = Right \text{ and } T_2 = \top) \\ [T_z \ddagger \ldots \ddagger \{T_2 \cup Tree\}] & \text{if } Branch = Right \text{ and } T_2 \neq \top \end{cases}$$
$L'_{Build} = L_{Build} \backslash \langle Q, Exs, Branch \rangle$

$$\frac{L_{Build}[\langle Q, Exs, Branch \rangle] \quad \texttt{stopCriteria}(Exs) = false}{\langle ag, C, M, T, L, BuildTree \rangle \rightarrow \langle ag, C, M, T, L, Cnds \rangle}(\textbf{Build3}_{rec})$$

Sometimes, the L_{Tree} stack has more than one element but there are no more elements for building inner nodes (e.g. the right side of a tree is deeper than the left one). In this cases, rule **Build3fit** flats the L_{Tree} stack adding the list on the top of the stack inside the one below until there is only one list in the stack.

$$\frac{L_{Tree}[T_2 \ddagger T_1] \quad T_2 \neq \top \quad L_{Build}[\top]}{\langle ag, C, M, T, L, BuildTree \rangle \to \langle ag, C, M, T, L', BuildTree \rangle}(\textbf{Build3}_{\textbf{fit}})$$

s.t. $L'_{Tree} = [T_z \ddagger \ldots \ddagger \{T_2 \cup T_1\}]$

Rule **Rho₁** generates the candidates to form part of the tree using the function rho() whose parameters are a query Q and the language bias. This rule updates the element L_{Cnds} when a non-empty set of candidates has been generated; otherwise, rule **Rho₂** moves the cycle forwards the *NoLearn* step.

$$\frac{LB = \{lb|lb = rmode(RM) \wedge ag_{lbs} \models lb\}}{\langle ag, C, M, T, L, Cnds \rangle \to \langle ag, C, M, T, L', BestTest \rangle}(\textbf{Rho}_\textbf{1})$$

s.t. $L'_{Cnds} = Candidates$
$\forall (cnd \in Candidates) \cdot cnd = at_1 \wedge \ldots \wedge at_n \quad (n \geq 1)$

$$\frac{LB = \{lb|lb = rmode(RM) \wedge ag_{lbs} \models lb\}}{\langle ag, C, M, T, L, Cnds \rangle \to \langle ag, C, M, T, L, NoLearn \rangle}(\textbf{Rho}_\textbf{2})$$

Rule **BestTst₁** evaluates iteratively each candidate in L_{Cnds} for selecting the candidate that maximizes gain ratio.

$$\frac{\texttt{gainRatio}(Head(L_{Cnds})) = G}{\langle ag, C, M, T, L, BestTest \rangle \to \langle ag, C, M, T, L', BestTest \rangle}(\textbf{BestTst}_\textbf{1})$$

s.t. $L'_{Bst} = \begin{cases} G & \text{if } G > L_{Bst} \\ L_{Bst} & \text{otherwise} \end{cases}$
$L'_{Cnds} = Tail(L_{Cnds})$

When all candidates have been evaluated, rule **BestTst₂** splits the training examples in those satisfying $Q \wedge L_{Bst}$ and those that do not. Two new *btp* tuples are added into the L_{Build} stack for building inner trees afterwards, and a new list is added into the L_{Tree}.

$$\frac{L_{Cnds} = \{\} \quad L_{Build}[\langle Q, Exs, Branch \rangle}{\langle ag, C, M, T, L, BestTest \rangle \to \langle ag, C, M, T, L', BuildTree \rangle}(\textbf{BestTst}_\textbf{2})$$

s.t. $bg = \{bs \in ag_{bs} | bs = at\text{:-}body \wedge body \neq \top\}$
$\quad Exs_L = \{Ex \in Exs | Ex = example(lbl, bs, class) \wedge (bs \cup bg) \models (Q \wedge L_{Bst})\}$
$\quad Exs_R = \{Ex \in Exs | Ex = example(lbl, bs, class) \wedge (bs \cup bg) \not\models (Q \wedge L_{Bst})\}$
$\quad L_{Build} = [btp_z \ddagger \ldots \ddagger \langle Q, Exs, Branch \rangle]$
$\quad L'_{Build} = [btp_z \ddagger \ldots \ddagger \langle Q, Exs_R, Right \rangle \ddagger \langle Q \wedge L_{Bst}, Exs_L, Left \rangle]$

Finally rule **Build3$_{end}$** indicates the end of the building process when there is no building tree parameters tuple in the stack L_{Build}. The flow of the cycle goes forward the step *Learning* for processing the learned hypothesis.

$$\frac{L_{Build}[\top]}{\langle ag, C, M, T, L, BuildTree \rangle \rightarrow \langle ag, C, M, T, L, Learning \rangle}(\textbf{Build3}_{\textbf{end}})$$

As mentioned before, once the Logical Decision Tree has been built, the learned hypothesis is used for updating the plans of the agent when more specific hypothesis is learned (see rule **Test$_1$**). If the learned hypothesis is either more general or similar to prior knowledge means that there was no learning, and therefore the reasoning cycle continues with its default operation.

5 Discussion and Future Work

The operational semantics presented in this paper defines Intentional Learning on *Jason*, which has served to create agents capable of learning new reasons for adopting intentions, when the executions of their plans failed. Learning is achieved through Top-Down Induction of Logical Decision Trees (TILDE), that has proved to be a suitable mechanism for supporting learning on Jason since the first-order representation of these trees is adequate to form training examples as sets of beliefs, while the obtained hypothesis is useful for updating the plans of the agents. Current work provides a formal and precise approach to incorporate Intentional Learning into BDI multi-agent systems, and a better understanding of the reasoning cycle of agents performing this type of learning. For reasons of space, a demonstration scenario showing the benefits of this approach is not presented here but can be found in [8], where we evaluate how agents improve their performance by executing Intentional Learning whenever the execution of a plan is failed.

The semantics presented in this paper paves the way for future research on Social Learning, as an alternative for recovering from individual learning failures. Social Learning has been defined as the phenomenon by means of which a given agent can update its own knowledge base by perceiving the positive or negative effects of any given event undergone or actively produced by another agent on a state of the world within which the learning agent has as a goal [6]. It would be interesting to identify the mechanisms that must be implemented at the agent level to enable them to learn from one another. A first answer is that the intentional level of the semantics presented in this paper is required to define distributed learning protocols as a case of collaborative goal adoption [5], where a group of agents sharing a plan has as social goal learning a new context for the plan in order to avoid possible future failures. As an example of learning protocol, agents in a group could share experiences (training examples) with the learner agent (the one that discovered the plan execution failure) to achieve this social goal.

Acknowledgements. This work has been jointly supported by the Spanish MICINN and the European Commission FEDER funds, under grant TIN2009-14475-C04. First author is supported by Conacyt doctoral scholarship number 214787. Third author is supported by Conacyt project number 78910.

References

1. Blockeel, H., Raedt, L.D., Jacobs, N., Demoen, B.: Scaling up inductive logic programming by learning from interpretations. Data Mining and Knowledge Discovery 3(1), 59–93 (1999)
2. Bordini, R.H., Hübner, J.F.: Semantics for the jason variant of agentspeak (plan failure and some internal actions). In: ECAI, pp. 635–640 (2010)
3. Bordini, R.H., Hübner, J.F., Wooldridge, M.: Programming Multi-Agent Systems in Agent-Speak using Jason. John Wiley & Sons Ltd. (2007)
4. Bratman, M.E.: Intention, Plans, and Practical Reason. Harvard University Press, Cambridge (1987)
5. Castelfranchi, C.: Modelling social action for ai agents. Artif. Intell. 103(1-2), 157–182 (1998)
6. Conte, R., Paolucci, M.: Intelligent social learning. Journal of Artificial Societies and Social Simulation 4(1) (2001)
7. Guerra-Hernández, A., El Fallah-Seghrouchni, A., Soldano, H.: Learning in BDI multi-agent systems. In: Dix, J., Leite, J. (eds.) CLIMA 2004. LNCS (LNAI), vol. 3259, pp. 218–233. Springer, Heidelberg (2004)
8. Guerra-Hernández, A., González-Alarcón, C.A., El Fallah Seghrouchni, A.: Jason Induction of Logical Decision Trees: A Learning Library and Its Application to Commitment. In: Sidorov, G., Hernández Aguirre, A., Reyes García, C.A. (eds.) MICAI 2010, Part I. LNCS, vol. 6437, pp. 374–385. Springer, Heidelberg (2010)
9. Guerra-Hernández, A., Ortíz-Hernández, G.: Toward BDI sapient agents: Learning intentionally. In: Mayorga, R.V., Perlovsky, L.I. (eds.) Toward Artificial Sapience, pp. 77–91. Springer, London (2008)
10. Rao, A.: AgentSpeak(L): BDI agents speak out in a logical computable language. In: Perram, J., Van de Velde, W. (eds.) MAAMAW 1996. LNCS, vol. 1038, pp. 42–55. Springer, Heidelberg (1996)
11. Rao, A., Georgeff, M.: Modelling rational agents within a BDI-Architecture. Technical Report 14, Carlton, Victoria (February 1991)
12. Rao, A.S., Georgeff, M.P.: Decision procedures for BDI logics. Journal of Logic and Computation 8(3), 293–342 (1998)
13. Singh, D., Sardina, S., Padgham, L.: Extending BDI plan selection to incorporate learning from experience. Robotics and Autonomous Systems 58(9), 1067–1075 (2010), Hybrid Control for Autonomous Systems
14. Singh, D., Sardiña, S., Padgham, L., James, G.: Integrating learning into a BDI agent for environments with changing dynamics. In: IJCAI, pp. 2525–2530 (2011)

A   Jason Semantic Rules

The following $AgentSpeak(L)$ operational semantic rules are relevant for the purposes of this paper, in particular for defining the relation with the rules defined for collecting training examples and the way that an achievement goal deletion event is triggered. For a detailed reviewing of this rules, is highly recommended to consult the text in [3,2].

$$\frac{T_\iota = i[head \leftarrow !at; h]}{\langle ag, C, M, T, ExecInt \rangle \rightarrow \langle ag, C', M, T, ProcMsg \rangle}\textbf{(AchvGl)}$$

s.t. $C'_E = C_E \cup \{\langle +!at, T_\iota \rangle\}, C'_I = C_I \setminus \{T_\iota\}$

$$\frac{T_\iota = [head \leftarrow \top]}{\langle ag, C, M, T, ClrInt \rangle \rightarrow \langle ag, C', M, T, ProcMsg \rangle}\textbf{(ClrInt}_1\textbf{)}$$

s.t. $C'_I = C_I \setminus \{T_\iota\}$

$$\frac{\langle a, i \rangle \in C_A \quad execute(a) = e}{\langle ag, C, M, T, ProcAct \rangle \rightarrow \langle ag, C', M, T, ProcAct \rangle}\textbf{(ExecAct)}$$

s.t.

$C'_A = C_A \setminus \{\langle a, i \rangle\}$
$C'_I = C_I \cup \{i'[te : ct \leftarrow h]\}$, if e
$C'_E = C_E \cup \{\langle -\%at, i \rangle\}$, if $\neg e \wedge (te = +\%at)$
with $i = i'[te : ct \leftarrow a; h]$ and $\% \in \{!, ?\}$

$$\frac{C_A = \{\} \vee (\neg \exists \langle a, i \rangle \in C_A \, . \, execute(a) = e)}{\langle ag, C, M, T, ProcAct \rangle \rightarrow \langle ag, C, M, T, ClrInt \rangle}\textbf{(ExecDone)}$$

SAT-Based Bounded Model Checking for Weighted Deontic Interpreted Systems*

Bożena Woźna-Szcześniak

IMCS, Jan Długosz University
Al. Armii Krajowej 13/15, 42-200 Częstochowa, Poland
b.wozna@ajd.czest.pl

Abstract. In this paper we present a SAT-based Bounded Model Checking (BMC) method for weighted deontic interpreted systems (i.e., Kripke structures where transitions carry a weight, which is an arbitrary natural number) and properties expressed in the existential fragment of a weighted temporal logic augmented to include knowledge and deontic components (WECTLKD). In particular, since in BMC both the system model and the checked property are translated into a Boolean formula to be analysed by a SAT-solver, we introduce a new Boolean encoding of the WECTLKD formulae that is particularly optimized for managing quantitative weighted temporal operators, knowledge operators, and deontic operators, which are typically found in properties of complex multi-agent systems in models of which we assume the possibility that agents may not behave as they are supposed to, and that acting (coordination, negotiation, cooperation, etc.) of agents may cost. We illustrate how the weighted deontic interpreted systems can be applied to the analysis of a variant of the standard bit transmission problem in which an agent may fail to do something it is supposed to do.

1 Introduction

Multi-agent systems (MASs)[13] are distributed systems in which *agents* are autonomous entities that engage in social activities such as coordination, negotiation, cooperation, etc. The formalism of *interpreted systems* (ISs) was introduced in [7] to model MASs and to reason about the agents epistemic and temporal properties. The formalism of *deontic interpreted systems* (DISs) [9] extends ISs to make possible reasoning about correct functioning behaviour of MASs. DISs provide a computationally grounded semantics on which it is possible to interpret deontic modalities as well as a traditional epistemic and temporal modalities. The mentioned deontic modalities differs from their standard meaning and application as one for indicating (modelling) obligation, prohibition and permission. Namely, they are used to represent the distinction between ideal (correct, acceptable) behaviour and actual (possibly incorrect, unacceptable) behaviour of the agents.

---

* Partly supported by National Science Center under the grant No. 2011/01/B/ST6/ 05317.

L. Correia, L.P. Reis, and J. Cascalho (Eds.): EPIA 2013, LNAI 8154, pp. 444–455, 2013.

By *weighted deontic interpreted systems* (WDISs) we mean an extension of DISs that allows to reason about agents quantitative properties as well as temporal, epistemic and deontic properties. WDISs generate weighted Kripke structures, i.e., transition systems where each transition holds a weight, which can be any integer value (including zero). Thus the difference between the Kripke structure generated by DIS and the weighted Kripke structure is that it is assumed that each transition is labelled not only by a joint action, but also by a natural number.

There are three main reasons why it is interesting to consider WDISs instead of DISs. First, if we assume that the weight represents elapsed time, then WDISs allow for transitions that take a long time, e.g. 100 time units. Such transitions could be simulated in DISs by inserting 99 intermediate states. But this increases the size of the model, and so it makes the model checking process more difficult. Second, WDISs allow transitions to have zero weight. This is very convenient in modelling the environment in which agents operate. Third, the transitions with the zero duration allow for counting specific events only and thus omitting the irrelevant ones from the model checking point of view.

The original motivation of bounded model checking (BMC) [3,12] methods was to take advantage of the success of SAT-solvers (i.e., tools implementing algorithms solving the satisfiability problem for propositional formulas), which during the last decade have immensely increased their reasoning power. The main idea of SAT-based BMC methods consists in translating the model checking problem [4,13] for a modal (temporal, epistemic, deontic, etc.) logic to the satisfiability problem of a propositional formula. Traditionally, this formula is the conjunction of two propositional formulae, the first of which encodes a part of the model and the second encodes the modal property. The usefulness of SAT-based BMC for error tracking and complementarity to the BDD-based symbolic model checking have already been proven in several works, e.g., [2,11].

To express the requirements of MASs various extensions of temporal logics [5] with epistemic (to represent knowledge) [7], doxastic (to represent beliefs) [8], and deontic (to represent norms and prescriptions) [9] components have been proposed. In this paper we consider an existential fragment of a deontic and epistemic extension of weighted CTL [10], which we call WECTLKD and we interpret over weighted Kripke structures generated by WDISs. WECTLKD allows for the representation of the quantitative temporal evolution of epistemic states of the agents, as well as their correct and incorrect functioning behaviour.

The original contributions of the paper are as follows. First, we define the WDIS as a model of MAS. Second, we introduce the WECTLKD language. Third, we propose a SAT-based BMC technique for WDISs and for the existential fragment of WECTLKD.

The structure of the paper is as follows. In Section 2 we introduce WDISs and the WECTLKD logic. In Section 3 we define a SAT-based BMC for WECTLKD interpreted over WDISs. In Section 4 we apply the BMC technique to an example close to the multi-agent systems literature: the weighted bit transmission problem with faults. In Section 5 we conclude the paper.

2   Preliminaries

Weighted Deontic Interpreted Systems (WDIS). We assume that a MAS consists of n agents (by $Ag = \{1, \ldots, n\}$ we denote the non-empty set of indices of agents) and a special agent $\mathcal{E}$ which is used to model the environment in which the agents operate. In the WDIS formalism, each agent $\mathbf{c} \in Ag$ is modelled using a set $L_{\mathbf{c}}$ of *local states*, a set $Act_{\mathbf{c}}$ of *possible actions*, a *protocol function* $P_{\mathbf{c}} : L_{\mathbf{c}} \to 2^{Act_{\mathbf{c}}}$ defining rules according to which actions may be performed in each local state, a (partial) *evolution function* $t_{\mathbf{c}} : L_{\mathbf{c}} \times Act_1 \times \cdots \times Act_n \times Act_{\mathcal{E}} \to L_{\mathbf{c}}$, and a *weight function* $d_{\mathbf{c}} : Act_{\mathbf{c}} \to \mathbb{N}$. Further, we assume that for each agent $\mathbf{c} \in Ag$, its set $L_{\mathbf{c}}$ can be partitioned into *faultless (green)* and *faulty (red)* states. Namely, for each agent $\mathbf{c} \in Ag$, a non-empty set of green states $\mathcal{G}_{\mathbf{c}}$ such that $L_{\mathbf{c}} \supseteq \mathcal{G}_{\mathbf{c}}$ is defined. The set of red states of agent $\mathbf{c}$, denoted by $\mathcal{R}_{\mathbf{c}}$, is defined as the complement of $\mathcal{G}_{\mathbf{c}}$ with respect to $L_{\mathbf{c}}$. Note that $\mathcal{G}_{\mathbf{c}} \cup \mathcal{R}_{\mathbf{c}} = L_{\mathbf{c}}$, for any agent $\mathbf{c}$.

Similarly to the other agents, $\mathcal{E}$ is modelled using a set of local states $L_{\mathcal{E}}$, a set of actions $Act_{\mathcal{E}}$, a protocol $P_{\mathcal{E}}$, an evolution function $t_{\mathcal{E}}$, and a weight function $d_{\mathcal{E}}$; it is assumed that local states for $\mathcal{E}$ are *public*. Further, for all agents including the environment, the sets $L_{\mathbf{c}}$ and $Act_{\mathbf{c}}$ are assumed to be non-empty, and the number of agents is assumed to be finite. Also, we do not assume that the sets $Act_{\mathbf{c}}$ (for all $\mathbf{c} \in Ag \cup \{\mathcal{E}\}$) are disjoint. Moreover, we define the set $Act = Act_1 \times \cdots \times Act_n \times Act_{\mathcal{E}}$ of joint actions, and with each joint action $a \in Act$ we associate a *weight* which is curried out by the action. This can be done by a function $d : Act \to \mathbb{N}$ defined, for example, as $d((a_1, \ldots, a_n, a_{\mathcal{E}})) = max\{d_{\mathbf{c}}(a_{\mathbf{c}}) \mid \mathbf{c} \in Ag \cup \{\mathcal{E}\}\}$ (note that this definition is reasonable, if we are interested in timing aspects of MASs that are simple enough to be described with durations on transitions. In the case the weight of a transition symbolises the duration time of the transition.), or as $d((a_1, \ldots, a_n, a_{\mathcal{E}})) = \sum_{\mathbf{c} \in Ag} d_{\mathbf{c}}(a_{\mathbf{c}}) + d_{\mathcal{E}}(a_{\mathcal{E}})$ (note that this definition is reasonable, if we are interested in MASs in which transitions carry some "cost", a notion more general than elapsed time).

Now for a given set of agents Ag, the environment $\mathcal{E}$ and a set of propositional variables $\mathcal{PV}$, we define a *weighted deontic interpreted system* (WDIS) as a tuple

$$(\{\iota_{\mathbf{c}}, L_{\mathbf{c}}, \mathcal{G}_{\mathbf{c}}, Act_{\mathbf{c}}, P_{\mathbf{c}}, t_{\mathbf{c}}, \mathcal{V}_{\mathbf{c}}, d_{\mathbf{c}}, \}_{\mathbf{c} \in Ag \cup \{\mathcal{E}\}})$$

where $\iota_{\mathbf{c}} \subseteq L_{\mathbf{c}}$ is a set of initial states of agent $\mathbf{c}$, and $\mathcal{V}_{\mathbf{c}} : L_{\mathbf{c}} \to 2^{\mathcal{PV}}$ is a valuation function for agent $\mathbf{c}$ that assigns to each local state a set of propositional variables that are assumed to be true at that state. Further, for a given WDIS we define: (1) a set of all *possible global states* $S = L_1 \times \ldots \times L_n \times L_{\mathcal{E}}$ such that $L_1 \supseteq \mathcal{G}_1, \ldots, L_n \supseteq \mathcal{G}_n, L_{\mathcal{E}} \supseteq \mathcal{G}_{\mathcal{E}}$; by $l_{\mathbf{c}}(s)$ we denote the local component of agent $\mathbf{c} \in Ag \cup \{\mathcal{E}\}$ in a global state $s = (\ell_1, \ldots, \ell_n, \ell_{\mathcal{E}})$; and (2) a *global evolution function* $t : S \times Act \to S$ as follows: $t(s, a) = s'$ iff for all $\mathbf{c} \in Ag$, $t_{\mathbf{c}}(l_{\mathbf{c}}(s), a) = l_{\mathbf{c}}(s')$ and $t_{\mathcal{E}}(l_{\mathcal{E}}(s), a) = l_{\mathcal{E}}(s')$. In brief we write the above as $s \xrightarrow{a} s'$. Finally, for a given WDIS we define a *weighted Kripke structure* (or *a model*) as a tuple $M = (\iota, S, T, \mathcal{V}, d)$, where

- $\iota = \iota_1 \times \ldots \times \iota_n \times \iota_{\mathcal{E}}$ is the set of all possible initial global state;
- S is the set of all possible global states as defined above;

- $T \subseteq S \times Act \times S$ is a transition relation defined by the global evolution function as follows: $(s, a, s') \in T$ iff $s \xrightarrow{a} s'$. We assume that the relation T is total, i.e., for any $s \in S$ there exists $s' \in S$ and an action $a \in Act \setminus \{(\epsilon_1, \ldots, \epsilon_n, \epsilon_{\mathcal{E}})\}$ such that $s \xrightarrow{a} s'$;
- $\mathcal{V} : S \to 2^{\mathcal{PV}}$ is the valuation function defined as $\mathcal{V}(s) = \bigcup_{c \in Ag \cup \{\mathcal{E}\}} \mathcal{V}_c(l_c(s))$.
- $d : Act \to \mathbb{N}$ is a weighted function associated to the WDIS.

Given a WDIS one can define the following relations: (1) $\sim_c \subseteq S \times S$ as an indistinguishability relation for agent c defined by: $s \sim_c s'$ iff $l_c(s') = l_c(s)$; and (2) $\bowtie_c \subseteq S \times S$ as a deontic relation for agent c defined by: $s \bowtie_c s'$ iff $l_c(s') \in \mathcal{G}_c$; Further, a *path* in M is an infinite sequence $\pi = s_0 \xrightarrow{a_1} s_1 \xrightarrow{a_2} s_2 \xrightarrow{a_3} \ldots$ of transitions. For such a path, and for $j \leq m \in \mathbb{N}$, by $\pi(m)$ we denote the m-th state s_m, by $\pi[j..m]$ we denote the finite sequence $s_j \xrightarrow{a_{j+1}} s_{j+1} \xrightarrow{a_{j+2}} \ldots s_m$ with $m - j$ transitions and $m - j + 1$ states, and by $D\pi[j..m]$ we denote the (cumulative) weight of $\pi[j..m]$ that is defined as $d(a_{j+1}) + \ldots + d(a_m)$ (hence 0 when $j = m$). By $\Pi(s)$ we denote the set of all the paths starting at $s \in S$ and by $\Pi = \bigcup_{s^0 \in \iota} \Pi(s^0)$ we denote the set of all the paths starting at initial states.

The logic WECTLKD. WECTLKD is an existential fragment of WCTL – which extends CTL with cost constraints on modalities – augmented with epistemic and deontic modalities. In the syntax of WECTLKD we assume the following: $p \in \mathcal{PV}$ is an atomic proposition, $c, c_1, c_2 \in Ag$ are three indices of agents, $\Gamma \subseteq Ag$ is a subset of indices of agents, and I is an interval in $\mathbb{N} = \{0, 1, 2, \ldots\}$ of the form: $[a, b)$ and $[a, \infty)$, for $a, b \in \mathbb{N}$ and $a \neq b$; In semantics we assume the following definitions of epistemic relations: $\sim_\Gamma^E \overset{def}{=} \bigcup_{c \in \Gamma} \sim_c$, $\sim_\Gamma^C \overset{def}{=} (\sim_\Gamma^E)^+$ (the transitive closure of $\sim_\Gamma^E$), $\sim_\Gamma^D \overset{def}{=} \bigcap_{c \in \Gamma} \sim_c$, where $\Gamma \subseteq Ag$.

The WECTLKD formulae are defined by the following grammar:

$$\varphi ::= \textbf{true} \mid \textbf{false} \mid p \mid \neg p \mid \varphi \wedge \varphi \mid \varphi \vee \varphi \mid \text{EX}\varphi \mid \text{E}(\varphi \text{U}_I \varphi) \mid \text{E}(\varphi \text{R}_I \varphi)$$

$$\mid \overline{\text{K}}_c \varphi \mid \overline{\text{E}}_\Gamma \varphi \mid \overline{\text{D}}_\Gamma \varphi \mid \overline{\text{C}}_\Gamma \varphi \mid \overline{\mathcal{O}}_c \varphi \mid \widehat{\text{K}}_{c_1}^{c_2} \varphi.$$

Intuitively, we have the existential path quantifier E, the temporal operators for "neXt time" (X), "weighted until" (U_I), and "weighted release" (R_I). Further, we have epistemic operators for "agent c does not know whether or not" ($\overline{\text{K}}_c$), and for the dualities to the standard group epistemic modalities representing distributed knowledge in the group Γ, everyone in Γ knows, and common knowledge among agents in Γ ($\overline{\text{D}}_\Gamma$, $\overline{\text{E}}_\Gamma$, and $\overline{\text{C}}_\Gamma$). Finally, we have the deontic operators $\overline{\mathcal{O}}_c$ and $\widehat{\text{K}}_{c_1}^{c_2}$ to represent the *correctly functioning circumstances of agents*. The formula $\overline{\mathcal{O}}_c \varphi$ is read as "there is a state where agent c is functioning correctly and φ holds". The formula $\widehat{\text{K}}_{c_1}^{c_2} \varphi$ is read as "under the assumption that agent c_2 is functioning correctly, agent c_1 does not know whether or not φ holds". Note that the combination of weighted temporal, epistemic and deontic operators allows us to specify how agent's knowledge or correctly functioning circumstances of agents evolve over time and how much they cost. Note also that the formulae for the "weighted eventually", and "weighted always" are defined as standard:

- $\mathrm{EF}_I\varphi \stackrel{def}{=} \mathrm{E}(\mathbf{true}\mathrm{U}_I\varphi)$ (it means that it is possible to reach a state satisfying φ via a finite path whose cumulative weight is in I),
- $\mathrm{EG}_I\varphi \stackrel{def}{=} \mathrm{E}(\mathbf{false}\mathrm{R}_I\varphi)$ (it means that there is a path along which φ holds at all states with cumulative weight being in I),

In this logic, we can express properties like: (1) "it is not true that every request is followed by an answer and it costs less than 5" (with the formula $\mathrm{EF}(request \wedge \mathrm{EG}_{[0,5)}answer)$; (2) "it is not true that agent $\mathbf{c}$ knows that every request is followed by an answer and it costs less than 5" (with the formula $\overline{\mathrm{K}}_{\mathbf{c}}(\mathrm{EF}(request \wedge \mathrm{EG}_{[0,5)}answer)))$; (3) "it is not true that agent $\mathbf{c}_1$ knows under the assumption that agent $\mathbf{c}_2$ is functioning correctly that every request is followed by an answer and it costs less than 5" (with the formula $\widehat{\underline{\mathrm{K}}}_{\mathbf{c}_1}^{\mathbf{c}_2}(\mathrm{EF}(request \wedge \mathrm{EG}_{[0,5)}answer)))$.

A WECTLKD formula φ is *true* in the model M (in symbols $M \models \varphi$) iff $M, s^0 \models \varphi$ with $s^0 \in \iota$ (i.e., φ is true at an initial state of the model M). The semantics (i.e., the relation $\models$) is defined inductively as follows (for simplicity, we write $s \models \varphi$ instead of $M, s \models \varphi$):

$s \models \mathbf{true}$, $s \not\models \mathbf{false}$, $s \models p$ iff $p \in \mathcal{V}(s)$, $s \models \neg p$ iff $p \notin \mathcal{V}(s)$,

$s \models \alpha \wedge \beta$ iff $s \models \alpha$ and $s \models \beta$,

$s \models \alpha \vee \beta$ iff $s \models \alpha$ or $s \models \beta$,

$s \models \mathrm{EX}\alpha$ iff $(\exists \pi \in \Pi(s))(\pi(1) \models \alpha)$,

$s \models \mathrm{E}(\alpha\mathrm{U}_I\beta)$ iff $(\exists \pi \in \Pi(s))(\exists m \geq 0)(D\pi[0..m] \in I$ and $\pi(m) \models \beta$ and $(\forall j < m)\pi(j) \models \alpha)$,

$s \models \mathrm{E}(\alpha\mathrm{R}_I\beta)$ iff $(\exists \pi \in \Pi(s))((\exists m \geq 0)(D\pi[0..m] \in I$ and $\pi(m) \models \alpha$ and $(\forall j \leq m)\pi(j) \models \beta)$ or $(\forall m \geq 0)(D\pi[0..m] \in I$ implies $\pi(m) \models \beta))$,

$s \models \overline{\mathrm{K}}_{\mathbf{c}}\alpha$ iff $(\exists \pi \in \Pi)(\exists i \geq 0)(s \sim_{\mathbf{c}} \pi(i)$ implies $\pi(i) \models \alpha)$,

$s \models \overline{\mathrm{Y}}_\Gamma\alpha$ iff $(\exists \pi \in \Pi)(\exists i \geq 0)(s \sim_\Gamma^Y \pi(i)$ and $\pi(i) \models \alpha)$, with $Y \in \{\mathrm{D, E, C}\}$,

$s \models \overline{\mathcal{O}}_{\mathbf{c}}\alpha$ iff $(\exists \pi \in \Pi)(\exists i \geq 0)(s \bowtie_{\mathbf{c}} \pi(i)$ and $\pi(i) \models \alpha)$,

$s \models \widehat{\underline{\mathrm{K}}}_{\mathbf{c}_1}^{\mathbf{c}_2}\alpha$ iff $(\exists \pi \in \Pi)(\exists i \geq 0)(s \sim_{\mathbf{c}_1} \pi(i)$ and $s \bowtie_{\mathbf{c}_2} \pi(i)$ and $\pi(i) \models \alpha)$.

3 Bounded Model Checking for WECTLKD

Bounded semantics is the backbone of each SAT-based bounded model checking (BMC) method, whose basic idea is to consider only finite prefixes of paths that may be witnesses to an existential model checking problem. A crucial observation is that, though the prefix of a path is finite (usually we restrict the length of the prefix by some bound k), it still might represent an infinite path if it is a loop. If the prefix is not a loop, then it does not say anything about the infinite behaviour of the path beyond its last state.

Let M be a model, and $k \in \mathbb{N}$ a bound. A k-*path* π_k in M is a finite sequence $s_0 \xrightarrow{\sigma_1} s_1 \xrightarrow{\sigma_2} \ldots \xrightarrow{\sigma_k} s_k$ of transitions (i.e., $\pi_k = \pi[0..k]$). $\Pi_k(s)$ denotes the set of all the k-paths of M that start at s, and $\Pi_k = \bigcup_{s^0 \in \iota} \Pi_k(s^0)$. A k-path π_k is a (k,l)-*loop* (or *loop*) iff $\pi_k(l) = \pi_k(k)$ for some $0 \leq l < k$. Note that if a

k-path π_l is a loop, then it represents the infinite path of the form uv^ω, where $u = (s_0 \xrightarrow{a_1} s_1 \xrightarrow{a_2} \ldots \xrightarrow{a_l} s_l)$ and $v = (s_{l+1} \xrightarrow{a_{l+2}} \ldots \xrightarrow{a_k} s_k)$. Moreover, since in the bounded semantics we consider finite prefixes of paths only, the satisfiability of the weighted release operator depends on whether a considered k-path is a loop. Thus, as customary, we introduce a function $loop : \bigcup_{s \in S} \Pi_k(s) \to 2^{\mathbb{N}}$ that is defined as: $loop(\pi_k) = \{l \mid 0 \leq l < k$ and $\pi_k(l) = \pi_k(k)\}$, and identifies these k-paths that are loops.

Bounded Semantics for WECTLKD. Let M be a model, $k \geq 0$ a bound, φ a WECTLKD formula, and $M, s \models_k \varphi$ denote that φ is k−true at the state s of M. The formula φ is k−true in M (in symbols $M \models_k \varphi$) iff $M, s^0 \models_k \varphi$ with $s^0 \in \iota$ (i.e., φ is k-true at an initial state of the model M). For every $s \in S$, the relation $\models_k$ (the bounded semantics) is defined inductively as follows:

$M, s \models_k$ **true**, $M, s \not\models_k$ **false**,

$M, s \models_k p$ iff $p \in \mathcal{V}(s)$, $M, s \models_k \neg p$ iff $p \notin \mathcal{V}(s)$,

$M, s \models_k \alpha \vee \beta$ iff $M, s \models_k \alpha$ or $M, s \models_k \beta$,

$M, s \models_k \alpha \wedge \beta$ iff $M, s \models_k \alpha$ and $M, s \models_k \beta$,

$M, s \models_k \mathrm{EX}\alpha$ iff $k > 0$ and $(\exists \pi \in \Pi_k(s))M, \pi(1) \models_k \alpha$,

$M, s \models_k \mathrm{E}(\alpha \mathrm{U}_I \beta)$ iff $(\exists \pi \in \Pi_k(s))(\exists 0 \leq m \leq k)(D\pi[0..m] \in I$ and
$\qquad\qquad\qquad\qquad M, \pi(m) \models_k \beta$ and $(\forall 0 \leq j < m)M, \pi(j) \models_k \alpha)$,

$M, s \models_k \mathrm{E}(\alpha \mathrm{R}_I \beta)$ iff $(\exists \pi \in \Pi_k(s))((\exists 0 \leq m \leq k)(D\pi[0..m] \in I$ and
$\qquad\qquad\qquad\qquad M, \pi(m) \models_k \alpha$ and $(\forall 0 \leq j \leq m)M, \pi(j) \models_k \beta)$ or
$\qquad\qquad\qquad\qquad (D\pi[0..k] \geq right(I)$ and $(\forall 0 \leq j \leq k)(D\pi[0..j] \in I$
$\qquad\qquad\qquad\qquad$ implies $M, \pi(j) \models_k \beta))$ or $(D\pi[0..k] < right(I)$ and
$\qquad\qquad\qquad\qquad (\exists l \in loop(\pi))((\forall 0 \leq j < k)(D\pi[0..j] \in I$ implies
$\qquad\qquad\qquad\qquad M, \pi(j) \models_k \beta)$ and $(\forall l \leq j < k)(D\pi[0..k]+D\pi[l..j+1] \in I$
$\qquad\qquad\qquad\qquad$ implies $M, \pi(j+1) \models_k \beta))))$,

$M, s \models_k \overline{\mathrm{K}}_\mathbf{c}\alpha$ iff $(\exists \pi \in \Pi_k)(\exists 0 \leq j \leq k)(M, \pi(j) \models_k \alpha$ and $s \sim_\mathbf{c} \pi(j))$,

$M, s \models_k \overline{\mathrm{Y}}_\Gamma \alpha$ iff $(\exists \pi \in \Pi_k)(\exists 0 \leq j \leq k)(M, \pi(j) \models_k \alpha$ and $s \sim_\Gamma^Y \pi(j))$,
$\qquad\qquad\qquad\qquad$ where $Y \in \{\mathrm{D}, \mathrm{E}, \mathrm{C}\}$,

$M, s \models_k \overline{\mathrm{O}}_\mathbf{c}\alpha$ iff $(\exists \pi \in \Pi_k)(\exists 0 \leq j \leq k)(M, \pi(j) \models_k \alpha$ and $s \bowtie_\mathbf{c} \pi(j))$,

$M, s \models_k \widehat{\mathrm{K}}_\mathbf{c}^d \alpha$ iff $(\exists \pi \in \Pi_k)(\exists 0 \leq j \leq k)(M, \pi(j) \models_k \alpha$ and $s \sim_\mathbf{c} \pi(j)$
$\qquad\qquad\qquad\qquad$ and $s \bowtie_d \pi(j))$.

The *bounded model checking problem* asks whether there exists $k \in \mathbb{N}$ such that $M \models_k \varphi$. The following theorem states that for a given model and a WECTLKD formula there exists a bound k such that the model checking problem $(M \models \varphi)$ can be reduced to the bounded model checking problem $(M \models_k \varphi)$. The theorem can be proven by induction on the length of the formula φ.

Theorem 1. *Let M be a model and φ a WECTLKD formula. Then, the following equivalence holds: $M \models \varphi$ iff there exists $k \geq 0$ such that $M \models_k \varphi$.*

Translation to SAT. Let M be a model, φ a WECTLKD formula, and $k \geq 0$ a bound. In BMC, in general, we define the propositional formula

$$[M, \varphi]_k := [M^{\varphi, \iota}]_k \wedge [\varphi]_{M,k} \tag{1}$$

that is satisfiable if and only if the underlying model M is the genuine model for the property φ. Namely, Formula (1) is satisfiable if and only if $M \models_k \varphi$ holds.

The definition of the formula $[M^{\varphi,\iota}]_k$ assumes that states of the model M are encoded in a symbolic way. Such a symbolic encoding is possible, since the set of states of M is finite. In particular, each state s can be represented by a vector $w = (\mathbf{w}_1, \ldots, \mathbf{w}_r)$ (called a *symbolic state*) of propositional variables (called *state variables*) whose length r depends on the number of local states of agents and the possible maximal value of weights appearing in the given MASs.

Further, since the formula $[M^{\varphi,\iota}]_k$ defines the unfolding of the transition relation of the model M to the depth k, we need to represent k-paths in a symbolic way. This representation is usually called a *j-th symbolic k-path π_j*. Moreover, we have to know how many symbolic k-paths should be considered in the propositional encoding. The number of k-paths that is sufficient to translate formulae of WECTLKD is given by the function f_k : WECTLKD $\to \mathbb{N}$ which is defined as follows: $f_k(\mathbf{true}) = f_k(\mathbf{false}) = f_k(p) = f_k(\neg p) = 0$, where $p \in \mathcal{PV}$; $f_k(\alpha \wedge \beta) = f_k(\alpha) + f_k(\beta)$; $f_k(\alpha \vee \beta) = max\{f_k(\alpha), f_k(\beta)\}$; $f_k(\mathrm{E}(\alpha \mathrm{U}_I \beta)) = k \cdot f_k(\alpha) + f_k(\beta) + 1$; $f_k(\mathrm{E}(\alpha \mathrm{R}_I \beta)) = (k+1) \cdot f_k(\beta) + f_k(\alpha) + 1$; $f_k(\overline{\mathrm{C}}_\Gamma \alpha) = f_k(\alpha) + k$; $f_k(Y\alpha) = f_k(\alpha) + 1$ for $Y \in \{\mathrm{EX}, \overline{\mathrm{K}}_\mathbf{c}, \overline{\mathcal{O}}_\mathbf{c}, \widehat{\overline{\mathrm{K}}}^d_\mathbf{c}, \overline{\mathrm{D}}_\Gamma, \overline{\mathrm{E}}_\Gamma\}$.

Given the above, the j-th symbolic k-path π_j is defined as the following sequence $((d_{0,j}, w_{0,j}), \ldots, (d_{k,j}, w_{k,j}))$, where $w_{i,j}$ are symbolic states and $d_{i,j}$ are *symbolic weights*, for $0 \le i \le k$ and $0 \le j < f_k(\varphi)$. The *symbolic weight* $d_{i,j}$ is a vector $d_{i,j} = (\mathbf{d}_{1,j}, \ldots, \mathbf{d}_{x,j})$ of propositional variables (called *weight variables*), whose length x depends on weighs appearing in the given MASs and the weight function d.

Let w and w' (resp., d and d') be two different symbolic states (resp., weighs). We assume definitions of the following auxiliary propositional formulae: $I_s(w)$ - encodes the state s of the model M, $\mathcal{T}((d, w), (d'w'))$ - encodes the transition relation of M, $p(w)$ - encodes the set of states of M in which $p \in \mathcal{PV}$ holds, $H(w, w')$ - encodes equality of two global states, $H_\mathbf{c}(w, w')$ - encodes the equivalence of two local states of agent $\mathbf{c}$. $HO_\mathbf{c}(w, w')$ - encodes the accessibility of a global state in which agent $\mathbf{c}$ is functioning correctly, $\mathcal{B}^I_k(\pi_n)$ - encodes that the weight represented by the sequence $d_{1,n}, \ldots, d_{k,n}$ of symbolic weights is less than $right(I)$, $\mathcal{D}^I_j(\pi_n)$ - encodes that the weight represented by the sequence $d_{1,n}, \ldots, d_{j,n}$ of symbolic weights belongs to the interval I, $\mathcal{D}^I_{k;l,m}(\pi_n)$ for $l \le m$ - encodes that the weight represented by the sequences $d_{1,n}, \ldots, d_{k,n}$ and $d_{l+1,n}, \ldots, d_{m,n}$ of symbolic weights belongs to the interval I.

The formula $[M^{\varphi,\iota}]_k$ encoding the unfolding of the transition relation of the model M $f_k(\varphi)$-times to the depth k is defined as follows:

$$[M^{\varphi,\iota}]_k \quad := (\bigvee_{s\in\iota} I_s(w_{0,0})) \wedge \bigwedge_{j=0}^{f_k(\varphi)-1} \bigwedge_{i=0}^{k-1} \mathcal{T}((d_{i,j}, w_{i,j}), (d_{i+1,j}, w_{i+1,j})) \quad (2)$$

For every WECTLKD formula φ the function f_k determines how many symbolic k-paths are needed for translating the formula φ. Given a formula φ and a set A of k-paths such that $|A| = f_k(\varphi)$, we divide the set A into subsets needed for translating the subformulae of φ. To accomplish this goal we need some auxiliary functions that were defined in [15]. We recall the definitions of these functions, but for more details see the paper [15].

The relation $\prec$ is defined on the power set of $\mathbb{N}$ as follows: $A \prec B$ iff for all natural numbers x and y, if $x \in A$ and $y \in B$, then $x < y$ (e.g., $\{1,2,3\} \prec \{5,6\}$, $\{1,2,5\} \not\prec \{3,6\}$). Now, let $A \subset \mathbb{N}$ be a finite nonempty set, and $n, e \in \mathbb{N}$, where $e \leqslant |A|$. Then,

- $g_l(A, e)$ denotes the subset B of A such that $|B| = e$ and $B \prec A \setminus B$ (e.g., $g_l(\{4,5,6,7,8\}, 3) = \{4,5,6\}$).
- $g_r(A, e)$ denotes the subset C of A such that $|C| = e$ and $A \setminus C \prec C$ (e.g., $g_r(\{4,5,6,7,8\}, 3) = \{6,7,8\}$).
- $g_\mu(A)$ denotes the set $A \setminus \{min(A)\}$ (e.g., $g_\mu(\{4,5,6,7,8\}) = \{5,6,7,8\}$).
- If n divides $|A| - e - 1$, then $h_n^U(A, e)$ denotes the sequence $(B_0, \ldots, B_n)$ of subsets of $A \setminus \{min(A)\}$ such that $\bigcup_{j=0}^n B_j = A \setminus \{min(A)\}$, $|B_0| = \ldots = |B_{n-1}|$, $|B_n| = e$, and $B_i \prec B_j$ for every $0 \leqslant i < j \leqslant n$. If $h_n^U(A, e) = (B_0, \ldots, B_n)$, then $h_n^U(A, e)(j)$ denotes the set B_j, for every $0 \leqslant j \leqslant n$. For example, if $A = \{1,2,3,4,5,6,7\}$, then $h_3^U(A, 0) = (\{2,3\}, \{4,5\}, \{6,7\}, \emptyset)$, $h_3^U(A, 3) = (\{2\}, \{3\}, \{4\}, \{5,6,7\})$, $h_3^U(A, 6) = (\emptyset, \emptyset, \emptyset, \{2,3,4,5,6,7\})$, $h_3^U(A, e)$ is undefined for $e \in \{0, \ldots, 7\} \setminus \{0, 3, 6\}$.
- If $n+1$ divides $|A| - e - 1$, then $h_n^R(A, e)$ denotes the sequence $(B_0, \ldots, B_{n+1})$ of subsets of $A \setminus \{min(A)\}$ such that $\bigcup_{j=0}^{n+1} B_j = A \setminus \{min(A)\}$, $|B_0| = \ldots = |B_n|$, $|B_{n+1}| = e$, and $B_i \prec B_j$ for every $0 \leqslant i < j \leqslant n+1$. If $h_n^R(A, e) = (B_0, \ldots, B_{n+1})$, then $h_n^R(A, e)(j)$ denotes the set B_j, for every $0 \leqslant j \leqslant n+1$.

Let φ be an WECTLKD formula, M a model, and $k \in \mathbb{N}$ a bound. The propositional formula $[\varphi]_{M,k} := [\varphi]_k^{[0,0,F_k(\varphi)]}$, where $F_k(\varphi) = \{j \in \mathbb{N} \mid 0 \leqslant j < f_k(\varphi)\}$, encodes the bounded semantics for WECTLKD, and it is defined inductively as shown below. Namely, let $0 \leqslant n < f_k(\varphi)$, $m \leqslant k$, $n' = min(A)$, $h_k^U = h_k^U(A, f_k(\beta))$, and $h_k^R = h_k^R(A, f_k(\alpha))$, then:

$[\mathbf{true}]_k^{[m,n,A]} := \mathbf{true}, \qquad [\mathbf{false}]_k^{[m,n,A]} := \mathbf{false},$

$[p]_k^{[m,n,A]} := p(w_{m,n}), \qquad [\neg p]_k^{[m,n,A]} := \neg p(w_{m,n}),$

$[\alpha \wedge \beta]_k^{[m,n,A]} := [\alpha]_k^{[m,n,g_l(A,f_k(\alpha))]} \wedge [\beta]_k^{[m,n,g_r(A,f_k(\beta))]},$

$[\alpha \vee \beta]_k^{[m,n,A]} := [\alpha]_k^{[m,n,g_l(A,f_k(\alpha))]} \vee [\beta]_k^{[m,n,g_l(A,f_k(\beta))]},$

$[\mathrm{EX}\alpha]_k^{[m,n,A]} := H(w_{m,n}, w_{0,n'}) \wedge [\alpha]_k^{[1,n',g_\mu(A)]},$ if $k > 0$; $\mathbf{false}$, otherwise,

$[\mathrm{E}(\alpha \mathrm{U}_I \beta)]_k^{[m,n,A]} := H(w_{m,n}, w_{0,n'}) \wedge \bigvee_{i=0}^k ([\beta]_k^{[i,n',h_k^U(k)]} \wedge \mathrm{D}_i^I(\boldsymbol{\pi}_{n'}) \wedge$
$\bigwedge_{j=0}^{i-1} [\alpha]_k^{[j,n',h_k^U(j)]}),$

$[\mathrm{E}(\alpha \mathrm{R}_I \beta)]_k^{[m,n,A]} := H(w_{m,n}, w_{0,n'}) \wedge (\bigvee_{i=0}^k ([\alpha]_k^{[i,n',h_k^R(k+1)]} \wedge \mathrm{D}_i^I(\boldsymbol{\pi}_{n'}) \wedge \bigwedge_{j=0}^i$
$[\beta]_k^{[j,n',h_k^R(j)]}) \vee (\neg \mathcal{B}_k^I(\boldsymbol{\pi}_{n'}) \wedge \bigwedge_{j=0}^k (\mathrm{D}_j^I(\boldsymbol{\pi}_{n'}) \to [\beta]_k^{[j,n',h_k^R(j)]}))$
$\vee (\mathcal{B}_k^I(\boldsymbol{\pi}_{n'}) \wedge \bigwedge_{j=0}^k (\mathrm{D}_j^I(\boldsymbol{\pi}_{n'}) \to [\beta]_k^{[j,n',h_k^R(j)]}) \wedge \bigvee_{l=0}^{k-1}$
$[H(w_{k,n'}, w_{l,n'}) \wedge \bigwedge_{j=l}^{k-1} (\mathrm{D}_{k;l,j+1}^I(\boldsymbol{\pi}_{n'}) \to [\beta]_k^{[j,n',h_k^R(j)]})])),$

$[\overline{\mathrm{K}}_\mathbf{c}\alpha]_k^{[m,n,A]} := (\bigvee_{s \in \iota} I_s(w_{0,n'})) \wedge \bigvee_{j=0}^k ([\alpha]_k^{[j,n',g_\mu(A)]} \wedge H_\mathbf{c}(w_{m,n}, w_{j,n'})),$

$$[\overline{O}_{\mathbf{c}}\alpha]_k^{[m,n,A]} := (\bigvee_{s \in \iota} I_s(w_{0,n'})) \wedge \bigvee_{j=0}^k ([\alpha]_k^{[j,n',g_\mu(A)]} \wedge HO_{\mathbf{c}}(w_{m,n}, w_{j,n'})),$$

$$[\widehat{K}_{\mathbf{c}}^d\alpha]_k^{[m,n,A]} := (\bigvee_{s \in \iota} I_s(w_{0,n'})) \wedge \bigvee_{j=0}^k ([\alpha]_k^{[j,n',g_\mu(A)]} \wedge \widehat{H}_{\mathbf{c}}^d(w_{m,n}, w_{j,n'})),$$

$$[\overline{D}_\Gamma\alpha]_k^{[m,n,A]} := (\bigvee_{s \in \iota} I_s(w_{0,n'})) \wedge \bigvee_{j=0}^k ([\alpha]_k^{[j,n',g_\mu(A)]} \wedge \bigwedge_{\mathbf{c} \in \Gamma} H_{\mathbf{c}}(w_{m,n}, w_{j,n'})),$$

$$[\overline{E}_\Gamma\alpha]_k^{[m,n,A]} := (\bigvee_{s \in \iota} I_s(w_{0,n'})) \wedge \bigvee_{j=0}^k ([\alpha]_k^{[j,n',g_\mu(A)]} \wedge \bigvee_{\mathbf{c} \in \Gamma} H_{\mathbf{c}}(w_{m,n}, w_{j,n'})),$$

$$[\overline{C}_\Gamma\alpha]_k^{[m,n,A]} := [\bigvee_{j=1}^k (\overline{E}_\Gamma)^j \alpha]_k^{[m,n,A]}.$$

The theorem below states the correctness and the completeness of the presented translation. It can be proven by induction on the complexity of the given WECTLKD formula.

Theorem 2. *Let M be a model, and φ a WECTLKD formula. Then for every $k \in \mathbb{N}$, $M \models_k \varphi$ if, and only if, the propositional formula $[M, \varphi]_k$ is satisfiable.*

Now, from Theorems 1 and 2 we get the following.

Corollary 1. *Let M be a model, and φ a WECTLKD formula. Then, $M \models \varphi$ if, and only if, there exists $k \in \mathbb{N}$ such that the propositional formula $[M, \varphi]_k$ is satisfiable.*

4   The Weighted Bit Transmission Problem with Faults

We adapted the scenario of *a bit transmission problem with faults* [1], and we called it the *weighted bit transmission problem with faults* (WBTP). The WBTP involves two agents, a sender $\mathfrak{S}$, and a receiver $\mathfrak{R}$, communicating over a possibly faulty communication channel (the environment), and there are fixed costs $c_{\mathfrak{S}}$ and $c_{\mathfrak{R}}$ associated with, respectively, sending process of $\mathfrak{S}$ and $\mathfrak{R}$. $\mathfrak{S}$ wants to communicate some information (e.g., the value of a bit) to $\mathfrak{R}$. One protocol to achieve this is as follows. $\mathfrak{S}$ immediately starts sending the bit to $\mathfrak{R}$, and continues to do so until it receives an acknowledgement from $\mathfrak{R}$. $\mathfrak{R}$ does nothing until it receives the bit; from then on it sends acknowledgements of receipt to $\mathfrak{S}$. $\mathfrak{S}$ stops sending the bit to $\mathfrak{R}$ when it receives an acknowledgement. Note that $\mathfrak{R}$ will continue sending acknowledgements even after $\mathfrak{S}$ has received its acknowledgement.

In this section we are interested in applying our SAT-based bounded model checking method for weighted deontic interpreted systems to verify a version of the scenario above where the receiver $\mathfrak{R}$ does not operate as it is supposed to. The non-weighted version of the scenario was first described in [1]. Specifically, we consider the possibility that $\mathfrak{R}$ may send acknowledgements without having received the bit.

Each agent of the scenario can be modelled by considering its local states, local actions, local protocol, local evolution function, local weight function, and local valuation function. For $\mathfrak{S}$, it is enough to consider four possible local states representing the value of the bit that $\mathfrak{S}$ is attempting to transmit, and whether or not $\mathfrak{S}$ has received an acknowledgement from $\mathfrak{R}$. Since we are not admitting the possibility of faults, its local states are all green. Thus, w have:

$\mathcal{G}_{\mathfrak{S}} = \{0, 1, 0\text{-}ack, 1\text{-}ack\}$, $\mathcal{R}_{\mathfrak{S}} = \emptyset$, and $L_{\mathfrak{S}} = \mathcal{G}_{\mathfrak{S}}$. Further, $\iota_{\mathfrak{S}} = \{0, 1\}$. For $\mathfrak{R}$, it is enough to consider six possible local states representing: the value of the received bit, the circumstance in which no bit has been received yet (represented by ϵ), the circumstance in which $\mathfrak{R}$ has sent an acknowledgement without having received the value of the bit (denoted by $\epsilon\text{-}ack$), and the circumstance in which $\mathfrak{R}$ has sent an acknowledgement having received the value of the bit (represented by $0\text{-}ack$ and $1\text{-}ack$). Thus, $L_{\mathfrak{R}} = \{0, 1, \epsilon, 0\text{-}ack, 1\text{-}ack, \epsilon\text{-}ack\}$ represents the set of local state of $\mathfrak{R}$, and $\iota_{\mathfrak{R}} = \{\epsilon\}$. Since we assume that the ack component can be viewed as a proof that one faulty acknowledgement was sent before the value of the bit was received, we classify the local states of $\mathfrak{R}$ as follow: $\mathcal{G}_{\mathfrak{R}} = \{0, 1, \epsilon\}$, and $\mathcal{R}_{\mathfrak{R}} = \{\epsilon\text{-}ack, 0\text{-}ack, 1\text{-}ack\}$. For the environment $\mathfrak{E}$, to simplify the presentation, we shall to consider just one local state: $L_{\mathfrak{E}} = \{\cdot\} = \iota_{\mathfrak{E}}$. Moreover, we assume that $L_{\mathfrak{E}} = \mathcal{G}_{\mathfrak{E}}$, $\mathcal{R}_{\mathfrak{E}} = \emptyset$. Now we can define the set of possible global states S for the scenario as the product $L_{\mathfrak{S}} \times L_{\mathfrak{R}} \times L_{\mathfrak{E}}$, and we consider the following set of initial states $\iota = \{(0, \epsilon, \cdot), (1, \epsilon, \cdot)\}$.

The set of actions available to the agents are as follows: $Act_{\mathfrak{S}} = \{sendbit, \lambda\}$, $Act_{\mathfrak{R}} = \{sendack, \lambda\}$, where λ stands for no action. The actions for $\mathfrak{E}$ correspond to the transmission of messages between $\mathfrak{S}$ and $\mathfrak{R}$ on the unreliable communication channel. The set of actions for $\mathfrak{E}$ is $Act_{\mathfrak{E}} = \{\leftrightarrow, \rightarrow, \leftarrow, -\}$, where $\leftrightarrow$ represents the action in which the channel transmits any message successfully in both directions, $\rightarrow$ that it transmits successfully from $\mathfrak{S}$ to $\mathfrak{R}$ but loses any message from $\mathfrak{R}$ to $\mathfrak{S}$, $\leftarrow$ that it transmits successfully from $\mathfrak{R}$ to $\mathfrak{S}$ but loses any message from $\mathfrak{S}$ to $\mathfrak{R}$, and $-$ that it loses any messages sent in either direction. The set $Act = Act_{\mathfrak{S}} \times Act_{\mathfrak{R}} \times Act_{\mathfrak{E}}$ defines the set of joint actions for the scenario. The local weight functions of agents are defined as follows: $d_{\mathfrak{S}}(sendbit) = n$ with $n \in \mathbb{N}$, $d_{\mathfrak{S}}(\lambda) = 0$, $d_{\mathfrak{R}}(sendack) = m$ with $m \in \mathbb{N}$, $d_{\mathfrak{R}}(\lambda) = 0$, and $d_{\mathfrak{E}}(\leftrightarrow) = d_{\mathfrak{E}}(\rightarrow) = d_{\mathfrak{E}}(\leftarrow) = d_{\mathfrak{E}}(-) = 0$. We assume zero-weight for the actions of $\mathfrak{E}$, since we wish to only count the cost of sending and receiving messages. The local protocols of the agents are the following: $P_{\mathfrak{S}}(0) = P_{\mathfrak{S}}(1) = \{sendbit\}$, $P_{\mathfrak{S}}(0\text{-}ack) = P_{\mathfrak{S}}(1\text{-}ack) = \{\lambda\}$, $P_{\mathfrak{R}}(0) = P_{\mathfrak{R}}(1) = \{sendack\}$, $P_{\mathfrak{R}}(\epsilon) = \{\lambda\}$, $P_{\mathfrak{R}}(0\text{-}ack) = P_{\mathfrak{R}}(1\text{-}ack) = \{sendack\}$, $P_{\mathfrak{R}}(\epsilon\text{-}ack) = \{\lambda\}$, $P_{\mathfrak{E}}(\cdot) = Act_{\mathfrak{E}} = \{\leftrightarrow, \rightarrow, \leftarrow, -\}$.

It should be straightforward to infer the model that is induced by the informal description of the scenario we considered above together with the local states, actions, protocols, and weighted functions defined above. In the model we assume the following set of proposition variables: $\mathcal{PV} = \{\mathbf{bit = 0}, \mathbf{bit = 1}, \mathbf{recack}\}$ with the following interpretation: $(M, s) \models \mathbf{bit = 0}$ if $l_{\mathfrak{S}}(s) = 0$ or $l_{\mathfrak{S}}(s) = 0\text{-}ack$, $(M, s) \models \mathbf{bit = 1}$ if $l_{\mathfrak{S}}(s) = 1$ or $l_{\mathfrak{S}}(s) = 1\text{-}ack$, $(M, s) \models \mathbf{recack}$ if $l_{\mathfrak{S}}(s) = 1\text{-}ack$ or $l_{\mathfrak{S}}(s) = 0\text{-}ack$.

Some properties we may be interested in checking for the example above are the following:

1. $EF_{[n+m,n+m+1)}(\mathbf{recack} \wedge \overline{K}_{\mathfrak{S}}(\overline{O}_{\mathfrak{R}}(K_{\mathfrak{R}} \neg (\mathbf{bit = 0}) \wedge \overline{K}_{\mathfrak{R}} \neg (\mathbf{bit = 1}))))$ - the property says that it is not true that forever in the future if an ack is received by $\mathfrak{S}$, then $\mathfrak{S}$ knows that in all the states where $\mathfrak{R}$ is functioning correctly, $\mathfrak{R}$ knows the value of the bit and the cost is $n + m$.

2. $EF_{[n+m,n+m+1]}(\overline{\mathcal{O}}_{\mathfrak{R}}(\overline{K}_{\mathfrak{S}}(\overline{K}_{\mathfrak{R}}\neg(\mathbf{bit} = \mathbf{0}) \wedge \overline{K}_{\mathfrak{R}}\neg(\mathbf{bit} = \mathbf{1}))))$ – the property says that it is not true that forever in the future in all the states where $\mathfrak{R}$ is functioning correctly, $\mathfrak{S}$ knows that $\mathfrak{R}$ knows the value of the bit and the cost is $n + m$.

3. $E(\overline{\mathcal{O}}_{\mathfrak{R}}(\overline{K}_{\mathfrak{R}}\neg(\mathbf{bit} = \mathbf{0}) \wedge \overline{K}_{\mathfrak{R}}\neg(\mathbf{bit} = \mathbf{1}))U_{[n+m,n+m+1]}\mathbf{recack})$ – the property says that at one point at the future an ack is received by $\mathfrak{S}$, the cost is $n+m$, and at all the preceding points in time where $\mathfrak{R}$ was operating as intended $\mathfrak{R}$ did not know the value of the bit.

To apply the BMC method we have to encode the local states of agents and the global weight function in the in binary form. Since the sender $\mathfrak{S}$ can be in 4 different local green states we shall need 2 bits to encode its state; we take: $(0,0) = 0$, $(0,1) = 1$, $(1,0) = 0\text{-}ack$, $(1,1) = 1\text{-}ack$. Since the receiver $\mathfrak{R}$ can be in 3 different local green states and in 3 different local red states, we shall need $2 + 2$ bits to encode its state; we take: $(1,0;0,0) = 0$, $(0,1;0,0) = 1$, $(0,0;0,0) = \epsilon$, $(0,0;1,0) = 0\text{-}ack$, $(1,1;0,0) = 1\text{-}ack$, $(1,1;1,0) = \epsilon\text{-}ack$. The modelling of the environment $\mathfrak{E}$ requires only one bit: $(0) = \cdot$ Therefore, a global state is modelled by a byte: $s = (s[1], s[2], s[3], s[4], s[5], s[6], s[7])$. For instance the initial states $s^0 = (0, \epsilon, \cdot)$ and $s^1 = (1, \epsilon, \cdot)$ are represented, respectively, by the tuples $(0,0,0,0,0,0,0)$ and $(0,1,0,0,0,0,0)$. Thus the propositional encoding of initial states is the following: $I_{s^0}(w_{0,0}) \vee I_{s^1}(w_{0,0}) = \bigwedge_{i=1}^{7} \neg w_{0,0}[i] \vee (\neg w_{0,0}[i] \wedge w_{0,0}[i] \wedge \bigwedge_{i=1}^{5} \neg w_{0,0}[i])$. Since the maximal weight associated to the joint actions of the scenario is $n + m$, to encode the weight we need $\lceil log_2(n + m)\rceil$ duration variables. Now we can encode the model of the example up to the depth $k \in \mathbb{N}$, but we do not do it here.

The translation of the propositions used in our formulae is the following: $(\mathbf{bit} = \mathbf{0})(w) := (\neg w[1] \wedge \neg w[2]) \vee (w[1] \wedge \neg w[2])$, which means that $(\mathbf{bit} = \mathbf{0})$ holds at all the global states with the first local state equal to $(0,0)$ or $(1,0)$. $(\mathbf{bit} = \mathbf{1})(w) := (\neg w[1] \wedge w[2]) \vee (w[1] \wedge w[2])$, which means that $(\mathbf{bit} = \mathbf{1})$ holds at all the global states with the first local state equal to $(0,1)$ or $(1,1)$. $\mathbf{recack}(w) := (w[1] \wedge \neg w[2]) \vee (w[1] \wedge w[2])$, which means that $\mathbf{recack}$ holds at all the global states with the first local state equal to $(1,0)$ or $(1,1)$.

The translation for the equality of the $\mathfrak{R}$-local states is as follows: $H_{\mathfrak{R}}(w, v) = \bigwedge_{i=3}^{6} w[i] \Leftrightarrow v[i]$, and the translation of an accessibility of a global state in which $\mathfrak{R}$ is running correctly is as follows: $HO_{\mathfrak{R}}(v) = (v[3] \wedge \neg v[4]) \vee (\neg v[3] \wedge v[4]) \vee (\neg v[3] \wedge \neg v[4]) \vee (\neg v[3] \wedge \neg v[4]) \vee (v[3] \wedge v[4])$. The translation of the equality of the $\mathfrak{S}$-local states is as follows: $H_{\mathfrak{S}}(w, v) = \bigwedge_{i=1}^{2} w[i] \Leftrightarrow v[i]$.

Having the above encoding, we can easily infer propositional formulae that encode all the properties mentioned above. Further, checking that the WBTP satisfies the properties 1–3 can now be done by feeding a SAT solver with the propositional formulae generated in the way explained above.

5   Conclusions

In this paper we redefined the methodology of SAT-based BMC for WECTLKD, presented in [14], by extending the MAS model from the deontic interleaved interpreted systems to the weighted deontic interpreted systems. The bounded model checking of the weighted deontic interpreted systems may also be performed by means of Ordered Binary Diagrams (OBDD). This will be explored in the future. Moreover, our future work include an implementation of the algorithm presented here, a careful evaluation of experimental results to be obtained, and a comparison of the OBDD- and SAT-based bounded model checking method for weighted deontic interpreted systems.

We would like to notice that the proposed BMC method can be used for solving some planing problems that can be formulated in therms of weighted automata. Namely, we can formalize the notion of the agent as a weighted automaton, and then apply our BMC methodology for weighted Kripke structure that are generated by a given network of weighted automata. A planning problem defined in therms of weighted automata was considered, e.g., in [6].

References

1. Lomuscio, A., Sergot, M.: Violation, error recovery, and enforcement in the bit transmission problem. In: Proceedings of DEON 2002. Imperial College Press (2002)
2. Cabodi, G., Camurati, P., Quer, S.: Can BDDs compete with SAT solvers on bounded model checking? In: Proceedings of DAC 2002, pp. 117–122. ACM (2002)
3. Clarke, E., Biere, A., Raimi, R., Zhu, Y.: Bounded model checking using satisfiability solving. Formal Methods in System Design 19(1), 7–34 (2001)
4. Clarke, E.M., Grumberg, O., Peled, D.A.: Model Checking. The MIT Press (1999)
5. Emerson, E.A.: Temporal and modal logic. In: van Leeuwen, J. (ed.) Handbook of Theoretical Computer Science, vol. B, ch. 16, pp. 996–1071. Elsevier Science Publishers (1990)
6. Fabre, E., Jezequel, L.: Distributed optimal planning: an approach by weighted automata calculus. In: Proceedings of CDC 2009, pp. 211–216. IEEE (2009)
7. Fagin, R., Halpern, J.Y., Moses, Y., Vardi, M.Y.: Reasoning about Knowledge. MIT Press, Cambridge (1995)
8. Levesque, H.: A logic of implicit and explicit belief. In: Proceedings of the 6th National Conference of the AAAI, pp. 198–202. Morgan Kaufman (1984)
9. Lomuscio, A., Sergot, M.: Deontic interpreted systems. Studia Logica 75(1), 63–92 (2003)
10. Markey, N., Schnoebelen, P.: Symbolic model checking of simply-timed systems. In: Lakhnech, Y., Yovine, S. (eds.) FORMATS/FTRTFT 2004. LNCS, vol. 3253, pp. 102–117. Springer, Heidelberg (2004)
11. Męski, A., Penczek, W., Szreter, M., Woźna-Szcześniak, B., Zbrzezny, A.: Two approaches to bounded model checking for linear time logic with knowledge. In: Jezic, G., Kusek, M., Nguyen, N.-T., Howlett, R.J., Jain, L.C. (eds.) KES-AMSTA 2012. LNCS, vol. 7327, pp. 514–523. Springer, Heidelberg (2012)
12. Penczek, W., Lomuscio, A.: Verifying epistemic properties of multi-agent systems via bounded model checking. Fundamenta Informaticae 55(2), 167–185 (2003)
13. Wooldridge, M.: An introduction to multi-agent systems. John Wiley (2002)
14. Woźna-Szcześniak, B., Zbrzezny, A., Zbrzezny, A.: The BMC method for the existential part of RTCTLK and interleaved interpreted systems. In: Antunes, L., Pinto, H.S. (eds.) EPIA 2011. LNCS, vol. 7026, pp. 551–565. Springer, Heidelberg (2011)
15. Zbrzezny, A.: Improving the translation from ECTL to SAT. Fundamenta Informaticae 85(1-4), 513–531 (2008)

Dynamics of Relative Agreement
in Multiple Social Contexts

Davide Nunes[1], Luis Antunes[1], and Frederic Amblard[2]

[1] GUESS / LabMAg / University of Lisbon
{davide.nunes,xarax}@di.fc.ul.pt
[2] Université Toulouse 1 Sciences Sociales
frederic.amblard@univ-tlse1.fr

Abstract. In real world scenarios, the formation of consensus is an self-organisation process by which actors have to make a joint assessment about a target subject being it a decision making problem or the formation of a collective opinion. In social simulation, models of opinion dynamics tackle the opinion formation phenomena. These models try to make an assessment, for instance, of the ideal conditions that lead an interacting group of agents to opinion consensus, polarisation or fragmentation. In this paper, we investigate the role of social relation structure in opinion dynamics using an interaction model of relative agreement. We present an agent-based model that defines social relations as multiple concomitant social networks and apply our model to an opinion dynamics model with bounded confidence. We discuss the influence of complex social network topologies where actors interact in multiple relations simultaneously. The paper builds on previous work about social space design with multiple contexts and context switching, to determine the influence of such complex social structures in a process such as opinion formation.

Keywords: opinion dynamics, consensus, bounded confidence, relative agreement, social contexts, social networks.

1   Introduction

Understanding trend and opinion spreading or consensus formation processes within a population is fundamental to construct coherent views and explanations for real-world events or phenomena. Examples of such processes include: joint assessments of a certain policy, the impact of a viral marketing campaign or even, in the context of economics and politics, the voting problem. This last problem was investigated in an early model proposed by Herbert Simon [20].

Formal opinion dynamics models try to provide an understanding if not an analysis of opinion formation processes. An early formulation of these models was designed to comprehend complex phenomena found empirically in groups [10]. In particular, the work on consensus building in the context of decision-making was initialised by DeGroot [8] and Lehrer [16]. Empirical studies of opinion formation in large populations have methodological limitations, as such, we use

L. Correia, L.P. Reis, and J. Cascalho (Eds.): EPIA 2013, LNAI 8154, pp. 456–467, 2013.
© Springer-Verlag Berlin Heidelberg 2013

simulation, in particular Multi-Agent Simulation (MAS), as a methodological framework to study such phenomena in a larger scale. Most opinion dynamics simulation models are based either on binary opinions [11,3] or continuous opinions [7,12,6,13]. In these models, agents update their opinions either under social influence or according to their own experience. For a detailed analysis over some opinion dynamics model analytical and simulation results, the reader can refer to [12].

In agent-based opinion dynamics models, agent interactions are guided by social space abstractions. In some models, dimensionality is irrelevant. Typically, all agents can participate in interactions with all other agents. Axelrod, takes a different approach in his model of dissemination of culture [4] and represents agents in a bi-dimensional grid which provides structure for agents to interact with each other. In Weisbuch's bounded confidence model with social networks [23], the agents are bound by different network topologies. A first definition for these types of bounded confidence models was given in [14].

In real-world scenarios, actors engage in a multitude of social relations different in kind and quality. Most simulation models don't explore social space designs that take into account the differentiation between coexisting social worlds. Modelling multiple concomitant social relations was an idea pioneered by Peter Albin in [1] and allows for the comprehension of a variety of real-world dynamics such as, e.g., the impact of on-line social media political campaigns or what it means socially to lose a job. Furthermore, such complex social structures are the basis for the formation of social identity [19,9] and play a decisive role in self-organised processes such as consensus formation [3,18].

This paper is aimed at extending the line of research regarding the representation of social spaces with explicit multiple concomitant social relations. This work, described in [3,18] presents interesting insights on how different complex social relation topologies influence consensus formation dynamics. We apply the notions of multiple social contexts to a model of continuous opinion dynamics called Relative Agreement (RA) model [6]. This model is an extension of the Bounded Confidence (BC) model [15,7,12].

The work in [3,18] explores multiple contexts applied to a simple game of consensus that can be seen as a simple binary opinion dynamics models. It is found that by considering coexisting social relations, the agent population converges to a global consensus both faster and more often. This happens due to what we call permeability between contexts. Permeability in multiple social contexts is created due to both social context overlapping and context switching [3]. Context switching models the existence of multiple distinct social relations from which each social agent switches to and from at different instances in time. As an example, take for instance the *work* and *family* relations.

As the RA model [6] is considerably more complex than the simple interaction game considered in [3,18], we perform a series of experiments to determine if this social space modelling methodology exerts a similar influence in this model. This will allow to understand if the multiple-context models present properties that are transversal to the interaction processes to which they are applied.

The paper is organised as follows. In section 2 we present the opinion dynamics model along with our social space design with multiple concurrent social networks. Section 3 describes the model of experiments presenting multiple simulation scenarios and indicators for the observations. Section 4 presents the results and the corresponding discussion. Finally, in section 5 we summarise up our findings and propose some future work guidelines.

2   The Proposed Model

The proposed model integrates both the multi-relational modelling approach [3] and the Relative Agreement (RA) interaction model [6]. We start by describing the multi-context model with context switching [3] and the continuous opinion dynamics model [6]. We then present the resulting ABSS model of continuous opinion formation with uncertainty, multiple social contexts and context switching dynamics.

2.1   A Model of Context Switching

The multi-context approach [3,18] considers a multitude of concomitant social relations to represent the complex social space of an agent. This setting can be seen in a simulation as a n-dimensional scenario where each dimension surface represents a different social relation (see figure 1) simulated with a social network model. Agents belong to distinct contexts (neighbourhoods) in these multiple relations.

Fig. 1. Multiplex social network structure forming the social space for our models of multiple concurrent social relations

In the particular model of *context switching* [3], a population of N agents populates multiple social networks. Each agent is active only in one context at a time. In each simulation step, the agents select a neighbour from their current context and update their opinion according to some rule. At the end of each interaction an agent switches to a different context with a probability ζ_c. For the sake of simplicity, the ζ_c probability is a parameter associated with each context c and it is valid for all the agents in that context. This allows for modelling of time spent in each context, in an abstract way. We can think of context switching as a temporary deployment in another place, such as what happens with temporary immigration.

2.2 Relative Agreement Interaction

We now describe the model of continuous opinion dynamics with relative agreement [6]. In this model, each agent i is characterised by two variables, its opinion x_i and its uncertainty u_i both being real numbers. The opinion values are drawn from a uniform distribution between -1 and 1.

This model can be seen as an extension of the Bounded Confidence (BC) model [15,7,12]. In the BC model, the agents have continuous opinions and the interactions are non-linear. The agents only exert influence on each other if their opinions are within a certain fixed threshold. The threshold can be interpreted as an uncertainty, or a bounded confidence, around the opinion [6]. It is assumed that agents do not take into account opinions out of their range of uncertainty.

The RA model differs from the BC model in the fact that the change in an opinion x_j of an agent j under the influence of an agent i, is proportional to the overlap between the agent opinion segments (the agreement), divided by the uncertainty of the influencing agent uncertainty u_i. Another difference is that the uncertainty is not fixed, the value of u_j is also updated using the same mechanism. The opinion and uncertainty updates are illustrated in figure 2.

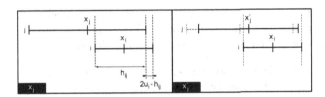

Fig. 2. Agent i (with the opinion x_i and the uncertainty u_i) influences agent j (with the opinion x_j and the uncertainty u_j). In this case, h_{ij} is the overlap between the agents and $2u_i - h_{ij}$ is the non-overlapping part. On the left is the representation of the opinion and uncertainty of agent j, on the right, the dashed lines represent the position of the segment before the interaction and the plain lines, the final values for the these two properties [6].

The opinion overlap h_{ij} is given by:

$$h_{ij} = min(x_i + u_i, x_j + u_j) - max(x_i - u_i, x_j - u_j) \qquad (1)$$

The opinion and uncertainty values are updated according to the following equations. As an example, the opinion x_j and the uncertainty u_j of agent j is updated according to equation 2 and 3 respectively, if $h_{ij} > u_i$.

$$x_j{'} = x_j + \mu \left(\frac{h_{ij}}{u_i} - 1 \right) (x_i - x_j) \qquad (2)$$

$$u_j{}' = u_j + \mu \left(\frac{h_{ij}}{u_i} - 1 \right) (u_i - u_j) \tag{3}$$

Where the μ is a constant parameter which amplitude controls the speed of the dynamics. For more details, refer to [6].

2.3 Context Switching with Relative Agreement

The proposed model integrates both the context switching (described in section 2.1) and the relative agreement models. In this model, agents are embedded in static social networks, interact using the opinion dynamics rules set by the RA model described in the previous section 2.2, and switch contexts (the agent neighbourhood in the network) according to a probability ζ_{C_k}, associated with each context C_k.

Our proposed simulation model behaves as follows. Consider a population of N agents distributed by M different social networks. The networks are static throughout the simulation. On each network an agent can be either *active* or *inactive* being that an agent can only be active in one network (context C_k) at a time.

On each simulation cycle, the N agents are schedule to execute their behaviour sequentially and in a uniform random order. The behaviour of the current agent i, located in the context C_k can be described as follows:

1. Choose an available neighbour (agent j) from the current context at random;
2. Update agent i and j opinions and uncertainties according to the equations 2 and 3 from the previous section 2.2;
3. Switch to a random distinct context C_l ($C_l \neq C_k$) with a probability ζ_{C_k}, which is a static parameter with different values for each context / network;

Note that although static complex social network models allow us to create abstract representation for social contexts can be relatively stable if we consider short to moderate periods of time, our social peers are not always available at all times and spend different amounts of time in distinct relations.

3 Experiment Design

The simulation experiments were created using the MASON framework [17] and executed in a grid environment described in [18]. In each experiment, a population of 300 agents interacts until 3000 cycles pass or the opinion values stabilise. We perform 30 simulation runs for each parameter combination considered. We have two main goals with the experiments presented in this paper. The first is to analyse the dynamics of opinion formation under the model of relative agreement described in the previous section. This model combines both the relative agreement interaction rules and the context switching social spaces with multiple contexts. The second goal is to analyse the influence of different network topologies in the formation of consensus in multi-agent societies.

In this paper, we present a set of experiments focused on the analysis of the dynamics induced by the context switching mechanism. We spanned the switching parameter (ζ_{C_i}) from 0 to 1 in intervals of 0.05 with two contexts. We also use different network topologies in these contexts. We then observe how different combinations of context switching probabilities and network structures affect the speed of convergence to stable opinion values.

The initial uncertainty parameter is set to $U = 1.4$. According to the previous work with relative agreement in single social networks [2], this value guaranteed the convergence in one central opinion value. We chose this parameter value to ensure that interactions are not heavily restricted by the uncertainty early in a simulation run. We want to study the influence of different network topologies so these are the structures that intrinsically guide such interactions early on.

We tested our model with three network models: regular networks, with the same number of connections for each agent; scale-free network, generated using the Barabasi-Albert (BA) model [5]; small-world network, generated using the Watts & Stroggatz (WS) model [22].

4   Results and Discussion

In this section we present and discuss the experimental results. We show how different values of switching between contexts influence the speed of convergence to stable opinion values. We also explore the interplay between the model of relative agreement presented in section 2.3 and different network topologies.

4.1   Context Switching with Regular Networks

In this set of experiments, we focus on the analysis of how the switching probabilities affect the opinion formation game. To do this, we construct several simulation scenarios where agents interact in two social relations. Each relation is associated with an abstract network model and has a context switching value ζ_{C_i}. This value corresponds to the probability of switching from a relation to another (as described in section 2.3).

We maintain homogeneous network structures and span the context switching values (ζ_{C_i}) from 0 to 1 in intervals of 0.05. Figure 3 depicts a landscape for this parameter span. In this case we create two contexts each one with a k-regular network with $k = 30$. Regular networks offer an easy way to model highly clustered populations of agents. For this type of networks each node is connected with $2k$ other nodes, where k is a parameter of the generative procedure. Regular networks also provide a convenient way to observe the influence of neighbourhood size in the opinion stabilisation process as the connectivity structure is equal for all the agents. They can also serve as models for highly clustered communities (although its structure is far from real world scenario topologies [22,5]).

In figure 3, we see that small probabilities ($0 \leq \zeta_{C_i} \leq 0.1$ proximately) in one of the contexts lead to a large numbers of encounters necessary to stabilise the opinion values. In extreme cases, stabilisation is never reached. We can also

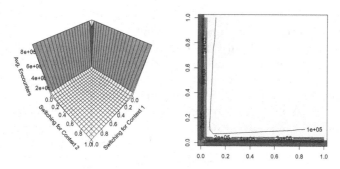

Fig. 3. Meetings to achieve convergence to stable opinion values with two 30-regular networks and $\zeta \in [0,1]$

observe the configuration $(\zeta_{C_1}, \zeta_{C_2}) = (0,0)$ is slightly better in these cases. This is because agents are isolated in each context and thus the opinions evolve separately the same way they would if the agents were placed within a single network. Also note that although not depicted in the figure, in this last case the opinions also evolve to two separate values.

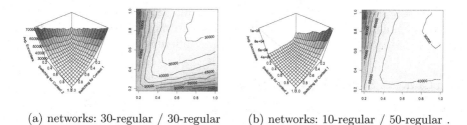

(a) networks: 30-regular / 30-regular   (b) networks: 10-regular / 50-regular .

Fig. 4. Meetings to achieve convergence to stable opinion values with for $\zeta \in [0.2, 1]$. In the first landscape (figure 4a) we use 30-regular networks in both contexts. In figure 4b we use a 10-regular and a 50-regular for context 1 and 2 respectively.

Figure 4a shows a zoom in the previous landscape (figure 3) with the ζ being between 0.2 and 1. This is the optimal zone in terms of encounters necessary to achieve stable opinions. Here we can see that the optimal values for switching with this regular configuration lies within $\zeta_{C_i} \in [0.8, 1]$ proximately. Moreover, if one of the contexts has a high switching probability, the other context should have a similar level of switching. Having one social relation with high switching while having the second with a low probability leads to a scenario where agent spend most of the time in one context but can still switch to another one. While they spend considerately less simulation time in this second context, this is enough to destabilise the opinion formation process.

In the next experiment we created a scenario to observe the effects of different connectivity levels for each context. Figure 4b depicts the span of the switching probability within the values $\zeta \in [0.2, 1]$. The first context is now a 10-regular network (each agent has 20 neighbours) while the second is a 50-regular network (each agent having a total of 100 neighbours).

As we can see the asymmetry in the connectivity has clear effects in the convergence to stable opinion values. In this case, we find that if an agent stays more time ($\zeta \in [0.2, 0.3]$) in the context with the lowest connectivity it seems to be important to switch less frequently from the highly connected and clustered social layer. Similarly to what was found in [18], a possible explanation is that in larger neighbourhoods, the probability of performing encounters with an agent with a very different opinion early in the simulation is considerable. The impact is clearly visible as disturbance in the convergence to stable opinions.

4.2   It's a Small World after All

One evidence of the importance of network structure can be found in the next results. We conducted experiments using the Watts & Strogatz (WS) model [22] to generate networks with small-world properties. These topologies are constructed by rewiring regular networks, introducing increasing amounts of disorder. Moreover, we can construct highly clustered networks, like regular lattices, yet with small characteristic path lengths, like random graphs. They are called small-world by analogy with the phenomenon [21], popularly known as six degrees of separation. This phenomena refers to the idea that everyone is on average approximately six steps away, by way of introduction, from any other person on Earth.

Figure 5 shows the results for a set-up with two WS networks with an initial $k = 30$ and a rewiring probability of $p = 0.1$ and $p = 0.6$. The value of $p = 0.1$ for the rewiring, introduces enough disorder in the network to lower the average path length without sacrificing the clustering coefficient too much. In figure 5a, we can see that the influence is very similar to the previous results with regular networks (see figure 4a) but the reduction in the path length causes the model to converge more rapidly for higher switching probabilities.

When we increase the level of disorder, for instance, to a value of $p = 0.6$, the network clustering coefficient is significantly reduced, while maintaining the low average path length. The results for this are depicted in figure 5b. Although the switching probability seems to have a more complex influence on the speed of convergence, globally, the number of encounters seem to be almost homogeneous throughout the switching vales $\zeta_{C_i} > 0.3$. Also, as the number of necessary encounters is a slightly lower, it seems that high values of switching are more important when the networks possesses highly clustered nodes.

4.3   Context Switching with Scale-Free Networks

In this section we briefly discuss the results for the experiments with the scale-free network models. We performed an experiment with two contexts each one with a scale-free network. In this network, each node has a minimum connectivity of 1, meaning that the preferential attachment mechanism only actuates once

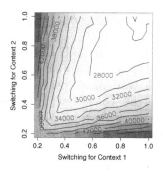

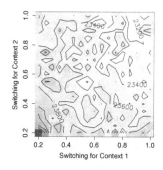

(a) WS network with $p = 0.1$.   (b) WS network with $p = 0.6$

Fig. 5. Meetings to achieve convergence to stable opinion values for two Watts & Strogatz small-world networks generated with initial degree $k = 30$ and rewiring probability $p = 0.1$ (5a) and $p = 0.6$ (5a). The switching values are $\zeta \in [0.2, 1]$.

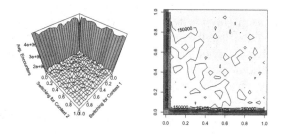

Fig. 6. Meetings to achieve convergence to stable opinion values for two scale-free networks with minimum degree $k = 1$. The switching values are $\zeta \in [0, 1]$.

each time a node is added to the network. This thus generates a network with a forest topology.

Figure 6 shows that although the majority of nodes has a very low connectivity (see, [5]), the small-world characteristics of this scale-free model provide means to achieve convergence to stable opinion values. This happens for switching probabilities approximately within $\zeta_{C_i} \geq 0.1$, much like what happens in the previously described experiments.

4.4 Results with a Single Network

For result comparison purposes we performed a series of experiments with single networks. Using a parameter k with the values $k = 1, 2, 3, 4, 5, 10, 20, 30, 40, 50$, we performed experiments with single k-regular, WS small-world and BA scale-free networks (see figure 7). For the regular networks the parameter k is the previously described connectivity with each agent having $2k$ neighbours. For the small-world networks, this parameter is used to construct the initial k-regular

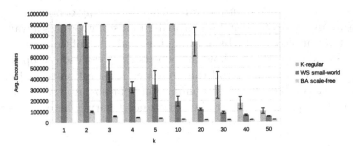

Fig. 7. Meetings to achieve convergence to stable opinion values with a single network context (without switching). Results for k-regular, WS small-world with $p = 0.1$ and BA scale-free networks.

structure, we used a rewiring probability $p = 0.1$ to keep these network highly clustered. For the BA scale-free networks the k is the minimum degree each agent will have upon generating the network instance.

Note that in figure 7, the maximum value of encounters is limited by the maximum number of simulation cycles allowed. In this case, the models that display the maximum number of encounters did not converge to stable opinion values.

In figure 7 we can see that for $k \geq 2$, scale-free networks seem to outperform the other models in terms of convergence speed. Also note that these results confirm that the switching mechanism allows the opinion formation process to converge both faster and more often. As an example, consider the results for scale-free networks (figure 6) where convergence was made possible by exposing the agents to two distinct contexts.

The results in this paper show that the usage of different network model structures plays an important role when modelling opinion or consensus formation processes. Context dynamics seems to be an advancement as a modelling methodology for complex real-world scenarios and has a deep influence in how simulation models behave. These are key points discussed both in the work of Antunes *et al.* [3,18] and Amblard *et al.* [6,2] from which this work stems from.

5   Conclusion and Future Work

The results in this paper corroborate the fact that multiple context structures play an important role in processes such as opinion formation. While complex network models are good for modelling real-world social relation scenarios, single network structures fail to capture the complexity of the multitude of existing relations. Social decision making and the phenomena associated with this processes are influenced in different ways by distinct kinds of social relations. Examples of this are found in real-world events such as contemporary political or marketing campaigns.

The model here presented, while abstract by nature, can unveil interesting dynamics that should be taken into account when modelling complex social spaces for simulation models. The switching probability also introduces a way to model interaction temporal dynamics by allowing the modelling of time agents dedicate to different social contexts and how this affects the formation of opinions.

For future work, we will extend the presented exploration to include heterogeneous context configurations, combining different social network models. We also consider to explore scenarios where the uncertainty is heterogeneous [2].

References

1. Albin, P.S.: The analysis of complex socioeconomic systems. Lexington Books, Lexington (1975)
2. Amblard, F., Deffuant, G.: The role of network topology on extremism propagation with the relative agreement opinion dynamics. Physica A: Statistical Mechanics and its Applications (January 2004)
3. Antunes, L., Nunes, D., Coelho, H., Balsa, J., Urbano, P.: Context Switching Versus Context Permeability in Multiple Social Networks. In: Lopes, L.S., Lau, N., Mariano, P., Rocha, L.M. (eds.) EPIA 2009. LNCS, vol. 5816, pp. 547–559. Springer, Heidelberg (2009)
4. Axelrod, R.: The Dissemination of Culture. Journal of Conflict Resolution 41(2), 203–226 (1997)
5. Barabási, A.L., Albert, R.: Emergence of scaling in random networks. Science 286(5439), 509 (1999)
6. Deffuant, G., Amblard, F., Weisbuch, G., Faure, T.: How can extremism prevail? A study based on the relative agreement interaction model. Journal of Artificial Societies and Social Simulation 5(4), 1 (2002)
7. Deffuant, G., Neau, D., Amblard, F., Weisbuch, G.: Mixing beliefs among interacting agents. Advances in Complex Systems 3(01n04), 87–98 (2000)
8. Morris, H.: DeGroot. Reaching a Consensus. Journal of the American Statistical Association 69(345), 118–121 (1974)
9. Ellemers, N., Spears, R., Doosje, B.: Self and social identity*. Annual Review of Psychology 53(1), 161–186 (2002)
10. French Jr., J.P.R.: A formal theory of social power. Psychological Review 63, 181–194 (1956)
11. Galam, S.: Rational group decision making: A random field ising model t = 0. Physica A: Statistical Mechanics and its Applications 238(1), 66–80 (1997)
12. Hegselmann, R., Krause, U.: Opinion dynamics and bounded confidence: models, analysis and simulation. J. Artificial Societies and Social Simulation 5(3), 2 (2002)
13. Jager, W., Amblard, F.: Uniformity, bipolarization and pluriformity captured as generic stylized behavior with an agent-based simulation model of attitude change. Computational & Mathematical Organization Theory 10, 295–303 (2004)
14. Krause, U.: Soziale Dynamiken mit vielen Interakteuren. Eine Problemskizze in Modellierung und Simulation von Dynamiken mit Vielen Interagierenden Akteuren. Modus. Universität Bremen (1997)
15. Krause, U.: A Discrete Nonlinear and Non-Autonomous Model of Consensus Formation. In: Communications in Difference Equations: Proceedings of the Fourth International Conference on Difference Equations, pp. 227–236. Gordon and Breach Pub., Amsterdam (2000)

16. Lehrer, K.: Social consensus and rational agnoiology. Synthese 31, 141–160 (1975)
17. Luke, S., Cioffi-Revilla, C., Panait, L., Sullivan, K., Balan, G.: MASON: A Multi-agent Simulation Environment. Simulation 81(7), 517–527 (2005)
18. Nunes, D., Antunes, L.: Consensus by segregation - the formation of local consensus within context switching dynamics. In: Proceedings of the 4th World Congress on Social Simulation, WCSS 2012 (2012)
19. Roccas, S., Brewer, M.B.: Social identity complexity. Personality and Social Psychology Review 6(2), 88–106 (2002)
20. Simon, H.A.: Bandwagon and underdog effects and the possibility of election predictions. The Public Opinion Quarterly 18(3), 245–253 (1954)
21. Travers, J., Milgram, S.: An experimental study of the small world problem. Sociometry 32(4), 425–443 (1969)
22. Watts, D.J., Strogatz, S.H.: Collective dynamics of 'small-world' networks. Nature 393(6684), 440–442 (1998)
23. Weisbuch, G.: Bounded confidence and social networks. The European Physical Journal B-Condensed Matter and Complex Systems 38(2), 339–343 (2004)

The Social Meaning of Physical Action

Frank Dignum[1], Virginia Dignum[2], and Catholijn M. Jonker[2]

[1] Utrecht University, The Netherlands
F.P.M.Dignum@uu.nl
[2] Delft University of Technology, The Netherlands
{m.v.dignum,c.m.jonker}@tudelft.nl

Abstract. Intelligent systems should be able to initiate and understand social interactions, which are always brought about by physical actions. Thus, all physical actions need to be interpreted for their social effects. This paper gives an analytical framework that identifies the main challenges in modelling social interactions. Such interpretation is subjective and might lead to misunderstandings between participants in the interaction. In real life, people use rituals, such as greetings, as a method for common grounding. Greeting as an essential prerequisite of joint activity is taken as case study to illustrate our approach.

1 Introduction

Intelligent systems will never play an important role in our lives, unless they acquire social intelligence. As social creatures, people are highly dependent on others in almost all aspects of life, not just for their physical well-being but especially for their psychological welfare. Well coordinated joint action is essential for success, which is only possible if there is enough common knowledge amongst the actors to anticipate each other's behaviour, thus increasing the effectiveness of the joint activity. Existing literature in agent technology on social interaction already covers many aspects of joint actions, coordination, cooperation and task dependencies but mainly treats these in terms of objectively observable effects. However, social interaction has essential effects on social reality, which can even outweigh the importance of the physical effects, but which are not directly nor objectively observable.

Our framework is based on the observation that joint activities are not just about physical actions, but also about the social interpretation of these actions, and the expectation that the other actors will continuously monitor and interpret these actions for their social meaning. Social interpretation is a private activity that is not objectively observable by others. However, social interpretation influences future actions that in turn might give others an indirect and subjective idea of the first social interpretation. For example, the forward motion of person A, might (wrongly) be interpreted by person B as a threat (social interpretation). As a result person B steps away. Which in turn person A might (wrongly) interpret as a "please come inside" gesture of person B. The smile with which person A enters B's house might convince B that A is no threat after all.

L. Correia, L.P. Reis, and J. Cascalho (Eds.): EPIA 2013, LNAI 8154, pp. 468–479, 2013.
© Springer-Verlag Berlin Heidelberg 2013

Considering the possible social meaning of any physical actions, might be confused with social interaction, as defined by [13], Chapter 9 as follows. *Social interactions are the acts, actions, or practices of two or more people mutually oriented towards each other's selves, that is, any behavior that tries to affect or take account of each other's subjective experiences or intentions.* In our framework we also consider the fact that people might attach social meaning to an action of another that did not intend to affect the other person's subjective experiences or intentions. For example, a door slammed shut because the wind pulled it out of the hand of a person A might be attributed as an angry action of A by another person B that had a somewhat conversation with A just a few minutes before. Person A might have had no such intention. Therefore, although the literature on social interaction is the most important source of inspiration for our work, social interaction is a subset of the actions studied in this paper, i.e., the set of all actions that can be attributed social meaning.

In this paper we introduce an analytical framework of the social meaning of physical action including the relevant concepts with which we can model social interactions, analyse where possible misunderstandings might take place, and what might be done to minimize these misunderstandings. Based on this framework we can design systems that perform more realistic interactions with possible social meaning. From this initial work we can infer that agent should keep an elaborate model of social reality in order to perform realistic social interactions. In order to illustrate the functioning of our framework we discuss the greeting ritual, as it contains all major elements of social interaction, i.e., physical coordination of joint activities, and their possible social interpretations and misunderstandings.

The rest of this paper is organized as follows. First some related work is presented in Section 2. In Section 3 our analytical framework for social interactions is presented. We illustrate the framework on a small case study of the greeting ritual in Section 4. In section 5 some conclusions are drawn.

2   Related Work

Social interactions have been extensively studied in social psychology and form an important subset of the set studied in this paper, i.e., the set of all physical actions that might have social meaning. The aspects that we distinguish in our framework are inspired by this area of research as exemplified by [2], [13], [15]. Moreover, we chose aspects that are not only important for describing social interactions but are also useful (necessary) for designing socially interacting systems. Within the AI community social interactions have a close relation to the research on joint actions, plans and goals. Joint action (and its related plans and goals) are per definition focussed on agents that are involved in activities with a joint goal and for reasons of performance, the actions of each participant are meant to have an effect on the other participants [8].

Especially we can build on the theoretical work from [7] and [11] that indicate the joint mental attitudes that are necessary to perform joint actions. In general this work provides a formal basis for the intuition of "common ground" necessary for joint action. These theories indicate that in order to succesfully complete a joint plan you need to have a common belief about this plan and its subplans (or the goal and subgoals). We use this part later on to check how much of this common belief is actually established by the context (e.g. culture, history, place, etc.) and what part needs to be established through the interaction itself. However, in more practical-oriented research, the emphasis is on how to distribute a given plan over a set of actors and how to coordinate the plan such that this joint activity is realized efficiently and effectively (see e.g. [8]). For this purpose GPGP [5] and its implementation platform TAEMS [10] define hierarchical plans with some specific relations that allow for easy distribution of parts of the plan to different actors. Teams will try to find optimal distributions of a plan over actors with their own capabilities. It also provides a monitoring capability to keep track of the plan during execution. However, in our situation we do not start with a known joint plan, but rather with an interaction of which not all steps might be fixed or known on forehand. Even if a ritual is performed (such as greeting) it is not sure that all parties of the ritual share the same idea about how the ritual should be performed and thus might have different expectations about the actions of the other parties. Besides this important difference social interactions also explicitly include the social interpretation of the actions of the parties. A party performs actions with the expectation that these actions will be socially interpreted by the others. However, it is difficult to check whether this interpretation is actually made (correctly). Thus we need mechanisms to cope with these differences.

A similar issue arises in communication. The effect of communication acts is also not (directly) observable. This has led to many debates about the semantics of (agent) communication [1]. All aspects identified in communication also play a role in the type of social interactions studied in this paper. Especially we will also use the work on common ground for communication (see e.g. [3]). However, we do not focus on the possible interpretations of the content of the message, instead we focus on the physical aspects and their social interpretation of the interactions. Finally, social interactions also are linked to research on Theory of Mind [12]. Theory of Mind tries to build up a model of the mental attitudes of a person based on the perceived actions. Although this model helps to establish common expectations and understanding of social relations, this is not the focus of our current research. In first instance we focus on the direct social intentions with which actions are performed rather than longer term social goals that a person might have. E.g. we do not need to find out an ulterior (social) goal of a greeting, but only want to infer whether a person wants to establish a peer-to-peer relation or establish himself as a superior.

Another important link between physical activity and social interpretation is context, i.e. timing or location, see e.g., [9],[4],[16]. This is also an issue in the research on Intelligent Virtual Agents (see e.g. [17]), but most of that work only focuses on one physical aspect at the time and does not consider the complete physical context in relation to the social interaction. In social interaction, these links are however essential. Taking again the greeting example, consider two persons, A and B, that meet for the first time. Person A (the initiator) extends a hand to the other person, looks B in the eyes and utters a greeting phrase. The timing of these activities is essential. From what distance does this greeting start? While walking up to the other we can already start extending our hand, while looking the other in the eye. However, when B starts to respond by also raising his hand, which A observes while maintaining eyecontact, at some point A needs to focus her eyes on the handshaking, and so should B. If not, A and B might poke each other, miss the hand completely, and so on. So proximity (distance), timing, continuous monitoring of the hands during a specific part of the greeting, and eye contact is important in this western culture greeting.

We conclude that a social interaction implies a common ground of the participants, knowledge about the possible actions in relation to context, knowledge about the possible social interpretations of those actions in relation to context. Social interaction is impossible without a form of monitoring of the activities and effects by all participants. Such monitoring has to include both the observation of the physical aspects of the activities as well as the social interpretation of physical activities and their effects.

3   Analytical Framework

Before listing the elements of our analytical framework we should say a few words about the concept of *social reality*. Although this concept has been used by several philosophers and sociologists (cf. [6],[14]), unlike physical reality, there is no general accepted definition or even common understanding about what it exactly constitutes. Intuitively it is the set of rules and believes that people have in common and that guide our social interactions. However, where for physical reality there are theories from physics, chemistry and biology to model reality from different perspectives and using connected models, the theories from social science about social reality are incomplete and disconnected. Thus creating models for physical reality for applications is easier than creating models for social reality. For the physical reality model we just have to decide how realistic the model has to be and choose the model delivering that amount of realism. I.e. for some video games realistically moving characters are important and they are modelled using bones and muscle structures, while in other games they can be cartoon-like. For social reality this choice is less obvious and depends mainly on the type of application. Realism depends more on which type of aspects are modelled than how many aspects are modelled.

We will show in the rest of the paper that modelling social reality in a similar way as physical reality and connecting the two in a principled way is essential for creating realistic social intelligent systems.

Using the literature discussed in the previous section, we identify the following concepts as being relevant for an analytical framework of social interactions:

- **Personal goal**
- **Physical context**
  - *Time & Space; Observability*
- **Physical action**
  - *Time-scale & Space (proximity) of the action; Monitoring ; Refinement of action to tune with the actions of the other participants; Dynamics*
- **Social context**
  - *Type of ritual; Relationships & Stakeholders; Expectations from stakeholders; Common understanding / xommon ground (e.g. History, Point of departure, Social purpose); Social consequences of success and failure*
- **Social action**
  - *Expectations; Trust; Predictability*

Our analysis of social interactions starts with identifying the **goal** of the social interaction. The goal depends on the context and determines the action. Both context and action have a physical, as well as a social aspect.

Social interaction is always realised through **physical actions**. I.e. there is a *counts-as* relation or interpretation relation that indicates the social effects of physical actions. However, physical actions can also have a functional purpose and effect. Thus the social effects of an action are determined by the way an action is performed. E.g. closing a door has a functional effect of the door being closed. If the door is closed very carefully, the social effect is that the person is seen as deferring towards the persons in the room. If the door is slammed closed, he might be seen as being angry or indifferent to the persons in the room.

Because the social action is connected to physical actions the **physical context** is important. Especially, because the social meaning of an action is usually conveyed by variations on how the action is performed rather than that this meaning can be derived from the physical purpose of the action. Thus shared space and time is needed in which all participants can observe the actions performed. Moreover, the observability of actions by the participants is important. The physical effect of closing a door is still observable after the action has been performed. However, the social effect needs to be established by direct observation of the performing of the act itself, because social effects are virtual and not (directly) observable afterwards. E.g. when a person slams a door, someone seeing the door closed 5 minutes later cannot derive anymore that the person closing the door was angry.

Further analysis of the physical context of the physical action itself concerns the time scale and proximity of the action as determinants social signals. By time scale we mean both the time it takes to complete an action with respect to the interaction with partners and the level of detail that is available in the (implemented) physical action. An action like "going out in town" has a long duration and can lead to a strong social relation (including a large common ground) between the participants. A greeting action is short and can only be used for basic social effects. Even with short actions it is important to know how much level of detail is available that can be used to convey social signals. E.g. when shaking hands, we can note how strong the handshake is, whether eye contact is made, whether the other hand remains static or moves to elbow or sholder of the partner, whether the upper body remains static or not. Shaking hands itself can denote a certain formal relation between the participants (opposed to giving a hug), but each of these details might convey whether there is a friendly or less friendly relation, who is considered the boss, etc. However, if we do not model these details but only have a standard greeting action available in our model none of these signals could be given or perceived.

Similarly the proximity plays a big role. When two people greet in a crowded room they might not be close enough to shake hands or kiss and thus wave hands if their history of social interaction allows for this option. However, in this situation having no physical contact does not indicate social distance, as it is enforced by the context.

Given the above elements it is clear that monitoring the physical actions and the rate with which the monitoring can be done also determines how well social signals can be perceived. This is directly linked with the following point; reactive behaviour is based on perceived social signals. Because the social interaction is by definition a joint interaction the partners should react as quick and accurate as possible on each others' parts of the interaction. This implies an active monitoring of the actions of the partners in order to check whether they conform to the expectations and if needed to perform a corrective action to give an immediate signal of possible misinterpretations. This monitoring and reaction to actions is in principle similar to what happens in any joint plan. However, in a joint plan you are not interested in every action of the other partners. You are usually only interested in the results of those actions in as far as your own actions or the joint goal depend on them. Usually this means that it suffices to monitor some key events or results. As stated before, in social interactions the social signals are carried in the actions themselves, thus a much closer level of monitoring is needed. Note that the level of monitoring depends on the level of expectation of the actions and the trust that they will be performed. E.g. when I greet someone in a well-known (common ground) situation with a handshake, I will not monitor the handshake very close (and might also miss some subtle signals). Because the close monitoring of every action of the partners in an interaction is (computationally) very expensive, we try to minimize the level of monitoring. This can be done by using more standard interactions and creating a known context. The more of these elements are known, the more accurate expectations we can form and the less points we have to monitor for deviations of these expectations.

Besides the physical context, the **social context** also determines which social signals can be conveyed and interpreted in an interaction. Firstly, we have to determine whether a ritual is used. Rituals are interaction patterns that can be assumed to be common knowledge within a group with the same culture. Often, rituals contain mainly actions that have little functional purpose and therefore are easily recognized and interpreted for their social effects. E.g. the handshake in a greeting in itself has no physical purpose except for synchronizing attention. Due to its limited functional purpose one can turn attention to the social interpretation of the handshake in the current context. Handshake rituals are commonly used to indicate group membership.

Of course, besides the ritual used the participants or stakeholders in the interaction and their social relations play a role in the social interpretation of the interaction. Do the partners know each other already? Do they have a peer-to-peer relation or is one dependent on the other? Do they have a formal relationship connected to a specific context (such as work, committee, sport, etc.) or are they general friends or family? Given such a starting point certain social signals of the interaction either confirm the relation, try to establish a relation or try to change (or end) it. Also, given certain roles of the stakeholders certain expectations of behaviour will arise within the interaction.

The next element that plays a role in the analysis of the social interaction is the common social ground of the partners. Do the partners have a shared history which determines whether the social signal they give confirms their relation or are they strangers and does the social signal establish such a relation? The historic relation is a part of the social common ground which is used to interpret the interaction. The second part follows from this history but adds the knowledge of the social departure point of the current social interaction. Do all partners belief that they are starting a certain ritual in order to establish a social relation or do they have different interpretations of their point of departure? E.g. one might expect a formal greeting at the start of a meeting, while the other only expects a head nod. The last part of this element is the social purpose of the interaction. This purpose might be perceived differently by all partners, but it does determine for each the way they interpret each social action and its purpose. E.g. a meeting can be seen by one participant as synchronizing and exchanging knowledge in order to coordinate further actions. However, another participant might see the meeting as a means to establish a power order between the participants that is used to make decisions later on.

A final aspect of the social context is the importance of success or failure of the social interaction. E.g. the consequence of a failed job interview is much larger than the consequence of a failed greeting of a person on the street who you wanted to ask for directions. When the consequences of failure are large the partners will perform a much closer monitoring function and also use many more aspects of the context in order to make optimal interpretations of the social signals and check whether they conform to the expectations of the other partners in order to create as much social common ground as possible. For less important interactions a crude interpretation of the social signal suffices.

The final component of the analysis consists of the **social action** itself. We consider three aspects that are impoortant for the interpretation: *expectation, trust and predictability.* If a social action is expected by all partners the interpretation of the action is easy and common to all partners. It follows a standard that apparently is part of the common ground (and which made all partners expect this action). When the partners trust each other they implicitly expect each other to conform to the expectations. Thus when the social interaction is trusted it will be less heavily monitored and standard interpretations will be used by default. Computationally this is very efficient. The predictability is also related to this aspect. If an action is completely predictable within a sequence of actions then there is no need to closely monitor it or do an extensive run time interpretation. The default interpretation can be used that is appropriate for the context and thus the social effects follow directly from the action. E.g. when a Dutch man greets another man in a business setting he will shake hands. This is completely predictable. However if the man greets a woman the action is less predictable. He might either kiss or shake hands. Thus this context requires closer monitoring and synchronization of actions.

4   Using the Analytical Framework

In this section, we provide a first, informal, description of an operational model for the social interaction framework. We don't take all aspects discussed in the previous sections into account but describe the most salient ones and provide some insights on how to extend and detail the framework. In figure 1, we describe how to refine the usual functional planning used by agents to account for social effects of actions. We assume that each possible functional goal corresponds to a specific method in the agent's plan. Each abstract method can be realized in possibly many different ways, each having a different social effect. Depending on the desired social effect, the agent will replace an abstract method in its functional plan by the concrete method most suitable for that social effect. We therefore assume that planning for social goals does not interfere with but refines planning for functional goals. This simplification suffices for many situations. Given this approach to planning, the main consequence of social reasoning is to determine which concrete method should be chosen given a social goal. In the following, we sketch a reasoning algorithm that can be implemented over an agent's reasoning engine. We assume that agent's belief base contains a social ontology describing the association between social and physical actions. A tuple *(pa, sa, v)* in this ontology, indicates the potential effect *v* for social action *sa* related to physical action *pa*. We can now introduce the relation *prefered social action,* ⪰, between action tuples. If the aim is to realize social action *sa*, then

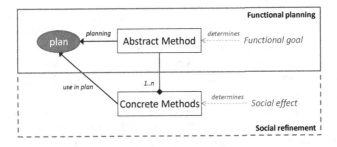

Fig. 1. Example of social refinement

of two actions $pa1$ is prefered over action $pa2$ when the potential value of $pa1$ is higher than that of $pa2$, and the amount of needed common ground to realise $pa1$ is less than that for $pa2$. Formally:

$$(pa_1, sa, v_1) \succeq (pa_2, sa, v_2) \text{ iff}$$
$$cg_a(c_i, pa_1) \subseteq cg_a(c_i, pa_2) \text{ and } v_1 \geq v_2$$

where $cg_a(c_i, pa)$ stands for the required common ground knowledge that is still lacking for agent a, in context c_i. Using this comparison, and assuming that

```
while p ∈ plan(a) do
    PossAlt =
        Select (pa, sa, v)
            From Soc_ontology(a)
                Where sa ∈ sg(a) ∧ pa ∈ concrete_methods(p);
        End
    (px, sx, vx) = max(PossAlt, ⪰);
    replace(p, px, plan(a));
end
```

Algorithm 1. Social refinement of agent plans

for each abstract physical method the agent may know several concrete methods with different possible social effects, Algorithm 1 describes the reasoning process of agent a to choose the most appropriate functional plan, given a social goal $sg(a)$. Informally, the reasoning is as follows. Given a component p of the agent's functional plan $plan(a)$, it first selects all possible tuples (pa,sa,v) from Soc-ontology(a) for which the social action, sa, contributes to the agent's social goal $sg(a)$, and for which the physical action pa is a concretization of p. From this list, the agent chooses the physical action px, with the highest *prefered social action*, $\succeq$, to replace p in $plan(a)$. The result is a social refinement of the original functional plan for the social goal of the agent.

Subsequently the perceiver is selects a social action that best corresponds to the physical action. However, it uses its own ontology, which might be different from that of the other agent. The social actions that are selected depend on the expected actions and also on how these social actions fit with the believed social goal of the other agent. Finally, the common ground of the agent is updated with the new perceived social fact. In the previous section we discussed the importance of the monitoring of the other agent during the social interaction. Three things are worth to mention here. The perceived physical action can be short and only a portion of a complete action when the monitoring is frequent. Secondly, the perception action itself $(perceive(pa))$ might also be observed by the other agent which has its own peception cycle. He might compare the perceived perception action $(perceive(perceive(pa)))$ with the expected perception of the action pa and interpret that again as having a specific social effect. Finally, an agent's past experiences with the other agents influence both the perception and the interpretation of the actions of these other agents.

while *true* **do**
  $perceive(pa)$;
  $PercSoc =$
    Select (pa, sa, v)
      From $Soc\_ontology(a)$
      Where $expected(sa) \wedge sa \in believed(sg(b))$;
    End
  $(px, sx, vx) = max(PercSoc, \succeq)$;
  $cg_a(c_{i+1}) = cg_a(c_i) \bigcup vx$;
end

Algorithm 2. Perceived effect by agent a of social action of b

Application Example . In order to show the applicability of the ideas proposed in this paper, we consider the (deceivingly) simple example of greeting another agent in the setting of an international conference. In this scenario, we assume agents to have different national backgrounds and be aware that others are possibly different. When greeting someone, people behave according to certain rituals. E.g. where it is usual in Latino countries for people of opposite sex to kiss each other (also in a first encounter), this would be unheard of in Asiatic cultures. Thus the default action taken when greeting is clearly determined by the cultural context. In this situation, we assume that the social refinement of plans with respect to greeting in professional contexts is depicted in Figure 2. We introduce two agents, *Alice* and *Bob*, each with a different social ontology, as in Figure 3. Given these ontologies, in an encounter between *Alice* and *Bob*, where both are assumed to have the same social goal, namely that of establishing a professional relation, within the overall functional plan of meeting participants in the conference, *Alice*'s reasoning would lead her to take the action 'Handshake' (cf. Algorithm 1) whereas *Bob* expects a 'Bow' (cf. Algorithm 2). When *Alice* starts with the handshake, *Bob* will monitor closely because he is uncertain in this environment whether his expectations are correct. Once he perceives the start of the handshake action he will again check his social ontology to find out whether there is a corresponding social action that still promotes the assumed social goal. In this case he finds that a handshake is also a formal greeting. At this point

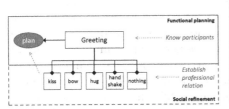

| Soc-ontology(Alice) | | |
pa	sa	v
Kiss	Formal-greeting	2
Handshake	Formal-greeting	4
Bow	Formal-greeting	3
Hug	Formal-greeting	1
Long handshake	friendly greeting	2

| Soc-ontology(Bob) | | |
pa	sa	v
Handshake	Formal-greeting	2
Bow	Formal-greeting	3
Do-nothing	Formal-greeting	1

Fig. 2. Social refinement model   **Fig. 3.** Social ontologies in greeting example

Bob recognizes that he and *Alice* are executing the ritual of a formal greeting and if *Bob* performs the next step in this ritual, which is extending his hand to shake A*lice*'s hand, he confirms to *Alice* his acknowledgement of taking part in the formal greeting ritual. The moment *Alice* perceives *Bob* extending his hand, which is the expected physical action within the ritual, she also acknowledges that she and *Bob* are jointly performing the formal greeting ritual.

As the ritual for both *Alice* and *Bob* will have a number of expected actions they will both monitor and expect these actions to take place in the sequel. However, also in these steps the exact expected actions can differ between *Alice* and *Bob*. E.g. *Bob* might expect the handshake to last 5 seconds, while *Alice* only expects it to last 2 seconds. The longer handshake of *Bob* can lead *Alice* to reevaluate and check her ontlogy and interpret now that *Bob* might try to give a friendly greeting. Another situation occurs when *Alice* would decide for a 'Kiss' as the refinement of Greeting, for instance because she sees it happening in the conference. In this case, *Bob* would not be able to interpret *Alice*'s action as a form of Formal greeting, and would need to reassess the situation as another type of social interaction.

5   Conclusions

In this paper we argued that it is essential to model social reality in order to create real intelligent interacting systems. It does not suffice to just add some protocols that implement social rituals as these might differ across cultures and persons in many subtle ways.

In Section 3 we have sketched aspects that are important in order to analyse and model social interactions. Because the social actions are performed through physical actions, both the physical as well as the social context have to be taken into account. Although our framework for the social meaning of physical actions can build upon the theories for joint plans and goals (in as far as the physical actions that execute the social interactions are joint plans) there are some additional considerations that influence the social interpretation of actions. First of all, the effects of social actions are purely mental. They change social reality, which exists based on (implicit) tacit agreements, but cannot be objectively checked. The effects are also achieved by interpretation of the receiver of the social signal. This interpretation again is subjective and internal and thus not observable by the performer of the social action. The interpretation of an action as a social signal depends on many factors from both the physical as well as the social context and upon the particular physical action that is used. Although we could not indicate all the formal connections in this paper we tried to give a flavour of what is involved to model this appropriately. In order to reduce the amount of aspects to be monitored rituals can be used. The main idea of a ritual is that the actions are easily recognizable as conveying a social signal and they can be assumed to be part of the common ground already, even if the specifics of the ritual might be unknown to some of the participants. Thus many aspects that influence the interpretation of the social actions in a ritual are standardized and can be assumed fixed.

Through the example in Section 4 we have shown that our framework can be used to design realistic social interactions, which is imperative for many applications. However, it also shows that adding social interactions to existing systems requires an extensive model of social reality and keeping track of social actions. In our future work we will expand upon this aspect. We are working towards a virtual reality application that helps people become aware of and know how to handle cultural differences in social interaction. For now the main recommendation for those developing agents for social interaction is to make a clear seperation in the actor between the physical actions and the social intent of that action, and within the other (observing) agent to make a clear separation between the objectively observable action and the subjective interpreation of an intended social effect and of the social effect on the observer.

References

1. Bentahar, J., Colombetti, M., Dignum, F., Fornara, N., Jones, A., Singh, M., Chopra, A., Artikis, A., Yolum, P.: Research directions in agent communication. ACM Transactions on Intelligent Systems and Technology 4(2), 1–27 (2013)
2. Argyle, M.: Social Interaction. Transaction Publishers Rutgers (2009)
3. Clark, H.: Using Language. Cambridge University Press (1996)
4. Condon, W.S., Ogston, W.D.: Sound film analysis of normal and pathological behavior patterns. J. Nervous Mental Disorders 143, 338–347 (1966)
5. Lesser, V., Decker, K.: Generalizing the partial global planning algorithm. International Journal of Intelligent Cooperative Information Systems 1(2), 319–346 (1992)
6. Durkheim, E.: The Rules of Sociological Method, 8th edn.
7. Grosz, B.J., Kraus, S.: Collaborative plans for complex group action. Artificial Intelligence 86(2), 269–357 (1996)
8. Jennings, N.R.: Controlling cooperative problem solving in industrial multi-agent systems using joint intentions. Artificial Intelligence 75(2), 195–240 (1995)
9. Kendon, A.: Movement coordination in social interaction: some examples described. Acta Psychologica 32, 101–125 (1970)
10. Decker, K., Wagner, T., Carver, N., Garvey, A., Horling, B., Neiman, D., Podorozhny, R., Nagendra Prasad, M., Raja, A., Vincent, R., Xuan, P., Lesser, V., Zhang, X.Q.: Evolution of the GPGP/TAEMS Domain-Independent Coordination Framework. International Journal of Autonomous Agents and Multi-Agent Systems 9(1), 87–143 (2004)
11. Cohen, P.R., Levesque, H.J., Nunes, J.H.T.: On acting together. In: Proceedings of the Eighth National Conference on Artificial Intelligence (AAAI 1990), pp. 94–99 (1990)
12. Pynadath, D., Marsella, S.: Psychsim: Modeling theory of mind with decision-theoretic agents. In: Proc. of the Internat. Joint Conference on Artificial Intelligence, pp. 1181–1186 (2005)
13. Rummel, R.J.: Understanding conflict and war, vol. 2. Sage Publications (1976)
14. Searle, J.R.: The Construction of Social Reality. Penguin (1996)
15. Thompson, L., Fine, G.A.: Socially shared cognition, affect and behavior: A review and integration. Personality and Social Psychology Review 3, 278–302 (1999)
16. Turner, J.H.: A theory of social interaction. Stanford University Press (1988)
17. Vilhjálmsson, H.H., Kopp, S., Marsella, S., Thórisson, K.R. (eds.): IVA 2011. LNCS (LNAI), vol. 6895. Springer, Heidelberg (2011)

Towards an Agent Based Modeling: The Prediction and Prevention of the Spread of the Drywood Termite *Cryptotermes brevis*

Orlando Guerreiro[1], Miguel Ferreira[2], José Cascalho[3], and Paulo Borges[1]

[1] Azorean Biodiversity Group (GBA, CITA-A) and Portuguese Platform for
Enhancing Ecological Research & Sustainability (PEERS),
Universidade dos Açores, Portugal
[2] Centro de Astrofísica, Universidade do Porto, 4150-762 Porto, Portugal
[3] Centro de Matemática Aplicada e Tecnologias de Informação (CMATI),
Universidade dos Açores, Portugal
orlandogue@gmail.com, {miguelf,jmc,pborges}@uac.pt

Abstract. We present initial efforts made to model the spread of the drywood termite in Angra do Heroísmo, Azores, using an agent based modeling approach. First we describe how a simple Cellular Automata (CA) model was created in Netlogo to simulate the spread of the species based on simple assumptions concerning the ecology of the species. A second step was taken by increasing the complexity of the initial CA approach, adding new specific characteristics to each cell, based again on ecology of the species and its behavior towards the environment. Finally, we add agents to the model in order to simulate the human intervention in fighting back the pest. This new model has become a two-level Agent-Based model. We also evaluated the costs of this intervention. These efforts were supported by field research which allowed a continuous cross-checking of the results obtained in the model with the field data.

Keywords: netlogo, pest control, agent based systems.

1 Introduction

Cryptotermes brevis is nowadays one of the worst pests in Azores being present in six of the nine islands of the archipelago and in its two major cities. This termite rapidly destroys the wood structures of houses with a huge economical and patrimonial costs. Here we describe the initial efforts made to model the spread of *Cryptotermes brevis* in Angra do Heroísmo, Azores, using agent based modeling approach with Netlogo [1]. The model was constructed in steps of increasing levels of complexity. The basic model was constructed within a Cellular Automata (CA) approach [2] [3] and then improved to simulate the efforts to control or eradicate the problem, through the addition of agents [4]. Agent-Based Modeling has been used to model pest control and management [5] [6] [7] [8] or to understand the behavior of endangered species [9]. Although we also have as

L. Correia, L.P. Reis, and J. Cascalho (Eds.): EPIA 2013, LNAI 8154, pp. 480–491, 2013.
© Springer-Verlag Berlin Heidelberg 2013

a *medium term goal*, the discussion of management issues, in this paper only an initial step is made towards that goal by adding the pest control agents.

The following main concerns guided the elaboration of the present work:

1. What can be the role of the simulation as a tool to predict the spread of this termite in a specific urban area?
2. How can a dynamic model using simulated agents that combat termite infestation help to identify strategies to reduce the presence of the termites, and to prevent them to spread into new areas?

The main contributions of this work are, first, to show how a simple model in Netlogo can successfully predict the spread of a termite taking into account a specific urban environment. Secondly, apply the model with pest control agents to understand which are the best strategies to control the spread of the pest, therefore opening the door to explore models of integrated management.

In the next section we briefly present how data about the termite required for the model was gathered. In section 3 we describe the basic model, present the results of the simulations and compare then with real data. We also study the effect of adding information about the buildings characteristics in the model. Finally, in section 4 we present preliminary results of simulations with pest control agents in the same environment and predict different scenarios for the next 40 years.

2   Collecting Data

The drywood termite individuals live their entire life in colonies inside wood structures in houses and furniture. However, as in other termite species, a swarming or short-term dispersal period occurs as part of its life-cycle.

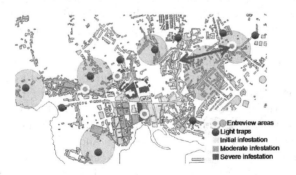

Fig. 1. Locations in Angra do Heroísmo where the data was collected. The red double arrow is the shortest distance from the main infested area to an isolated spot of infestation.

In this part of the life cycle, that usually occurs in the summer/warmer season, the young alates leave the parents nest and fly in search for a partner to start a new colony. It is in this period that new houses become infested. Once the termites form a new colony, it takes at least 5 years until this colony becomes a source of new infestations. The flight capacity of this termite was not known.

Different methodologies were applied during and after this swarming period in order to obtain the maximum possible information. The following sources of data were selected (Fig.1):

- Interviews to termite pest control agents;
- Interviews to inhabitants living in certain areas of the city, mostly in the border between infested and non-infested areas;
- Placement of UV Light traps to capture the young winged *C. brevis* individuals.

From these different sources it was possible to obtain: a) that this termite has been present in Angra do Heroísmo for at least 40 years; b) the alates have a flight capacity of the order of 100m or higher; c) a more accurate map of the areas infested and their degree of infestation and d) the most likely location of the first infested houses (see [10] for details).

Based on the termites biology described above, one infers that the pest can only advance at each 5 years interval. Assuming that the infestation started 40 years ago one obtains that the average distance *Cryptotermes brevis* alates had to fly on each swarming period was about 125 meters [1], in agreement with the 100m minimum distances obtained with the light traps experiments.

3   The Basic Model

3.1   Netlogo as a Tool to Predict the Spread of the Pest

Netlogo is a multistage tool, which has been used for a large number of experiments across several domains (see the link to the library of applied models in "http://ccl.northwestern.edu/netlogo/models/"). The entities in Netlogo are patches, turtles and the observer. The observer controls the experiment in which the other two entities participate. The patches are static identities in the environment, which corresponds to cells in an Cellular Automata environment or a kind of stationary "agents" [11] in a Agent-Based Modeling perspective. The environment is a grid of patches that perceive the place where they are and interact with other patches as well as the turtles. Turtles are agents that are able to move, to be created, to die and to reproduce and interact as well with the patches in environment and the other agents. Agents can also be organized in groups as different *breeds* that co-exist in the world. These *breeds* can be viewed as specialized agents that assume specific tasks as it is the case presented in this paper.

---

[1] This disregards human active contribution to its dispersion and assumes a single initial source of infestation.

The rational behind the selection of Netlogo to produce an initial study of the spread of this pest in Angra do Heroísmo was twofold:

- To supply additional information about the dimension of the infestation, based on some of the evidences collected during the field work and the knowledge about the ecology of the species;
- To give clues about main factors that influence the spread of the pest through all the city and to predict new places where the species could be already present.

3.2 The Basic Properties of the Model

In this problem we are interested in the houses and how the infestation propagates, and not in the termites *per se*. The city is divided into patches or cells. The patches either represent buildings or immune structures (e.g. streets, car parks, fields). We added the map of the city from a converted GIS map.

Table 1. A set of basic principles of the model

Cell states	There are four possible states that are represented in figure 2 and explained in the main text below.
Patch size	Each patch is 10x10m. A map of the city of Angra do Heroísmo was imported and all non black patches represent the buildings in the city.
Time step	The unit of time of the simulation is one year.
All buildings are born equal	In similar circumstances all cells representing buildings have equal probability of being infested.
Dispersion Radius	This is the flight distances of the alates. The neighbors of a cell are all cells within a dispersion radius.
Probability of infestation	This is the probability of a house becoming infested if it has one infested neighbor. All infested neighbors have the same probability of infesting it. This probability is a parameter of the model.

In Table 1 we describe the model properties. These are essentially some of the basic principles of the ODD protocol [4].

The probability of a cell becoming infested increases with the number of infested neighbors according to the rules of probability.

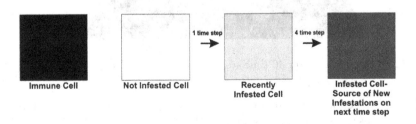

Fig. 2. The different states acessible to a cell and how they can evolve in time

The possible states of each cell are:

- *Imune*(IM). These represent everything that is not a building and is not susceptible to infestation.
- *Not infested* (NI). These correspond to houses not yet infested;
- *Recently infested* (RI). These are recently infested houses which are not a source of infestation yet. The cells only remain on this state for 4 time steps becoming source of infestation at the beginning of the 5th time step;
- *Infested* (IF). These are infested houses that are sources of infestation to other houses.

3.3   Results of the Basic Model

Here we present the results of a simulation with the following parameters and assumptions:

- Probability of infestation 0.25.
- Dispersion radius 10 (corresponding to 100 meters).
- Two places were selected as initial infestation sources (two initial red cells). These locations were selected according to the interviews made in the city center (cf. Sect. 2).
- The simulations were run for 40 timesteps, from the initial infestation up to the present.

In [10] we studied in detail the influence of different radius and probability in the results.

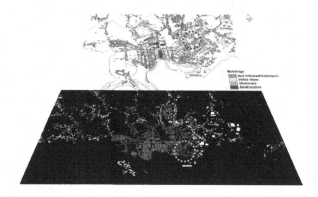

Fig. 3. Simulation after 40 time steps compared to the map of infestation [12]

For instance, we studied how the number of infested houses and recently infested houses vary in time as well as the degree of asymetry in the infested region that can occur due to the stochastic nature of the model. We found that the final results of the model are insensitive to the values of the probability unless it takes very small values ($p << 0.1$). This possibility was excluded based on what is known about the infestation process. For example, with $p = 0.1$ we found that the total number of infested houses had a coefficient of variation of $CV = 0.008$ for 20 simulations and was very similar to the deterministic case.

To better analyze the results we show a map where we overlap the infested areas of Angra do Heroísmo and the outcome of one of the simulations (Fig. 3).

The results of the simulations are in general agreement to what was known in 2009. The simulations predict that virtually no houses inside the infested region can remain uninfested, even when the probability of infestation is relatively low, and this is indeed what is observed in the field. The predicted region of infestation is similar to the known map of infestation, but a closer look reveals some differences. The simulation forecasts the spread of the pest into areas that are not present in that map such as the areas which are surrounded by green and blue dotted circles in the map. Also, the map of infestation is far from being symmetric with respect to the first infested houses. This significant asymmetric growth of the infestation is not obtained with this model. A possible explanation is that our assumption that all houses are "equal" could be strongly violated. In fact, a large number of houses in these areas are more recent and have less wood than the traditional houses in the city center. Another explanation presented in [10] was that the model is essentially correct and the infestation had already reached these locations but was undetected. These results and questions motivated us to pursue further investigations and led to changes in the basic model as explained in the following section.

3.4 Adding Complexity to the Model

Recently an updated and more accurate map of the infested areas and their degree of infestation was obtained [13][14] (Fig.4).

Fig. 4. New infestation map [13][14]

This map shows that one of the hypothesis raised was partially correct and the infestation had indeed reached some of the locations predicted by the model. Here we test whether taking into account that houses are not all equal leads to a better fit of the model with the data.

We consider three different types of buildings representing different wood structures of the buildings:

- A *full wood structure (FWS)*: These are the cases in which patches are associated to houses with a wooden roof or another wood structure. In these cases a patch jumps from RI stage to IF stage after fourth year of infestation as defined in the basic model.
- A *partial wood structure (PWS)*: In these cases patches jump from RI stage to IF stage less often than in PWS cases. The rationale behind this is that the number of colonies in PWR are necessarily smaller and so the spread of the pest is smaller too. The probability of a patch going from stage RI to IF is taken as 0.5.
- *No wood structure(NWS)*: In this case, we consider that these patches belong to houses in which wood exists in windows, doors, or furniture. The probability to jump to a IF stage is smaller and is taken as 0.25.

The buildings were taken to be in one of these three different categories according to where they were located. The buildings in the historical city centre were considered as FWS, those far from the centre and more recent were considered as NWS, and those at an intermediate distance from the centre as PWS. As it can be seen in figure 5 a kind of a 'corridor' for termites propagation is set by the distribution of the different kind of houses in the city. Although this distribution was based mainly on suppositions, it is known, by historical reasons, that the center of the city has more buildings with a wooden structure.

These new definitions, together with a radius of dispersion of 150 meters, changed the way the spread of infestation proceeded and the results of the simulations reproduce closely the recent data maps (Fig. 5).

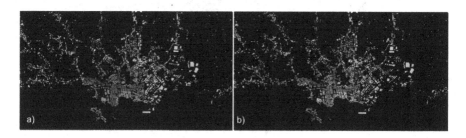

Fig. 5. Image *a)* shows the different building structures. The darker patches are the FWS cells, the lighter patches are the NWS cells and the intermediate shades are the PWS cells. Image *b)* shows the result of a simulation obtained after 41 time steps.

4 Adding Pest Control Agents to the Model

In recent years several companies have been actively fighting the pest, using different methods with different results in terms of efficacy.

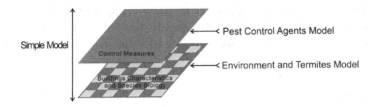

Simple Model

Pest Control Agents Model

Environment and Termites Model

Fig. 6. Adding pest-control agents to simulate the desired fight againts the pest

This leads us to redirect our research towards another question: "How can a dynamic model using simulated agents that combat termite infestation, help to identify strategies to reduce the presence of the termites and to prevent them to spread into new areas?" We now aim to simulate future scenarios in which efforts are made by citizens or public institutions to control the spread of the pest.

Table 2. The different pest control methods and the associated agents

Pest control method	Efficacy	Price (€/m^2)	Model action	Rational
Chemical treatment	kill 60 to 70% of termites	44	– The cell becomes *yellow*, in the first year of infestation (RI state); – No changes in building structure.	With this method the cell is still infested and in four years it can become a source of infestation again (IF state).
Heat treatment	kill 100% of termites	60	– The cell becomes white (NI state); – No changes in building wood structure.	This method will eliminate the infestation in the cell. The cycle of infestation can start again, depending on the number of infested neighbor cells
Replacement of wood by non-wood materials	The termites are permanently eliminated.	150	– The cell becomes *white* (NI state). – Changes in building wood structures to NWS.	This method eliminates the wood structure in the cell. The probability of being a source of infestation turns to zero. It is similar to a NWS building, but with a probability to jump to a IF stage equal to zero.

Adding pest control agents changed the model from a one layer to a two layers model, as shown in Figure 6.

Based on different on-field observations we extracted some data that guided us to define three different type of pest control agents, presented in the table 2. They are in Netlogo three different breeds, one for each type of agent. These pest control agents move all over the environment and act as described in the table. Our goal is threefold:

- To understand how many agents are needed to control the pest;
- To determine how important are the details that rule the agents action;
- To estimate, by large, the costs associated to that control through the following years.

4.1   Experiments

Table 3 shows the number of agents and options for the pest control agents in two different experiments.

Table 3. The two different experiments using pest control agents

Experiments	Number of agents	Scheduling
Random agents action	heat(100); chemicals(100); rebuilders(20)	The different breeds were created initially and spread randomly through the environment. They then move randomly to a yellow (RI) or red (IF) patch in a radius of 150m.
Coordinated agents action	heat(100); chemicals(100); rebuilders(20)	The different breeds were created initially and spread randomly through the environment. They then move randomly to a red patch (IF) in a radius of 150m and only if there is no red patch they search for a yellow patch (RI).

We consider image b) in the figure 5 as the starting point for these new experiments.

In the first experiment pest-control agents select one of the *red* or *yellow* patches randomly at predefined maximum distance, while in the second they select the *red* patches first.

The former was intended to model individual action made by citizens to apply treatment to their houses. While the latter was intended to simulate coordinated actions in which a priority was defined to treat first the houses source of infestation and only then the recently infested ones.

Figure 7 presents the results of the simulations starting at the present and ending in 40 years. What strikes most about the figure is the huge difference between images b) and c). If the agents are not coordinated the pest can hardly be controlled. Once they act in the slightest coordinated way, selecting the spots with the highest infestation, the results are impressively different. The pest is completely controlled.

Figure 8 clarifies the differences between the two experiments. We see clearly that the number of red cells decreases considerably in a linear way for the best scenario[2]. This reveals that it is possible to control the pest in just a few years.

---

[2] Note that in fig. 8 b) the increment of yellow cells is explained by the fact that, in the model, all cells can become infested although some of them will never be a source of infestation again.

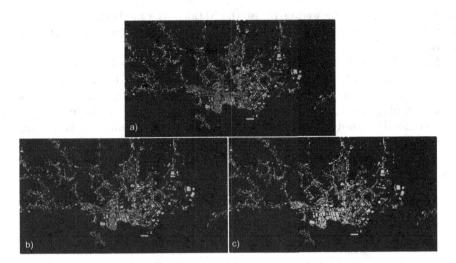

Fig. 7. Three different scenarios in the year 2053,: the image a) without pest control agents, the image b) the 'Random Agents Actions' experiment and finally the image c) the 'Coordinated Agents Action' experiment

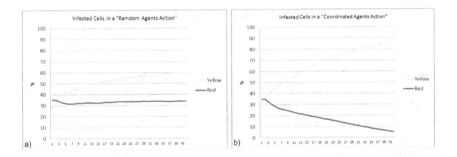

Fig. 8. The percentage of infested cells along the time steps for the two experiments: a) Random Agents Action b) Coordinated Agents Action

A rough estimate of the cost based on the number of 'moves' and their costs for each type of agents in the simulation scenario was made (see table 2). The values obtained are very high, about 15 millions euros in the first 15 years or 1 million euro/year.

5   Discussion and Conclusions

In this paper we describe a simple CA model of the spread of the infestation by *Cryptotermes brevis* in the urban environment of Angra do Heroísmo. The results of the model were validated with field data. This initial model was then improved by increasing its complexity with the introduction of simulated pest control agents. This allowed us to use the Netlogo as a predicting tool for future scenarios in which the pest was fighted back by three different pest control methods.

The preliminary results obtained suggest that coordination is important to provide a solution to the problem in a medium term. But it also demonstrates the huge investment that is needed. Most importantly it suggests that the way pest control agents act can result in a lot of "money down the drain". However, a more detailed study is required before definite conclusions can be obtained. For example, it will be necessary to explore several combinations of the different kind of agents and determine the optimal combination in terms of costs and results in controlling the problem.

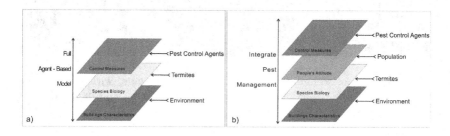

Fig. 9. Next steps to a full Agent-Based model and to study Integrated Urban Pest Management modeling

Changing the modeling method from a CA approach to a Agent-Based Modeling opens up the strides for future research steps. Figure 9 suggests that another two levels could be added to the model presented in this work. The first one is the level of the termites while the second one is the level of the citizens attitude towards the pest. In the former we search for a more complex model of the pest behavior including relevant details about the ecology of the species. In the latter we search for an integrated urban pest management model, for which there is some research in progress [15].

Acknowledgements. This study was partly supported by grant M221-I-002-2009 TERMODISP (DRCT, Azores, Portugal). O. Guerreiro was supported by a grant from Azorean Government (SFRCT - DRCT - M3.1.5/F/003/2010) and holds currently a Ph.D. Grant from Azorean Government (DRCT - M3.1.2/F/2011).

References

1. Willensky, U.: Netlogo: Center for connected learning and computer based modeling. Northwestern University (1999)
2. Langton, C.G.: Studying artificial life with cellular automata. Physica D: Nonlinear Phenomena 22(1-3), 120–149 (1986); Proceedings of the Fifth Annual International Conference

3. Bone, C., Dragicevic, S., Roberts, A.: A fuzzy-constrained cellular automata model of forest insect infestations. Ecological Modelling 192(1-2), 107–125 (2006)
4. Railsback, S., Grimm, V.: Agent-Based and Individual-Based Modeling: A Practical Introduction. Princeton University Press (2011)
5. Rebaudo, F., Dangles, O.: An agent-based modeling framework for integrated pest management dissemination programs. Environmental Modelling and Software 45, 141–149 (2012)
6. Rebaudo, F., Crespo-Pérez, V., Silvain, J.F., Dangles, O.: Agent-based modeling of human-induced spread of invasive species in agricultural landscapes: Insights from the potato moth in ecuador. Journal of Artificial Societies and Social Simulation 14(3) (2011)
7. Railsback, S.F., Johnson, M.D.: Pattern-oriented modeling of bird foraging and pest control in coffee farms. Ecological Modelling 222(18), 3305–3319 (2011)
8. Chadli, A., Tranvouez, E., Bendella, F.: Combining agent-based participatory simulation and technology enhanced learning for rodent control. In: Proceedings of the 2009 Summer Computer Simulation Conference, SCSC 2009, pp. 377–382. Society for Modeling & Simulation International, Vista (2009)
9. Falbo, K.R.: An individual based larval dispersion model for the hawaiian hawskbill sea turtle in the hawaiian archipelago. Master's thesis, Humboldt State University, Environmental Systems: Mathematical Modeling (2011)
10. Guerreiro, O.: Contribution to the management of the drywood termite cryptotermes brevis (walker, 1853) in the azorean archipelago. Master's thesis, University of Azores (2009)
11. Sakellariou, I., Kefalas, P., Stamatopoulou, I.: Enhancing netlogo to simulate bdi communicating agents. In: Darzentas, J., Vouros, G.A., Vosinakis, S., Arnellos, A. (eds.) SETN 2008. LNCS (LNAI), vol. 5138, pp. 263–275. Springer, Heidelberg (2008)
12. Borges, P.A., Lopes, D., Simoes, A., Rodrigues, A., Bettencourt, S., Myles, T.: Determinação da distribuição e abundancia de térmitas (isoptera) nas habitações do concelho de angra do heroísmo. Technical report, Universidade dos Açores (2004)
13. Borges, P.A., Guerreiro, O., Borges, A., Ferreira, F., Bicudo, N., Ferreira, M.T., Nunes, L., Sao Marcos, R., Arroz, A.M., Scheffrahn, R.H., Myles, T.G.: As térmitas no arquipélago dos açores: monitorização e controle dos voos de dispersão e prevenção da colonização nas principais localidades afectadas com ênfase na térmita de madeira seca Cryptotermes brevis (walker). Technical report, Universidade dos Açores (2011)
14. Borges, P.A., Guerreiro, O., Borges, A., Ferreira, F., Bicudo, N., Ferreira, M.T.: As térmitas no arquipélago dos açores: monitorização e controle dos voos de dispersão e prevenção da colonização nas principais localidades afectadas com ênfase na térmita de madeira seca Cryptotermes brevis (walker). Technical report, Universidade dos Açores (2012)
15. Arroz, A., Marcos, R., Neves, I., Guerreiro, O., Gabriel, R., Borges, P.A.: Relatório final da campanha: SOS térmitas - unidos na prevenção. Technical report, Universidade dos Açores (2012)

Automatic Extraction of Explicit and Implicit Keywords to Build Document Descriptors

João Ventura* and Joaquim Silva

CITI/DI/FCT, Universidade Nova de Lisboa,
Campus de Caparica, 2829-516 Caparica, Portugal
joao_ventura@netvisao.pt, jfs@di.fct.unl.pt

Abstract. Keywords are single and multiword terms that describe the semantic content of documents. They are useful in many applications, such as document searching and indexing, or to be read by humans. Keywords can be explicit, by occurring in documents, or implicit, since, although not explicitly written in documents, they are semantically related to their contents. This paper presents a statistical approach to build document descriptors with *explicit* and *implicit* keywords automatically extracted from the documents. Our approach is language-independent and we show comparative results for three different European languages.

1 Introduction

Keywords are semantically relevant terms that are used to reflect the core content of documents. Some of the first works related to the automatic extraction of keywords were addressed in [1], [2] and [3]. However, in many applications, as in library collections, the extraction of keywords remains mainly a manual process.

In this paper we propose a statistical and language-independent method to build document descriptors. Each *global* document descriptor is made of two distinct descriptors: an *explicit descriptor* containing explicit keywords, and an *implicit descriptor* with implicit keywords. Both types of keywords are automatically extracted from the documents.

Our approach starts by extracting the concepts from each document using *ConceptExtractor*. Then, it applies *Tf−Idf* to the extracted concepts and ranks them by their relevance to the content of the document. The first ranked concepts are selected as explicit keywords and will form the explicit descriptor.

However, there are other meaningful concepts that, although not occurring in a document, are semantically related to its content. We call these the implicit keywords. They may, for instance, provide a user of a search engine the access to documents that may not contain these keywords, but are semantically related to them. Concepts such as "car emissions", "Toxicology" and "acid rains" may be useful if automatically added as implicit keywords of a document about "air pollution", in case those terms do not occur explicitly in that document.

---

* Supported by FCT-MCTES PhD grant SFRH/BD/61543/2009.

L. Correia, L.P. Reis, and J. Cascalho (Eds.): EPIA 2013, LNAI 8154, pp. 492–503, 2013.
© Springer-Verlag Berlin Heidelberg 2013

To extract the implicit keywords of a document, the Semantic Proximity is calculated between each concept extracted from the corpus and each keyword of the document's explicit descriptor. The first ranked concepts, according to a metric, are selected as the document's implicit keywords and form the document implicit descriptor.

In the next section we analyze the related work. A brief explanation of ConceptExtractor is presented in Section 3 where we also propose a modification. In Sections 4 and 5 we present our methodology for building explicit and implicit descriptors, and the conclusions are in Section 6.

2 Related Work

There are different methodologies for the extraction of keywords. Linguistic approaches, such as [4], [5], [6], [7], [8], [9] tend to use one or more language-specific tools, as Part-Of-Speech (POS) taggers, syntactic patterns, stemmers, stop word lists or ontologies (Wordnet). Despite the good results of some of these approaches, their generalization for other languages is not guaranteed, since those linguistic tools may not be available for other languages.

Regarding statistical approaches, for instance in [10] the LocalMaxs algorithm is used to extract relevant multiword expressions (MWEs). The top ranked MWEs are selected as keywords and the best results were obtained using a metric based on the median length of words. However, LocalMaxs does not extract unigrams, which impoverishes results for languages such as German and Dutch where many concepts tend to be agglutinated in a single word. Also, in [11], we find a comparison of metrics for the extraction of keywords. For multiwords, it also uses LocalMaxs, which performs less than ConceptExtractor [12]. For single words, they apply a 6 character filter to improve results which may leave out important concepts (such as "dad" or "car") or even acronyms (e.g. "RAM" or "CEO") that may be relevant in some documents.

Hybrid approaches, such as [8], use *Tf–Idf* and Support Vector Machines, in addition to POS filters. Other approaches, such as [13], require either domain specific knowledge or training samples, which imply human interaction and need to be adapted to the specific application domain.

Regarding structure-based approaches, for instance in [14] an approach for Web document summarization is addressed which considers the textual content of all documents explicitly linked to it. The efficiency of this approach depends on the existence of *html* links to the target document. In [15] the authors propose a keyword extraction method based on the exploration of some properties on Wikipedia articles such as *inlink, outlink, category* and *infobox*. This type of approaches, including [16] and [17], are dependent on the existence of links on the Web regarding the topics and the language of the target documents.

The alternative approach we present needs no structured data, since it works with raw text. Also, by not using language-specific tools, we believe our method can be generalized for other languages than the ones tested in this paper.

3   Extraction of Concepts

As we show in Section 4, the keywords of documents are essentially the most meaningful concepts in them. By extracting concepts from documents we are in fact reducing the search space from all possible sequences of words/multiwords in a document to a much smaller set of terms that are semantically meaningful.

For the task of the extraction of concepts we use ConceptExtractor [12]. Other approaches, such as [18], [19] and [20], were also considered but they are limited for our needs. The works in [18] and [19] only extract multiwords, and [20] reports low precision values. ConceptExtractor is able to extract single and multiword concepts, and presents Precision/Recall values above 0.85. For paper self-containment, we briefly present this extractor here.

3.1   ConceptExtractor

The ConceptExtractor is a statistical and language independent method for the extraction of concepts from corpora. It is based on the following assumptions: (a) concepts are words with semantic meaning (e.g. *government* vs. *the*); (b) concepts can join with other concepts to form compound and more specific concepts (e.g. (*President, Republic*) form *President of the Republic*); (c) compound concepts begin and end with concepts (e.g. *President of the Republic* vs. *President of the*); (d) strong compound concepts tend to have fixed distances between the single word concepts that form them; (e) concepts can be more or less specific (e.g. *bike* vs. *mountain bike*).

For this extractor, it is very important to measure the tendency of any pair of words to occur at fixed distances. Thus, for each single word w in a corpus, a list of all neighbor words $B = [b_1, b_2, .., b_m]$ is obtained. Each neighbor b_i may occur at different positions relative to w, where w is at the center of a window of size s. For each pair (w, b_i), $X_{(w,b_i)}$ is a list of co-occurrence frequencies:

$$X_{(w,b_i)} = [x_{-\frac{s}{2}}, \ldots, x_{-1}, x_1, \ldots, x_{\frac{s}{2}}] , \qquad (1)$$

where x_j is the co-occurrence frequency of word b_i at position j relative to the center word w. Pairs (w, b_i) that show preference to occur at fixed positions present higher $Rel\_var(.)$ values than pairs that usually occur scattered:

$$Rel\_var(X_{(w,b_i)}) = \frac{1}{s(s-1)} \sum_{j=1}^{s} \left(\frac{x_j - \bar{x}}{\bar{x}} \right)^2 \qquad \bar{x} = \frac{1}{s} \sum_{j=1}^{s} x_j . \qquad (2)$$

Being $B = [b_1, \ldots, b_m]$ the list of all m neighbors of a word w, the specificity of a single word w is measured by:

$$Spec(w) = Rel\_var([Rel\_var(X_{(w,b_1)}), \ldots, Rel\_var(X_{(w,b_m)})]) , \qquad (3)$$

where $X_{(w,b_i)}$ is the list of the co-occurrence frequencies of b_i near w. The underlying idea with $Spec(w)$ is that a single word w is more specific if it is strongly associated with just some neighbors and weakly with all others.

For a multiword W consisting in a sequence of words $(w_1, \ldots, w_n)$, its specificity is measured by (4).

$$SpecM(W) = \frac{1}{\binom{n}{2}} \sum_{\substack{i,j \in \{1\ldots n\} \\ \wedge\, i<j}} uq(w_i, w_j) \cdot pq(w_i, w_j) \,. \tag{4}$$

$$uq(w_i, w_j) = \sqrt{Spec(w_i) \cdot Spec(w_j)} \qquad pq(w_i, w_j) = \frac{x_{j-i}}{\sum_{k \in Pos} x_k} \,.$$

Basically, the specificity of a multiword W is given by the average influence of each possible single word pair in W in terms of two factors: (a) the average specificity of their isolated single words, given by $uq(w_i, w_j)$ (*unigram quality*), and (b) the tendency for the pair to co-occur at fixed positions, i.e., to form a compound concept, which is given by $pq(w_i, w_j)$ (*pair quality*). x_{j-i} is the number of co-occurrences of w_j at position $j-i$ relative to w_i; $Pos = \{-\frac{s}{2}, \ldots, -1, 1, \ldots, \frac{s}{2}\}$ is the set of all relative positions in the window of size s.

The more specific a term is, the more likely it represents a concept. So, by applying a fixed threshold for specificities, concepts can be extracted (further details in [12]).

3.2 Improving the Extraction of Multiword Concepts

On [12], the Precision and Recall values for multiwords are well below the results for single word concepts. We think it is due to the fact that the rule *Compound concepts start and end with concepts* is not actually being applied. In fact, $SpecM(.)$ only measures the average specificity of the pairs (w_i, w_j) that form a multiword. Being W a multiword $(w_1, w_2, \ldots, w_n)$, we propose the following alternative to $SpecM(.)$ for measuring the specificity of multiwords:

$$SpecM_2(W) = SpecM(W) \cdot min(Spec(w_1), Spec(w_n)) \,. \tag{5}$$

Basically, the $SpecM(.)$ value of W is multiplied by the minimum $Spec(.)$ value of the words that start/end the multiword. This has the effect of harming the multiwords that start or end with function or meaningless words.

To test the results of $SpecM_2(.)$, we built three corpora of random Wikipedia documents for three different languages (English, Portuguese and German). Table 1 summarizes the information about the corpora.

Table 1. Basic statistics about the Wikipedia-based corpora used in the experiments

Corpus	EN	PT	DE
Number of documents	2 714	1 811	4 682
Total words	12 176 000	11 974 000	11 305 000
Words by document (avg)	4 486	6 611	2 414

300 multiwords were automatically and randomly extracted from each corpus and manually classified as being or not concepts. Figure 1(a) uses the English test-set to show that Precision and Recall invert their tendency for different $SpecM_2(.)$ thresholds. There is a F-value peak for thresholds near 3×10^{-7}.

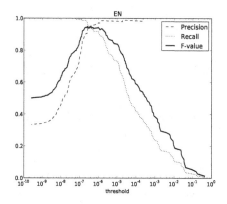 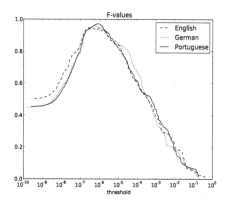

Fig. 1. (a) Precision, Recall and F-measure values for different $SpecM_2(.)$ thresholds – English test-set (log. X-axis); (b) F-values for different $SpecM_2(.)$ thresholds for the three different languages test-sets (log. X-axis)

Figure 1(b) shows the F-values by $SpecM_2(.)$ threshold for each language test-set. The behavior is similar for each language. Table 2 compare the results between the proposed $SpecM_2(.)$ and the original $SpecM(.)$, for the thresholds that maximize the F-values of each language test-set.

Table 2. $SpecM$ vs $SpecM_2$ comparative results for multiword concepts

	EN	PT	DE	EN	PT	DE
	$SpecM$	$SpecM$	$SpecM$	$SpecM_2$	$SpecM_2$	$SpecM_2$
Precisiom	0.87	0.82	0.87	0.93	0.95	0.91
Recall	0.89	0.82	0.82	0.95	0.97	0.97
Threshold	5.3×10^{-4}	7.4×10^{-4}	11.0×10^{-4}	2.7×10^{-7}	4.3×10^{-7}	7.6×10^{-7}

From $SpecM(.)$ to $SpecM_2(.)$ there is an average increase of 7.6% for Precision and of 12% for Recall values. The improvement of the results comes from the fact that non-concepts that have strongly connected single-words (e.g. "President of the", "by the", etc.) are now harmed by starting/ending with function words. The difference between concepts and non-concepts are more clearly reflected by $SpecM_2(.)$ than by $SpecM(.)$.

Given the $SpecM_2(.)$ thresholds for the three languages, we propose a generic threshold of 4.9×10^{-7} (an average from those values) as the new threshold for multiword concepts, when applying $SpecM_2(.)$ to extract multiword concepts.

4   The Explicit Descriptor

The explicit descriptor is a set of keywords which occur explicitly in documents. For our purposes, the explicit descriptor of a document is formed by 20 keywords: the 10 best scored single words and the 10 best scored multiwords. To extract the keywords from a document, we compute the *Tf-Idf* to its extracted concepts.

Tf-Idf (Term frequency−Inverse document frequency) is a statistical metric often used in IR to evaluate how important a term W (word or multiword) is for a document d_j in a corpus $\mathcal{D}$. It is defined as:

$$Tf\text{-}Idf(W, d_j) = \frac{f(W, d_j)}{size(d_j)} \cdot \log \frac{\|\mathcal{D}\|}{\|\{d : W \in d\}\|} . \tag{6}$$

$\|\mathcal{D}\|$ means the number of documents of corpus $\mathcal{D}$; $\|\{d : W \in d\}\|$ is the number of documents containing term W, and $size(d_j)$ the number of words of d_j. To prevent bias towards longer documents, we use the probability $(f(W, d_j)/size(d_j))$ of term W in document d_j instead of the absolute frequency $(f(W, d_j))$.

To evaluate the results of *Tf-Idf* applied to the extracted concepts, we used the same corpora as in Table 1. We built the explicit descriptors of 10 documents randomly taken from each corpus and then, three independent reviewers, who had full access to the documents, evaluated the quality of each descriptor. Reviewers were instructed to consider as keywords the concepts which describe, or are strongly related, with one or more sections of a document, including the whole document. Table 3 shows the titles of the documents for each language and the reviewers' agreement rate, which was obtained by measuring the number of "correct" keywords in which all three reviewers agreed on.

Table 3. Reviewers' agreement on keyword classification

EN		PT		DE	
Doc. Title	Agr.	Doc. Title	Agr.	Doc. Title	Agr.
Abortion	.95	Descobrimentos	.85	Adolph Hitler	.85
Brain	.90	Al-Andalus	.95	Genetik	.70
Nostradamus	.75	Direitos animais	.85	Demokratie	.75
Dog	.95	ADN	.90	G. Rossini	.75
Saint Peter	.75	Hist. Espanha	.90	Immunsystem	.85
Imagism	.90	Gato	.90	Kairo	.75
Monopoly	.90	W. A. Mozart	.90	Microsoft	.80
Desert	.90	Teosofia	.70	Papageien	.50
Plate Tectonics	1	Vasco da Gama	.85	Pflicht	.90
History	.75	Nazismo	.85	Wolga	.75

The average agreement was about 0.83, although it was lower for German because two reviewers had to rely on automatic translators for this language.

Table 4 shows the explicit descriptor for the English *Brain* document. Although some terms may not be accepted as correct keywords ("phenomena are identical", "central nervous"), most describe the core content of the document.

Table 4. Explicit descriptor – *Brain* document

	Ranked list (#1, #2, .. #10)
Single words	brain, neurons, disease, animals, nervous, cells, brains, intelligence, body, vertebrates.
Multiwords	spinal cord, cerebral cortex, artificial intelligence, optical lobes, olfactory bulb, central nervous, brain stems, Parkinson's disease, simple reflexes, phenomena are identical.

Table 5 shows the evaluation of the approach. Precision gives the average rate of correct keywords in each descriptor; Recall measures the rate of concepts that did not need to be exchanged by "better" keywords outside the descriptor. Since the reviewers' agreement was not 100%, Precision and Recall values were measured by assuming the majority of choices among the three reviewers.

Table 5. Precision and Recall for explicit keywords

Corpus	Single words		Multiwords	
	Precision	Recall	Precision	Recall
English	0.89	0.80	0.87	0.79
Portuguese	0.88	0.86	0.91	0.83
German	0.89	0.89	0.85	0.80

The results are similar for the three languages, despite some slight differences, and show that the combination of *Tf–Idf* with ConceptExtractor is quite able to extract keywords to build explicit descriptors.

We have also compared the usage of *Tf-Idf* with concepts (referred as *Explicit*) with other extraction methods. Table 6 compares it with *Tf–Idf* without the usage of concepts, while Table 7 compares it with *LeastCv, LeastRvar* and *Mk[2.5]*, as described in [10] and [11].

Table 6. Comparison of methods for explicit document descriptors – single words

Language	*Explicit*		*Tf-Idf*	
	Precision	Recall	Precision	Recall
English	**0.89**	**0.80**	0.87	0.79
Portuguese	**0.88**	**0.86**	0.86	0.86
German	**0.89**	**0.89**	0.87	0.88

Our *Explicit* method scores higher than the others. Although *Tf-Idf* (without concepts) shows results similar to our method for single words, it scores poorly for multiwords. This happens because *Tf-Idf* tends to assign high values to rare sequences, such as "do ADN" and "ADN é" ("of DNA" and "DNA is", respectively) which in this case occurs only in the Portuguese *ADN* document.

Table 7. Comparison of methods for explicit document descriptors – multiwords

Language	Explicit		Tf-Idf		LeastCv		LeastRvar		Mk[2.5]	
	P.	R.	P.	R.	P.	R.	P.	R.	P.	R.
English	**0.87**	**0.79**	0.50	0.35	0.62	0.61	0.65	0.63	0.76	0.72
Portuguese	**0.91**	**0.83**	0.52	0.38	0.61	0.60	0.64	0.61	0.75	0.72
German	**0.85**	**0.80**	0.49	0.37	0.62	0.60	0.64	0.63	0.76	0.72

5   The Implicit Descriptor

The implicit descriptor of a document is a set of keywords that do not occur explicitly in a document but whose meanings are semantically related with the document's content. For instance, a document may focus on topics such as "air pollution", "carbon monoxide" and "ground level ozone", but concepts such as "lung cancer" or "water cycle", even not occurring explicitly in the document, could enrich the *global* document descriptor, since they are semantically related with its contents. A richer descriptor provides an extended semantic scope that can be useful in IR and Web Search applications.

Basically, the implicit keywords of a document are concepts from other documents of a corpus which have strong *Semantic Proximity* values with most of the keywords of the document's explicit descriptor.

5.1   Semantic Proximity

We define the semantic proximity between two terms as a combination of two factors: how correlated the terms are in all documents (*Inter-document Proximity*) and how the terms relate inside a document (*Intra-document Proximity*).

The idea behind the Inter-document Proximity is that, if two terms A and B tend to occur in the same documents, probably they are semantically close and their correlation is strong. We use $Corr(A, B)$ to measure the correlation between terms A and B:

$$Corr(A, B) = \frac{Cov(A, B)}{\sqrt{Cov(A, A)} . \sqrt{Cov(B, B)}} . \tag{7}$$

$$Cov(A, B) = \frac{1}{\|\mathcal{D}\| - 1} \sum_{d_i \in \mathcal{D}} d(A, d_i) . d(B, d_i) . \tag{8}$$

$$d(A, d_i) = p(A, d_i) - p(A, .) \qquad d(B, d_i) = p(B, d_i) - p(B, .)$$

$$p(A, d_i) = \frac{f(A, d_i)}{size(d_i)} \qquad p(A, .) = \frac{1}{\|\mathcal{D}\|} \sum_{d_i \in \mathcal{D}} p(A, d_i)$$

$Corr(A, B)$ is based on Pearson's correlation coefficient. $\|\mathcal{D}\|$ is the number of documents of corpus $\mathcal{D}$; d_i is the i-th document in $\mathcal{D}$; $size(d_i)$ is its number of words and $f(A, d_i)$ the frequency of term A in d_i. $Corr(A, B)$ ranges from

Table 8. Correlation values for some pairs of terms in the English corpus

Term A	suanpan	supply	Microsoft	airport	electricity
Term B	Chinese abacus	demand	Windows	automobile	Mozart
$Corr(A,B)$	1.000	0.810	0.791	0.011	0.006

-1 to $+1$: it gets negative results when A tends to occur and B not, zero when the correlation is weak, and close to $+1$ when the correlation is strong. Table 8 shows some $Corr(A,B)$ values for pairs of terms from the English corpus.

Although the correlation assesses how strong some pairs are related through-out a set of documents, it is not sensitive to their positions inside the documents. For instance, the correlation between "suanpan" and "Chinese abacus" (as in Table 8) is the same as the correlation between "suanpan" and "Babylonian abacus", since all terms occur only in the English *Abacus* document. However, since "suanpan" is a Chinese abacus, it should be, say, *more strongly related* with "Chinese abacus" than with "Babylonian abacus". This leads us to the Intra-document Proximity.

The idea with the Intra-document Proximity is that two terms are more strongly related if they tend to occur near each other in a document. In fact, in the *Abacus* document, "Chinese abacus" and "suanpan" occur in the same section while "Babylonian abacus" occurs two sections before. We define $IP(A, B)$ (Intra-document Proximity) between terms A and B as:

$$IP(A, B) = 1 - \frac{1}{\|\mathcal{D}^*\|} \sum_{d \in \mathcal{D}^*} \frac{dist(A, B, d)}{farthest(A, B, d)} . \tag{9}$$

$$dist(A, B, d) = \sum_{o_i \in Occ(A,d)} nearest(o_i, B, d) + \sum_{o_k \in Occ(B,d)} nearest(o_k, A, d) . \tag{10}$$

$\mathcal{D}^*$ is the set of documents containing terms A and B, $Occ(A, d)$ stands for the set of all occurrences of A in document d, and $nearest(o_i, B, d)$ gives the distance from occurrence o_i to the nearest occurrence of B in d, in number of words; distances are positive. $dist(A, B, d)$ represents a global distance between A and B, considering all occurrences of both terms in d. This distance is normal-ized by the maximum global distance between A and B considering all possible distributions of occurrences in d, which is given by $farthest(A, B, d)$ in (9). This *extreme case* happens when all occurrences of one term are located at the be-ginning of d and the occurrences of the other, at the end. $farthest(A, B, d)$ is given by:

$$farthest(A, B, d) = C_1 - C_2 + C_3 - C_4 . \tag{11}$$

$$C_1 = f(A, d) . (size(d) - f(B, d)) \qquad C_2 = \frac{(f(A, d) - 1)^2 + f(A, d) - 1}{2}$$

$$C_3 = f(B, d) . (size(d) - f(A, d)) \qquad C_4 = \frac{(f(B, d) - 1)^2 + f(B, d) - 1}{2} , \tag{12}$$

where $f(A, d)$ and $f(B, d)$ are the number of occurrences of A and B in d and $size(d)$ is the number of words of d. $IP(.,.)$ is sensitive to document distances where $IP(\text{suanpan}, \text{Chinese abacus}) = 0.97$ and $IP(\text{suanpan}, \text{Babylonian abacus}) = 0.67$. Due to lack of space, equations C_1, C_2, C_3 and C_4 are not proved.

Finally, we define the Semantic Proximity between two terms A and B as the multiplication of $Corr(A, B)$ by $IP(A, B)$. However, our tests confirmed that the $Corr(A, B)$ factor should have more weight in the Semantic Proximity calculation, hence the square root on $IP(A, B)$:

$$SemProx(A, B) = Corr(A, B) \cdot \sqrt{IP(A, B)} \ . \tag{13}$$

5.2   Ranking Implicit Concepts

For a document d, let k_i be the i-th ranked keyword of its explicit descriptor. If C is a concept not occurring in d, we propose to measure C as implicit keyword of d by the following metric:

$$score(C, d) = \sum_{i=0}^{n} \frac{SemProx(C, k_i)}{i} \ , \tag{14}$$

where n is the size of the explicit descriptor of d, which we set to 20 as referred. In the calculation of $score(C, d)$, $SemProx(C, k_i)$ is weighted by the inverse of the k_i ranking, which is $1/i$. Good implicit keywords should be more semantically related to the top explicit keywords, rather than the bottom ones.

5.3   Experimental Conditions and Results

For the evaluation, we used the same corpora and documents as mentioned in tables 1 and 3. For each document d, and for each extracted concept C not occurring in d, $score(C, d)$ was calculated. The first 20 concepts ranked by $score(., d)$ formed the implicit descriptor of d. Tables 9 and 10 show the top ranked concepts of the implicit descriptors for an English and a German document.

Table 9. Implicit descriptor (the top part) for the *Brain* English document

$score(.,.)$	Keyword	$score(.,.)$	Keyword	$score(.,.)$	Keyword
1.465	peripheral nervous sys.	0.666	Purkinje	0.663	granule cells
1.277	transverse nerves	0.664	Purkinje cells	0.661	cerebellar nuclei
1.276	CNS	0.663	cerebellar cortex	0.650	cerebellar

For each implicit descriptor we have evaluated Precision results. The criterion followed by the reviewers was that an implicit keyword should be accepted as correct only if they recognized that, although not occurring in the document, the keyword was semantically close to its contents. Recall was not evaluated since it would be impractical to find concepts in the about 2000 other documents of the same corpus that could be considered better than some of the implicit keywords. Table 11 shows the measured Precision results.

Table 10. Implicit descriptor (the top part) for the *Immunsystem* German document

score(.,.)	Keyword	score(.,.)	Keyword
0.530	Komponenten des Immunsystems	0.410	Lymphozyten
0.509	Reaktion des Immunsystems	0.395	eukaryotischen Zellen
0.431	Eukaryoten	0.358	Zellteilung
0.348	Emil Adolf von Behring	0.327	Hormone

Table 11. Precision values for the implicit descriptors

	English	Portuguese	German
Precision	0.84	0.87	0.83

Although slightly lower than those obtained for the explicit descriptors, we believe that these are good results, considering that implicit descriptors are able to extend the semantic scope of the documents.

6   Conclusions and Future Work

In this paper we have presented a language-independent method for the automatic building of document descriptors formed by explicit and implicit keywords.

We start by identifying concepts on the documents which we then use as explicit keywords. We have shown that, for this task, *Tf-Idf* returns better results when using concepts. We have also proposed metrics to identify semantic relations between terms in order to measure the relevance of a concept as implicit keyword of a document. Implicit keywords offers an extended semantic scope to the global descriptors of documents, with great applicability. Ultimately, the combination of 20 explicit keywords with 20 implicit keywords allows for a greater *semantic completeness* of the global descriptors.

Our methodology is independent of any language-specific tools, as we tried to show by obtaining similar results for the different languages. In the future our work will focus mainly on the improvement of the results of implicit keywords, which, as far as we know, is not being currently addressed.

Acknowledgements. Prof. Dr. Maria Francisca Xavier of the Linguistics Department of FCSH/UNL, Ms. Carmen Matos and Eng. Amarílis Jones are kindly acknowledged for providing their expertise in the manual evaluation processes.

References

1. Luhn, H.P.: The automatic creation of literature abstracts. IBM Journal of Research and Development 2, 159–168 (1958)
2. Jones, K.S.: A statistical interpretation of term specificity and its application in retrieval. Journal of Documentation 28, 11–21 (1972)

3. Salton, G., Yang, C.: On the specification of term value in automatic indexing. Journal of Documentation 29(4), 351–372 (1973)
4. Cigarrán, J.M., Peñas, A., Gonzalo, J., Verdejo, M.F.: Automatic selection of noun phrases as document descriptors in an fca-based information retrieval system. In: Ganter, B., Godin, R. (eds.) ICFCA 2005. LNCS (LNAI), vol. 3403, pp. 49–63. Springer, Heidelberg (2005)
5. Hulth, A.: Enhancing linguistically oriented automatic keyword extraction. In: Proceedings of Human Language Technology - North American Association for Computational Linguistics, pp. 17–20 (2004)
6. Alani, H., Sanghee, K., Millard, D.E., Weal, M.J., Lewis, P.H., Hall, W., Shadbolt, N.: Automatic extraction of knowledge from web documents. In: Proceedings of Workshop of Human Language Technology for the Semantic Web and Web Services, 2nd International Semantic Web Conference (2003)
7. Ercan, G., Cicekli, I.: Using lexical chains for keyword extraction. Information Processing and Management: An International Journal Archive 6, 1705–1714 (2007)
8. Zhang, K., Xu, H., Tang, J., Li, J.: Keyword extraction using support vector machine. In: Yu, J.X., Kitsuregawa, M., Leong, H.-V. (eds.) WAIM 2006. LNCS, vol. 4016, pp. 85–96. Springer, Heidelberg (2006)
9. Mihalcea, R., Tarau, P.: Textrank: Bringing order into texts. In: Proceedings of the Conference on Empirical Methods in Natural Language Processing (EMNLP 2004), vol. 2 (2004)
10. Silva, J.F., Lopes, G.P.: Towards automatic building of document keywords. In: COLING 2010 The 23rd International Conference on Computational Linguistics, pp. 1149–1157 (2010)
11. Teixeira, L.F., Lopes, G.P., Ribeiro, R.A.: An extensive comparison of metrics for automatic extraction of key terms. In: Proceedings of 4th International Conference on Agents and Artificial Intelligence, pp. 55–63 (2012)
12. Ventura, J., Silva, J.F.: Mining concepts from texts. In: International Conference on Computer Science (2012)
13. Suzuki, Y., Fukumoto, F., Sekiguchi, Y.: Keyword extraction of radio news using term weighting with an encyclopedia and newspaper articles. In: SIGIR (1998)
14. Delort, J.Y., Bouchon-Meunier, B., Rifqi, M.: Enhanced web document summarization using hyperlinks. In: Proceedings of the Fourteenth Association for Computing Machinery Conference on Hypertext and Hypermedia (2003)
15. Xu, S., Yang, S., Lau, F.C.: Keyword extraction and headline generation using novel word features. In: Proceedings of the Twenty-Fourth AAAI Conference on Artificial Intelligence (AAAI 2010) (2010)
16. Mihalcea, R., Csomai, A.: Wikify!: linking documents to encyclopedic knowledge. In: CIKM 2007: Proceedings of the 16th ACM Conference on Information and Knowledge Management, vol. 2 (2010)
17. Milne, D., Witten, I.H.: An effective, low-cost measure of semantic relatedness obtained from wikipedia links. In: Proceedings of Wikipedia and AI Workshop at the AAAI 2008 Conference (WikiAI 2008) (2008)
18. Silva, J.F., Lopes, G.P.: A local maxima method and a fair dispersion normalization for extracting multiword units. In: Proceedings of the 6th Meeting on the Mathematics of Language, pp. 369–381 (1999)
19. Frantzi, K., Ananiadou, S.: Extracting nested collocations. In: The 16th International Conference on Computational Linguistics (COLING 1996), pp. 41–46 (1996)
20. Yoshida, M., Nakagawa, H.: Automatic term extraction based on perplexity of compound words. In: Dale, R., Wong, K.-F., Su, J., Kwong, O.Y. (eds.) IJCNLP 2005. LNCS (LNAI), vol. 3651, pp. 269–279. Springer, Heidelberg (2005)

Compact and Fast Indexes
for Translation Related Tasks

Jorge Costa[1], Luís Gomes[1], Gabriel P. Lopes[1],
Luís M.S. Russo[2], and Nieves R. Brisaboa[3]

[1] Faculdade de Ciências e Tecnologia - Universidade Nova de Lisboa
Caparica, Portugal
jorge.costa@campus.fct.unl.pt, luismsgomes@gmail.com, gpl@fct.unl.pt
[2] Instituto Superior Técnico – Universidade Técnica de Lisboa
Lisboa, Portugal
luis.russo@ist.utl.pt
[3] Database Lab, University of A Coruña
A Coruña, Spain
brisaboa@udc.es

Abstract. Translation tasks, including bilingual concordancing, demand an efficient space/time trade-off, which is not always easy to get due to the usage of huge text collections and the space consuming nature of time efficient text indexes. We propose a compact representation for monotonically aligned parallel texts, based on known compressed text indexes for representing the texts and additional uncompressed structures for the alignment. The proposed framework is able to index a collection of texts in main memory, occupying less space than the text size and with efficient query response time. The proposal supports any type of alignment granularity, a novelty in concordancing applications, allowing a flexible environment for linguistics working in all phases of a translation process. We present two alternatives for self-indexing the texts, and two alternatives for supporting the alignment, comparing the alternatives in terms of space/time performance.

Keywords: Text compression, Machine Translation, bilingual concordancer, parallel text alignment, alignment granularity.

1 Introduction

The process of translating adequately a text or a sentence from one language to another is a difficult task for humans, namely selecting the best translation for each word or multi-word expression from several possible translation candidates. A bilingual concordancer is an important tool for human translators, helping them in the tasks of translation, post-editing machine-made translations [13] and other related tasks, as it allows the user to look at the context of a translation candidate in the parallel corpus from where it was extracted [1].

A bilingual concordancer allows querying for a pattern or a pair of patterns (translation candidates), with one or more tokens, over the parallel corpora that

L. Correia, L.P. Reis, and J. Cascalho (Eds.): EPIA 2013, LNAI 8154, pp. 504–515, 2013.
© Springer-Verlag Berlin Heidelberg 2013

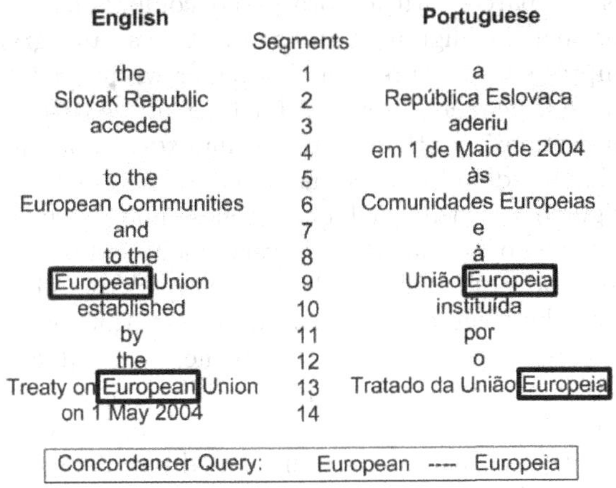

English		Portuguese
	Segments	
the	1	a
Slovak Republic	2	República Eslovaca
acceded	3	aderiu
	4	em 1 de Maio de 2004
to the	5	às
European Communities	6	Comunidades Europeias
and	7	e
to the	8	à
European Union	9	União Europeia
established	10	instituída
by	11	por
the	12	o
Treaty on European Union	13	Tratado da União Europeia
on 1 May 2004	14	

Concordancer Query: European ---- Europeia

Fig. 1. Sub-sentence alignment fragment of text in English and Portuguese, with a simplified example of a concordancer result

have been aligned beforehand. Following the aligned text fragment in Figure 1, a search for the pair "European" – "Europeia" would report two co-occurrences on the 9th and 13th aligned segments of both languages. A co-occurrence is a pair of occurrences that occur in aligned (or nearby) segments. On the other hand, searching for "on 1 May 2004" – "em 1 de Maio de 2004" might not show the 4th and 14th segments as a co-occurrence given their distance of 10 segments apart (which is typically a large distance).

Concordancers like TransSearch [15] and Sketch Engine [11] use a fixed-grain alignment (either word or sentence), which can be a limitation for users. In our study, we use a monotonic sub-sentence alignment with variable coarseness, ranging from near word-level [8] as in Figure 1, in cases of closely related languages such as Spanish and Portuguese, to coarse alignments when considering hard-to-align languages, resulting in segments consisting on sentences or paragraphs. We typically refer these two levels of granularity as fine and coarse respectively.

Additionally, dealing with large text collections is a challenge, as texts may have sizes in the order of the Gigabyte per language. Typically, time efficient text indexes, like suffix arrays [16] and suffix trees [2], lead to high space consumptions [5], while efficient space performances can lead to considerably slower solutions. For an application such as concordancer, which demands efficient querying over parallel corpora, this space/time trade-off is a major concern.

In this paper we propose a compressed bilingual concordancer representation, usable for any language pairs and alignment granularity, providing answers in real time and occupying less space in memory than the text sizes themselves.

The text representation is based on compressed indexes for representing the aligned parallel texts, namely a byte-oriented wavelet tree [6]. Compressed indexes achieve remarkable compression ratios, in order of about 25% to 40% of the text size [6], with logarithmic search times as a typical indexing technique.

This approach is compared with a word-based compressed suffix array [5] in Section 5. To represent the alignment and support a variable granularity, we use additional uncompressed structures. In this paper we present two approaches, bitmaps and arrays of integers, prepared for finer and coarser alignments, and compare them in Section 5. Both compressed indexes and additional structures are explained in further detail in Sections 3 and 4.

Supporting alignments with variable coarseness makes our concordancer usable in a broader range of scenarios, from very finely and precisely aligned corpora to situations where (for any number of reasons) the alignments are very coarse. Additionally, by using this compact representation, we are able to index huge amounts of text collections in the main memory, reducing the access to secondary memory, hence obtaining efficient query response times. These factors allied to a generic API, make the proposed application easy to use and to integrate with other related tasks, based on aligned parallel corpora or bilingual lexica. Such an application is also useful for validators of freshly extracted translation pairs[1] [3], as they can use the concordancer for any translation candidate being evaluated and change the alignment granularity on demand.

2   Bilingual Concordancer

We consider any word, punctuation mark or separator as a token, as all are considered words in the text indexes.

Parallel text alignment is the process of identification of the matching segments between texts in two different languages. Figure 1 has an example snapshot of a finely aligned parallel text, in English and Portuguese, where the aligned segments appear at the same line, including void (or null) segments, as is the case of the 4th segment in English and the 14th in Portuguese. A coarser segment typically contains several fine segments [8] or a whole sentence.

By using aligned parallel text as context, a concordancer query response shows the segments where the pattern occurs and the respective aligned segments in the other language. When the query is bilingual the application returns the co-occurrences of the pair, together with the surrounding context. Two patterns co-occur if $|off_1 - off_2| \leq MaxDist$, where off_1 and off_2 are the offsets of the language 1 and 2 occurrences of the patterns and $MaxDist$ is the maximum distance allowed between co-occurring expressions, in terms of segments. This distance can be changed on demand, allowing more flexibility when using corpora with different alignment coarseness. If two patterns co-occur often, then the probability of being a correct translation candidate is higher.

In Figure 2 we show a typical bilingual concordancer result, in this particular case the two co-occurrences of the pair "European – Europeia", alongside with the context surrounding each co-occurrence. With $MaxDist = 2$, the pair "European – Europeia" co-occur two times, as shown in Figure 2. The surrounding context is controlled by a window. In Figure 2, the window is 3, so the

---

[1] According to our experience, validation of automatically extracted translations enables more precise alignments later.

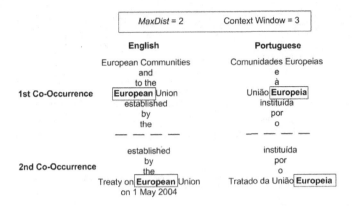

Fig. 2. Example of concordancer result

three segments before and after the occurrence are shown, without exceeding the coarse (sentence) segment limits. The second co-occurrence context exceeds the sentence limit, thus only one more segment is shown after the occurrence.

Our bilingual concordancer uses two layers: the compression layer based on the compressed indexes and the alignment layer supported by bitmaps or arrays.

3 Compression Layer

The compression layer of the proposed concordancer framework uses a compressed self-index to represent the texts in the main memory. The proposals are Byte-Oriented Wavelet Trees (BOC-WT) and Word-Based Compressed Suffix Arrays (WCSA). With this layer, the application obtains the offsets of the occurrences of a pattern in the respective text, in terms of tokens.

3.1 Byte-Oriented Wavelet Tree

Brisaboa et. al. [6] proposed a reordering of the bytes in the codewords of the compressed text following a wavelet tree [10] like strategy, instead of having a simple sequence of concatenated codewords, where each codeword represents a token. This enables self-synchronization independently of the encoding used and allows logarithmic search times over the text, as a typical indexing technique.

The Byte-Oriented Wavelet Tree is a multi-ary wavelet tree in which the bytes of the codewords are placed in different nodes (see Figure 3). The root contains the first byte of all codewords, following the text order, thus having as many bytes as words in the text. The root has as many children as the number of different first bytes of the codewords. Each of these nodes in the second level has the second byte of the codewords represented, following the order from the upper level. The same process is done for every following level of the tree. The length of the codewords depends on the frequency of the word, with the most frequent words having smaller length codewords.

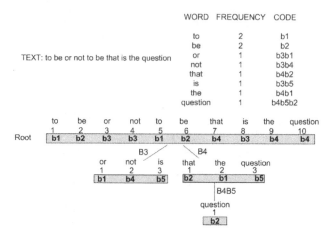

Fig. 3. Example of a Byte-Oriented Wavelet Tree

Searching for a token or multi-token pattern is heavily based on level-wise *rank* and *select* queries. With S as a sequence of symbols, $rank_b(S, i) = y$ if the symbol b appears y times in $S_{1,i}$ and $select_b(S, j) = x$ if the jth occurrence of the symbol b in S appears at position x. *Rank* and *select* queries are the core of the main functionalities of the index, which are: *count* the number of occurrences of a pattern, *locate* a pattern and *extract* a pattern in a given position of the text. *Rank* is used for locating a i-th token of the text, while *select* is used for finding the i-th occurrence of a pattern, as explained in more detail in [6].

As both *rank* and *select* are executed locally in each level of the tree [6], a search query is more efficient against a search regarding the whole text, because it depends on the height of the tree and not on the size of the text.

3.2   Word-Based Compressed Suffix Array

A different alternative for compressing the text and reorganizing the bytes is to use a form of compressed index. A Word Based Compressed Suffix Array [5] is an adaptation of a typical Compressed Suffix Array (CSA), like Sadakane's CSA implementation [18], to a word-based alphabet, which is much larger than a character-based one.

A WCSA consists of two layers (see Figure 4). One is VA, an array with the tokens of the source text following a lexicographic order, mapped to a correspondent unique integer *id*. The other is an integer-based CSA (iCSA) for indexing the mentioned *ids*. The main idea is to replace the source text for the respective integer sequence *Sid*, with the concatenated identifiers of the tokens, following the order of the text. This sequence is indexed on iCSA and discarded after.

The iCSA layer provides the typical self-indexing operations like *count*, *locate* and *extract*, but for integer patterns. Thus, when querying for a string pattern, it is necessary to find the respective *ids* of each token of the pattern in VA

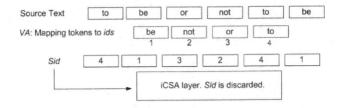

Fig. 4. General Structure of a Word-Based Compressed Suffix Array

[5]. Then, with all the integers obtained from VA concatenated, we can use the mentioned functionalities of the iCSA layer, in order to find the location of the patterns or counting how many times they occur in the text.

4 Alignment Layer

The alignment layer uses the offsets obtained from the compression layer, in order to find out on which aligned segments these offsets occur, independently of the granularity. This is important to determine the co-occurrences as detailed in Section 2, as $MaxDist$ is based on the number of segments.

4.1 Arrays of Integers and Bisection

Each position of the arrays has the zero-based offsets of the start of a segment. As we can have different levels of coarseness, we need five arrays per language pair: two for coarse segments (CA), two for fine segments (FA) and one for relating the coarse and fine segments (RA), as shown in Figure 5.

In each position of a CA, we store the absolute offset[2] of the beginning of a segment. In Figure 5, for language 1 the coarse segments start at offsets 0, 10, 19 and 27. In a FA, the offsets in each position are relative to the coarse segment and not absolute, which leads to using only 2 bytes per each fine segment offset, instead of the 4 bytes necessary for representing the absolute offsets. In Figure 5, $FA_1[0] = FA_1[4] = FA_1[7] = FA_1[10] = 0$, indicate the start of a new fine segment, but as it coincides with the start of a coarse segment, the offset is 0.

We need an auxiliary array to relate the absolute offsets in each CA with the offsets stored in the respective FA. Thus, RA stores the positions in both FA where the coarse segments begin. In Figure 5, $RA[1] = 4$, thus the second coarse segment starts at position 4 on FA_1 and FA_2. Only one RA is needed as two aligned coarse segments always have the same number of fine segments.

For searching for the alignment segment of a particular offset, we apply the bisection algorithm over the CA of the respective language. Considering Figure 5, $bisect(CA_1, 10) = 1$, thus the offset 10 occurs in the second coarse segment. For finding the fine segment, the algorithm continues with $RA[1] = 4$ and $RA[1+1] =$

---

[2] An absolute offset considers the whole text.

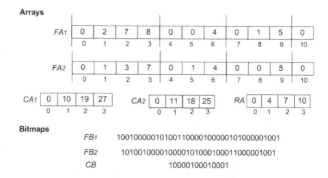

Fig. 5. Example of a representation of aligned segments, with arrays and bitmaps

7. Then, a bisection is performed over FA_1 in the interval $[4, 7]$, to find for the offset $off - off_c = 10 - 10 = 0$, where off is the absolute offset and off_c is the offset of the start of coarse segment (the second in this particular example). $FA_1[5] = 0$, thus the offset 10 is at the 6th fine segment ($FA_1[4] = 0$ as well, but it is a void segment as $FA_1[5] - FA_1[4] = 0$).

4.2 Bitmaps and *Rank* and *Select*

With bitmaps we use a codification where each 1 represents the start of a new segment and each 0 a token of the segment. Figure 5 shows the bitmap layer for the same example presented for arrays. FB_1 and FB_2 are the bitmaps for the segments of FA_1 and FA_2 respectively. The first fine segment has two tokens, so the bitmap starts with a 1 marking the start of the segment and two 0s, one for each token. The next segment has five tokens, introducing a 1 and five 0s and the same occurs for the remaining segments. The void segments also start with a 1, but do not have any 0.

For representing the alignment we need three bitmaps: two for fine segments and one for coarse segments. In the bitmap for coarse segments each 0 represents a fine segment. We need only one bitmap to represent the coarse segments, as segments with such granularity have the same number of fine segments, independently of the language. Consider the coarse bitmap in Figure 5. Between the start of the first coarse segment, at offset 0, and the start of the next one at offset 10, there are 4 fine segments, thus we add a 1 for marking the start of the coarse and four 0s. The same idea prevails for the remaining bits.

For finding out in which segment a pattern occurs, we use an efficient implementation of *rank* and *select* over the bitmaps [9]. We start at the fine segments and do a $rank_1(FB_1, select_0(FB_1, off))$, where off represents the offset. *Select* operation indicates in which position of the bitmap is the segment, while *rank* indicates which segment we are looking for. Picking the example on Figure 5, consider searching for the offset 10: $select_0(FB_1, 10) = 16$, $rank_1(FB_1, 16) = 6$, thus the offset occurs at the 6th fine segment. For finding the coarse segment, the idea is similar but using the fine segment found. With the fine segment 6 of the example, $select_0(CB, 6) = 7$, $rank_1(CB, 7) = 2$. Thus, the offset 10 occurs at the 6th fine segment and 2nd coarse segment.

5   Experimental Results

For the experimental results, we tested the bilingual concordancer with all possible index combinations: $WT + bsect$, $WT + bmap$, $CSA + bsect$ and $CSA + bmap$, with $bsect$ indicating the approach based on arrays and $bmap$ the approach using bitmaps. Figure 6 shows the results for memory consumption in Megabytes and time performance in seconds, for the total values, and microseconds for the average time consumption per occurrence. We used BOC-WT with the codification Plain Huffman [6] and WCSA with Delta Codes [5] codification.

We considered four parallel corpora of different sizes: European Commission's Directorate-General for Translation (DGT)[3] with 804 Mb, Europarl [12] with 394 Mb, European Medicines Agency corpora (EMEA) [19] with 194 Mb and the European Constitution corpora (Euconst) [19] with 1.6Mb. For the query time evaluation we used 699,500 bilingual *locate* queries. The results were obtained using a computer with 8 GB of memory and i7-2670QM 2.20GHz processor, using a 32-bit Ubuntu operative system.

Memory Consumption

Corpora	Alignment Layer		Text Layer		Total / Compression Percentage							
	bmap	bsect	WT	CSA	WT + bmap		WT + bsect		CSA + bmap		CSA + bsect	
Euconst	0.25	0.42	0.66	0.53	0.91	57%	1.1	68%	0.78	49%	0.95	60%
EMEA	12	64	63	33	75	39%	127	65%	45	23%	97	50%
Europarl	22	104	130	110	152	39%	234	59%	132	34%	214	54%
DGT	49	236	267	173	316	39%	503	62%	222	28%	409	51%

Query Time Consumption

Corpora	Number of Occurrences	WT Average Time per Occurrence		WT Total Time		CSA Total Time	
		bsect	bmap	bsect	bmap	bsect	bmap
Euconst	2,213,760	15 μs	18 μs	34 s	41 s	1017 s	1022 s
EMEA	166,794,611	5 μs	9 μs	860 s	1619 s		
Europarl	468,721,150	8 μs	13 μs	3717 s	6203 s		
DGT	938,024,690	7 μs	11 μs	6662 s	9762 s		

Fig. 6. Spatial and temporal performance of the indexes

5.1   Memory Consumption

Analyzing the results from Figure 6, displayed in Megabytes, we can see that any solution occupies less space than the text itself, which is of extreme importance, as texts can have sizes in the order of Gigabyte. This happens mainly because of

---

[3] DGT available here: http://ipsc.jrc.ec.europa.eu/index.php?id=197

the compressed indexes, but show as well that both alternatives for the alignment layer maintain acceptable sizes for the application's space demands, even without using compressed bitmaps or arrays (which is a possibility for future tests).

The memory results show as well that the smaller text has the higher compression percentage, due to the less amount of repetition present in the text. The other three corpora used show very similar spatial performance, with percentages that vary at most 5%. However, even the largest percentage is lower than 70% of the text size.

The most space consuming approach is the $WT + bsect$. This happens for two reasons. The wavelet trees are more space consuming than WCSA, mainly because of the information that needs to be kept in each level and node. Additionally, the bitmaps occupy less space than the arrays of integers, because every bit on the bitmaps occupies less space than each integer in the array (4 bytes), even considering that the bitmaps need more elements (0s and 1s) than the arrays to represent the alignment. The arrays need one integer per segment, while the bitmaps need 1 bit $+ n$ bits, with n as the segment size in tokens.

5.2   Query Time

Figure 6 shows that BOC-WT is much faster than WCSA. We only experimented WCSA for the smaller corpora (Euconst) because it showed a considerably worst temporal performance when compared with BOC-WT. With these results, even considering that WCSA has a better space consumption, the difference does not compensate the considerable difference in time performance.

The results for the alignment layer, in Figure 6, show that the approach based on arrays is much faster than the one supported by bitmaps. This happens for two reasons: the usage of bisection over the arrays, which have fewer elements than the bitmaps, and hence it needs less iterations for finding the segments. As an example, for finding the fine segment where an offset occurs, only a single bisection is needed over the arrays for finding that offset. While using bitmaps, we need to do a *select* and a *rank* for each offset. Additionally, the coarse segments in the bitmaps depend on the fine segments, making it slower to find the coarse segments, as we need to query over the fine ones first, which are in much more quantity than the coarse segments.

The time increases significantly when the text size increases, namely because the number of occurrences increases as well. 699,500 queries in a file with 1.6 Mb have much less occurrences than in an 800 Mb text, so as more occurrences are reported, more time is spent by the application. Another interesting fact is that the average time spent per occurrence are quite similar between texts. The deviation between the reported values occurs because a larger text has more repetition, thus more occurrences of a single pattern, making it faster to find them. For instance, "the" appears more than 2 million times in DGT and it is faster to find these 2 million occurrences, than the same number of occurrences divided over 699,500 different queries.

Considering the gains in terms of memory, especially when compared to an approach based on non-compressed indexes, such as suffix arrays which take 4 times the text size, the query performance shows positive results, leading to the desired efficient space/time tradeoff. The efficient implementations of the bisection algorithm and of *rank* and *select*, together with the logarithmic complexity of the compressed index queries, are the main contributors for these results.

6   Related Work

TransSearch [15] and Sketch Engine [11] are well known and studied concordancer systems, with important improvements over the years, like for instance the translation spotting in TransSearch [4]. Translation spotting is typically interesting for one sided queries, using only one language, as it indicates the respective translation for the pattern in the query. In our approach, translation spotting is done in a different manner. For every word or multi-word shown in the context of the occurrences or co-occurrences returned by the concordancer, if it has a known translation, the spotting is done automatically just by selecting the word or multi-word in the text. However, there are more relevant differences between these approaches and our proposal.

First, we wanted a space/time efficient concordancer application, easy to merge with other Machine Translation tasks. As an example, when a user is evaluating a translation pair candidate, just by pressing a button, the results of the concordancer request appear. If the text source of the translation candidate is in the main memory, the result only takes the query time, which is low as shown in Section 6. Thus with efficient space consumption, more texts can be indexed at the same time in memory, avoiding rebuilding the index any time a new text is used and speeding up the application. Second, our approach can be used for several alignment granularity, with the possibility of changing the granularity on demand, without modifying the application.

Callison-Burch et. al. [7] used suffix arrays to store and search efficiently for phrase pairs, enabling the finding of the best matches for a pattern. This method was evaluated for 120 queries and 50,000 sentence pairs, while we tested 699,500 queries for different sizes of corpora. Additionally, suffix arrays occupy 4 times the text size, which is too space demanding. Lopez [14] used suffix arrays as well, for developing lookup algorithms over hierarchical phrase-based translation with good computational time results, but inefficient in terms of space.

Lucene[4] is a Java-based information retrieval software used for full text indexing, just like the proposed compressed indexes, with efficient searching capability and with space consumption around 20% to 30% of text size. This system is commonly used in search engines, but it could be adapted to support our needs, by having an empty list of stopwords (we want to look for a whole phrase and not keywords as in standard search engines). However, Lucene uses compressed inverted indexes, which were proved to be better than suffix arrays for a considerable amount of occurrences, namely 5000 [17]. For concordancing purposes, the vast majority of queries consider terms or phrases that occur only once or twice in the corpora, which are around 75% of the terms in the texts.

---

[4] http://lucene.apache.org/

7   Conclusions

In this paper we presented a compact framework for supporting a bilingual concordancer application, based on compressed indexes and additional structures. This framework has four main objectives: index a huge text collection in the main memory (in the order of Gigabytes), maintain an immediate query response for the users, support various levels of alignment granularity and provide a generic interface for supporting other Machine Translation tasks based on parallel corpora or similar textual information.

These goals were achieved by using a compressed byte-oriented wavelet tree for indexing the texts and two alternatives based on uncompressed bitmaps, or arrays of integers, for supporting the text alignment. All these solutions led to a space consumption lower than the text size itself, with the most space consuming approach taking less than 70% of the text size, using the smallest corpora, while it maintains an efficient query response (in microseconds). In terms of space, the combination wavelet tree + bitmap is the most compressed one, while the fastest is the wavelet trees + arrays combination. The wavelet tree was compared against an alternative based on word-based compressed suffix arrays, however the slowdown demonstrated by the latter in the experimental results comparatively, does not compensate the gains verified in terms of memory.

This application was designed for supporting a bilingual concordancer application, however due to the space and time results and the support of variable grain alignment, we can use this framework for supporting other Machine Translation tasks, supporting bilingual lexica or determination of language models, which involve textual information similar to parallel corpora. Additionally, we want to extend our approach to support alignments with gaps. One example of such phrases is "shall V_1 and V_2", where V_1 and V_2 are two verbal forms, representing phrases like "shall analyze and report", "shall look and verify", etc, whose translations will not be word by word translations. These patterns can have one or more variables (typically no more than five) and are extremely important for finding new and correct translations. With more known translations, the translation tasks based on bilingual lexicons can obtain better results.

Acknowledgments. The authors would like to acknowledge Susana Ladra for providing the byte-oriented wavelet tree implementation and FCT/MCTES for supporting this work with the ISTRION project (ref. PTDC/EIA-EIA/114521/2009) and with personal scholarships for Jorge Costa (ref. SFRH/BD/78390/2011) and Luis Gomes (ref. SFRH/BD/64371/2009).

References

1. Aires, J., Lopes, G., Gomes, L.: Phrase translation extraction from aligned parallel corpora using suffix arrays and related structures. In: Lopes, L.S., Lau, N., Mariano, P., Rocha, L.M. (eds.) EPIA 2009. LNCS, vol. 5816, pp. 587–597. Springer, Heidelberg (2009)

2. Apostolico, A.: The myriad virtues of subword trees. In: Combinatorial Algorithms on Words. NATO ISO Series, pp. 85–96. Springer (1985)
3. Bilbao, V.D., Lopes, J.P., Ildefonso, T.: Measuring the impact of cognates in parallel text alignment. In: Portuguese Conference on Artificial Intelligence, EPIA 2005, pp. 338–343. IEEE (2005)
4. Bourdaillet, J., Huet, S., Langlais, P., Lapalme, G.: Transsearch: from a bilingual concordancer to a translation finder. Machine Translation 24(3), 241–271 (2010)
5. Brisaboa, N.R., Fariña, A., Navarro, G., Places, Á.S., Rodríguez, E.: Self-indexing natural language. In: Amir, A., Turpin, A., Moffat, A. (eds.) SPIRE 2008. LNCS, vol. 5280, pp. 121–132. Springer, Heidelberg (2008)
6. Brisaboa, N., Fariña, A., Ladra, S., Navarro, G.: Reorganizing compressed text. In: Proceedings of the 31st Annual International ACM SIGIR Conference on Research and Development in Information Retrieval, pp. 139–146. ACM (2008)
7. Callison-Burch, C., Bannard, C.: A compact data structure for searchable translation memories. In: EAMT 2005 (2005)
8. Gomes, L., Aires, J., Lopes, G.: Parallel texts alignment. In: Proceedings of the New Trends in Artificial Intelligence, 14th Portuguese Conference in Artificial Intelligence, EPIA 2009, Aveiro, pp. 513–524. Universidade de Aveiro (October 2009)
9. González, R., Grabowski, S., Mäkinen, V., Navarro, G.: Practical implementation of rank and select queries. In: Poster Proc. Volume of 4th Workshop on Efficient and Experimental Algorithms, pp. 27–38. CTI Press and Ellinika Grammata (2005)
10. Grossi, R., Gupta, A., Vitter, J.: High-order entropy-compressed text indexes. In: Proceedings of the Fourteenth Annual ACM-SIAM Symposium on Discrete Algorithms, pp. 841–850. Society for Industrial and Applied Mathematics (2003)
11. Kilgarriff, A., Rychly, P., Smrz, P., Tugwell, D.: Itri-04-08 the sketch engine. Information Technology 105, 116 (2004)
12. Koehn, P.: Europarl: A parallel corpus for statistical machine translation. In: MT summit. 5 (2005)
13. Kremer, G., Hartung, M., Padó, S., Riezler, S.: Statistical machine translation support improves human adjective translation. Translation: Computation, Corpora, Cognition 2(1) (2012)
14. Lopez, A.: Hierarchical phrase-based translation with suffix arrays. In: Proc. of EMNLP-CoNLL, pp. 976–985 (2007)
15. Macklovitch, E., Simard, M., Langlais, P.: Transsearch: A free translation memory on the world wide web. In: Second International Conference On Language Resources and Evaluation, LREC, vol. 3, pp. 1201–1208 (2000)
16. Manber, U., Myers, G.: Suffix arrays: a new method for on-line string searches. SIAM Journal on Computing 22(5), 935–948 (1993)
17. Puglisi, S.J., Smyth, W.F., Turpin, A.: Inverted files versus suffix arrays for locating patterns in primary memory. In: Crestani, F., Ferragina, P., Sanderson, M. (eds.) SPIRE 2006. LNCS, vol. 4209, pp. 122–133. Springer, Heidelberg (2006)
18. Sadakane, K.: New text indexing functionalities of the compressed suffix arrays. Journal of Algorithms 48(2), 294–313 (2003)
19. Tiedemann, J.: News from opus-a collection of multilingual parallel corpora with tools and interfaces. Recent Advances in Natural Language Processing 5, 237–248 (2009)

Using Clusters of Concepts to Extract Semantic Relations from Standalone Documents

João Ventura* and Joaquim Silva

CITI/DI/FCT, Universidade Nova de Lisboa,
Campus de Caparica, 2829-516 Caparica, Portugal
joao_ventura@netvisao.pt, jfs@di.fct.unl.pt

Abstract. The extraction of semantic relations from texts is currently gaining increasing interest. However, a large number of current methods are language and domain dependent, and the statistical and language-independent methods tend to work only with large amounts of text. This leaves out the extraction of semantic relations from standalone documents, such as single documents of unique subjects, reports from very specific domains, or small books.

We propose a statistical method to extract semantic relations using clusters of concepts. Clusters are areas in the documents where concepts occur more frequently. When clusters of different concepts occur in the same areas, they may represent highly related concepts.

Our method is language independent and we show comparative results for three different European languages.

Keywords: Concepts, clusters, semantic relations, statistics.

1   Introduction

Natural language texts are undoubtedly one of the most common and reliable sources of knowledge ever invented. Even with the widespread use of computer networks such as the Internet, large quantities of unstructured texts are still being created, containing knowledge waiting to be made available by computers. Semantic relations between concepts are examples of such knowledge and they are used with several degrees of success in various NLP applications, such as word sense disambiguation [1], query expansion [2], document categorization [3], question answering [4] and semantic web applications [5].

However, most methodologies for the extraction of semantic relations from texts have scalability issues. For instance, while some methods extract semantic relations by exploring syntactic patterns in texts, others use external semantic lexicons such as thesauri, ontologies or synonym dictionaries. These kind of approaches are deeply language and domain dependent, since not all languages share the same syntactic patterns, and lexical databases are not available for all languages and domains. On the other hand, most statistical methods are

---

* Supported by FCT-MCTES PhD grant SFRH/BD/61543/2009.

L. Correia, L.P. Reis, and J. Cascalho (Eds.): EPIA 2013, LNAI 8154, pp. 516–527, 2013.
© Springer-Verlag Berlin Heidelberg 2013

language-independent but tend to have the need for large amounts of text in order to be effective.

This poses a problem for the extraction of semantic relations from standalone documents. Standalone documents are, essentially, isolated or single documents, such as documents of unique subjects or domains, reports from very specific fields of expertise or even small books. The specificity of some fields of expertise in some of these documents may imply that no external ontologies exist for those domains, and given the small amount of text in those documents, statistical methods, with their correlation-like metrics, are not efficient. It is undeniable, however, that these isolated and autonomous documents are also a source of knowledge, so a local, more document-centric analysis is required.

In this paper we propose a statistical and language-independent method for the extraction of semantic relations between concepts in standalone documents. Given a pair of concepts occurring in a single document, we are able to measure how semantically related they are. Specifically, for a single document, we start by extracting its concepts. Then, for each concept, we analyze the local clusters that it may form. Finally, since relevant concepts on a document tend to form clusters in certain areas, clusters occurring in the same areas may represent highly related concepts.

We propose methods to identify clusters of concepts in documents, measure the degree of intersection between clusters and to quantify semantic relations. Although we are able to measure the degree of semantic "relatedness" (or semantic closeness) between concepts, we cannot infer the type of relation. Moreover, we also briefly show that statistical methods using the correlation for the extraction of semantic relations from collections of documents, may benefit from this local analysis.

This paper is structured as follows: in the next section we review the related work. Section 3 proposes the method for the identification of clusters and for the extraction of semantic relations from them. Section 4 shows the results of this approach. In Section 5 we briefly show how our methodology may work on collections of documents and Section 6 presents the conclusions.

2   Related Work

Current surveys ([7], [8]) have identified four different classes of approaches for the extraction of semantic relations: (1) based on distributional similarity and pattern recognition; (2) as classification tasks by means of feature training; (3) approaches which extends other semantic lexicons; (4) statistical approaches.

Pattern recognition approaches, such as [9] and [10], tend to consider that two terms are semantically related when they share the same patterns. For instance, in [9] a semantic relation exists between two terms if their neighbor terms up to a certain distance are the same, while for [10], the terms must occur in the same sentence. Also, a great amount of work has been done to explore the fixed structure of some on-line resources, such as [11] where the authors explore the explicit links on each Wikipedia article to derive semantic relations. However, these kind of approaches tend to be extremely dependent on the structure of a specific language, tool or resource.

Works such as [12] and [13] are representative of the classification area. For instance, in [12] the authors are interested in assigning semantics to data-mining database fields. They use a web crawler to extracts names, *urls* and prices of apparel, then manually classify a set of features such as age group, functionality, price point, degree of sportiness, etc. These methods are domain specific and may be difficult to adapt to other domains.

Approaches like [14], [15] and [16] are examples of methodologies which use external semantic lexicons such as FrameNet [17] and WordNet [18]. For instance, in [16] the authors identify lexical patterns that represent semantic relations between WordNet concepts. However, these external lexicons tend to be manually maintained and are not available for many languages and domains.

In the statistics field, most approaches are aimed at the extraction of semantic relations from collections of documents. For instance, Latent Semantic Indexing [19] identifies semantic relations using matrix calculations. However, the text is represented as an unordered collection of words and can not handle multiword units. In [20] and [21] correlation metrics are used, but in [21], a method to compute correlations in fixed-sized windows is proposed which could work for standalone documents. However, from our observations, the distances between related concepts may be more than the 16 words the authors propose.

3   Clusters of Concepts – Extracting Semantic Relations

Clusters of concepts occur when the distances between successive occurrences of a concept are less than what would be expected by chance. In other words, a cluster is a specific area on a text where a concept is relevant and tends to occur rather densely. For instance, consider the following paragraph from the English Wikipedia article *Arthritis*:

> **Gout** is caused by deposition of **uric acid** crystals in the joint, causing inflammation. (...) The joints in **gout** can often become swollen and lose function. (...) When **uric acid** levels and **gout** symptoms cannot be controlled with standard **gout** medicines that decrease the production of **uric acid** (e.g., allopurinol, febuxostat) or increase **uric acid** elimination from the body through the kidneys (e.g., probenecid), this can be referred to as refractory chronic **gout** or RCG.

Fig. 1. A paragraph from the *Arthritis* article – English Wikipedia

This paragraph is the only place, in the *Arthritis* article, where *gout* and *uric acid* occur. Since both concepts occur rather densely only in this paragraph, each one forms a cluster here. And since both concepts form a cluster in the same area, we consider the concepts highly related. Undoubtedly, *gout* and *uric acid* are related concepts ("gout is caused by deposition of uric acid crystals") and highly relevant in this paragraph.

3.1   Identifying Clusters of Concepts

In a formal way, a cluster of a concept exists where the distances between some of its successive occurrences are less than what would be expected by chance. So, the question is how to define the expected behavior of a concept on a document. Be $L_c = \{t_1, t_2, \cdots, t_m\}$ the list of the t_i positions where a concept c occurs in a document of size n. From L_c, we can obtain $\hat{u}_a$ (see (1)) which measures the average separation that would exist if c occurred uniformly (or randomly) on the document:

$$\hat{u}_a = \frac{n+1}{m}. \tag{1}$$

The underlying idea is that, for two successive occurrences (t_i, t_{i+1}) of c, if their separation is less than $\hat{u}_a$, both are part of a cluster, else, they are not.

Unfortunately, $\hat{u}_a$, as it is, tends to favor rare words. For instance, a concept which occurs 4 times in a document of size 2000 will have $\hat{u}_a \approx 500$. If the occurrences are spread over 4 successive paragraphs of size 200, the distances will be always less than 500 – the maximum distance would be 400, for one occurrence in the beginning of one paragraph and the next occurrence in the end of the following paragraph. Thus, this rare concept will always form a cluster, but, instead of being highly concentrated on a single paragraph or two, the concept is weakly scattered over four paragraphs. To allow clusters for great distances may be too much, so we propose an upper limit for such rare cases.

Figures 2, 3 and 4 show, on the left side, the number of paragraphs (y-axis) by paragraph size (x-axis, in words), and on the right side, the average number of words in a paragraph (y-axis) by document size (x-axis/10).

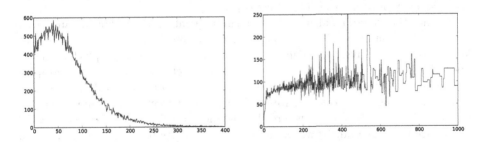

Fig. 2. Paragraph analysis on a corpus of English documents

As it is evident in the figures, the behavior of the paragraphs tends to be quite similar for all languages tested. On the left side, we can see that most paragraphs have about 50 words, and that 95% have less than 150 words. On the right side, except for small document with less than 100–200 words, the average paragraph length is independent of the size of documents. For our purposes, since clusters are somewhat associated with paragraphs (or parts of paragraphs), and since

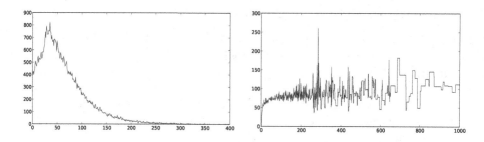

Fig. 3. Paragraph analysis on a corpus of Portuguese documents

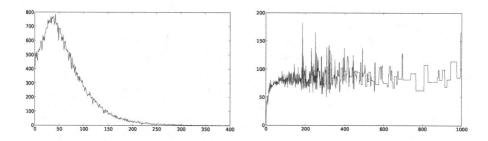

Fig. 4. Paragraph analysis on a corpus of German documents

95% of paragraphs have less than 150 words, we propose an **upper limit of 150 words**. This means that no cluster may be formed where the distance between successive occurrences of a concept is greater than 150 words, independently of the frequency of occurrence of the concept.

On the other hand, $\hat{u}_a$ also tends to harm frequent concepts. For instance, in a typical document of size 2000, a very frequent concept (usually the most relevant keyword) occurs in average 60 times ($\hat{u}_a \approx 33$). Since most paragraphs have a size about 50 words, a frequent concept may not form clusters in those paragraphs, for instance, if it occurs only 2 times in the paragraph, but in distinct edges. Considering this, we propose a **lower limit of 50 words**. This means that a cluster will always be formed where the distance between successive occurrences of a concept is less than 50 words, independently of its frequency of occurrence.

Formally, being $L_c = \{t_1, t_2, \cdots, t_m\}$ the list of the positions where concept c occurs, (2) measures the new proposed average separation to consider whether c occurs randomly in a document:

$$\hat{u} = \begin{cases} 150 & \text{, if } \hat{u}_a > 150 \\ 50 & \text{, if } \hat{u}_a < 50 \\ \hat{u}_a & \text{, otherwise} \end{cases} . \tag{2}$$

The next step consists in the calculation of the cohesions between successive occurrences of c, given by equation (3):

$$coh(t_i, t_{i+1}) = \frac{\hat{u} - d(t_i, t_{i+1})}{\hat{u}} . \tag{3}$$

$$d(t_i, t_{i+1}) = t_{i+1} - t_i . \tag{4}$$

Basically, the cohesion measures the distance between successive occurrences (t_i, t_{i+1}), proportional to $\hat{u}$. If the distance is small, the cohesion will tend to 1.0, else, it will tend to values less than zero. Zero stands as the frontier case, where the distance will be equal to $\hat{u}$.

The final step consists in traversing the L_c list and join together occurrences belonging to the same clusters, since a concept may form more than one cluster (or none). Figure 5 shows a pseudo-code sample for finding clusters in L_c.

```
def findClusters(Lc):
    clusterList = ClusterList()
    currCluster = Cluster()
    for (ti, ti+1) in Lc:
        if (coh(ti, ti+1) > 0):
            // Add the pair to the cluster
            currCluster.addPair(ti, ti+1)
            currCluster.addCohesion(coh(ti, ti+1))
        else:
            // If cluster is not empty
            if (currCluster.numberPairs() > 2):
                currCluster.computeAverageCohesion()
                clusterList.add(currCluster)
            // start a new empty cluster
            currCluster = Cluster()
    return clusterList
```

Fig. 5. Pseudo-code for finding clusters in L_c

The final cohesion value for each cluster is the arithmetic average of the cohesion values for the successive occurrences of the concept in the cluster. Also, although not required, in our tests we enforce that a cluster, to be valid, must have at least 3 occurrences of the concept (or 2 pairs as in the pseudo-code).

3.2 Intersection and Semantic Closeness of Clusters

As we said previously, the underlying idea is that a pair of concepts is highly related if they tend to make clusters in the same areas of a document. Thus, the purpose behind the intersection is to find whether two clusters occupy the same area of a text. So, for two clusters $C_A = \{p_{A1}, p_{A2}, \cdots, p_{An}\}$ and $C_B = \{p_{B1}, p_{B2}, \cdots, p_{Bm}\}$, where p_{Xi} is a position where concept X occurs in the text, we measure the intersection using (5):

$$intersection(C_A, C_B) = \frac{span(C_A, C_B)}{spanMin(C_A, C_B)} , \tag{5}$$

$$span(C_A, C_B) = min(p_{An}, p_{Bm}) - max(p_{A1}, p_{B1}) , \tag{6}$$

$$spanMin(C_A, C_B) = min(p_{An} - p_{A1}, p_{Bm} - p_{B1}) . \tag{7}$$

Basically, $spanMin(C_A, C_B)$ measures the size of the smallest cluster while $span(C_A, C_B)$ measures the size of the real intersection between clusters. The result is how much of the smallest cluster is intersected. Equation (5) returns values between $-\infty$ and 1.0, where 1.0 occurs when one cluster is completely inside the other, and values less than 0.0 occur when the clusters do not intersect.

Since we are now able to measure intersections between clusters, we measure the Semantic Closeness for a pair of concepts (A, B) using (8).

$$SC(A, B) = AvgIntersection(A, B) * AvgCoh(A) * AvgCoh(B) , \tag{8}$$

where $AvgIntersection(A, B)$ is the average of all positive intersections between clusters of A and B (i.e., only when $intersection(C_A, C_B) > 0$), and $AvgCoh(X)$ is the average of all cohesions for all clusters of X. Pairs of concepts for which their clusters are strongly intersected and the individual clusters are cohesive, are highly related. Tables 1, 2 and 3 show some results of Semantic Closeness between concepts from some documents of the tested corpora (see section 4).

Table 1. Semantic Closeness for terms in the *Arthritis* article - EN Wikipedia

Term A	Term B	$SC(A, B)$
gout	gouty arthritis	0.671
gout	uric acid	0.604
rheumatoid arthritis	osteoarthritis	0.472
medications	exercise	0.067
rheumatoid arthritis	psoriatic arthritis	0.000
systemic	history	0.000

The tables clearly show that the results are quite balanced among all languages. Top results are for pairs of concepts whose relations are pretty obvious in the respective documents. For instance, in the English *Arthritis* article, *gout* is synonym of *gouty arthritis* and *uric acid* causes *gout*. In the Portuguese article, Aminoacyl-tRNA (*aminoacil-trna*) is an enzyme to which an amino acid (*aminoácido*) is cognated, and insuline (*insulina*) is a hormone to process glucose, where glycogen (*glicogénio*) is glucose stored in cells.

Table 2. Semantic Closeness for terms in the *Metabolismo* article - PT Wikipedia

Term A	Term B	$SC(A, B)$
aminoacil-trna	aminoácidos	0.768
insulina	glicogénio	0.627
glicose	gluconeogénese	0.443
ácidos gordos	ácidos tricarboxílicos	0.282
via	energia	0.049
álcool	ferro	0.000

Table 3. Semantic Closeness for terms in the *Autismus* article - DE Wikipedia

Term A	Term B	$SC(A, B)$
intelligenz	sprachentwicklung	0.657
frühkindlichen autismus	atypischer autismus	0.512
autismus	sprachentwicklung	0.264
intelligenz	autismus	0.208
autismus	begriff	0.048
wissenschaftler	diagnosekriterien	0.000

Bottom results are essentially for pairs which are not usually related, such as *systemic* and *history*. However, there are also cases for which, although the pair seems related, the relation is not explicit in the document. For instance, although *rheumatoid arthritis* and *psoriatic arthritis* are two types of arthritis, they are different types of arthritis, with different causes and different symptoms, thus, not related on a low-level (rheumatoid arthritis affects tissues and organs while psoriatic arthritis affects people who have the chronic skin condition, psoriasis).

4   Experimental Conditions and Results

For evaluating the results of the proposed approach, we started by building three Wikipedia document-based corpus for three different European languages, namely English, Portuguese and German. We used *Catscan*[1] to extract titles of documents from the *Medicine* category down to a certain depth (number of subcategories). Then we extracted an *XML dump file*[2] with the contents of the articles and, finally, we used WikiCorpusExtractor[3] to parse the XML and build the corpora files. Table 4 shows some information about the number of documents, sizes and depth for each corpus.

From each corpus we extracted 10 random documents with a minimum of 2000 words. Then, for each document, we used *ConceptExtractor* [6] to obtain their concepts. Table 5 shows the titles of the selected documents.

Finally, for each document, we extracted 30 pairs of concepts and computed their *Semantic Closeness* (see (8)). It resulted in a list with 300 pairs of concepts

---

[1] http://toolserver.org/~magnus/catscan_rewrite.php

[2] http://en.wikipedia.org/wiki/Special:Export for the case of EN Wikipedia.

[3] https://github.com/joaoventura/WikiCorpusExtractor

Table 4. Statistics about the corpora based on Wikipedia *Medicine* articles

Corpora	English	Portuguese	German
N. documents	4 160	4 066	4 911
Tot. words	4 657 053	4 153 202	4 337 068
Avg. words/doc	1 120	1 022	884
Depth	2	4	2

Table 5. Random documents extracted from the EN, PT and DE Wikipedia

English	Portuguese	German
Arthritis	Esclerose tuberosa	Schuppenflechte
Orthotics	Ácido desoxirribonucleico	Homopathie
Pediatric ependymoma	Transtorno mental	Keratokonus
Long-term effects of benzodiazepines	Cinética enzimática	Nosokomiale Infektion
Mutagen	Sistema imunitário	Tuberkulose
Canine reproduction	Bactéria	Phagentherapie
Schizophrenia	Antidepressivo	Krim-Pfingstrose
Menopause	Terapia genética	Verhaltenstherapie
Glucose meter	Micronutriente	Oberkieferfraktur
Rabbit haemorrhagic disease	Sistema circulatório	Sexualwissenschaft

for each language, indexed by document title, which were manually classified as being related or not. The criterion for the classification was the following: a pair of concepts should only be classified as related if those concepts were explicitly related in their document of origin. This implies that the documents had to be available for reading. As an example of the criterion, Table 6 shows the classified results for the article *Pediatric ependymoma*.

Table 6. Classification results for the article *Pediatric ependymoma*

Pair		Pair	
0.697 gene expression – telomerase	X	0.000 occur – tend	
0.657 mutations – ependymoma	X	0.000 arise – kit	
0.554 tumor suppressor – nf2	X	0.000 favorable – frequently	
0.492 classification – ependymoma	X	0.000 intracranial – correlated	
0.333 tumors – ependymomas	X	0.000 inversely – supratentorial	
0.327 genes – notch		0.000 significantly – remains	
0.312 expression – pediatric ependymomas	X	0.000 loss – down-regulation	
0.226 suppressor genes – mutations	X	0.000 loss – tyrosine	
0.204 pathway – pediatric ependymomas	X	0.000 men1 – inversely	
0.189 tumor suppressor – ependymomas	X	0.000 remains – candidate genes	
0.132 genes – p53	X	0.000 mmp14 – ependymomas	X
0.065 progression – p53		0.000 mmp2 – lethargy	
0.000 location – neurofibromatosis		0.000 mutations – mmp14	
0.000 chromosome – genomic hybridization	X	0.000 outcome – myxopapillary	

Since the extracted lists were sorted by rank, in order to obtain Precision and Recall values we had to force a threshold, such that above the threshold a pair was to be *automatically* considered relevant, and below, non-relevant. We set that threshold empirically on 0.1. Table 7 shows the results.

Table 7. Precision and Recall results

Language	Precision	Recall
English	0.91	0.83
Portuguese	0.92	0.85
German	0.89	0.79

As it can be seen, the proposed approach is quite balanced for all languages tested. *Precision* measures how many of the pairs above the threshold are indeed related while *Recall* measures how many of the really related pairs (the ones classified with an 'X') are correctly above the threshold. As expected, recall results were lower than Precision results: given the lack of statistical information in one single document, our approach is not able to correctly identify all possible relations. For instance, in table 6, the pair (*mmp14– ependymomas*) is an example. *MMP14* is an enzyme related with *ependymomas*. However, since *mmp14* only occurs 2 times in the document, and both occurrences are relatively distant, it never forms a cluster. Rare, scattered concepts, are problematic for our approach. However, for most applications, higher precision values are more relevant.

5   On Collections of Documents

As already mentioned, because of the ability to do a local analysis on a document, we believe that our method can aid other methods when dealing with collection of documents. As a brief example, table 8 shows the Pearson correlation values for some concepts co-occurring with *gout* in the documents of the English corpus.

Table 8. Pearson correlation values for concepts co-occurring with *gout* –Eng. corpus

Concept	$Corr(.)$	Concept	$Corr(.)$
lawrence c. mchenry	0.544	christopher smart	0.544
dr johnson	0.544	gouty arthritis	0.352
hester thrale	0.544	arthritis	0.257
samuel swynfen	0.544	uric acid	0.198

In this example, the higher correlated concepts are person's names. They come from a document that relates the health of these persons with gout. By being rare in the corpus, these names are extremely valued by correlation metrics. However, specially for applications such as the creation of thesauri, this type of knowledge may have little interest. As an exercise, in table 9 we show the

Table 9. Concepts co-occurring with *gout* in the English corpus

Concept	$Corr(.)$	$SC(.)$	$Avg(.)$	Concept	$Corr(.)$	$SC(.)$	$Avg(.)$
gouty arthritis	0.352	0.67	0.511	dr johnson	0.544	0.00	0.272
uric acid	0.198	0.60	0.399	hester thrale	0.544	0.00	0.272
arthritis	0.257	0.36	0.301	samuel swynfen	0.544	0.00	0.272
lawrence c. mchenry	0.544	0.00	0.272	christopher smart	0.544	0.00	0.272

same concepts, but including the results of the *Semantic Closeness*, as well as the average value between the correlation and the Semantic Closeness.

Gouty arthritis, uric acid and *arthritis* are concepts explicitly related with *gout* in some documents of the English corpus. Sorting by the average value allows them to appear in the first positions of the ranking. As we mentioned, this type of knowledge may be of interest for some applications.

6 Conclusions

In this paper we have presented a method for the extraction of semantic relations from standalone documents. These are documents that, given their specific domains and text size, external ontologies may not exist and standard statistical methods such as the correlation may not work.

Our methodology works by identifying clusters in order to measure the Semantic Closeness between pairs of concepts. By measuring the intersection between clusters of different concepts, we are able to measure their semantic relatedness. We justify our method by presenting results for three different European languages. We have also shown with a small example, that the local analysis done by this approach may aid statistical methods, such as those based on correlations, when extracting semantic relations from collections of documents.

In general, although precision results are quite encouraging, we are only able to extract semantic relations which are explicit in the texts. This is shown by the lower recall results. Future work may be done to address this issue.

References

1. Patwardhan, S., Pedersen, T.: Using WordNet-based context vectors to estimate the semantic relatedness of concepts. In: Proceedings of the EACL 2006 Workshop Making Sense of Sense, pp. 1–12 (2006)
2. Hsu, M.-H., Tsai, M.-F., Chen, H.-H.: Query expansion with conceptNet and word-Net: An intrinsic comparison. In: Ng, H.T., Leong, M.-K., Kan, M.-Y., Ji, D. (eds.) AIRS 2006. LNCS, vol. 4182, pp. 1–13. Springer, Heidelberg (2006)
3. Tikk, D., Yang, J.D., Bang, S.L.: Hierarchical text categorization using fuzzy relational thesaurus. KYBERNETIKA-PRAHA 39(5), 583–600 (2003)
4. Yousefi, J., Kosseim, L.: Using semantic constraints to improve question answering. In: Kop, C., Fliedl, G., Mayr, H.C., Métais, E. (eds.) NLDB 2006. LNCS, vol. 3999, pp. 118–128. Springer, Heidelberg (2006)

5. Sheth, A., Arpinar, I.B., Kashyap, V.: Relationships at the heart of semantic web: Modeling, discovering, and exploiting complex semantic relationships. In: Nikravesh, M., Azvine, B., Yager, R., Zadeh, L.A. (eds.) Enhancing the Power of the Internet. STUDFUZZ, vol. 139, pp. 63–94. Springer, Heidelberg (2003)
6. Ventura, J., Silva, J.F.: Mining concepts from texts. In: International Conference on Computer Science (2012)
7. Biemann, C.: Ontology Learning from Text: A Survey of Methods. LDV-Forum Journal 20(2), 75–93 (2005)
8. Gmez-Prez, A., Manzano-Macho, D.: Deliverable 1.5: A survey of ontology learning methods and techniques. Ontology Based Information Exchange for Knowledge Management and Electronic Commerce 29243 (2003)
9. Grefenstette, G.: Evaluation techniques for automatic semantic extraction: comparing syntactic and window based approaches. In: Corpus Processing for Lexical Acquisition, pp. 205–216. MIT Press, Cambridge (1996)
10. Akbik, A., Broß, J.: Wanderlust: Extracting Semantic Relations from Natural Language Text Using Dependency Grammar Patterns. In: Proceedings of the 18th International World Wide Web Conference, Madrid, Spain (2009)
11. Nakayama, K., Hara, T., Nishio, S.: Wikipedia Link Structure and Text Mining for Semantic Relation Extraction. In: SemSearch 2008, CEUR Workshop Proceedings (2008) ISSN 1613-0073
12. Ghani, R., Fano, A.: Using Text Mining to Infer Semantic Attributes for Retail Data Mining. In: Proceeding of the 2nd IEEE International Conference on Data Mining (ICDM 2002), Maebashi, Japan, pp. 195–203 (2002)
13. Snow, R., Jurafsky, A., Ng, A.: Learning syntactic patterns for automatic hypernym discovery. In: Advances in Neural Information Processing Systems (NIPS 2004), Vancouver, British Columbia (2004)
14. Mohit, B., Narayanan, S.: Semantic Extraction with Wide-Coverage Lexical Resources. In: Proceedings of the North American Chapter of the Association for Computational Linguistics - Human Language Technologies, Edmonton, Canada, pp. 64–66 (2003)
15. Gildea, D., Jurafsky, D.: Automatic Labeling of Semantic Roles. Computational Linguistics 28(3), 245–288 (2002)
16. Ruiz-Casado, M., Alfonseca, E., Castells, P.: Automatic extraction of semantic relationships for wordNet by means of pattern learning from wikipedia. In: Montoyo, A., Muñoz, R., Métais, E. (eds.) NLDB 2005. LNCS, vol. 3513, pp. 67–79. Springer, Heidelberg (2005)
17. Baker, C.F., Fillmore, C.J., Lowe, J.B.: The Berkeley FrameNet project. In: Proceedings of the 17th International Conference on Computational Linguistics, Montreal, Canada, pp. 86–90 (1998)
18. Miller, G.A.: WordNet: A Lexical Database for English. Communications of the ACM 38(11), 39–41 (1995)
19. Deerwester, S., Harshman, R., Dumais, S., Furnas, G., Landauer, T.: Improving Information Retrieval with Latent Semantic Indexing. In: Proceedings of the 51st Annual Meeting of the American Society for Information Science, pp. 36–40 (1988)
20. Panchenko, A., Adeykin, S., Romanov, A., Romanov, P.: Extraction of Semantic Relations between Concepts with KNN Algorithms on Wikipedia. In: Proceedings of the 10th International Conference on Formal Concept Analysis, Leuven, Belgium (2012)
21. Terra, E., Clarke, C.L.A.: Frequency estimates for statistical word similarity measures. In: Proceedings of the Human Language Technology and North American Chapter of Association of Computational Linguistics Conference 2003 (HLT/NAACL 2003), pp. 244–251 (2003)

Rule Induction for Sentence Reduction

João Cordeiro[1,3], Gaël Dias[2], and Pavel Brazdil[3]

[1] University of Beira Interior, DI – Hultig,
Rua Marquês d'Ávila e Bolama, Covilhã 6200 – 001, Portugal
jpaulo@di.ubi.pt
[2] University of Caen Basse-Normandie, GREYC Hultech,
Campus Côte de Nacre, F-14032 Caen Cedex, France
gael.dias@unicaen.fr
[3] University of Porto, INESC – TEC,
Rua Dr. Roberto Frias, 378, 4200 – 465 Porto, Portugal
{jpcc,pbrazdil}@inescporto.pt

Abstract. Sentence Reduction has recently received a great attention from the research community of Automatic Text Summarization. Sentence Reduction consists in the elimination of sentence components such as words, part-of-speech tags sequences or chunks without highly deteriorating the information contained in the sentence and its grammatical correctness. In this paper, we present an unsupervised scalable methodology for learning sentence reduction rules. Paraphrases are first discovered within a collection of automatically crawled Web News Stories and then textually aligned in order to extract interchangeable text fragment candidates, in particular *reduction cases*. As only positive examples exist, *Inductive Logic Programming* (ILP) provides an interesting learning paradigm for the extraction of sentence *reduction rules*. As a consequence, reduction cases are transformed into first order logic clauses to supply a massive set of suitable learning instances and an ILP learning environment is defined within the context of the *Aleph* framework. Experiments evidence good results in terms of irrelevancy elimination, syntactical correctness and reduction rate in a real-world environment as opposed to other methodologies proposed so far.

1 Introduction

The task of Sentence Reduction (or Sentence Compression) can be defined as summarizing a single sentence by removing information from it [8]. Therefore, the compressed sentence should retain the most important information and remain grammatical. But a more restricted definition is usually taken into account and defines sentence reduction as dropping any subset of words from the input sentence while retaining important information and grammaticality [9]. This formulation of the task provided the basis for the noisy-channel and decision tree based algorithms presented in [9], and for virtually most follow-up work on data-driven sentence compression [1,11,16]. One exception can be accounted for [2,3], who consider sentence compression from a more general perspective and generate abstracts rather than extracts.

Sentence reduction has recently received a great deal of attentions from the Natural Language Processing research community for the number of applications where

L. Correia, L.P. Reis, and J. Cascalho (Eds.): EPIA 2013, LNAI 8154, pp. 528–539, 2013.

sentence compression can be applied. One of its (direct) applications is in automatic summarization [8,17,20]. But there exist many other interesting issues for sentence reduction such as automatic subtitling [18], human-computer interfaces [6] or semantic role labeling [19].

In this paper, we present an unsupervised scalable methodology for learning sentence reduction rules following the simplest definition of sentence compression. This definition makes three important assumptions: (1) only word deletions are possible and no substitutions or insertions allowed, (2) the word order is fixed and (3) the scope of sentence compression is limited to isolated sentences and the textual context is irrelevant. In other words, the compressed sentence must be a subsequence of the source sentence, which should retain the most important information and remain grammatical.

The proposed methodology is based on a pipeline. First, Web news stories are crawled and topically clustered. Then, paraphrase extraction and alignment are respectively performed based on text surface similarity measures and biologically-driven alignment algorithms. Then, *reduction cases*, which can be seen as interchangeable text fragments are extracted and transformed into first order logic clauses, eventually enriched with linguistic knowledge, to supply a massive set of suitable learning instances for an *Inductive Logic Programming* framework (*Aleph*). Experiments evidence good results in terms of irrelevancy elimination, syntactical correctness and reduction rate in a real-world environment as opposed to existing methodologies proposed so far.

2 Related Work

One of the first most relevant work is proposed by [9], who propose two methods. The first one is a probabilistic model called the noisy channel model in which the probabilities for sentence reduction $P(t|s)$ (where t is the compressed sentence and s is the original sentence) are estimated from a training set of 1035 (s, t) pairs, manually crafted, considering lexical and syntactical features. The second approach learns syntactic tree rewriting rules, defined through four operators: SHIFT, REDUCE, DROP and ASSIGN. Sequences of these operators are learned from a training set and each sequence defines a transformation from an original sentence to its compressed version. The results are interesting but the training data set is small and the methodology relies on deep syntactical analysis.

To avoid language dependency, [11] propose two sentence reduction algorithms. The first one is based on template-translation learning, a method inherited from the machine translation field, which learns lexical reduction rules by using a set of 1500 (s, t) pairs selected from a news agency and manually tuned to obtain the training data. Due to complexity difficulties, an improvement is proposed through a stochastic Hidden Markov Model, which is trained to decide the best sequence of lexical reduction rules that should be applied for a specific case. This work proposes an interesting issue but the lack of a large data set of learning pairs can not really assess its efficiency.

To avoid supervised learning, a semi-supervised approach is presented in [16], where the training data set is automatically extracted from the Penn Treebank corpus to fit a noisy channel model. Although it proposes an interesting idea to automatically provide new learning instances, it is still dependent upon a manually syntactically-labeled data set, in this case the Penn Treebank corpus.

[1] propose an hybrid system, where the sentence compression task is defined as an optimization of an integer programming problem. For that purpose, several constraints are defined, through language models, linguistic and syntactical features. Although this is an unsupervised approach, without using any parallel corpus, it is completely knowledge driven, as a set of hand-crafted rules and heuristics are incorporated into the system to solve the optimization problem.

More recently, [2,3] propose a tree-to-tree transduction method for sentence compression. Their model is based on synchronous tree substitution grammar, a formalism that allows local distortion of the tree topology and can thus naturally capture structural mismatches. Their experimental results bring significant improvements over a state-of-the-art model, but still rely on supervised learning and deep linguistic analysis.

In this paper, we propose a real-world scalable unsupervised ILP learning framework based on paraphrase extraction and alignment. Moreover, only shallow linguistic features are introduced and propose an improvement over surface text processing.

3   The Overall Architecture

Our system autonomously clusters topically related Web news stories, from which paraphrases are extracted and aligned. From these aligned sentence pairs, structures called reduction cases are extracted and transformed into learning instances, which feed an ILP framework called *Aleph* appropriately configured for our problem. The learning cycle is then computed and a set of sentence reduction rules is generated, as a result.

Fig. 1. The main architecture of our system

A reduction rule obtained from the learning process states a number of conditions over a candidate elimination segment (X), as well as its relative left (L) and right (R) contextual segments[1]. In particular, the input Web news stories are splitted into sentences, which are then part-of-speech tagged and shallow parsed[2]. For that purpose, the *OpenNLP* library[3] has been used. So, given sentence (1),

```
Pressure will rise for congress to enact a massive fiscal stimulus package (1)
```

its shallow parsed counterpart is defined in sentence (2).

```
[NP Pressure/NNP] [VP will/MD rise/VB] [PP for/IN] [NP congress/NN]
[VP to/TO enact/VB] [NP a/DT massive/JJ fiscal/JJ stimulus/NN package/NN] (2)
```

---

[1] A segment means just a sequence of words.

[2] A shallow parser is a non-hierarchical sentence parser.

[3] http://opennlp.apache.org/

In the learning process, three types of learning features are considered: words, part-of-speech tags and chunks[4]. As a consequence, a learned rule may state conditions involving these three types. Formula 1, expresses such rule in a high level notation.

$$eliminate(X) \; <= |X| = 2 \; \wedge \; L_c = \text{VP} \; \wedge \; X_2 = \text{NN} \; \wedge \; R_1 = \text{to} \; \wedge \; R_c = \text{VP} \quad (1)$$

The rule should be read as follows: *eliminate segment X if its size is two words long, its second word is a noun (NN), the left and right context segments are verb phrases (VP), and the first word of the right context is the word "to"*. We can see that this rule applies to sentence (2), resulting in the elimination of the `"for congress"` segment thus giving rise to the compressed sentence (3).

`Pressure will rise to enact a massive fiscal stimulus package (3)`

3.1 Paraphrase Extraction and Alignment

The first process called the *Gauss Selection* consists in the automatic identification and extraction of paraphrases from topically related Web news stories. It is based on lexical unigram exclusive overlaps in a similar way to some approaches in the literature [5,7]. In the work of [5], a significant comparison of paraphrase identification functions is proposed and results over standard test corpora reveal that some of their proposed functions achieve good performances. After several experiments, we noticed that the Gaussian functions could be adequately parametrized for the extraction of assymetrical paraphrases, thus satisfying our practical goal. A Gaussian function has the general form described in Equation 2.

$$f(x) = a \cdot e^{-\frac{(x-b)^2}{2 \cdot c^2}} \quad (2)$$

In our case, the Gaussian function computes the likelihood of two sentences being paraphrases. So, (x) is the relative number of lexical unigrams in exclusive overlap between a pair of sentences. So, the more exclusive links two sentences share, the more likely they will be paraphrases. As the measure must be a real value in the unit interval a is settled to 1. Moreover, in order to learn reduction rule we must find paraphrases where one sentence is smaller than the other one. As a consequence, the b parameter allows to tune this dissimilarity degree, which in our case is equal to 0.5. An example is given below where sentences (4) and (5) are paraphrases.

`Pressure will rise for congress to enact a massive fiscal stimulus package (4)`
`Pressure will rise to enact a fiscal package (5)`

After paraphrase extraction, a combination of Biology-based sequence alignment algorithms [12,14] is proposed in [4] to word-align any paraphrasic sentence pair. As a consequence, the alignment process over sentences (4) and (5) gives rise to the aligned sentences (6) and (7). For that purpose, a specific mutation matrix based on a modified version of the edit distance is computed as in [4].

---

[4] A sentence segment of related words grouped by a shallow parser.

```
Pressure will rise for congress to enact a massive fiscal stimulus package (6)
Pressure will rise ___ _____ to enact a _____ fiscal _____ package (7)
```

3.2   Learning Instance Selection

After identifying relevant paraphrases and aligning them at word level, specific text segments that evidence local sentence reduction cases, which can then be used as reduction
instances in the learning process, must be defined. We call these structures *reduction
cases*. For example, by looking at sentences (6) and (7), we can observe three reduction
cases: (a) "for congress", (b) "massive", and (c) "stimulus".

So, first, in order to *consider* a reduction case, one must have a segment aligned with
an empty sequence (lets say a middle segment, represented by X), surrounded by left
(L) and right (R) contexts of commonly aligned words. For example, in segment (a) we
have L = "Preasure will rise" and R ="to enact a", and so as a consequence,
the triple $\langle L, X, R \rangle$ is selected as a reduction case. In that same alignment, we have two
more reduction cases, with X="massive" and X="stimulus".

Then, in order to *select* a relevant reduction case, an evaluation function is defined as
in Equation 3. In particular, the $value(\langle L, X, R \rangle)$ function gives preference to reduction cases having long contexts relatively to the misaligned segment X. The longer the
contexts the higher the linguistic evidence indicating a true reduction case. A threshold
must be set for the selection decision. We have pick one ensuring that the length of the
contexts outweighs the length of the misaligned segment, i.e. $|X| \leq |L| + |R|$.

$$value(\langle L, X, R \rangle) = 1 - \frac{|X|}{|L| + |R| + \frac{1}{2}} \tag{3}$$

Following our example, the first selected reduction case is given in sentences (8) and
(9). Note that all the other reduction cases would be selected although with different
strength values.

```
Pressure will rise for congress to enact (8)
Pressure will rise ___ _____ to enact (9)
```

3.3   ILP Learning

After extracting sentence reduction cases, our final step is to transform them into learning instances in order to build a learning model. Within this context, one interesting
advantage of ILP[5] is the possibility to learn exclusively from positive instances, con-
trarily to what is required by most supervised learning models. In our problem, this
turns out to be a key aspect, since negative examples are difficult to obtain or even not
available.

Another important characteristics of ILP is the way in which learning features can be
defined, normally through first-order logic predicates. Indeed, most learning paradigms
require a complete and exact feature set specification, before starting the learning process. With ILP, we can afford to simply define a broad set of "possible features" that can
be selected by the system during the learning process. This characteristics is particularly

---

[5] Inductive Logic Programming.

interesting as the set of all the exact learning features can be huge and as consequence lead to data sparseness in a classical learning paradigm.

We have considered three feature categories: *words*, *part-of-speech tags* and *chunks*. Each reduction case is transformed into a first-order logic representation, involving these features. An example related with the reduction case of section 3.2 is given below:

```
reduct(1, t(2,0),
    [pressure/nnp/np, will/md/vp, rise/vb/vp],
    [for/in/pp, congress/nn/np] ---> [],
    [to/to/vp, encat/vb/vp, a/dt/np]).
```

Each reduction case is represented by a 5-ary *Prolog* term "reduct/5". The first argument is a sequential number[6]. The second one contains a 2-ary term, which represents the reduction dimensionality, where its first argument is the misaligned segment size ($|X|$) and the second argument the kind of transformation that is applied, e.g. 0 means that there is a complete deletion of the misaligned segment. The third, fourth and fifth arguments contain *Prolog* lists representing respectively the L, X, and R segments. Each list element is a triple with the form of "WORD/POS/CHUNK".

4   The Aleph Framework

The learning process has been perfomed with an ILP system called *Aleph* [15], that we have specifically configured for our task. Aleph is a machine learning system written in *Prolog* and was initially designed to be a prototype for exploring ILP ideas. It has become a mature ILP implementation used in many research projects ranging form Biology to Natural Language Processing. In fact, Aleph is the successor of several and "more primitive" ILP systems such as: Progol [10] and FOIL [13], among others, and may be appropriately parametrized to emulate any of those older systems.

4.1   Configuring *Aleph*

Before starting any learning process in Aleph, a set of several specifications must be defined which will direct the learning process. Those involving the definition of the concept to be learned, the declaration of the predicates that can be involved in the rule formation, the definition of the learning procedure, optionally the definition of rule constraints and a set of learning parameters, among other details. In this subsection, we describe the most relevant settings, defined for our learning problem.

In Aleph, the learning instances are divided in three files: the background knowledge (BK) file (*.b) and two files containing the positive (*.f) and negative (*.n) learning instances . This last one is optional and was not used in our case, as explained before. Hence, our positive instances file contains the set of all sentence reduction cases extracted from the aligned paraphrasic sentences and transformed into first-order logic predicates.

The BK file contains the learning configurations including their associated predicates and parameters. We start by showing an excerpt of the head of our BK file, which contains the *settings*, *modes*, and *determinations*, for our learning task.

---

[6] It simply means the instance identifier.

```
%-------------------------------------------
% SETTINGS
:- set(minpos, 5).
:- set(clauselength, 8).
:- set(evalfn, user).
:- set(verbosity, 0).
%-------------------------------------------
% DECLARATIONS
:- modeh(1, rule(+reduct)).
:- modeb(1, transfdim(+reduct, n(#nat,#nat))).
:- modeb(3, chunk(+reduct, #segm, #chk)).
:- modeb(*, inx(+reduct, #segm, #k, #tword)).
:- determination(rule/1, transfdim/2).
:- determination(rule/1, chunk/3).
:- determination(rule/1, inx/4).
```

The first block specifies the learning parameters we have been using, where `minpos` is the minimum coverage and `clauselength` is the maximum number of clauses (conditions) a rule can have. The `evalfn` parameter establishes that the rule evaluation function is defined by the "user", meaning that we are defining our own evaluation function in the BK file. The verbosity parameter is simply related with the level of output that is printed during the learning process.

The second block of the BK file header contains the main procedures for rule construction. The `modeh/2` function defines the "learning concept", which is called as `rule`. The `modeb/2` and `determination/2` directives establish the predicates that can be considered for rule construction, as well as the way they can be used, like number of times that a given predicate can occur in the rule (first argument of `modeb/2`).

In particular, we defined three predicates that can be used in the rule formation process: `transfdim/2`, `chunk/3`, and `inx/4`. The first one states the transformation dimensionality (e.g. a reduction from 2 to 0 words), the second one states the chunk type for a specific text segment and the third predicate states a positional[7] word or part-of-speech (POS) occurrence. Note that in the mode directives, the predicate arguments starting with # represent data types, which are also defined in the BK file. For instance, #nat and #k represent natural numbers, #segm a text segment, #chk a chunk tag and #tword represents either a word or a POS tag. In order to understand the kind of rules being produced, the following example codifies Formula 1 presented in section 3.

```
rule(A) :-
    transfdim(A,n(2,0)), chunk(A,left,vp),
    inx(A,center:x,2,pos(nn)),
    inx(A,right,1,to),
    chunk(A,right,vp).
```

From the rule body, we have `transfdim(A,n(2,0))` as the first literal, stating that it is a two word elimination rule. The second and fifth literals respectively state that the left and right textual contexts must be verb phrases (vp). The third literal states that the second word from the elimination segment (center:x) must be a noun (pos(nn)) and the fourth literal obliges that the first word in the right context must be "to". With this example we can see that different linguistic aspects (lexical, morpho-syntactical and shallow-syntactical) can be mingled into a single rule.

---

[7] In a relative index position (third argument: #k).

It is important to point at the fact that special concern has been dedicated to the misaligned segment i.e. literals of the form chunk(A, center:x, *), as it can be formed by multiple chunks. Thus, only for this segment (center:x), we let rules with multiple chunk types to be generated. Two structures can be formed: XP-YP and XP*YP, with XP and YP representing chunk tags. In particular, the first structure means a sequence of exactly two chunks and the second structure represents a sequence of three or more chunks, with the first one being XP and the last one YP. For example, pp*np represents a sequence of three or more chunks starting with a prepositional phrase (pp) and ending with a noun phrase (np). This would match chunk sequences like "pp np np" or "pp np vp np".

We have set a *user-defined cost function* and a number of *integrity constraints* as a strategy to better shape and direct the learning process [15]. The cost function shown as follows combines the rule coverage with a given distribution length, giving preference to rules having four and five literals. The 17047 value is the number of learning instances used. For each training set, this value is automatically defined by the *Java* program that generates the *Aleph* learning files.

```
cost(_, [P,_,L], Cost) :-
    value_num_literals(L, ValueL),
    Cost is P/17047 * ValueL.

value_num_literals(1, 0.10).  %        |
value_num_literals(2, 0.25).  % 1.0 -              _
value_num_literals(3, 0.50).  %        |           _    _
value_num_literals(4, 1.00).  %        |           _    _    _    _
value_num_literals(5, 0.60).  %        |           _    _    _    _    _    _
value_num_literals(6, 0.40).  %        |      _    _    _    _    _    _    _
value_num_literals(7, 0.20).  %        ------------------------------------------>
value_num_literals(_, 0.00).  %             1    2    3    4    5    6    7
```

The set of integrity constraints was designed to avoid certain undesired rule types, such as reduction rules without any condition over one of the three textual segments (left, center:x and right). This is achieved through the constraint shown below, where the countSegmRestr/5 predicate counts the number of conditions on each segment.

```
false :-
    hypothesis(rule(_), Body, _),
    countSegmRestr(Body, NL, NX, NY, NR),
    not_valid(NL, NX, NY, NR).

not_valid( _, 0, _, _).  %--> the center:x segment is free
not_valid( 0, _, _, _).  %--> left segment is free.
not_valid( _, _, _, 0).  %--> right segment is free.
```

As a consequence of several experimental iterations taken, we have decided that it would be better to include constraints for avoiding the generation of a kind of overgeneral rules, which are likely to yield bad reduction results. For example, rules that just constrain on chunks. This is exactly what the following two integrity constraints are stating:

```
false :-
    hypothesis(rule(_), Body, _),
    Body = (chunk(_,_,_), chunk(_,_,_), chunk(_,_,_)).
false :-
    hypothesis(rule(_), Body, _),
    Body = (transfdim(_,_), chunk(_,_,_), chunk(_,_,_), chunk(_,_,_)).
```

4.2   Learned Rules

The output of a learning run produces a set of sentence reduction rules similar to the ones illustrated in section 4.1. In particular, we will discuss the results of the quality of the set of learned reduction rules by applying them on new raw sentences and measuring their correctness with different measures in section 5. It is important to keep in mind that the learning model can generate thousands of reduction rules and in Table 1 we show only four of them, as well as their application on new sentences[8].

Table 1. Four examples of learned rules applied to new text

1	$L_1 = \text{IN} \wedge X_1 = \text{DT} \wedge R_1 = \text{NN} \wedge \|X\| = 4$	*for all the iraqi people and for **all those who love** iraq*	✓
2	$L_1 = \text{NNS} \wedge X_3 = \text{NN} \wedge R_1 = \text{IN}$	*we need new faces **and new blood** in politics*	✓
3	$L_c = \text{VP} \wedge X_1 = \text{NN} \wedge R_1 = \text{to} \wedge \|X\| = 1$	*my comment has **everything** to do with the way the*	✓
4	$L_1 = \text{NNS} \wedge X_2 = \text{NN} \wedge R_1 = \text{IN}$	shia and kurdish parties **took control** of parliament	✗

From these four examples, we can see that three rules were positively applied and the rule from case 4 was badly applied. This case illustrates one of the main difficulties that still persists: the generation of too general rules. Indeed, a good learning model must be balanced in terms of specificity and generality. In fact, specific rules may be very precise but seldom apply, while general rules may have high coverage but low precision. These issues can be evidenced by the kind of extracted rules. For example, rules 2 and 4 are similar and both state constraints only on morpho-syntactical information. As such, they are general rules On the contrary, rule 3 is more specific by stating that the right context R of a possible deletion of size one ($\|X\| = 1$) must contain the word "to" immediately after the deleted segment ($R_1 = \text{to}$). Therefore, it is much less error prone.

5   Experimental Results

To estimate the quality of the produced reduction rules, we followed an empirical experiment using a data set of Web news stories collected along a period of 90 days over Google News API. This data set is called T90Days and contains 90 XML files, one per day, covering the most relevant news events from each day.

In each given file, the news are grouped by events or topics, where each one contains a bunch of related documents[9]. The T90Days corpus contains a total of 474MB of text data and a total number of 53 986 aligned paraphrases extracted through the method described in subsection 3.1. From these aligned paraphrases, a total of 13 745 reduction cases were selected and transformed into learning instances following the methodology described in subsection 3.2. Finally, the induction process yielded an amount of 2507 sentence reduction rules. It is important to notice that all these data sets and results are

---

[8] To fit space constraints, we only show the relevant sentence fragment and not the overall sentence. Moreover, the marked segment is the deleted one.

[9] Usually from 20 to 30 Web news stories.

available online[10], in order to provide the research community with a large scale golden data set compared to the existing ones so far.

The evaluation of the induced rule set was performed over news set, different from the one used for learning. For the sake of this evaluation, we applied the best rule to each sentence and compared it with the baseline, which consists in directly applying the reduction cases to the sentences i.e. only lexical information is taken into account.

In particular, we had to define what is the "best" rule to apply to a given sentence. For that purpose, rule optimality was computed by combining rule support, number of eliminated words and the application of a syntactical 4-gram language model applied to the context of the reduction segment. While rule support guarantees some confidence in the rule application and the number of eliminated words must be maximized, the idea of the syntactical language model is to guarantee the syntactical correctness of the sentence after the application of the deletion. As a consequence, only the reduction rules, which can guarantee that the compressed sentence will be more syntactically correct than the longer one will be applied. For that purpose, we trained a syntactical 4-gram language model over a part-of-speech tagged corpus to evaluate the syntactical complexity of any given sentence by a sequence probability as defined in Equation 4. Here, $F = [t_1, t_2, ..., t_m]$ is the sequence of part-of-speech tags for a given sentence with size m. In particular, $P(t) > P(s)$ is the condition that triggers the application of the sentence reduction rule, where t is the compressed version of sentence s.

$$P(F) = \Big(\prod_{k=4}^{m-4} P(t_k \mid t_{k-1}, ..., t_{k-4}) \Big)^{\frac{1}{m}} \qquad (4)$$

So, for each one of the two evaluations (baseline and best rule application), a random sample of 100 reductions was manually cross-evaluated by two human annotators. In order to take into account irrelevancy elimination and syntactical correctness, each reduction had to be scored with a value of 1 (incorrect reduction), 2 (semantically correct but incorrect syntactically) and 3 (semantically and syntactically correct). Additionally, each score was weighted by the number of eliminated words in order to give more importance to longer reductions. The results are presented in Table 2 for a *Cohen's K* value for inter-rater agreement of 0.478, meaning "moderate agreement".

Table 2. Results with four evaluation parameters

Test	Mean Rank	Precision	Mean $|X|$	Rules/Sentence
Baseline	1.983	66.09%	1.15	0.042
ILP Rules	2.056	68.54%	1.78	5.639

In particular, column 2 (**Mean Rank**) presents the average value of both annotators. Column 3 contains the average size of the eliminated segment (in words) and column 4 evaluates the ratio of the number of rules applied by sentence. In fact, columns 3 and 4 evidence the *utility* of a rule set in terms of the number of eliminated words and the number of reduced sentences. As stated before, on one hand the baseline test consists

---

[10] http://hultig.di.ubi.pt/

in the direct use of the reduction cases, as they are in T90Days, on sentence reduction, meaning that no induction was used. This approach resembles to what is done in [11]. On the other hand, the ILP Rule test implies the application of the best learned rule and shows an improvement both in terms of quality and reduction size, although both results still need to be improved. In the final section, we will propose new perspectives to improve our approach.

6   Conclusions and Future Directions

In this paper, we described an ILP learning strategy that learns sentence reduction rules from massive textual data automatically collected from the Web. After paraphrase identification based on Gaussian functions and alignment through a combination of biology-based sequence alignment algorithms, sentence reduction cases are selected and prepared to be used in the learning process, handled by an ILP framework called *Aleph* [15]. Different aspects distinguish our system from existing works. First, it relies on its exclusive automation. Each step takes place in a pipeline of tasks, which is completely automatic. As a result, the system can process huge volumes of data compared to existing works. Second, it is based on shallow linguistic processing, which can easily be adapted to new languages. Finally, we propose a freely available golden data set to the research community in order to apply existing techniques to larger data sets than existing ones [9].

However, improvements still need to be performed. As the overall strategy is based on a pipeline, different errors tend to accumulate step after step. So, each stage must be individually improved. In particular, we noticed from the results that many errors were due to incorrect text tokenization. As a consequence, we believe that the identification of multiword units will improve the quality of rule induction. Moreover, we will propose to automatically tune the generalization process so that we can avoid the induction of over-generalized reduction rules.

Acknowledgements. This work is funded by the ERDF – European Regional Development Fund through the COMPETE Programme (operational programme for competitiveness) and by National Funds through the FCT – Fundação para a Ciência e a Tecnologia (Portuguese Foundation for Science and Technology) within project "FCOMP - 01-0124-FEDER-022701".

References

1. Clarke, J., Lapata, M.: Constraint-based Sentence Compression: An Integer Programming Approach. In: Proceedings of the 21st International Conference on Computational Linguistics and 44th Annual Meeting of the Association for Computational Linguistics (2006)
2. Cohn, T., Lapata, M.: Sentence Compression Beyond Word Deletion. In: Proceedings of the 22nd International Conference on Computational Linguistics (2008)
3. Cohn, T., Lapata, M.: Sentence Compression as Tree Transduction. Journal of Artificial Intelligence Research 34(1), 637–674 (2009)

4. Cordeiro, J., Dias, G., Cleuziou, G., Brazdil, P.: Biology Based Alignments of Paraphrases for Sentence Compression. In: Proceedings of the Workshop on Textual Entailment and Paraphrasing associated to the 45th Annual Meeting of the Association for Computational Linguistics Conference (2007)
5. Cordeiro, J., Dias, G., Brazdil, P.: New Functions for Unsupervised Asymmetrical Paraphrase Detection. Journal of Software 2(4), 12–23 (2007)
6. Corston-Oliver, S.: Text Compaction for Display on Very Small Screens. In: Proceedings of the Workshop on Automatic Summarization associated to the 2nd North American Chapter of the Association for Computational Linguistics Conference (2001)
7. Dolan, W.B., Quirck, C., Brockett, C.: Unsupervised Construction of Large Paraphrase Corpora: Exploiting Massively Parallel News Sources. In: Proceedings of 20th International Conference on Computational Linguistics (2004)
8. Hongyan, H., McKeown, K.R.: Cut and Paste based Text Summarization. In: Proceedings of the 1st North American Chapter of the Association for Computational Linguistics Conference (2000)
9. Knight, K., Marcu, D.: Summarization beyond sentence extraction: A probabilistic approach to sentence compression. Artificial Intelligence 139(1), 91–107 (2002)
10. Muggleton, S.: Inductive Logic Programming: Issues, Results and the Challenge of Learning Language in Logic. Artificial Intelligence 114(1-2), 283–296 (1999)
11. Le Nguyen, M., Horiguchi, S., Ho, B.T.: Example-based Sentence Reduction using the Hidden Markov Model. ACM Transactions on Asian Language Information Processing 3(2), 146–158 (2004)
12. Needleman, S.B., Wunsch, C.D.: A General Method Applicable to the Search for Similarities in the Amino Acid Sequence of Two Proteins. Journal of Molecular Biology 48(3), 443–453 (1970)
13. Quinlan, J.R.: Learning Logical Definitions from Relations. Machine Learning 5(3), 239–266 (1990)
14. Smith, T.F., Waterman, M.S.: Identification of Common Molecular Subsequences. Journal of Molecular Biology 147, 195–197 (1981)
15. Srinivasan, A.: The Aleph Manual. Technical Report. Oxford University, UK (2000)
16. Turner, J., Charniak, E.: Supervised and Unsupervised Learning for Sentence Compression. In: Proceedings of the 43rd Annual Meeting of the Association for Computational Linguistics Conference (2005)
17. Siddharthan, A., Nenkova, A., McKeown, K.: Syntactic Simplification for Improving Content Selection in Multi-Document Summarization. In: Proceedings of the 20th International Conference on Computational Linguistics (2004)
18. Vandeghinste, V., Pan, Y.: Sentence Compression for Automated Subtitling: A Hybrid Approach. In: Proceedings of the Workshop on Text Summarization Branches Out associated to the 44th Annual Meeting of the Association for Computational Linguistics Conference (2004)
19. Vickrey, D., Koller, D.: Sentence Simplification for Semantic Role Labeling. In: Proceedings of the 46th Annual Meeting of the Association for Computational Linguistics Conference (2008)
20. Zajic, D.M., Dorr, B.J., Lin, J.: Single-Document and Multi-Document Summarization Techniques for Email Threads using Sentence Compression. Information Processing and Management 44(4), 1600–1610 (2008)

Author Index